Invertebrate Immunity

ADVANCES IN EXPERIMENTAL MEDICINE AND BIOLOGY

Recent Volumes in this Series

Volume 700
REGULATION OF MICRORNAS
Helge Großhans

Volume 701
OXYGEN TRANSPORT TO TISSUE XXXII
Duane F. Bruley and J.C. LaManna

Volume 702
RNA EXOSOME
Torben Heick Jensen

Volume 703
INFLAMMATION AND RETINAL DISEASE
John D. Lambris and Anthony P. Adamis

Volume 704
TRANSIENT RECEPTOR POTENTIAL CHANNELS
Md. Shahidul Islam

Volume 705
THE MOLECULAR IMMUNOLOGY OF COMPLEX CARBOHYDRATES-3
Albert M. Wu

Volume 706
ADHESION-GPCRS: STRUCTURE TO FUNCTION
Simon Yona and Martin Stacey

Volume 707
HORMONAL AND GENETIC BASIS OF SEXUAL DIFFERENTIATION DISORDERS AND HOT TOPICS IN ENDOCRINOLOGY
Maria I. New and Joe Leigh Simpson

Volume 708
INVERTEBRATE IMMUNITY
Kenneth Söderhäll

A Continuation Order Plan is available for this series. A continuation order will bring delivery of each new volume immediately upon publication. Volumes are billed only upon actual shipment. For further information please contact the publisher.

Invertebrate Immunity

Edited by

Kenneth Söderhäll
Department of Comparative Physiology, Uppsala University, Uppsala, Sweden

Springer Science+Business Media, LLC

Landes Bioscience

Springer Science+Business Media, LLC
Landes Bioscience

Springer Science+Business Media, LLC, 233 Spring Street, New York, New York 10013, USA
http://www.springer.com

Please address all inquiries to the publishers:
Landes Bioscience, 1806 Rio Grande, Austin, Texas 78701, USA
Phone: 512/ 637 6050; FAX: 512/ 637 6079
http://www.landesbioscience.com

The chapters in this book are available in the Madame Curie Bioscience Database.
http://www.landesbioscience.com/curie

Invertebrate Immunity, edited by Kenneth Söderhäll. Landes Bioscience / Springer Science+Business Media, LLC dual imprint / Springer series: Advances in Experimental Medicine and Biology.

ISBN: 978-1-4419-8058-8

Library of Congress Cataloging-in-Publication Data

Invertebrate immunity / edited by Kenneth Söderhäll.
p. ; cm. -- (Advances in experimental medicine and biology ; v. 708)
Includes bibliographical references and index.
ISBN 978-1-4419-8058-8
1. Invertebrates--Immunology. I. Söderhäll, Kenneth. II. Series: Advances in experimental medicine and biology ; v. 708. 0065-2598
[DNLM: 1. Invertebrates--immunology. W1 AD559 v.708 2010 / QL 362.85]
QL362.85.I58 2010
571.9'612--dc22

2010042470

DEDICATION

This volume is dedicated to all my present and former PhD students, post-docs and fellow scientists who worked in my laboratory and who have contributed to our research on invertebrate immunity.

PREFACE

The total number of animal species range up to around 1,000,000 species, of which invertebrates constitute more than 95%, including a vast diversity of organisms from unicellular protozoans to the much more complex echinoderms and protochordates. Insects are by far the largest group of animals within the invertebrates and have received a large amount of research interest due to their importance as vectors for human and animal diseases, such as malaria, and the serious harm done by insects to crops and food.

Thus a long-standing research interest in diseases, pathogens and immune responses of insects there has been. The finding by the late Professor Hans G. Boman and colleagues that Drosophila responded to a challenge with dead or live bacteria by the synthesis of antibacterial peptides initiated an intense research interest on the mechanism of this induction. In 1996 Jules Hoffmann, Bruno Lemaitre and colleagues published a paper in which they for the first time showed that Toll was involved in the production of the antifungal peptide drosomycin. Subsequently, this led to the discovery of Toll-like receptors (TLRs) in vertebrates and their importance in immune responses, particularly in mammals. Research on the role of TLRs in immune responses is very intense and the importance of these receptors is probably greater in mammals than in most invertebrates. The finding that Toll was involved in induction of an antimicrobial peptide opened the possibility of performing detailed genetic studies of the signaling pathways involved in the production of antimicrobial peptides. The completion of the Drosophila genome in 2000 made it possible to carry out powerful molecular genetic analysis of the immune system of this insect. For the past 10 years several genomes of insects and other invertebrates have been sequenced, making comparisons between invertebrates possible, not only between Drosophila and mammals. Several chapters in this book deal with immune responses in different groups of insects. It is evident that the immune responses are very similar between insects, but that there are also differences (Chapters 8-12). Interesting research is now being performed on mosquitoes because they are vectors for many human diseases. One important aspect, for example, is how malaria avoids any immune responses while entering into and being inside the vector mosquitoes (Chapter 12). Interesting studies are now carried out on developmental biology and innate immunity in Hydra (Chapter 1), and there is now interest in studying immune reactions in other invertebrate groups, especially in the model organism *Caenorhabiditis elegans* (Chapter 6), but also in leeches and earthworms (Chapters 4 and 5).

For the past 10 years research has increasingly been focused on crustaceans (shrimp, crabs and crayfish) (see Chapter 13) and mollusks (scallop, oysters and mussels) (see Chapters 2-3), mainly because of their importance as farmed species for consumption. In these aquatic animals antibacterial peptides have a greater variation and each peptide appears to be produced in several isoforms induced by different bacterial species, indicating that there seems to be some sort of specific response to different bacterial species. The ways in which hemocytes are synthetized have also been studied in detail in crayfish, and the finding of a family of astakines similar to prokineticins, which are involved in hematopoiesis, show that these animals may be well suited for studies on hemocytes and their synthesis (Chapter 13).

Recent research has shown that there are great differences in immune responses between different invertebrate groups, but of course also that there are many similarities. Toll receptors are present in some invertebrate groups but so far the importance of their function has been mainly studied in insects, while sea urchins, which have more than 200 Toll receptors still await further studies in toll receptor function (Chapter 14). The significance of another group of molecules, the so-called Dscams, which seem to be restricted to insects and crustaceans and are also present in vertebrates, is difficult to define. Their exact role in immunity is unkown at this moment.

To further emphasize the diversity between different invertebrate animal groups, the clotting reaction can serve as a good example and is described in two chapters (Chapters 7 and 13). In crustaceans it comprises a transglutaminase and a clotting protein, whereas in horse-shoe crabs clotting it is induced by microbial polysaccharides (LPS and beta-1,3-glucans) so that a proteolytic cascade is activated and terminates with coagulogen being cleaved by the proclotting enzyme to form the clot (Chapter 7). There is no similarity between the clotting proteins of crustaceans and horse-shoe crabs, the process is totally different. The only similarity is that transglutaminase is involved in both groups of animals.

The melanization reaction is an important innate immune response and which is present in most invertebrates. It was first noted by Söderhäll and Unestam in 1977 that this process, i.e., activation of prophenolocxidase, was induced in the presence of beta-1,3-glucans. Subsequently, this was shown in insects and several other invertebrates. Recently Lee and colleagues managed to show that the prophenoloxidase activating system and the induction of the Toll pathway in an insect share the same proteolytic cascade (Chapter 9). Interestingly, melanization is a highly conserved immune response which is present in nearly all vertebrates where it provides protection against UV-light and other stressors. Melanization in vertebrates is catalysed by tyrosinase, which has no homology with prophenoloxidase except for the copper binding sites. However, the reactions tyrosinase and phenoloxidase catalyze are exactly the same, converting phenols to quinones and subsequently melanin. Melanization is responsible for skin color in humans, and in other animals, and it is also an important component in the brain and eyes.

Another innate immune system which is present in both vertebrates and invertebrates is the complement system, and to date a complement-like system seems to be operable in echinoderms (Chapter 14), horseshoe crabs (Chapter 7) and tunicates (Chapter 15). In insects the so-called thiolester-containing proteins are proposed to be complement-like and have been shown to function as opsonic proteins in mosquitoes (Chapter 12). It remains to be shown whether true complement proteins are present in other protostome groups and not only in horse-shoe crabs.

A growing interest in invertebrate immunity is the study of the link between physiology and immunity. Surely diet, ageing, reproduction, reproductive behavior, time of day, and use of pathogens are all likely to have an effect on immunity and immune studies. The time of day for immune and challenge studies are important, and both in Drosophila and crayfish it has been shown that the immune system varies in efficiency during a day. This means that scientists should be aware of these facts when planning or making experiments.

In this book I have gathered scientists who are working with different invertebrates, and it can be seen that the insects are the still attracting most research and researchers. However, an increasing interest is emerging to study new invertebrate groups, especially those where the genome is known, as seen in Chapters 1, 5, 14 and 15. Even though Drosophila has been and still is an excellent model for immune studies, it is now clear that there are great differences between immune responses in Drosophila and that of several other invertebrates, which indeed calls for more research on other invertebrates.

Kenneth Söderhäll
Department of Comparative Physiology, Uppsala University
Uppsala, Sweden

ABOUT THE EDITOR...

KENNETH SÖDERHÄLL studied at Uppsala University and obtained his MSc degree in 1972 and PhD in 1978. He was promoted to associate professor (Docent) in 1980. After receiveing his PhD he worked in many different laboratories as a post-doc or as a guest scientist at the University of Montpellier/St.Christol, the Medical Cell Biology department at Tromsö University, the Millport Marine Biological Station in Scotland, University College of Wales in Swansea, the Marine Biological Laboratory at Woods Hole and John Hopkins University. In 1986 he held a Royal Society Fellowship at Swansea. In collaboration with Valerie J. Smith, he developed a new method to isolate and separate blood cells from invertebrates,which was based on using an anticoagulant with a low pH and EDTA, a method now used to isolate most invertebrate blood cells. When he returned to Uppsala he took a position as a researcher at the Swedish Science Research Council in 1987, obtaining a chair and appointment to professor and head of department at Uppsala University in 1989. He continued with research mainly on the proPO-system in arthropods. His group was first to clone proPO from an invertebrate and he has worked with this so-called melanization reaction for several years. His research team has also deciphered the clotting reaction in crustaceans and shown that it consists of a clotting protein present in plasma and a clotting enzyme, a transglutaminase present in the blood cells. This clotting system is distinct from that of a horseshoe crabs' in which a proteinase cascade demonstrates the great diversity in immune processes in invertebrates. He has published 230 original papers and 65 reviews in journals and books, and he has edited 4 books. He is presently on the editorial board on *Fish and Shellfish Immunology*, and is the editor of *Journal of Experimental Biology*. Dr. Söderhäll previously served on *Journal Experimental Zoology*, *Animal Biology*, *Journal of Invertebrate Pathology* and *Developmental and Comparative Immunology*. Since 2000 he has coedited the journal *Developmental and Comparative Immunology* and beginning in 2009 he is editor-in-chief for this journal. He was president for the International Society of Developmental and Comparative Immunology from 2006-2009. Currently he heads the Department of Comparative Physiology at Uppsala University (http://www.fu.uu.se/jamfys/is.html).

PARTICIPANTS

René Augustin
Zoological Institute
Christian-Albrechts-University
Kiel
Germany

Martin Bilej
Institute of Microbiology
Academy of Sciences
of the Czech Republic
Prague
Czech Republic

Thomas C.G. Bosch
Zoological Institute
Christian-Albrechts-University
Kiel
Germany

Katherine M. Buckley
Department of Medical
Biophysics and Department
of Immunology
Sunnybrook Health Sciences Centre
University of Toronto
Toronto, Ontario
Canada

Veronika Burešová
Institute of Parasitology
Biology Centre
Academy of Sciences
of the Czech Republic
University of South Bohemia
České Budějovice
Czech Republic

Lage Cerenius
Department of Comparative
Physiology
Uppsala University
Uppsala
Sweden

Jun-Ho Chai
National Research Laboratory
of Defense Proteins
College of Pharmacy
Pusan National University
Busan
Korea

Lori A. Clow
US Patent and Trademark Office
Alexandria, Virginia
USA

Sirlei Daffre
Departamento de Parasitologia
Instituto de Ciências Biomédicas
Universidade de São Paulo
São Paulo
Brazil

Nolwenn M. Dheilly
Department of Biological
Sciences
Macquarie University
Sydney
Australia

Ilka Engelmann
Centre d'Immunologie de Marseille-Luminy
Université de la Méditerranée
Marseille
France

Julie Ghosh
Department of Biological Sciences
George Washington University
Washington, DC
USA

Nam-Chul Ha
National Research Laboratory of Defense Proteins
College of Pharmacy
Pusan National University
Busan
Korea

Ondřej Hajdušek
Institute of Parasitology
Biology Centre
Academy of Sciences of the Czech Republic
University of South Bohemia
České Budějovice
Czech Republic

Tor Haug
Norwegian College of Fishery Science
University of Tromso
Tromso
Norway

John H. Henson
Department of Biology
Biochemistry and Molecular Biology Program
Dickinson College
Carlisle, Pennsylvania
USA

Julián F. Hillyer
Department of Biological Sciences
Institute for Global Health
Vanderbilt University
Nashville, Tenesseee
USA

Hyun-Ok Hwang
National Research Laboratory of Defense Proteins
College of Pharmacy
Pusan National University
Busan
Korea

Haobo Jiang
Department of Entomology and Plant Pathology
Oklahoma State University
Stillwater, Oklahoma
USA

Pikul Jiravanichpaisal
Molecular Aquatic Biology and Genetic Laboratory
National Center for Genetic Engineering and Biotechnology
National Science and Technology Development Agency
Rajdhevee, Bangkok
Thailand

Radka Josková
Institute of Microbiology
Academy of Sciences of the Czech Republic
Prague
Czech Republic

Michael R. Kanost
Department of Biochemistry
Kansas State University
Manhattan, Kansas
USA

Shun-ichiro Kawabata
Department of Biology
Kyushu University
Fukuoka
Japan

Chan-Hee Kim
Division of Infectious Disease
Department of Medicine
University of Massachusetts Medical School
Worcester, Massachusetts
USA

Petr Kopáček
Institute of Parasitology
Biology Centre
Academy of Sciences
of the Czech Republic
University of South Bohemia
České Budějovice
Czech Republic

Shoichiro Kurata
Graduate School
of Pharmaceutical Sciences
Tohoku University
Sendai
Japan

Kenji Kurokawa
National Research Laboratory
of Defense Proteins
College of Pharmacy
Pusan National University
Busan
Korea

Bok Luel Lee
National Research Laboratory
of Defense Proteins
College of Pharmacy
Pusan National University
Busan
Korea

Chun Li
Norwegian College
of Fishery Science
University of Tromso
Tromso
Norway

Hai-peng Liu
State Key Laboratory of Marine
Environmental Science
College of Oceanography
and Environmental Science
Xiamen University
Fujian
China

Eric S. Loker
Center for Evolutionary
and Theoretical Immunology
Department of Biology
University of New Mexico
Albuquerque, New Mexico
USA

Cheng Man Lun
Department of Biological
Sciences
George Washington University
Washington, DC
USA

Audrey J. Majeske
Department of Biological
Sciences
George Washington University
Washington, DC
USA

Valeria Matranga
Consiglio Nazionale
delle Richerche
Instituto di Biomedicina
e immunologia Molecolare
"Albert Monroy"
Palermo
Italy

Sham V. Nair
Department of Biological
Sciences
Macquarie University
Sydney
Australia

Masaru Nonaka
Department of Biological
Sciences
Graduate School of Science
The University of Tokyo
Hongo, Tokyo
Japan

Ji-Won Park
National Research Laboratory
of Defense Proteins
College of Pharmacy
Pusan National University
Busan
Korea

Keun-Hwa Park
National Research Laboratory
of Defense Proteins
College of Pharmacy
Pusan National University
Busan
Korea

Petra Procházková
Institute of Microbiology
Academy of Sciences
of the Czech Republic
Prague
Czech Republic

Nathalie Pujol
Centre d'Immunologie
de Marseille-Luminy
Université de la Méditerranée
Marseille
France

Limei Qiu
Key Laboratory of Experimental
Marine Biology
Institute of Oceanology
Chinese Academy of Sciences
Qingdao
China

David A. Raftos
Department of Biological
Sciences
Macquarie University
Sydney
Australia

Jonathan P. Rast
Department of Medical
Biophysics
Department of Immunology
Sunnybrook Health Sciences Centre
University of Toronto
Toronto, Ontario
Canada

Mattias Roth
Department of Biological
Sciences
Macquarie University
Sydney
Australia

Jiang Rui
National Research Laboratory
of Defense Proteins
College of Pharmacy
Pusan National University
Busan
Korea

Kyung-Hwa Ryu
National Research Laboratory
of Defense Proteins
College of Pharmacy
Pusan National University
Busan
Korea

Sandro Sacchi
Department of Biological
Sciences
George Washington University
Washington, DC
USA

Michel Salzet
Université Lille Nord De France
Laboratoire de Neuroimmunologie
et Neurochimie Evolutives
Université Lille
Lille
France

Honoo Satake
Suntory Institute for Bioorganic
Research
Wakayamadai, Shimamoto, Mishima
Osaka
Japan

Catherine S. Schrankel
Department of Biological
Sciences
George Washington University
Washington, DC
USA

Marcela Šilerová
Institute of Microbiology
Academy of Sciences
of the Czech Republic
Prague
Czech Republic

L. Courtney Smith
Department of Biological Sciences
George Washington University
Washington, DC
USA

Irene Söderhäll
Department of Comparative Physiology
Uppsala University
Uppsala
Sweden

Kenneth Söderhäll
Department of Comparative Physiology
Uppsala University
Uppsala
Sweden

Linsheng Song
Institute of Oceanology
Chinese Academy of Sciences
Qingdao
China

Klara Stensvåg
Norwegian College of Fishery Science
University of Tromso
Tromso
Norway

Aurélie Tasiemski
Université Lille Nord De France
Laboratoire de Neuroimmunologie et Neurochimie Evolutives
Université Lille
France

Andreas Vilcinskas
Institut für Phytopathologie und Angewandte Zoologie
Justus-Liebig-Universität Gießen
Heinrich-Gießen
Germany

Lingling Wang
Key Laboratory of Experimental Marine Biology
Institute of Oceanology
Chinese Academy of Sciences
Qingdao
China

Huan Zhang
Key Laboratory of Experimental Marine Biology
Institute of Oceanology
Chinese Academy of Sciences
Qingdao
China

CONTENTS

ACKNOWLEDGEMENTS

I wish to thank all who reviewed the chapters of this book and the Swedish Science Research Council for continous support of my research.

CHAPTER 1

CNIDARIAN IMMUNITY:
A Tale of Two Barriers

René Augustin and Thomas C.G. Bosch*
Zoological Institute, Christian-Albrechts-University Kiel, Kiel, Germany
**Corresponding Author: Thomas C.G. Bosch—Email: tbosch@zoologie.uni-kiel.de*

Abstract: The phylum Cnidaria is one of the earliest branches in the animal tree of life providing crucial insights into the early evolution of immunity. The diversity in cnidarian life histories and habitats raises several important issues relating to immunity. First, in the absence of specific immune cells, cnidarians must have effective mechanisms to defend against microbial pathogens. Second, to maintain tissue integrity, colonial forms have to rely on their capacity of self/nonself discrimination to rapidly detect approaching allogeneic cells as foreign and to eliminate them. And third, since cnidarians are colonized by complex bacterial communities and in many cases are home to algal symbionts, successful growth means for cnidarians to be able to distinguish between beneficial symbionts and pathogenic intruders. The aim of this chapter is to review the experimental evidence for innate immune reactions in Cnidaria. We show that in these diploblastic animals consisting of only two cell layers; the epithelial cells are able to mediate all innate immune responses. The endodermal epithelium appears as a chemical barrier employing antimicrobial peptides while the ectodermal epithelium is a physicochemical barrier supported by a glycocalix. Microbial recognition is mediated by pattern recognition receptors such as Toll- and Nod-like receptors. Together, the data support the hypothesis that the establishment of epithelial barriers represents an important step in evolution of host defense in eumetazoan animals more than 600 million years ago.

INTRODUCTION: CNIDARIA PROVIDE INFORMATION ABOUT THE EVOLUTION OF IMMUNITY

Cnidarians are among the earliest known phyletic lineages and are sister taxon to the Bilateria (Fig. 1).[1-4] Cnidarians possess most of the gene families found in bilaterians[1,5-7] and

---

Invertebrate Immunity, edited by Kenneth Söderhäll.

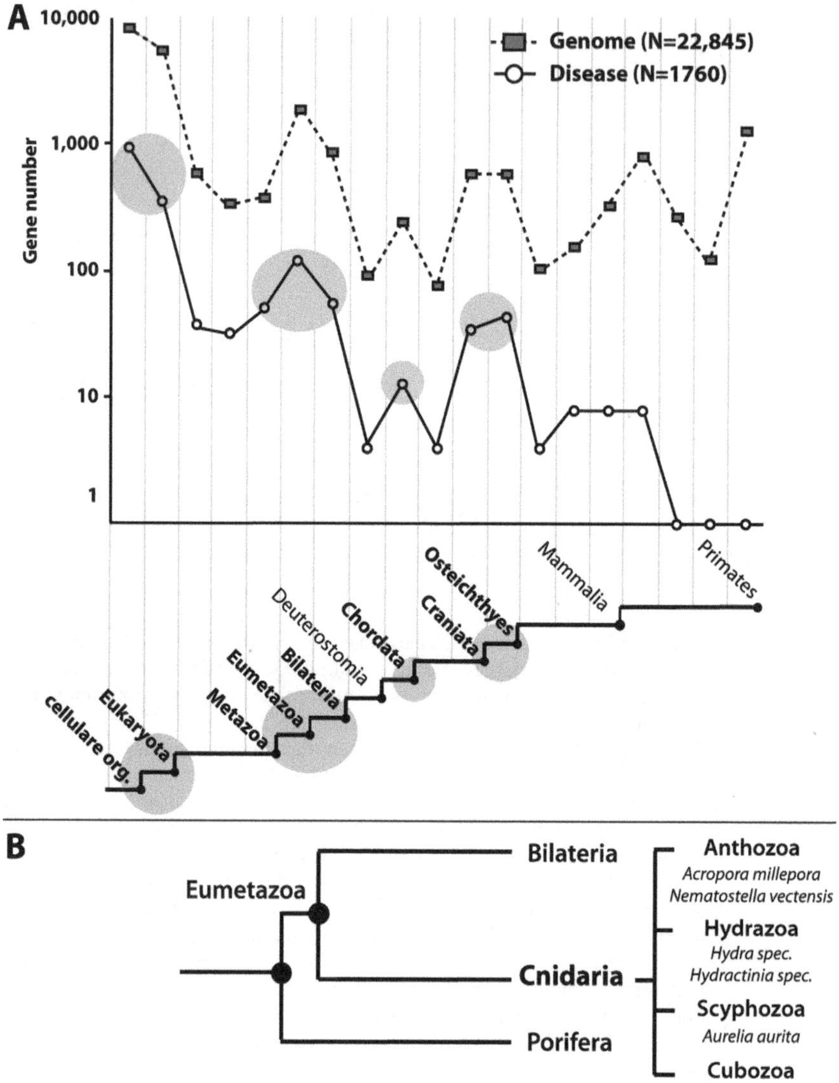

Figure 1. Phylostratigraphy of all human genes and disease-causing genes. A) The total number of human genes found in different phyla is plotted (dashed line—boxed). Distribution respectively first appearance of evaluated human disease genes in animal phylogeny (N = 1,760 black line—open circle) (Panel A modified with permission from Domazet-Loso et al,[18] ©2008 Oxford University Press). B) Cnidarians are among the earliest known phyletic lineages and are sister taxon to the Bilateria. The four classes of Cnidaria (Hydrozoa, Scyphozoa, Cubozoa, Anthozoa) contain popular model organisms for comparative immunology such as the freshwater polyp *Hydra* or the colonial Hydrozoan *Hydractinia.*

have retained many ancestral genes that have been lost in *D. melanogaster* and *C. elegans.*[7,8] Since their genome organization and content is remarkably similar to that of morphological complex bilaterians, they provide insight into the content of the "genetic tool kit" present in the Cnidarian–bilaterian ancestor. The four classes of Cnidaria (Hydrozoa, Scyphozoa,

Cubozoa, Anthozoa) contain popular model organisms for comparative immunology such as the freshwater polyp *Hydra* or the colonial Hydrozoan *Hydractinia* (Fig. 1B).

For analytical purposes, an important technical breakthrough in Cnidarian immunology was the development of a transgenic procedure allowing efficient generation of transgenic *Hydra*[9] and *Nematostella*[10] lines by embryo microinjection. The application of morpholino techniques in *Nematostella* embryonic tissue to knock down gene expression offers a way to analyse the function of novel immuno-candidate genes during development.[11,12] Furthermore, computational approaches have become increasingly powerful and sophisticated with each basal metazoan genome that has been completely sequenced[1-4,13] and with each advance in the technologies used for genome analysis.[14]

The ultimate goal of using bioinformatics in the comparative immunology field is to be able to identify the genomic basis for key immunological characteristics appearing in our metazoan ancestor. This goal has yet to be reached, but there has been considerable progress in this area. For instance, the origin of major regulators of innate and adaptive immune response, the interferon regulatory factors family and the Rel/NFκB family, was traced back to the appearance of multicellular organisms.[15-17] Moreover, rich sequence data from numerous species of basal metazoans allowed to determine the evolutionary origin of mutations leading to Mendelian genetic diseases. In a phylostratigraphic analysis[18] of about 1700 human genes that have been linked to a heritable genetic disease, their evolutionary origin could be traced back to the origin of the first cells. This study also showed that other large groups of disease genes emerged more than one billion years ago around the first appearance of multicellular animals as well as at the time of origin of bony fishes about 400 million years ago. Surprisingly, almost no disease associated gene emerged after the origin of mammals (Fig. 1A). The study suggests that over 90% of the disease genes have emerged before the bilaterian radiation. The results are important because they show that many diseases have ancient origin and that functional knowledge gained about such disease genes from remote model organisms such as Cnidarians is relevant for understanding the genes in humans. The identification of genes responsible for human diseases affecting biological barriers (e.g., skin or intestinal mucosa) often does not in itself provide a clue to etiopathogenesis or therapeutic targets, as the interaction of a suite of genes in a complex system such as the human is difficult to understand. Likewise the involved pathways that ultimately lead to the development of the disease phenotype are unclear. Searching for the evolutionary origin of the disease-causing genes and characterizing the variation in such genes under known evolutionary pressures may provide insights into the development of diseases in humans and identify new targets for therapy or prevention. "*To understand the context of the biological processes in which a gene is involved, it may be advisable to use model organisms that represent the evolutionary level at which these genes emerged*" (cited from ref. 18, pp 2706).

CNIDARIAN IMMUNE RESPONSES ARE PERFORMED BY TWO EPITHELIAL BARRIERS

Cnidarians are diploblastic consisting of an ectodermal and an endodermal epithelium. While both layers are separated by an extracellular matrix (mesoglea) a true mesoderm is missing. In the freshwater polyp *Hydra* (Fig. 2A) epithelial cells in both layers (see Fig. 2B), are multifunctional having both secretory and phagocytic activity.[19,20] A combined biochemical and transcriptome analysis approach revealed that in *Hydra* most innate

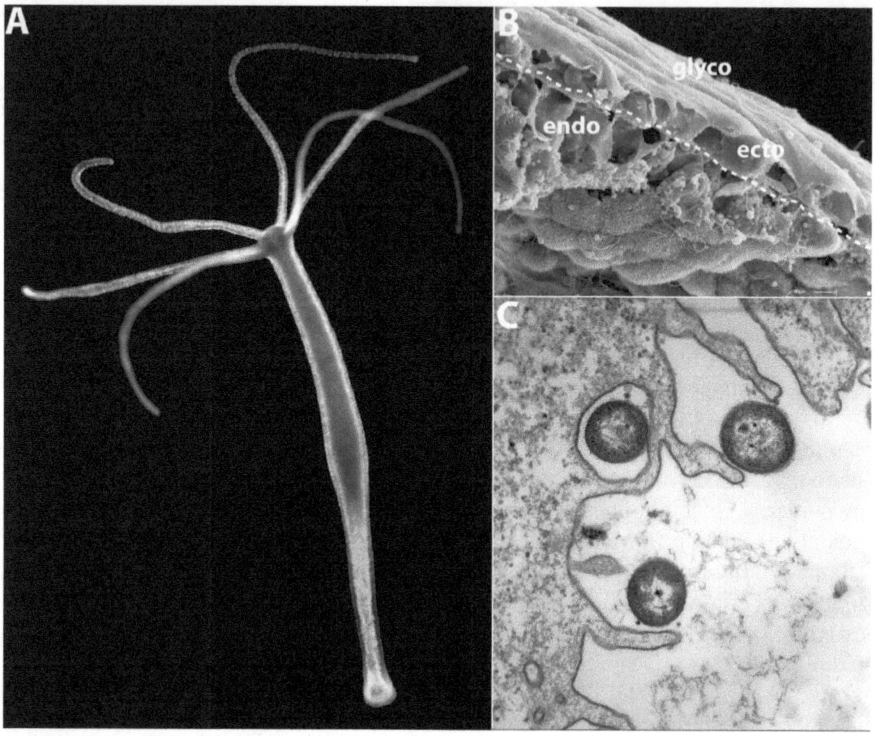

Figure 2. Cnidarians are diploblastic animals. A) Live image of *Hydra oligactis* (photo by S. Fraune). B) Raster electron micrograph showing the ectodermal (ecto) and endodermal epithelium (endo); which are separated by an extracellular matrix (mesoglea—dashed line); a true mesoderm is missing. The apical part of hydra ectodermal epithelial cells is covered by a glycocalix layer (glycol) (photo by F. Anton-Erxleben). C) Transmission electron micrographic of endodermal epithelial cells phagocyte bacteria from the gastric lumen. (Panel C modified from Bosch et al,[20] ©2010, with permission from Elsevier)

immune responses are mediated by epithelial cells.[20,21] Endodermal epithelial cells in *Hydra* not only contribute to digestion and uptake of food but also are capable to phagocytose and destroy bacteria present in the gastric cavity (Fig. 2C). Within the endodermal layer gland cells in addition to epithelial cells contribute to innate immune reactions by producing potent antimicrobial serine protease inhibitors.[22] There are no motile immune effector cells or phagocytes in sensu strictu (i.e., specialized cells that engulf and ingest other cells or particles) known in *Hydra.* In the sea fan coral *Gorgonia ventalina* granular amoebocytes were reported[23] to increase in number after contact with a fungal pathogen (*Aspergillus sydowii*). Interestingly, at the site of infection a high prophenoloxidase (melanisation) activity was observed which seems to have its origin in the surrounding amoebocytes.[23] Thus, this protective strategy seems to protect a whole colony rather than the individual polyp. Some cnidarians have a remarkable capability of regeneration. In *Hydra*, for example, gross damage to the tissue is quickly repaired due to the presence of continuously proliferating stem cells.[24] Since cells infected or damaged by pathogens such as bacteria or fungi are quickly removed by apoptosis[19] and replaced by noninfected cells, this enormous regeneration capacity may be considered an additional arm of the innate immune defense.

CNIDARIANS DO NOT LIVE ALONE BUT ARE INTIMATELY ASSOCIATED WITH SYMBIONTS

The principal function of the immune system is to mediate interactions of the host with the associated or colonizing microbiota.[25] Cnidarian tissue is loaded with microbes. Phylogenetic analyses as well as molecular techniques including FISH[26] and metagenomic approaches have demonstrated that in Cnidarians such as corals the microbial communities are highly complex.[27,28] The communities are distinct from those in the water column, are coral species-specific, are spatially and temporally stable and most likely have multiple roles in the physiological function of the coral host.[26,29-34] Disturbances in the balance between corals and colonizing microbiota appear to facilitate emergence of coral disease.[35] When investigating the microbiotic world in the freshwater polyp *Hydra*, different species were found to be associated with species-specific microbiota (Fig. 3A).[36] Since microbial phylotypes were identical in polyps taken directly from the wild and polyps cultured under constant laboratory conditions, the epithelia appear to select their species-specific microbes. Interestingly, disturbing the cellular composition of the epithelium results in rapid changes of the associated microbiota.[37] The molecular basis of this "interkingdom" host-microbe communication is currently under investigation. While the site of residence of most of the polyp-associated microbes is not yet known, in *Hydra oligactis* endodermal epithelial cells were discovered to contain intracellular symbionts belonging to the α-Proteobacteria clade (Fig. 3B).

In addition to being intimately associated with and colonized by microbes, some Cnidarian species contain symbiotic algae. In fact, Cnidaria are the phylogenetically oldest Eumetazoa phylum known to form symbiotic relationships with unicellular algae. In Anthozoans most intracellular symbionts belong to the Zooxanthellae. They are coevolved and important for long term survival of the solitary polyp or the coral colony. Various forms of stress including increase of temperature[38] or infection by pathogens cause cnidarian polyps to dispel their symbionts.[39,40]

To understand and ultimately prevent this "coral bleaching", the cellular and molecular interactions underlying this interaction are currently investigated in several coral species with particular emphasis on the establishment, maintenance and breakdown of these cooperative partnerships.[41-44] Most coral species must acquire symbionts anew with each generation and therefore must engage complex recognition and specificity processes, that results in the establishment of a stable symbiosis.

Another model organism for studying the set-up of long-term symbiotic interactions is the freshwater polyp *Hydra*. *Hydra viridis* forms a stable symbiosis with intracellular green algae of the *Chlorella* group.[45] The symbionts are located in endodermal epithelial cells (Fig. 3C). Each alga is enclosed by an individual vacuolar membrane resembling a plastid of eukaryotic origin at an evolutionary early stage of symbiogenesis.[46] Proliferation of symbiont and host is tightly correlated. The photosynthetic symbionts provide nutrients to the polyps enabling *Hydra* to survive extended periods of starvation.[45,47] Symbiotic *Chlorella* is unable to grow outside the host indicating a loss of autonomy during establishment of the intimate symbiotic interactions with *Hydra*. During sexual reproduction of the host, *Chlorella* algae are translocated into the oocyte giving rise to a new symbiont population in the hatching embryo.[48] How the cnidarian host not only tolerates but also supports the continuous presence of eukaryotic symbionts within its cells is one most of the fascinating questions in biology.

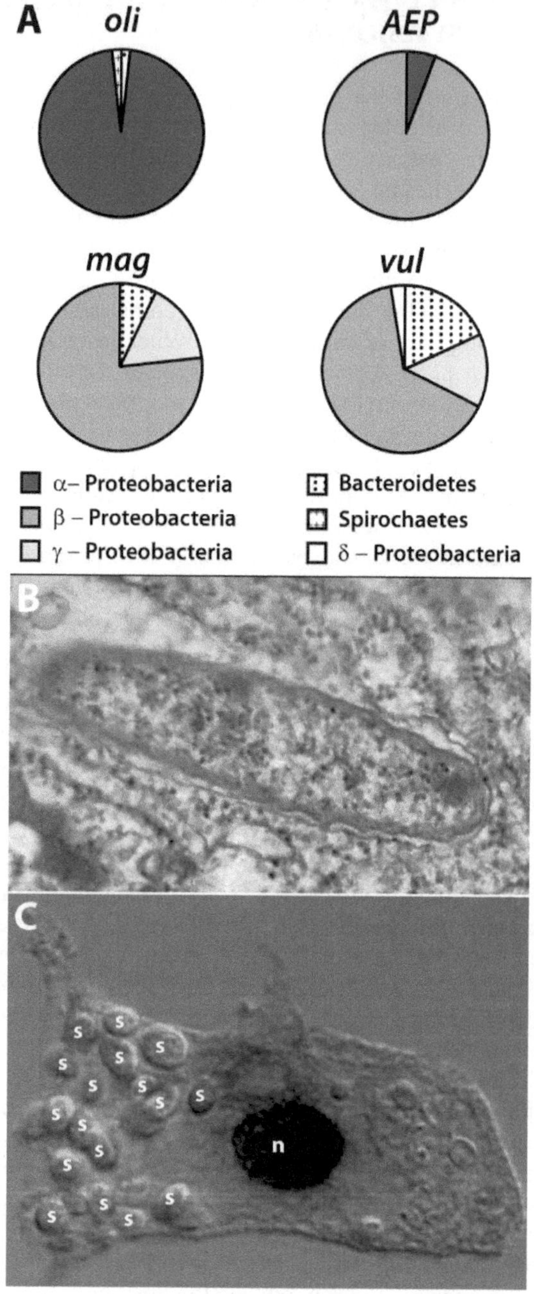

Figure 3. Cnidarians are intimately associated with symbionts. A) Bacterial communities identified from different hydra species. B) Transmission electron micrograph of bacterial endosymbiont in the cytoplasm of an ectodermal epithelial cell of *Hydra oligactis* (Panel B modified from S. Fraune et al,[36] ©2007, with permission from the National Academy of Sciences, USA). C) Phase contrast micrograph of a single mazerated endodermal epithelial cell containing symbiotic algae (s) in the basal part below the nucleus (n) (Panel C modified from Habetha et al[48]).

CNIDARIANS RECOGNIZE MICROBIAL ASSOCIATED MOLECULAR PATTERNS THROUGH GERM LINE ENCODED PATTERN RECOGNITION RECEPTORS (PRRs)

Among the most conserved cell surface receptors for microbial associated molecular patterns (MAMPs) are the Toll-like receptors (TLRs). While recognition of MAMPs by a prototypical TLR is mediated by leucine rich regions (LRRs) in the extracellular part, signal transduction requires the Toll-interleukin-like receptor (TIR) domain in the cytosolic region of the receptor. In vertebrates, different TLRs recognize different MAMPs.[49] Screening of the *Hydra* transcriptome and genome lead to the unexpected discovery that a bona fide TLR gene is not present.[15] However, data bank searches revealed that in *Hydra* there are two transmembrane proteins with LRRs similar to LRRs present in vertebrate TLRs. These proteins, HyLRR-1 and HyLRR-2, lack any intracellular domain and, therefore, also do not contain the typical TIR domain (Fig. 4A). The screening approach also uncovered two genes, HyTRR-1 and HyTRR-2, encoding for transmembrane proteins with a canonical TIR domain but lacking any conspicuous extracellular domains. Functional assays involving HEK cells co-expressing HyTRR-1 and HyLRR-2 receptors indicated that both transmembrane proteins function together to recognize and transduce the bacterial flagella component flagellin.[20] Thus, in *Hydra* the TLR function is present but allocated on two different genes. Since the anthozoan *Nematostella vectensis* contains a prototypical TLR,[15] MAMP recognition by TLRs seems to be a very ancestral feature. This is supported by the finding that in addition to the TLR transmembrane receptors also TLR-specific signal transduction components such as MyD88 and NFκB are present (Fig. 4A). The findings imply that TLR-mediated innate immune responses are ancient, originating in the common ancestor of bilaterian animals. Besides flagellin which is sensed by HyTRR-1–HyLRR-2 receptor complex also LPS was found to induce innate immune reactions.[20] How the LPS signal is transmitted into the cell and if the TLR like pathway is involved, is currently under investigation.

In all organisms there is an elaborate intracellular detection array for internal invasion. Intracellular MAMP recognition is mediated by Nod-like receptors (NLRs). As summarized elsewhere,[50] NLR genes encode for cytosolic proteins that comprise a trimodular domain structure, characterized by a central NACHT, a N-terminal DEATH-fold-like effector binding domain, e.g., a Pyrin domain (PYD), DEATH or caspase recruitment domain (CARD) and a C-terminal leucine rich repeat region (LRR). Upon ligand sensing via the C-terminal LRRs, which can be regarded as the intracellular counterpart of the ectodomain of the TLRs, the molecules have propensity to form self-oligomers, thereby recruiting proximity activated binding partners. This leads to either activation of proinflammatory signalling pathways such as NF-κB or direct activation of proinflammatory caspases. How do Cnidarians detect microbial invaders inside their cells? As indicated in Figure 4B, a typical *Hydra* HyNLR consists of a DEATH and a NACHT domain. Interaction with other DEATH domain containing proteins such as HyDODE may result in activation of a caspase cascade leading to programmed cell death (Fig. 4B). Homologs of the NLRs (e.g., R genes) have been discovered throughout the plant and animal kingdoms.[50] Although the Nod-receptor family has many unresolved features, the high evolutionary conservation of the NLRs underlines their significance in host defense. In contrast to vertebrates, Cnidaria such as *Acropora*, *Nematostella* and *Hydra* have numerous NLR-related receptors of different molecular configuration.[51] Since several NLRs have the ability to activate the inflammasome complex and cell death,[52] in Cnidarians with their enormous regeneration

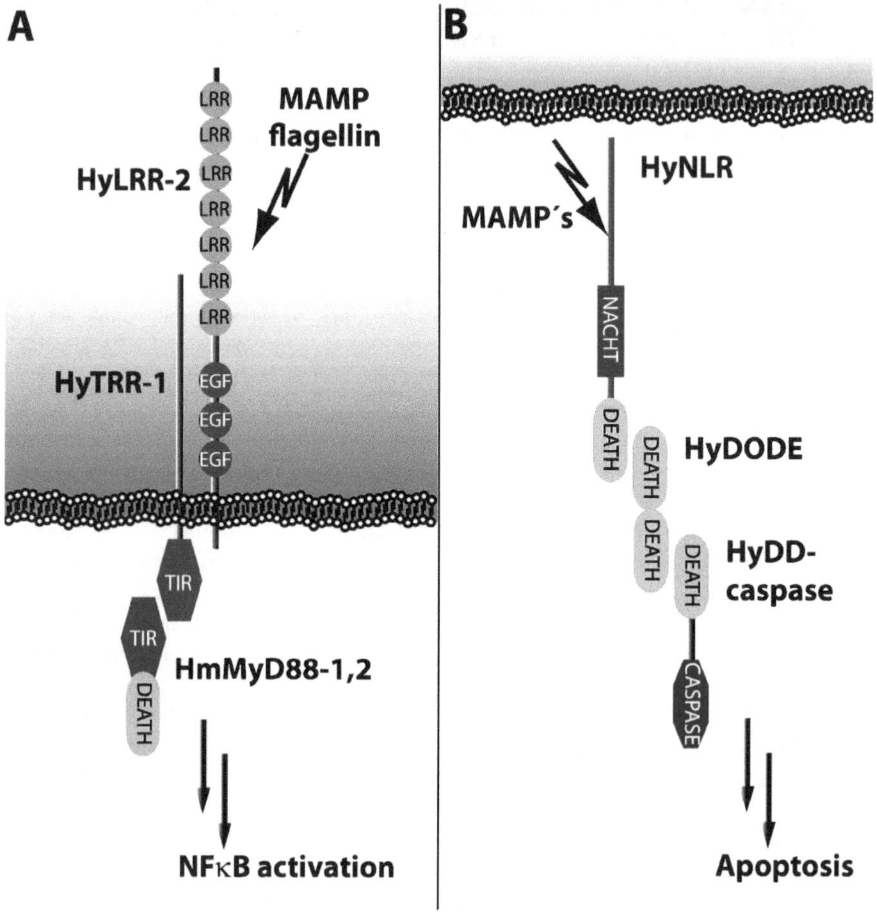

Figure 4. Microbial associated molecular pattern recognition in *Hydra*. Scheme representing the molecules responsible for A) extracellular and B) cytosolic recognition and transmission of MAMPs in *Hydra* A). The MAMP flagellin is detected by HyLRR-2 and via HyTRR-1 transmitted into the cell. The signal is transduced further by conserved TLR-signalling pathway components. B) Cytosolic MAMP´s are sensed by HyNLR´s and by homotypic interaction of death domain containing proteins programmed cell death is initiated (modified form Lange et al[51]).

capacity, apoptosis appears to be an important part of innate immunity. Unravelling the signaling pathways and exact molecular functions of these ancient NLRs will shed light on the primary cellular defense programs directed against invading microbiota.

Interestingly not only MAMPs are able to induce a defence reaction in Cnidarians. A much larger group of inducers are "danger associated molecular patterns" (DAMPs). These are substances normally buried inside the cell and therefore not detectable by neighbouring cells. In case of cell injury and/or cell death, these substances are released and suggested danger to other intact cells within the tissue. Two examples for DAMPs which are easy to apply in experimental procedures are extracellular nucleic acid and monosodium urate. In *Hydra*, increased concentrations of extracellular nucleic acid as well as monosodium urate stimulate an innate immune response.[20] The receptors for these danger signals in *Hydra* or other Cnidarians are currently not known.

CNIDARIANS PRODUCE ANTIMICROBIAL PEPTIDES WHICH ARE EFFECTIVE EVEN AGAINST HUMAN PATHOGENS

Silencing of HyTRR-1 and HyLRR-2 leads to drastic reduction in antimicrobial activity[20] and indicates that in Cnidarians—as in other organisms—activation of MAMP receptors induces the production of a plethora of different antimicrobial peptides (AMPs). A detailed biochemical characterization of the antimicrobial activity in *Hydra* revealed the presence of both conserved AMPs such as Hydramacin-1 (Fig. 5A) and novel or taxon-specific AMPs such as Arminin 1a and Periculin-1 which have no counterpart in the transcriptomes of any other organisms. We proposed elsewhere[53] that these taxon-specific genes, in combination with rewiring of the genetic networks of conserved regulatory genes, accomplish fine adaptations of Cnidarians to constantly changing ecological conditions. An intriguing example of such a taxonomically restricted antimicrobial peptide is Periculin-1, termed due to its rapid response to a wide variety of bacterial and tissue ''danger'' signals.[20] Analysis of the deduced amino acid sequence of Periculin-1 and the charge distribution within the molecule revealed an anionic N-terminal region and an 8 cysteine residues containing cationic C-terminal region.[20] No identifiable orthologues were found in any sequence database outside the genus *Hydra*. Periculin-1 has a bactericidal activity and is expressed in the endodermal epithelium as well as in a subpopulation of ectodermal interstitial cells. Screening other taxa indicates that each animal species contains a significant number of such "orphan" genes encoding potent antimicrobial peptides. The novel antimicrobial peptide Aurelin is another examples of such orphan genes (Fig. 5A) found in *Aurelia aurita*, one of the most common and widely recognized type of jellyfish found near the coasts in the Atlantic, Arctic and Pacific Oceans.[54] Aurelin, an antimicrobial peptide with a molecular mass of 4.3 kDa was shown to exhibit activity against Gram-positive and Gram-negative bacteria. The primary structure of its molecular precursor consists of a canonical signal peptide, anionic propiece and a mature cationic part.[54]

Functional characterization of Hydramacin-1 indicated a novel mode of antimicrobial action. As shown in Figure 5B, exposure of bacteria to low amounts of Hydramacin-1 caused bacteria to aggregate. Soon thereafter bacteria display typical features of programmed cell death with cytosolic proteolysis but intact cell membranes (Fig. 5B). Determination of the three dimensional structure[21] revealed that the molecule possesses two short α-helices at the N-terminus which are separated by a long flexible loop (Fig. 5C). The C-terminal region contains two β-strands in an antiparallel arrangement separated by a long flexible loop. This predicted structure (Fig. 5C) indicates that Hydramacin-1 functions as "glue" causing clumping of bacterial cells followed by programmed cell death. Moreover, when used in liquid growth inhibition assays, recombinant Hydramacin-1 was capable of killing a large number of human Gram-negative pathogens, including the extended spectrum beta-lactamase (ESBL) strains of *E.coli*, *Klebsiella oxytoca* and *Klebsiella pneumoniae*, which are resistant to penicillin derivates. Based on the observation of Hydramacin-induced clumping of bacteria followed by cell death (Fig. 5B) it is tempting to speculate that Hydramacin-1 acts as efficient AMP by interfering with the quorum sensing system of *Hydra's* pathogenic bacteria.

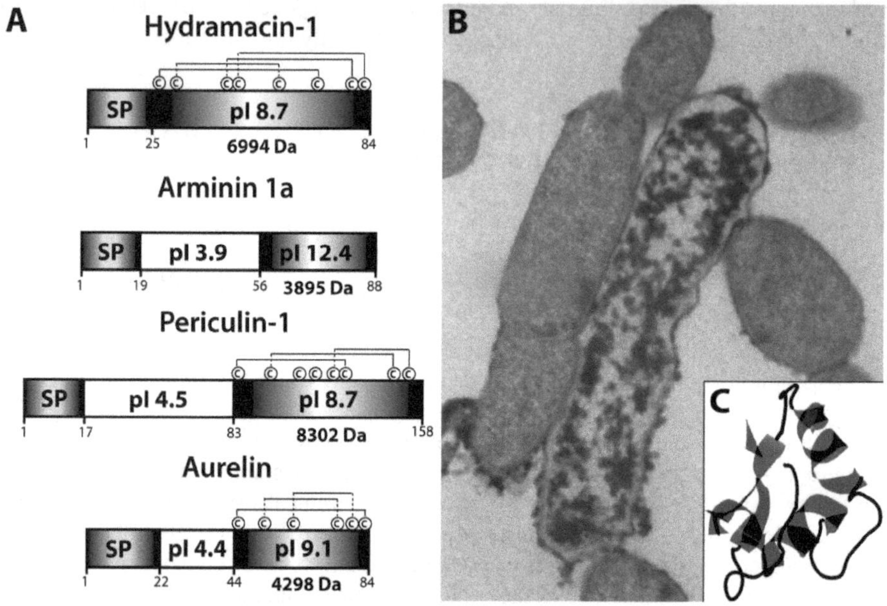

Figure 5. Cnidaria produce highly efficient antimicrobial peptides. A) Schematic representation of four antimicrobial peptides from *Hydra magnipapillata* and *Aurelia aurita*. All peptides contain a signal peptide (SP) and a cationic C-terminal active part (note the pI-value). In addition, Arminin 1a, Periculin-1 and Aurelin are made as precursor peptides, which need to be activated by the cleavage off the anionic N-terminal part. B) Transmission electron micrograph of Hydramacin-1 treated *Escherichia coli*. (photo by F. Anton-Erxleben) C) Ribbon cartoon representation of the average three dimensional structure of Hydramacin-1.

Since the last decade has seen the inexorable proliferation of a host of antibiotic-resistant bacteria including a particularly pernicious strain of bacteria known as methicillin-resistant *Staphylococcus aureus* (MRSA), we started to screen *Hydra* tissue systematically for the presence of novel antibiotics (Augustin and Bosch, unpubl.). Soon after the discovery of novel antimicrobial peptide Arminin 1a[55] we have developed the molecule "ARM-C31-N". ARM-C31-N consists of 31 amino acids which are amidated at the C-terminus. The molecule has a calculated mass of 3896 Da and an isoelectric point of 12.1. When used in liquid growth inhibition assays, ARM-C31-N is capable of killing a large number of resistant bacteria including methicillin resistant *S. aureus* and vancomycin resistant strains of *E. faecalis* and *E. faecium*.[55] More generally speaking, antimicrobial peptides from basal metazoans which are only distantly related to humans may provide interesting leading-structures to design a novel generation of antibiotics since human pathogens share little or no evolutionary history with Cnidarian-associated microbes and, therefore, appear to be particularly vulnerable by Cnidarian-antimicrobial molecules.

Aurelin, Periculin-1 and Arminin 1a are made as precursors. For their activation a negatively charged N-terminal part is cleaved to release a highly positively charged C-terminal part. In some, but not all AMPs this cationic C-terminal region is rich in cysteines indicating that this domain requires a distinct three dimensional structure for activity. The regulatory mechanisms promoting constitutive and inducible antimicrobial peptide (AMP) expression in the epithelial barriers are not yet well understood.

Cnidarian tissue is not only packed with strong antimicrobial molecules but also with serine protease inhibitors with strong antimicrobial activity. For example, when biochemically analyzing *Hydra magnipapillata* tissue for antistaphylococcal activity we discovered a kazal-type serine protease inhibitor, Kazal-2 produced in endodermal gland cells.[22] In liquid growth inhibition assays, native Kazal-2 protein has potent antistaphylococcal activity which most likely is due to the inhibition of a bacterial specific serine protease.[22]

Interestingly all peptide or protease inhibitors isolated so far in *Hydra*, Hydramacin-1, Arminin 1a, Periculin-1, as well as Kazal 2, are expressed in the endodermal layer.[20,22] We, therefore, assume that they contribute to the chemical defense properties of this layer.

HOW DO CNIDARIA DISTINGUISH SELF FROM NONSELF?

Alloimmune specificity and histocompatibility, driven by genetic polymorphism, are ancient determinants of self-/nonself–recognition.[56] *Hydractinia* is a colonial marine Cnidarian composed of a limited number of repeating structural units, polyps and stolons. Stolons are vascular-type canals which join the asexually proliferating polyps. *Hydractinia* maintain self-perpetuating stem cell lineages throughout their life history with the interstitial stem cells giving rise to the germ line and to several other cell types. After metamorphosis the planula larva forms the first polyp of a new *Hydractinia* colony. Whenever two or more planulae recruited to the same shell, the colonies may grow into contact and allorecognition interactions start. Allogeneic contacts have two major classes of outcomes: rejection and fusion.[57-59] Lange et al have shown that all contacts follow a similar sequence of events.[60] When the leading edge of two asexually expanding colonies comes into contact, a large number of nematocysts are transported to the regions in contact (Fig. 6A). Once a threshold number of nematocysts have accumulated, they either disperse in fusion interactions or fire and damage the allogeneic tissue in rejection interactions.[60] These allorecognition responses play a fundamental role in maintaining the genetic and physiological integrity of the colony [61-63] because the germ line is not sequestered and because interstitial cells migrate within *Hydractinia* colonies. Thus, depending on the outcomes of allorecognition reactions, there is the risk of losing access to the germ line through "somatic cell parasitism".[59,64] More than 50 years ago Hauenschild postulated a model in which allorecognition and the ability to fuse between stolons of different colonies is under the control of one polymorphic locus.[58,65,66] Five decades after Hauenschild's pioneering experiments, an extensive inbreeding program[67] demonstrated that allorecognition segregate in a single chromosomal region but contains two closely linked loci, *alr1* and *alr2* within 1.7 cM.[68,69] Individual *Hydractinia* that share at least one allele at both loci undergo fusion. In contrast, if no alleles are shared, a rejection process is initiated.

Analysis of the region of the *Hydractinia* genome corresponding to the *alr2* locus led Nicotra et al[70] recently to the identification of a transmembrane protein, CDS7, which has three polymorphic domains in the extracellular region (Fig. 6B). Although a functional characterization of this receptor is still missing, Nicotra et al[70] found that polymorphism in the first domain of CDS7 is an obligatory element of histocompatibility. These observations for the first time have uncovered a likely candidate for allorecognition in a basal metazoan. Since this receptor is unrelated to allorecognition receptors in tunicates[71] or vertebrates (MHC), this report in addition supports the view that mechanisms used

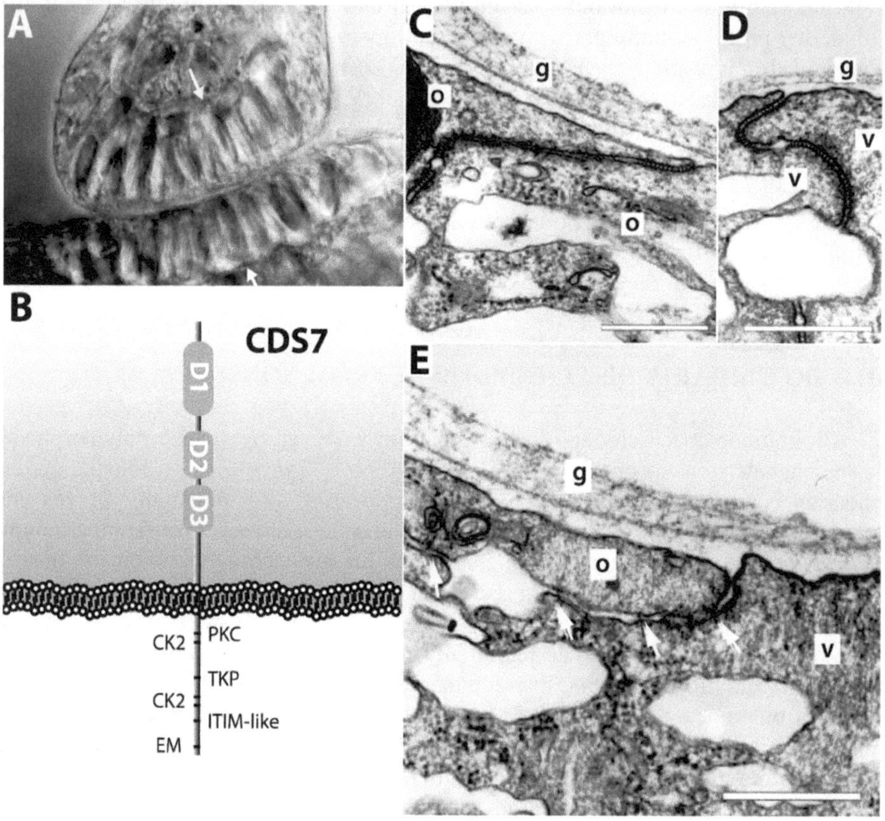

Figure 6. Cnidaria discriminate between self and nonself. A) Magnification of a stolon tip and stolon flank from rejecting *Hydractinia symbiolongicarpus* colonies, showing recruitment of nematocytes (arrowheads) to the contact point (Panel A modified from Lange et al,[60] ©2010, with permission from Elsevier). B) Schematic representation of a transmembrane receptor (CDS7) with three polymorphic domains (D1-D3) proposed from allorecognition locus *alr2* by Nicotra et al.[70] C-D) Transmission electron micrographs from ectoderm of grafting experiments between two animals of C) *Hydra oligatis* (o), D) *Hydra vulgaris* (v) and E) *Hydra vulgaris* (v) and *Hydra oligatis* (o). White arrows point to the heterotypic contact with a few irregular septae and spaces between cell membranes (compare septae in C) and D)). Note the species-specific differences in thickness of glycocalix (g). C-E) Modified from Kuznetsov et al.[73]

for self/nonself recognition evolved independently in different organisms.[72] Intensive searches for identifying molecules involved in allorecognition in solitary cnidarians such as *Hydra* were unsuccessful up to now.[73,74] There is, however, ample evidence for xenorecognition based on grafting experiments involving different Cnidarian species.[75] Close examination of such xenografts revealed strong histoincompatibility due to the inability to establish functional cell-cell contact structures such as septate junctions (Fig. 6C to E).[73]

CONCLUSON AND PERSPECTIVE—WHERE WILL THE TALE LEAD US?

In the absence of an adaptive immune system, Cnidarians employ an elaborate innate immune system to detect and eliminate nonself using their two cell layers as efficient defense barriers. In the absence of additional protective structures the endoderm appears to function as a chemical barrier relying exclusively on the activity of endodermally produced antimicrobial substances. In contrast to the endoderm, the ectoderm in *Hydra* is covered by a glycocalix (Fig. 2B, Fig. 6C-E). The fact that in conditions where the glycocalix is destroyed, the epithelium is highly vulnerable to pathogenic bacteria and fungi (Augustin and Bosch, pers. observation) strongly indicates that the ectoderm may function as a physicochemical ectodermal barrier protecting against invaders from the environment. In both barriers, multifunctional epithelial cells serve for pathogen recognition and defense. Their unique properties allow each epithelial barrier to deal with specific microenvironments and control the microbiota of each milieu. An important point for understanding the maintenance of homeostasis in our ancient multicellular ancestor is that neither barrier functions alone, but within the context of an organism and thus, the function of each epithelial barrier is likely to influence each other. As researchers look more closely at innate immune responses in a wide spectrum of animal taxa, it becomes clear that in all eumetazoan animals the first responders to pathogens are often the infected host epithelial or endothelial cells, rather than the arsenal of "professional" innate immune cells (macrophages and dendritic cells).[76] Thus, strict maintenance of epithelial barriers is certainly an ancestral and essential feature of innate immunity.

In future we expect Cnidarians to contribute to at least three different areas of comparative immunology. First, the recent accumulation of genomic, phylogenetic and functional data on components of innate immunity in Cnidarians will allow to uncover the innate immune repertoire in the eumetazoan ancestor. Scientists have traditionally devoted considerably more energy to understanding how immune systems work than to how they have evolved. Devoting energy for unravelling the logic of the immune system in Cnidarians, however, seems to be indispensable to understand how it all started at the dawn of evolution of multicellular animals and—maybe even more important—to reveal the evolutionary conserved mechanisms which allows the adaptation of organismal defense to environmental conditions. Second, Cnidarians are always associated with a complex microbiota. Understanding the Cnidarian host—microbe interactions and the communication molecules involved, will not only contribute to understanding the interactions of Cnidarian epithelia with microbial communities but will also provide a window into the evolution of beneficial microbiomes. Which roles do microbial communities play in the health of Cnidarians? Does the microbial flora associated with Cnidaria contribute to the antimicrobial defense of the animal? What principles govern the assembly and maintenance of the Cnidarian microbiome? Do antimicrobial peptides not only kill bacteria but are also involved in keeping the structure of the microbial community in balance? Answers to these questions will considerably improve our understanding of the role that tissue-associated microbial communities play in health and disease. This type of basic information in simple model systems is badly needed in light of the fact that disturbance of host microbe interactions in man results in severe inflammatory diseases such as bowl diseases. Third, in human medicine, the increasing prevalence of antibiotic-resistant microbes requires the development of new antimicrobial compounds. Antimicrobial peptides of animal origin may be an effective alternative or additive of conventional antibiotics for therapeutic use. The recent characterisation of highly active antimicrobial peptides in *Hydra* and *Aurelia* show that antimicrobial peptides from marine and freshwater cnidarians may represent a largely unexploited resource to design new antibiotics with broad-spectrum antimicrobial activity.

Much remains to be learned about the mechanisms of microbe recognition and defence in Cnidarians. It is clear that we are only beginning to understand the diversity and details of the mechanisms that are used to regulate their interaction with the microbial world, as well as the factors involved. Recent technological advances, particular in genomic analysis, however, suggest that the next few years are likely to be filled with many exciting and unanticipated discoveries that could rapidly reveal the mysteries of the two barriers.

ACKNOWLEDGEMENT

Research in our laboratory is supported in part by grants from the Deutsche Forschungsgemeinschaft (DFG) and grants from the DFG Cluster of Excellence programs "The Future Ocean" and "Inflammation at Interfaces".

REFERENCES

1. Putnam NH, Srivastava M, Hellsten U et al. Sea anemone genome reveals ancestral eumetazoan gene repertoire and genomic organization. Science 2007; 317:86-94.
2. Dunn CW, Hejnol A, Matus DQ et al. Broad phylogenomic sampling improves resolution of the animal tree of life. Nature 2008; 452:745-9.
3. Schierwater B, Eitel M, Jakob W et al. Concatenated analysis sheds light on early metazoan evolution and fuels a modern "urmetazoon" hypothesis. PLoS biology 2009; 7:e20.
4. Philippe H, Derelle R, Lopez P et al. Phylogenomics revives traditional views on deep animal relationships. Curr Biol 2009; 19:706-12.
5. Kortschak RD, Samuel G, Saint R et al. EST analysis of the cnidarian Acropora millepora reveals extensive gene loss and rapid sequence divergence in the model invertebrates. Curr Biol 2003; 13:2190-5.
6. Kusserow A, Pang K, Sturm C et al. Unexpected complexity of the Wnt gene family in a sea anemone. Nature 2005; 433:156-60.
7. Miller DJ, Ball EE, Technau U. Cnidarians and ancestral genetic complexity in the animal kingdom. Trends Genet 2005; 21:536-9.
8. Technau U, Rudd S, Maxwell P et al. Maintenance of ancestral complexity and nonmetazoan genes in two basal cnidarians. Trends Genet 2005; 21:633-9.
9. Wittlieb J, Khalturin K, Lohmann JU et al. Transgenic Hydra allow in vivo tracking of individual stem cells during morphogenesis. Proc Natl Acad Sci USA 2006; 103:6208-11.
10. Renfer E, Amon-Hassenzahl A, Steinmetz PR et al. A muscle-specific transgenic reporter line of the sea anemone, Nematostella vectensis. Proc Natl Acad Sci USA 2010; 107:104-8.
11. Saina M, Genikhovich G, Renfer E et al. BMPs and chordin regulate patterning of the directive axis in a sea anemone. Proc Natl Acad Sci USA 2009; 106:18592-7.
12. Rentzsch F, Fritzenwanker JH, Scholz CB et al. FGF signalling controls formation of the apical sensory organ in the cnidarian Nematostella vectensis. Development 2008; 135:1761-9.
13. Chapman JA, Kirkness EF, Simakov O et al. The Dynamic Genome of Hydra. Nature 2010; 464(7288):592-6
14. Hemmrich G, Bosch TC. Compagen, a comparative genomics platform for early branching metazoan animals, reveals early origins of genes regulating stem-cell differentiation. Bioessays 2008; 30:1010-8.
15. Miller DJ, Hemmrich G, Ball EE et al. The innate immune repertoire in Cnidaria—ancestral complexity and stochastic gene loss. Genome Biology 2007; 8
16. Reitzel AM, Sullivan JC, Traylor-Knowles N et al. Genomic survey of candidate stress-response genes in the estuarine anemone Nematostella vectensis. Biol Bull 2008; 214:233-54.
17. Nehyba J, Hrdlickova R, Bose HR. Dynamic evolution of immune system regulators: the history of the interferon regulatory factor family. Mol Biol Evol 2009; 26:2539-50.
18. Domazet-Loso T, Tautz D. An ancient evolutionary origin of genes associated with human genetic diseases. Mol Biol Evol 2008; 25:2699-707.

19. Bosch TCG, David CN. Immunocompetence in Hydra: Epithelial cells recognize self-nonself and react against it. J Exp Biol 1986; 238:225-34.
20. Bosch TC, Augustin R, Anton-Erxleben F et al. Uncovering the evolutionary history of innate immunity: the simple metazoan Hydra uses epithelial cells for host defence. Dev Comp Immunol 2009; 33:559-69.
21. Jung S, Dingley AJ, Augustin R et al. Hydramacin-1, structure and antibacterial activity of a protein from the basal metazoan Hydra. J Biol Chem 2009; 284:1896-905.
22. Augustin R, Siebert S, Bosch TC. Identification of a kazal-type serine protease inhibitor with potent anti-staphylococcal activity as part of Hydra's innate immune system. Dev Comp Immunol 2009; 33:830-7.
23. Mydlarz LD, Holthouse SF, Peters EC et al. Cellular responses in sea fan corals: granular amoebocytes react to pathogen and climate stressors. PloS one 2008; 3:e1811.
24. Bosch TC. Why polyps regenerate and we don't: towards a cellular and molecular framework for Hydra regeneration. Dev Biol 2007; 303:421-33.
25. McFall-Ngai M, Nyholm SV, Castillo MG. The role of the immune system in the initiation and persistence of the Euprymna scolopes-Vibrio fischeri symbiosis. Semin Immunol 2009.
26. Ainsworth TD, Fine M, Blackall LL et al. Fluorescence in situ hybridization and spectral imaging of coral-associated bacterial communities. Appl Environ Microbiol 2006; 72:3016-20.
27. Dinsdale EA, Edwards RA, Hall D et al. Functional metagenomic profiling of nine biomes. Nature 2008; 452:629-32.
28. Dinsdale EA, Pantos O, Smriga S et al. Microbial ecology of four coral atolls in the Northern Line Islands. PloS one 2008; 3:e1584.
29. Breitbart M, Bhagooli R, Griffin S et al. Microbial communities associated with skeletal tumors on Porites compressa. FEMS Microbiol Lett 2005; 243:431-6.
30. Frias-Lopez J, Zerkle AL, Bonheyo GT et al. Partitioning of bacterial communities between seawater and healthy, black band diseased and dead coral surfaces. Appl Environ Microbiol 2002; 68:2214-28.
31. Rohwer F, Breitbart M, Jara J et al. Diversity of bacteria associated with the Caribbean coral Montastraea franksi. Coral Reefs 2001; 20:85-91.
32. Rohwer F, Seguritan V, Azam F et al. Diversity and distribution of coral-associated bacteria. Mar Ecol Prog Ser 2002; 243:1-10.
33. Yokouchi H, Fukuoka Y, Mukoyama D et al. Whole-metagenome amplification of a microbial community associated with scleractinian coral by multiple displacement amplification using phi 29 polymerase. Environ Microbiol 2006; 8:1155-63.
34. Littman RA, Willis BL, Bourne DG. Bacterial communities of juvenile corals infected with different Symbiodinium (dinoflagellate) clades. Ma Ecol Prog Ser 2009; 389:45-59.
35. Ritchie KB. Regulation of microbial populations by coral surface mucus and mucus-associated bacteria. Mar Ecol Prog Ser 2006; 322:1-14.
36. Fraune S, Bosch TCG. Long-term maintenance of species-specific bacterial microbiota in the basal metazoan Hydra. Proc Natl Acad Sci USA 2007; 104:13146-51.
37. Fraune S, Abe Y, Bosch TCG. Disturbing epithelial homeostasis in the metazoan Hydra leads to drastic changes in associated microbiota. Environ Microbiol 2009; 11:2361-9.
38. Dunn SR, Thomason JC, Le Tissier MD et al. Heat stress induces different forms of cell death in sea anemones and their endosymbiotic algae depending on temperature and duration. Cell Death Diff 2004; 11:1213-22.
39. Weis VM. Cellular mechanisms of Cnidarian bleaching: stress causes the collapse of symbiosis. J Exp Biol 2008; 211:3059-66.
40. Palmer CV, Bythell JC, Willis BL. Levels of immunity parameters underpin bleaching and disease susceptibility of reef corals. FASEB J.
41. Rodriguez-Lanetty M, Phillips WS, Weis VM. Transcriptome analysis of a cnidarian-dinoflagellate mutualism reveals complex modulation of host gene expression. BMC Genomics 2006; 7:23.
42. Barneah O, Benayahu Y, Weis V. Comparative proteomics of symbiotic and aposymbiotic juvenile soft corals. Mar Biotechnol 2006; 8:11-6.
43. Voolstra CR, Schwarz JA, Schnetzer J et al. The host transcriptome remains unaltered during the establishment of coral-algal symbioses. Mol Ecol 2009; 18:1823-33.
44. Weis VM, Allemand D. What Determines Coral Health? Science 2009; 324:1153-5.
45. Muscatine L, Lenhoff HM. Symbiosis of hydra with algae. J Gen Microbiol 1963; 32:6.
46. O´Brien TL. Inhibition of vacuolar membrane-fusion by intracellular symbiotic algae in hydra-viridis (florida strain). J Exp Zool 1982; 223:211-8.
47. Thorington G, Margulis L. Hydra-Viridis—Transfer of Metabolites between Hydra and Symbiotic Algae. Biol Bull 1981; 160:175-88.
48. Habetha M, Anton-Erxleben F, Neumann K et al. The Hydra viridis/Chlorella symbiosis. Growth and sexual differentiation in polyps without symbionts. Zoology 2003; 106:101-8.

49. Beutler B. Innate immunity: an overview. Mol Immunol 2004; 40:845-59.
50. Rosenstiel P, Philipp EER, Schreiber S et al. Evolution and Function of Innate Immune Receptors—Insights from Marine Invertebrates. J Innate Immun 2009; 1:291-300.
51. Lange C, Hemmrich G, Klostermeier U et al. Complex repertoire of NOD-like receptors at the base of animal evolution 2010:submitted.
52. Schroder K, Zhou RB, Tschopp J. The NLRP3 Inflammasome: A Sensor for Metabolic Danger? Science 2010; 327:296-300.
53. Khalturin K, Hemmrich G, Fraune S et al. More than just orphans: are taxonomically-restricted genes important in evolution? Trends Genet 2009; 25:404-13.
54. Ovchinnikova TV, Balandin SV, Aleshina GM et al. Aurelin, a novel antimicrobial peptide from jellyfish Aurelia aurita with structural features of defensins and channel-blocking toxins. Biochem Biophys Res Commun 2006; 348:514-23.
55. Augustin R, Anton-Erxleben F, Jungnickel S et al. Activity of the novel peptide arminin against multiresistant human pathogens shows the considerable potential of phylogenetically ancient organisms as drug sources. Antimicrob Agents and Chemother 2009; 53:5245-50.
56. Dishaw LJ, Litman GW. Invertebrate Allorecognition: The Origins of Histocompatibility. Current Biology 2009; 19:R286-R8.
57. Buss LW, Grosberg RK. Morphogenetic Basis for Phenotypic Differences in Hydroid Competitive Behavior. Nature 1990; 343:63-6.
58. Hauenschild C. Genetische und entwichlungphysiologische Untersuchungen ueber Intersexualitaet und Gewebevertraeglichkeit bei Hydractinia echinata Flem. Wilhelm Roux's Arch Entwicklungsmech Org 1954; 147:1-41.
59. Buss LW, Shenk MA. Hydroid Allorecognition Regulates Competition at Both the Level of the Colony and at the Level of the Cell Lineage. Defense Molecules 1990; 121:85-105
60. Lange R, Plickert G, Muller WA. Histoincompatibility in a Low Invertebrate, Hydractinia-Echinata—Analysis of the Mechanism of Rejection. J Exp Zool 1989; 249:284-92.
61. Yund PO, Cunningham CW, Buss LW. Recruitment and Postrecruitment Interactions in a Colonial Hydroid. Ecology 1987; 68:971-82.
62. Buss LW, Yund PO. A Comparison of Recent and Historical Populations of the Colonial Hydroid Hydractinia. Ecology 1988; 69:646-54.
63. Yund PO, Parker HM. Population-Structure of the Colonial Hydroid Hydractinia Sp-Nov C in the Gulf of Maine. J Exp Mar Biol Ecol 1989; 125:63-82.
64. Buss LW. Somatic-Cell Parasitism and the Evolution of Somatic Tissue Compatibility. Proc Natl Acad Sci USA 1982; 79:5337-41.
65. Hauenschild C. Uber Die Vererbung Einer Gewebevertraglichkeits-Eigenschaft Bei Dem Hydroidpolypen Hydractinia-Echinata. Zeitschrift Fur Naturforschung Part B-Chemie Biochemie Biophysik Biologie Und Verwandten Gebiete 1956; 11:132-8.
66. Ivker FB. Hierarchy of Histo-Incompatibility in Hydractinia-Echinata. Biol Bull 1972; 143:162-174.
67. Mokady O, Buss LW. Transmission genetics of allorecognition in Hydractinia symbiolongicarpus (Cnidaria:Hydrozoa). Genetics 1996; 143:823-7.
68. Cadavid LF. Self-discrimination in colonial invertebrates: genetic control of allorecognition in the hydroid Hydractinia. Dev Comp Immunol 2004; 28:871-9.
69. Powell AE, Nicotra ML, Moreno MA et al. Differential effect of allorecognition loci on phenotype in Hydractinia symbiolongicarpus (Cnidaria: Hydrozoa). Genetics 2007; 177:2101-7.
70. Nicotra ML, Powell AE, Rosengarten RD et al. A Hypervariable Invertebrate Allodeterminant. Current Biology 2009; 19:583-9.
71. Nyholm SV, Passegue E, Ludington WB et al. fester, A candidate allorecognition receptor from a primitive chordate. Immunity 2006; 25:163-73.
72. Khalturin K, Panzer Z, Cooper MD et al. Recognition strategies in the innate immune system of ancestral chordates. Mol Immunol 2004; 41:1077-87.
73. Kuznetsov SG, Anton-Erxleben F, Bosch TCG. Epithelial interactions in Hydra: apoptosis in interspecies grafts is induced by detachment from the extracellular matrix. J Exp Biol 2002; 205:3809-17.
74. Kuznetsov SG, Bosch TCG. Self/nonself recognition in Cnidaria: contact to allogeneic tissue does not result in elimination of nonself cells in Hydra vulgaris. Zoology 2003; 106:109-16.
75. Theodor JL. Distinction between Self and Not-Self in Lower Invertebrates. Nature 1970; 227:690-692.
76. Thomas PG, Doherty PC. New approaches in immunotherapy. Science 2010; 327:249.

CHAPTER 2

GASTROPOD IMMUNOBIOLOGY

Eric S. Loker
Center for Evolutionary and Theoretical Immunology, Department of Biology, University of New Mexico, Albuquerque, New Mexico, USA
Email: esloker@unm.edu

Abstract: Over their 500 million year history, gastropods have radiated into marine, freshwater and terrestrial environments and adopted life styles ranging from herbivory to carnivory to endoparasitism to symbiont-mediated chemoautotrophy. They contend with many pathogens, including several lineages of specialized eukaryotic parasites. Their immunobiology is as yet poorly known, in part because most studies focus on a very small segment of gastropod diversity. Gastropod genome sequences are now forthcoming but synthetic overviews of the gastropod immunome are not yet available. Most immunological studies focus on interactions between gastropods and the larval stages of digenetic trematodes (digeneans) such as the medically important schistosomes. Digeneans elicit demonstrable and relevant snail defense responses and provide insights, augmented by the recently available schistosome genome sequences, for how gastropod responses are subverted. Survival of digeneans in snails depends at least in part on their ability to mimic host glycotopes, to overcome the immediate attack of reactive oxygen and nitrogen species produced by host hemocytes, and to induce long-term down-regulation of immune functions. Gastropods can mount distinct responses to different categories of pathogens, and can orchestrate effective elevated secondary responses under certain circumstances. Defense responses of at least one gastropod species, *Biomphalaria glabrata*, involve hemolymph lectins that are diversified by a variety of processes, including somatic diversification. Such observations have played a role in revising our general concept of invertebrate defense to include the possibility of more sophisticated and diversified responses beyond the production of limited repertoires of invariant pattern recognition molecules. The study of gastropod immunobiology is thus of basic interest and has several applied uses as well, including our need to conserve imperiled gastropod diversity.

Invertebrate Immunity, edited by Kenneth Söderhäll.

INTRODUCTION: AN HOMAGE TO GASTROPOD ANTIQUITY AND DIVERSITY

Any attempt to develop an overview of gastropod immunobiology should be framed by a consideration of gastropod antiquity and diversity of life styles. Gastropods were unarguably extant by the Upper Cambrian (488-501 mya) and perhaps even earlier, in the preCambrian, more than 542 million years ago.[1,2] They have since undergone dramatic radiations, numbering today between 40,000 and 150,000 living species,[3] ranging in size from less than 1 mm to nearly a meter in length.[4] A single New Caledonian coral reef lagoon can harbor over 2,000 gastropod species, and for some families, more than 80% of the diversity may yet to be recorded.[5] Gastropods are the most speciose class of animals to inhabit marine environments, and have also colonized land and freshwater repeatedly. Some species have life spans measured in decades[6,7] though many are annuals or have life histories marked by long periods of dormancy.[8] Most gastropod species are shelled and benthic, but some have abandoned the shell and some have adopted fully pelagic life styles. Some live in extreme habitats such as hydrothermal vents or sulfide seepages. Most are herbivorous though many are carnivorous, some are kleptoparasites, and some are nearly unrecognizable, highly modified endoparasites of other marine invertebrates. Although all gastropods probably depend to some extent on symbionts for their survival, in some this dependence is extraordinary, such that the gastropod host effectively "farms" its symbionts on its gills for eventual consumption,[9] or the gastropod/symbiont unit approaches the status of a photosynthetic[10] or chemoautotrophic organism.[11]

INFECTIOUS CHALLENGES TO GASTROPODS

Another important step in comprehending gastropod immunity is to gain an appreciation for the rogue's gallery of pathogens with which they must contend. Like any group of organisms, gastropods have viruses[12,13] yet the extent of challenge posed by viruses, and the nature of the gastropod antiviral response, are almost completely unknown. Gastropods in marine habitats are confronted with generalized[14-16] and specialized bacterial pathogens[17] which can cause general die-offs, especially in commercially grown species like abalones, particularly when stressed.[18] For most gastropods, natural die-offs resulting from exposure to viruses or bacteria are very likely to go unnoticed, so the pathogen burden borne by gastropods may be higher than presently appreciated. The increased application of culture-free metagenomics methods to the study of gastropod microbiology[19] will help clarify this issue.

Along with the usual background of viral or bacterial challenge, gastropods face other pathogens that are unequivocally gastropod specialists, the best known being the digenetic trematodes, also known as digeneans or "flukes". There are ~18,000 nominal species of digeneans, nearly all of which are parasites of molluscs, and most of which are obligate gastropod parasites.[20] Digeneans typically exhibit considerable host specificity with respect to their molluscan hosts. They undergo a complex developmental program in their chosen host, one that involves intimate contact with host tissues and extensive proliferation of larval stages that culminates, often weeks after initial infection, in production and release of large numbers of cercariae (Fig. 1). Infection often persists until the death of the host and in long-lived gastropods, infections can last for decades.[7] Digenean infection imposes a steep fitness cost as the host is partially or completely

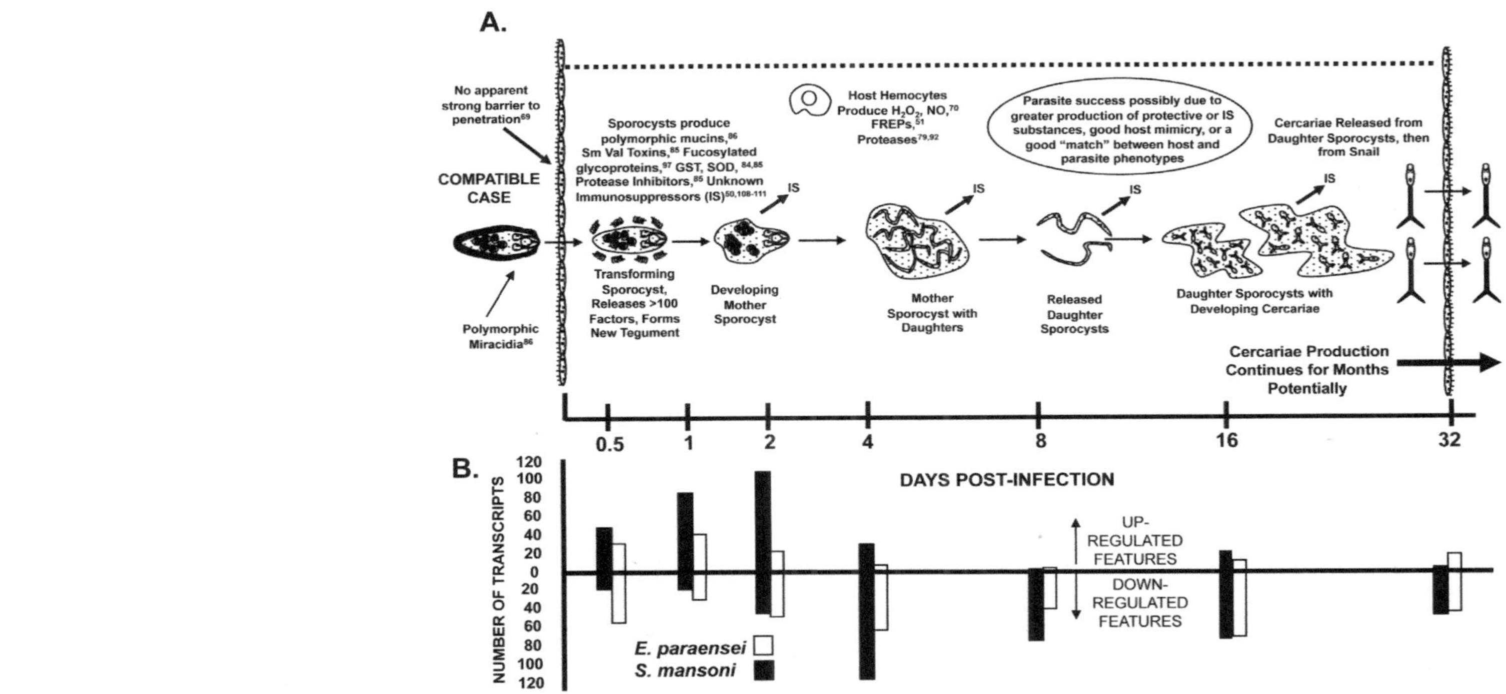

Figure 1. A) A timeline to show the sequence of events involved in the development of a digenetic trematode such as *Schistosoma mansoni* in the snail Biomphalaria glabrata, an example of a digenean life cycle involving sporocyst stages and no rediae. Infection is initiated by penetration of the snail epithelium by a miracidium, followed by the sequence of events noted, culminating in the production of cercariae approximately one month later. Cercariae cross the snail's epithelium to become free-swimming; their production can persist for several months. Indicated are key references that relate to events occurring early in infection, believed to be the critical period when successful infection is established. B) Shown in the same timeframe are the results of a microarray study indicating that following exposure of *B. glabrata* to either *S. mansoni* or *Echinostoma paraensei*, there is an early preponderance of host immune-related features that are up-regulated (especially for *S. mansoni*). Thereafter, throughout the long period required to complete development, a sustained predominance of down-regulated features occurs for both species.[50] This is indicative of the need for these parasites to induce a long-term state of immunosuppression while at the same time massively proliferating in their host, all without causing its premature demise.

castrated.[21] Digenean infections are known from a broad spectrum of gastropods ranging from basal limpet species to derived heterobranchs. Some gastropod species play host to dozens of digenean species, and simultaneous infection of individual gastropods with two or more digenean species is not uncommon.

A nonexhaustive list of additional pathogens containing at least some members with a significant degree of specialization on gastropods includes apicomplexans,[22] microsporidans,[23,24] ichthyosporeans,[25] nematodes,[26] copepods[27] and hematophagous mites.[28]

It must be noted that the vast majority of gastropod species have never been studied in an immunological context: our current information is based on just a few species, biased towards those of medical or commercial significance (Fig. 2). Furthermore, we should resist the temptation to assume that all gastropod defense systems are the same. It seems probable, especially given the age of the gastropod lineage, that the unique challenges posed by different habitats, feeding styles, and levels of commitments to symbiotic associations will have selected for diverse kinds of immune systems. For example, species heavily besieged by digeneans may have adopted very different immune "portfolios" than species living in dense colonies where directly-transmitted microparasites may be more of a concern. These portfolios might feature expansions of different gene families or the emergence of entirely novel defense mechanisms to deal with their respective immune challenges. The study of

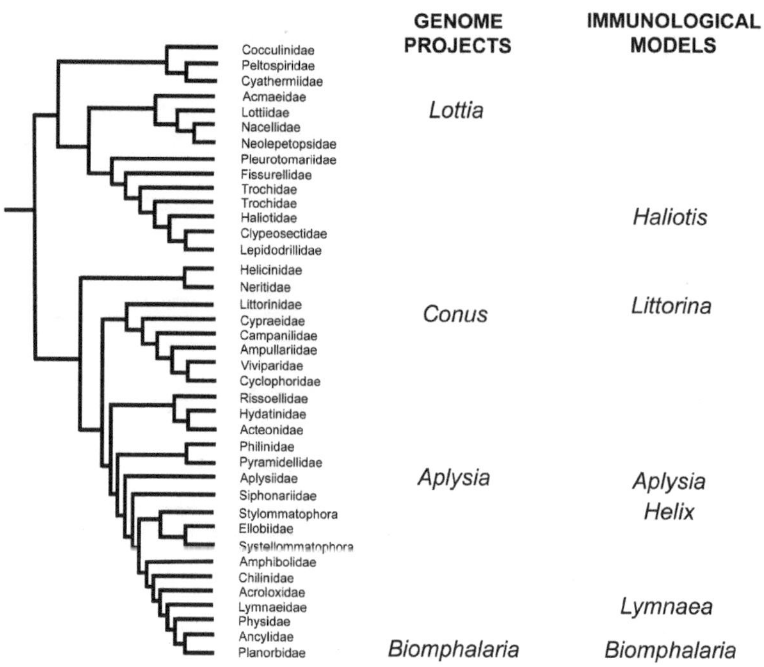

Figure 2. An overview of gastropod phylogeny, with emphasis on taxa most relevant to the study of gastropod immunobiology. This tree should not be viewed as a formal hypothesis of gastropod relationships but as an overview to assist readers in placing the approximate phylogenetic positions among gastropods prominently figuring in studies of gastropod immunology. Relationships outlined in trees presented by Aktipis et al (2008)[4] and Klussmann-Kolb et al (2008)[137] were used to delineate the relationships among the groups singled out for discussion.

alternative models of gastropod immunity should be encouraged, and Table 1 highlights some of the recent results forthcoming from the study of other than freshwater pulmonates.

GASTROPOD GENOME PROJECTS—AWAITING THE DELUGE

The completion of genome sequencing projects has revolutionized our understanding of both ecdysozoan[29,30] and deuterostome[31] invertebrate immunobiology. Recent years have also seen the initiation of gastropod genome projects, two of which have now been substantially completed with a third well underway. The first gastropod, and molluscan, genome sequence to be obtained is that of the owl limpet *Lottia gigantea* (*Patellogastropoda, Lottiidae*). The 359.5 Mbp *L. gigantea* genome is also the smallest known molluscan genome, and has been sequenced to 8.87X coverage and both automated and custom annotation provided (http://genome.jgi-psf.org/Lotgi1/Lotgi1.home.html). The initial shotgun sequence (7X coverage) and assembly of the 1.8Gbp genome of the California sea hare *Aplysia californica* (Opisthobranchia, Aplysiidae) has also been completed (http://www.broadinstitute.org/ science/projects/mammals-models/vertebrates-invertebrates/aplysia/aplysia-genome-sequencing-project), and sequencing of the 931 Mbp sequence of the medically important gastropod *Biomphalaria glabrata* (Pulmonata, Planorbidae) is well underway (http://biology.unm.edu/biomphalaria-genome/index.html.). A genome project emphasizing the venoms (venomics) of the cone snail *Conus consors* (www.conco.eu) is also underway. Fortuitously, these genome projects provide representation for diverse parts of the gastropod family tree (Fig. 2). As yet there has been no overarching synthetic or comparative interpretations for any of these genomes, particularly with respect to elaboration of the immunome, though a deluge of new information will be forthcoming in the near future, one that will surely also revolutionize our understanding of gastropod immunobiology.

THE DISTINCTIVE ARCHITECTURE OF GASTROPOD IMMUNE SYSTEMS OFFERS MANY OPPORTUNITIES FOR STUDY

The soft, moist body surface of gastropods is protected by a ciliated, mucus-producing epithelium that provides an initial physical trap and barrier to colonization by pathogens, one that is at present under-studied with respect to its role in defense.[32] The isolation of achacin from the body mucus of the giant African land snail *Achatina fulica* suggests mucus also provides a chemical barrier. Achacin is an L-amino oxidase that generates H_2O_2 from oxidative deamination of L-lysine and L-arginine and has antibacterial and tumoricidal properties.[33] The regular production and shedding of body mucus is also likely advantageous in cleansing the surface of pathogens like bacteria which are stimulated to grow in biofilms in mucus trails.[34] The composition, turnover rates, and dynamics of pathogen shedding of snail mucus are all worthy of more study.

Hemocytes—Multifunctional Guardians of the Snail Body

The open circulatory system of gastropods is populated with circulating defense cells called hemocytes (formerly often called amebocytes) with well appreciated defensive roles in phagocytosis and encapsulation reactions. These cells wander into tissue spaces as well. Gastropods also possess "fixed" defensive cells that are found around internal

Table 1. Important models to increase the phylogenetic breadth of the study of gastropod immunobiology

	Gastropods Studied				
Function	*Lottia*	*Haliotis*	*Littorina*	*Aplysia*	*Helix*
Immune cell and system description		hemocyte description[15]	hemocyte description[38,56] Signaling pathways[139]	hemocyte description[39] Albumen gland source of AMPs[140] Aplysianin in eggs[141] Ink as source of tumorcidal L-AA[142] 2 FREPS[127]	hemocyte description[138] Agglutinins from albumen gland[74]
Genome Mining	G-protein coupled recptors[62] FBG-encoding genes[95]				
Immune Gene Discovery		Perforin-like[143,144] NF-KB homolog[67] Pattern recog[145] B-thymosin[146] Cytidine deaminase[147] Abhisin AMP[148] TNFa homolog[149] GST cloned[150]			

continued on next page

Table 1. Continued

Function	Lottia	Haliotis	Gastropods Studied Littorina	Aplysia	Helix
Immune Gene Discovery		Fas ligand[151] C-type lectin[152] Mx protein[153]			
Immune-Pathogen Interactions		Vibriosis[154] Withering syndrome[17] Stress review[47] Vibrio and p38[63] Pedal mucus and bacteria[34] Bacteria for trt of vibriosis[156] Global warming effects[18] Herpesvirus[157] Gene regulation study[158] Tagged vibrios[14] Stress markers[151]	Signalling and trematodes[155] Hemocytes and digeneans[111] Hematopoiesis and digeneans[106]		

spaces and trap and/or phagocytose particulate objects.[35] Also present are rhogocytes (also known as pore cells) which are involved in synthesis or processing of respiratory proteins,[36] but have also been implicated in ingestion of small foreign particles.[37] The relative contribution of fixed cells to defense is not well understood.

In addition to their more familiar roles, hemocytes are also involved in wound healing, nerve repair, shell formation and repair, tissue remodeling possibly including dissolution of gonads after spawning, and metabolite and nutrient movement.[38-40] Gastropod hemocytes also engage in diapedesis, the movement of ingested foreign materials to and across external or gut epithelia.[37] The extent to which this occurs and influences production of new hemocytes or potentially increases susceptibility to new pathogen challenges is unknown.

We are far from a comprehensive understanding of where hemocytes are produced, how many distinct lineages of hemocytes are produced, their life spans and how functionally diversified they are: these parameters will almost certainly be shown to vary among different gastropod lineages. A distinct hematopoietic organ, called the amebocyte producing organ or APO, has been identified in some snails like *B. glabrata* and *Lymnaea truncatula*.[41,42] The APO of *B. glabrata*, if removed from snails of a strain resistant to *Schistosoma mansoni* infection and implanted into snails of a susceptible strain will successfully transfer a degree of resistance.[43,44] Others have doubted whether this organ alone can provide sufficient numbers of hemocytes for the entire snail, and have postulated that hemocyte production also occurs in peripheral vascular locations.[45] In other pulmonates, both individual circulating and tissue-dwelling hemocytes can divide,[46] and circulating blast-like cells in *Littorina littorea* are also capable of division.[38] For abalones, a hematopoietic organ has yet to be localized.[47]

The pathways involved in activating hematopoiesis are poorly known and may involve direct stimulation by pathogens or indirect stimulation via mitogenic hemocyte cytokines.[48] Excretory/secretory products of sporocysts of the digenean *Echinostoma paraensei* stimulated enlargement of the APO of *B. glabrata*,[49] and extracts of *S. mansoni* stimulated an increase in mitosis in excised APOs prompting the suggestion that the parasite provides a direct mitogenic or nutritive effect on hematopoiesis.[48] APOs treated with phorbol myristate acetate (PMA), a stimulator of protein kinase C (PKC), showed increased mitotic activity.[48] Recent studies have identified from *B. glabrata* potential mitogen-related proteins including TGF-β Type 1 receptors and epidermal growth factor (EGF)-related proteins, which are up-regulated early in the course of digenean infection and thus are promising candidates as hematopoiesis-promoting factors.[50]

A monograph could be written on the classification and description of gastropod hemocytes,[15,39] and suffice it to say here that gastropod blood typically contains hemocytes that do not spread or spread minimally when placed on artificial surfaces (often called round cells, or blast-like cells), whereas the majority of cells typically spread avidly, forming conspicuous filopodia or lamellopodia in the process, and are believed to be the mature effectors in phagocytosis and encapsulation responses. Such spreading cells have been referred to as hyalinocytes or as granulocytes, the difference in terminology relating to the numbers of "granules" within the cells, a parameter likely to vary among gastropod taxa and as a function of hemocyte ontogeny. Round/blast-like cells are frequently considered to be of an earlier stage in hemocyte ontogeny, and in *L. littorea* such cells were shown to have incorporated 5-bromo-2'-deoxyuridine indicative of recent cell division and are believed to differentiate into mature effector cells.[38]

Related subjects awaiting further study are how long gastropod hemocytes live, and whether mature effector hemocytes are terminally differentiated, although it has been noted hemocytes participating in encapsulation and wound healing responses can still divide.[46] In comparison to other invertebrates, hemocytes of the small number of gastropods so far studied are relatively few in distinctive morphological types, generally lack large aggregates of conspicuous granules or pigments, lack flagella, and do not have an obvious propensity to lyse when removed from the host or presented with an antigenic stimulus. Gastropod hemocytes are characteristically sticky and readily form aggregates, a tendency noted to be more pronounced in marine than freshwater species.[38]

Hemocyte Signaling Pathways

Hemocytes are responsible for synthesis and release of several defense-related factors such as reactive oxygen species (ROS),[39] lectins such as fibrinogen-related proteins (FREPs)[51] and antimicrobial peptides (AMPs).[52] As compared to many other invertebrates, the study of gastropod AMPs is in its infancy. Hemocyte membrane-associated lectins have been implicated in detection of intruders. For example, a galectin present on the surface of ~60% of *B. glabrata* hemocytes has been characterized, and in recombinant form binds to the tegument of *S. mansoni* sporocysts in a carbohydrate-inhibitable manner, suggesting it is a hemocyte-bound pattern recognition molecule.[53] It was also suggested there may be "counter receptors" on hemocytes, such as integrin-like molecules,[54] that could be bound by soluble forms of galectin, such that the galectin could also serve in cross-bridging hemocytes to a parasite surface.[53]

Key to understanding hemocyte effector functions are the intracellular signal transduction pathways likely to be activated by exposure of hemocytes to exotic stimuli. PMA stimulation of hemocytes, acting on PKC, results in H_2O_2 generation in *B. glabrata*,[55] of superoxide anions in *L. littorina*,[56] and of NO in Lymnaea stagnalis.[57] PKC activation is likely to result in activation via phosphorylation of mitogen activated protein kinases (MAPKs) like ERK or p38 because inhibitors of MAPKs also prevent hemocyte spreading or H_2O_2 production.[58,59] More natural stimuli such as laminarin also activate PKC and H_2O_2 production in *L. stagnalis* hemocytes.[60] A role for phosphatidylinositol 3-kinase in controlling phagocytic activity has been shown in *L. stagnalis* hemocytes,[61] and G-protein coupled membrane receptors have also recently been reported from *L. littorea*.[62] The notion that components of gastropod signaling pathways such as p38 can be targeted by pathogens, as they often are with pathogens of plants, has recently been documented in abalones contending with *Vibrio harveyi*.[63]

Toll-like receptors (TLRs) and associated Toll pathway components such as MyD88 are known from cephalopods[64] and bivalves.[65,66] Homologs of TLRs are present in the *Lottia* genome, a Rel-like NF-kB transcription factor is known from abalones,[67] and both Dorsal-like and Rel-like transcriptional factors occur in *B. glabrata* (Zhang, personal communication) so additional Toll pathway homologs are likely to be present in gastropods, although their functional relevance is as yet unknown. It is somewhat surprising that the numerous studies focused on digenean-snail interactions have not as yet turned up TLRs or transcription factors of canonical Toll-signaling pathways. Acting upstream of Toll-signaling pathways in other invertebrates are pattern recognition molecules like PGRPs (peptidoglycan recognition proteins) and GNBP (gram negative binding protein, or B-1-3 glucan recognition/binding protein or LGBP). Both short and long form PGRP-encoding genes are present in *B. glabrata*[68] and the latter generates different transcripts via alternative splicing, some of which may have amidase activity. At least three different GNBPs are also known from *B. glabrata*, and two of these group with other GNBPs that may preserve glucanase activity.[68] Short form PGRP and GNBP were down-regulated at 6 hours after exposure to 3 types of microbes and neither responded early in the course of digenean infection, whereas the former was modestly up-regulated after two weeks of digenean infection. Long-form PGRP expression was consistently low and not modified by exposure to microbes or digeneans.[68]

A Plethora of Immune-Relevant Molecules Awaiting Functional Validation

One of the conceptual cornerstones of gastropod immunobiology is that hemocytes interact in concert with several different categories of molecular defense factors, many of them circulating in the hemolymph, together more effectively recognizing and disabling pathogens.[39,56] Among the prominent categories of immune-relevant molecules either produced by, or working in concert with, hemocytes are ROS, lectins, AMPs, PGRPs, GNBPs, proteases, protease inhibitors, and complement-like proteins. Much of the current focus of gastropod immunobiology is on revealing and identifying these factors, many of which are alluded to in the following sections of this document, and have been recapped in recent reviews[69,70] and in Table 1 and Figure 3. As just a few recent examples of intriguing immune-relevant molecules that have also proven to be responsive to microbial or digenean stimulation on a *B. glabrata* oligo-based microarray,[71] are a complement C1q-like protein with a lectin domain, serpins, several signaling pathway components including an NF-κB p105 subunit, several FREPs, a FREM which is a unique juxtaposition of fibrinogen and EGF domains, and a macrophage migration inhibition-like factor. All await further study. Also most worthy of further consideration are the many "unknown" sequences (up to 60%) responsive to immune challenge that are often recovered from gene discovery efforts. Since they do not fit established paradigms, further study of immune responsive unknowns may lead to the most novel future discoveries of gastropod defenses.

Some Additional Comments on Gastropod Internal Defense Systems

Gastropod blood does not obviously coagulate upon removal from the animal or after exposure to pathogens. However, tubular helical filaments visible with electron microscopy have been found in hemolymph,[39,72] and plasma from *B. glabrata* infected with *E. paraensei* often forms precipitates.[51] It has been suggested that matrilin in *B. glabrata* plasma might participate in formation of filaments which could serve to keep hemocytes from contacting digenean sporocysts in susceptible snails.[73] Further study is thus warranted to determine if gastropods have as yet unappreciated abilities to alter their hemolymph in ways that might be easily overlooked and are relevant for defense.

The albumen gland, well-known for its role in providing nutrients in the perivitelline fluids packaged around each developing embryo in an egg mass, has received increasing attention for its immunological roles. A unique hexameric "H-type" lectin is produced by the albumen gland of *Helix pomatia* and *Cepaea hortensis* and is believed to protect developing embryos from bacteria.[74] A recent proteomics study revealed that 16 of 20 identified polypeptides in *B. glabrata* egg mass fluids have a defensive role, and include protease inhibitors, compounds with possible phenoloxidase activity, Cu-Zn superoxide dismutase, C-type lectins, GNBP, aplysianin/achacin type proteins, and LBP/BPI or lipopolysaccharide binding protein/bacterial permeability increasing protein.[75] Defensive

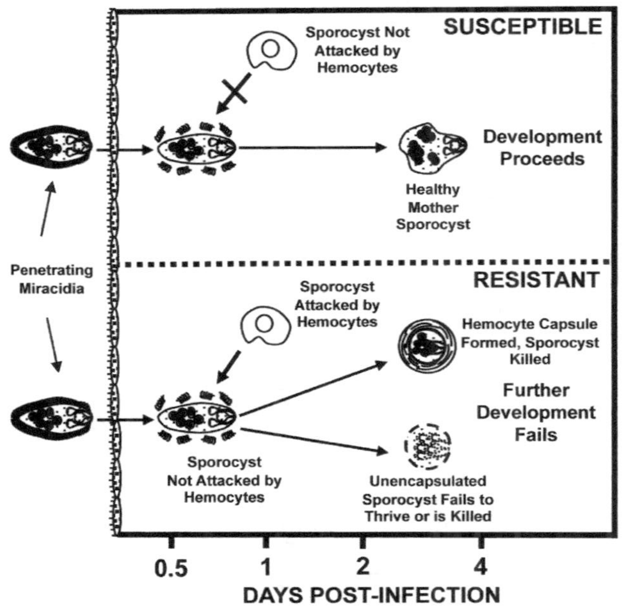

CONSTITUTIVE DIFFERENCES BETWEEN SUSCEPTIBLE AND RESISTANT SNAILS

Resistant Snails have more hemocytes than susceptible snails
- Bg-Sm: [176]

Resistant hemocytes have different behavior or properties
- Bg-Sm: [177]
- Bt-Sm: [178]

Resistant Snails produce more parasite-toxic reactive oxygen species (ROS)
- Bg-Sm: [70,101,179-182]

Resistant Snails have novel recognition/killing capacity in their plasma
- Bt-Sm: [183,184]
- Bg-Sm: [44]

Innate differences in the levels of potentially immune relevant factors
- Bg-Sm: [185]
- Bg-Ec: [73,76,94,186,187]
- Bt-Sm: [188]

Resistant snails have greater ability to resist parasite offensive strategies, including signaling pathway targets
- Bg-Sm: [48,116]
- Bg-Ec: [110]

Resistant and susceptible snails may differ in the ease with which they are mimicked by parasites
- Bg-Sm: [97]

RESPONSE CAPACITIES FOLLOWING EXPOSURE TO INFECTION

Resistant Snails are more stimulated to make hemocytes following exposure
- Bg-Sm: [189]

Resistant snails respond more prominently or differently to infection than susceptible snails
- Bg-Sm: [92,100,102,103,185,189-191]
- Bg-Ec: [73,76,77,94,187]
- Bt-Sm: [188]

Figure 3. Explanations for the phenomenon of resistance of snails to digenean infection. At least three different experimental model systems have been developed to examine the phenomenon of resistance of snails to trematode infection: Bg-Sm: This is the most prominent and involves *Schistosoma mansoni* and its interactions with *Biomphalaria glabrata* strains either selected for resistance (such as 10-R2 or 13-16-R1 strains) or that are naturally resistant (BS-90 snails, also known as Salvador snails). Bg-Ec: This model involves *B. glabrata* and strains selected for resistance or susceptibility to *Echinostoma caproni*. Bt-Sm: A third model gaining recent prominence involves *S. mansoni* and *Biomphalaria tenagophila* isolates that are either naturally susceptible (Cabo Frio) or resistant (Taim) to infection. Some of the prominent explanations for the underlying raison d'etre for resistance derived from these models and recent applicable references are outlined in the attached list. The more recent papers will lead readers to additional older papers on similar topics.

compounds produced by the albumen gland are probably distributed systemically with the plasma to protect the adult snail as well.[76,77]

One surprising feature of gastropod immunity is the limited extent to which snails so far have been shown to rely on phenoloxidase and the eventual production of melanin to kill or wall off pathogens, at least in the most commonly studied gastropod models.[39, 78] PPO activity is reported from snails,[79] and the egg mass fluids they produce,[80] and it is conceivable that particular groups of gastropods, once investigated, will have a greater reliance on melanization and associated reactions for defense than seen thus far. A molluscan hallmark, the ability to produce the stacked, compartmentalized structure of aragonite crystals to make shell nacre, can also be exploited in defense by encasing macroscopic pathogens such as trematode metacercariae. The molecular basis of nacre formation is being revealed in bivalves[81] and homologs should be sought in gastropods.

SNAILS AND DIGENEANS AS MODELS TO STUDY SPECIFIC, INTIMATE AND LONG-TERM HOST-PARASITE SYSTEMS

The recent study of gastropod immunology has been dominated by the use of freshwater pulmonate gastropods, especially the planorbid snail *B. glabrata*, as study subjects. This is understandable because such snails serve as the intermediate hosts for digeneans of medical or veterinary significance, such as *S. mansoni* in the case of *B. glabrata*. Digeneans are conveniently available, natural pathogens of gastropods that can be used as probes to elicit relevant host immune responses then amenable to study. Snail-digenean systems are excellent models to study the "give and take" involved in intimate, long-term host-parasite associations. The use of digeneans as a means to study gastropod defenses is more fruitful now than ever because genome sequences have been recently forthcoming for *S. mansoni* and the related *S. japonicum*.[82,83] These have already been of use in illuminating aspects of the intramolluscan biology of digeneans[84,85] that in turn provide important clues for how the snail response might be orchestrated.[86,87] Furthermore, a number of important tools and resources have been developed that not only allow further dissection of the outcome of digenean-snail encounters but also enable exploration of responses to other pathogens, or to environmental perturbations. These include the ongoing *B. glabrata* genome project[88] and associated development of BAC libraries[89] and EST databases,[90] the development of microarrays for both digenean[91] and snail,[71,92] the use of echinostome flukes such as *Echinostoma paraensei* or *E. caproni* to elicit host responses that can be compared and contrasted with those engendered by *S. mansoni*[69] and the development of strains of *B. glabrata* susceptible or resistant to infection with either *S. mansoni* or *E. caproni* (Fig. 3). Recent reviews can be consulted for more detail,[21,69,70] particularly with respect to the specific identity of a number of candidate molecules identified for further study.

Use of both EST datasets and a *B. glabrata* microarray have shown that the gastropod response to different insults—wounding, injection of gram positive or negative bacteria, or exposure to infection with *S. mansoni* or *E. paraensei*—is not "one size fits all" [50, 71] and that snails can discriminate among different kinds of immunological stimuli, a phenomenon also noted by others.[90,93-95] Wounding alone provoked rather mild responses, the two kinds of bacteria upregulated only partially overlapping subsets of array features, and the response to the two digeneans not only were different, but also easily separable from those elicited by bacteria.[71]

With respect to the more specific issue of the determinants of compatibility between digeneans and snails, several recent productive lines of investigation have been developed, many scrutinizing differences between schistosome-susceptible and -resistant snails (Fig. 3). None of the explanations generated should be ruled out and all may apply to one extent or the other. With respect to the role of ROS, hemocytes from snails resistant to *S. mansoni* infection produce higher levels of H_2O_2 than do susceptible snails.[70] Hydrogen peroxide is particularly lethal to *S. mansoni* sporocysts which have also been shown to be naturally vulnerable to high oxygen tensions.[96] Furthermore, resistant snails possess a different allelic form of the enzyme Cu/Zn superoxide dismutase involved in hydrogen peroxide production than do susceptible snails,[70] though it is unclear if this differences or other gene regulatory steps are responsible for the different levels of hydrogen peroxide achieved. On the parasite side, proteomic studies of the larval transformation proteins (LTPs) of both *S. mansoni* and *E. paraensei* sporocysts[84,85] reveal molecules involved in scavenging of ROS to be prominent, highly suggestive of their importance in protection from hemocyte attack. Also found among the LTPs are intriguing venom-related proteins[85] and glycoconjugates,[97] many of which bear fucose residues and make excellent candidates for host-mimicking molecules as host molecules are also known to bear fucose residues.[98,99] As noted by Wu et al,[85] a veritable "cloud" of parasite produced factors are released, at least 99 of which have been characterized.[85] Among these molecules are highly polymorphic mucins and it has been speculated that these parasite produced molecules might serve as a smokescreen to divert attack by humoral factors such as FREPs,[86,87] at least some of which have a known binding preference for fucose residues.[51] Thanks to the availability of the *S. mansoni* genome, we can gain a much better perspective on the kinds of molecules released by transforming parasites, or by developing daughter sporocysts, and can begin to formulate hypotheses for how these might confuse or suppress the host defense response.

Although protective and potentially obfuscating parasite molecules are produced, it is clear that *B. glabrata* is nonetheless able to detect the presence of either *S. mansoni* or *E. paraensei* infection within hours of penetration.[50,92,100-103] Thus unless subverted, a growing host response could increasingly jeopardize the developing parasite. A number of lines of evidence stemming from the original work of Lie and colleagues[104] suggest immunosuppression, or "interference" in the parlance of the snail-digenean literature, is mediated by larval digeneans, and, importantly, that this effect can be implemented within one hour after penetration[105] and likely persists for the duration of the infection.[69] Hemocytes from digenean-infected snails often show diminished ability to adhere to surfaces, engage in phagocytosis or to encapsulate trematode larvae[106,107] effects that can be reproduced by exposure of hemocytes to LTP from echinostome sporocysts or rediae.[108-111] A recent microarray study documenting the responses of *B. glabrata* from 0.5 to 32 days following exposure to either *S. mansoni* or *E. paraensei* revealed that both parasites, after a brief early preponderance of upregulated array features, provoked a persistent excess of down-regulated features relative to up-regulated features after 2 dpe (Fig. 1).[50] Insofar as this study examined whole body extracts from exposed snails it is not unreasonable to assume that significant down-regulation occurred even earlier among host cells in the immediate vicinity of developing parasites, in accordance with earlier work documenting trematode-mediated interference in *B. glabrata*.[109,112] Also, it should be emphasized that although interference has been most associated with echinostome development, this recent study, also in agreement with Lie's classical studies,[112] suggests that schistosomes also engage in a strategy of immunosuppression though it seems to take somewhat longer to be manifested.[50] Indeed, based on results obtained from other snail-digenean studies where evidence consistent with interference has been obtained,[111] it seems likely that a strategy of immunosuppression may be a general requirement for parasites like digeneans with a long developmental program and persistent infections.

The specific digenean-produced factors responsible for inducing interference, and their exact mechanisms of action are unknown, but more detailed information regarding the in vitro larval digenean proteome[84,85] provide good starting points for testing candidates, though the actual in vivo proteome of larval digeneans may be somewhat different. One likely possibility is that digeneans produce factors that alter the activity of hemocyte signaling pathways, likely in more than one way.[113,114] Exposure of adherent *B. glabrata* hemocytes to LTP of *E. paraensei* sporocysts or rediae caused both induction of calcium waves and rounding, not always both in the same cell, suggestive of at least two distinct effects of LTP.[113] Following exposure to LTP or to whole *S. mansoni* sporocysts, ERK-dependent signaling was impaired in hemocytes from *B. glabrata* susceptible to infection, but not in schistosome-resistant snails. As ERK regulates important hemocyte effector functions, it was suggested that disruption of ERK signaling, possibly via carbohydrates expressed on parasite surfaces or in LTP, facilitated survival of *S. mansoni* in susceptible snails.[115,116] It was also noted that components in plasma may protect hemocytes from the impact of parasite-mediated suppression of hemocyte signaling pathways,[115] providing a potential explanation for why some plasma factors, for instance FREP4, are persistently upregulated in snails that have been successfully infected with digeneans.[50,117]

In contemplating the study of schistosome-snail interactions, including those studies outlined in Figure 3, it should be kept in mind that the relative amount of genetic diversity present in both digenean and snail laboratory populations is low, and has been skewed by artificial selection. Furthermore, natural populations of schistosome intermediate hosts such as *B. glabrata* from South America, or of *Biomphalaria pfeifferi* from Africa, typically do not exhibit wholesale resistance to infection with most or all *S. mansoni* miracidia. In contrast, in more natural settings, the success or failure of a particular schistosome-snail encounter depends on the genetic constitution of each participant,[118,119] such that some combinations are compatible and some are not, e.g., parasite and host phenotypes are matched or mismatched. Thus the results of laboratory studies must be interpreted cautiously, and more studies using fresh field isolates encouraged. The underlying basis for phenotype matching remains unknown and building on the intriguing results of Roger et al[86,87] should be a priority for future study.

ADDRESSING FUNDAMENTAL ISSUES IN IMMUNOLOGY USING GASTROPOD MODELS

Gastropods and their immune systems can also serve as useful models to consider important conceptual issues applicable to all organisms and the problems they confront with pathogens. The standard paradigm emphasizes the view that invertebrates elaborate a basic set of pathogen pattern recognition receptors coupled to standard responses like activation of phagocytosis or production of protective factors such as ROS, lectins or AMPs. An emerging challenge to this paradigm, in which studies of gastropods have figured prominently, has emphasized that the threats posed by the ubiquity, diversity and rapid evolvability of pathogens require a greater degree of sophistication from invertebrate defenses than previously considered.[120] It also notes that many invertebrates, including some gastropods live for decades, and must be able to contend with pathogens that could "lock on" and decimate a host with a fixed immune repertoire.[121,122]

Central to this discussion is the issue of the production of molecules with diverse recognition capability, such that broader spectra of pathogens, including those that may actively alter their own surface molecules or secretions, could be achieved. Lectins play an important role in nonself recognition in gastropods.[123] Vasta et al[124] note that even in the absence of unusual genetic rearrangements, a substantial degree of recognition diversity can be achieved because lectins exist in multi-gene families, can engage in alternative splicing, and the carbohydrate recognition domains of lectins exhibit considerable "plasticity" in their interactions with ligands and so might still bind multiple pathogens with slightly altered surface carbohydrates. A cursory examination of the *Lottia* genome indicates that several C-type lectins are present, and the same is almost certainly true for the *B. glabrata* genome, so these considerations are likely to apply.

Studies of fibrinogen-related proteins, or FREPs, from *B. glabrata*, which were first identified following exposure to digeneans, also exemplify some of the points of Vasta et al,[124] however, they differ in one important regard, namely the presence of an unforeseen mechanism of genetic rearrangement. FREPs are circulating hemolymph proteins that as presently understood are comprised of either one or two N-terminal immunoglobulin domains connected via a short interceding region to a C-terminal fibrinogen domain.[117] They exist in a large family comprised of 14 identifiable sub-families. FREPs have calcium-dependent lectin activity, and bind soluble digenean antigens, or to the surfaces of digenean sporocysts and bacteria.[95] Recent microarray studies suggest they are highly responsive to digenean infection, some being up-regulated, and others down-regulated. In our experience, the persistent up-regulation of FREP4 is effectively a marker of digenean infection in *B. glabrata*.[50]

In addition to their multiple subfamily representation, FREPs are further diversified by alternative splicing, and because at least some FREPs exist in their native configuration as multimers,[95] this represents a further opportunity for diversification, as has been noted for fibrinogen-related defense molecules in mosquitoes.[125] Additionally, study of the FREP3 subfamily surprisingly provided evidence for somatic diversification resulting from a combination of point mutations and putative gene conversion events acting on a relatively small number of source sequences,[126] the latter process favored by tandem arrangement of FREP3 genes, as has been noted from recent studies of BAC architecture (Adema, personal communication). Interestingly, this diversification process has either not been noted or is far less prominent for members of other FREP subfamilies such as FREP2 and 4, which show levels of variation more in line with expectations of normal allelic variation or representation from a small number of germline loci. This further suggests the FREP3 diversification noted is not artifactual (Adema, personal communication). One noteworthy aspect of FREP3 somatic diversification is that it provides a potential mechanism to generate diversity among different individual hemocytes[117] and raises the possibility that not all gastropod hemocytes "are created equal", offering a potential means to greatly increase the immune recognition repertoire.

The specific combination of IgSF and fibrinogen domains has been found in other planorbid snails such as *Helisoma trivolvis*, and recently two bonafide FREPs have been reported from *Aplysia californica*,[127] extending their representation to include another major lineages within the heterobranch clade. The *Lottia* genome contains an estimated 70 fibrinogen-encoding genes,[95] but they are not believed to be associated with IgSF domains. In general, gastropod studies have helped draw attention to the importance of the fibrinogen domain which is proving to be common among other invertebrate groups as well, frequently being implicated in defense responses.[117]

Recent studies of *S. mansoni* miracidia and sporocysts indicate they have a remarkable system for generating diversified mucin molecules that are released only in the molluscan host, such that each individual miracidium is potentially unique in the spectrum of mucins produced.[86,87] Heterogeneity in glycotope expression within and among *S. mansoni* miracidia has also been noted by Peterson et al.[97] These findings serve to accentuate the point that even metazoan parasites can present a highly diversified face to the gastropod defense system. Although the function of diversified mucins is presently unknown, it has been speculated they could potentially interact with FREPs, possibly serving as a smokescreen to divert host recognition efforts away from the parasite, thus potentially offering an explanation for why schistosome-snail compatibility depends on the specific genetic composition of the schistosome being matched to that of the snail.[86,-87] Although this is an attractive hypothesis, much more work is needed to determine if digenean mucins and host FREPs actually interact with one another, and how these systems might work in the more relevant test bed of naturally occurring snails and schistosomes.

A number of other lines of evidence also indicate that gastropods can produce diversified defense molecules. Bouchut et al[73] in a study of genes involved in hemocyte adherence or extracellular filament formation in *B. glabrata* found evidence for point mutations in six of the genes studied, but only if transcripts originated from hemocytes as opposed to whole bodies, and concluded "it raises the question of the existence of a gene wide diversification process taking place preferentially (or exclusively) in hemocytes of *B. glabrata*." Ittiprasert et al[102] noted that *B. glabrata* mucin transcripts were heterogeneous and commented they could match the heterogeneity of *S. mansoni*-produced mucins. In their study of *B. glabrata* egg mass fluids, Hathaway et al[75] noted that the most abundant protein LBP/BPI existed in variant forms, and at least six similar yet variant C-type lectins were present. Zhang et al[95] noted the presence of multiple forms of both PGRPs and GNBPs from *B. glabrata*, and in a study of hemocyte-mediate nerve repair in *L. stagnalis*, Hermann et al[40] speculated that different integrins may exist in different combinations on hemocytes thus creating diverse populations. In a similar vein, recent studies of the bivalve *Mytilus edulis* have revealed that one category of AMPs, myticin C, exhibits extraordinary variation within and among individuals, and suggested that all the variants were generated from more common sequences by an as yet unknown mechanism.[128] Perhaps the presence of molluscan mobile genetic elements such as *Nimbus*[129] can help to explain some of the genetic rearrangements these results imply. It is important to remain cautious about the real implications of all the studies documenting diversification in recognition capacity as much more study is needed to document the exact mechanisms involved, and to validate their actual relevance in terms of protection from pathogen challenge.

Gastropods also provide an excellent opportunity to teach us about another fundamental aspect of invertebrate defense: by virtue of prior exposure to immunological stimuli, can defense responses be induced to quantitatively or qualitatively different levels that diminish susceptibility to subsequent pathogen challenge in a manner actually relevant to the animal? This topic is fraught with controversy among invertebrate immunologists and caution is required to assure that outcomes attributable to immunological priming or memory actually have an immunological basis.[130] One of the most compelling set of studies addressing this general topic was provided by Lie et al[131] who showed that by first exposing *B. glabrata* to irradiated echinostome miracidia, those same snails would subsequently be resistant upon challenge with normal miracidia up to 10 days later. By contrast, control snails exposed only to normal snails readily became infected. The study was all the more remarkable for showing a degree of specificity: snails with acquired resistance to one echinostome species had partial resistance to other echinostomes, but remained fully susceptible to *S. mansoni*. Snails sensitized twice and then challenged showed an even faster time course of parasite destruction, and a role of humoral factors in immobilizing parasites and making them vulnerable to encapsulation was invoked.[132] They concluded acquired resistance was "not due to increased numbers of circulating amebocytes, but rather to intrinsic changes in the recognition and probably also killing capacity of amebocytes".[104] This model system awaits further study and has the potential to help us define the limits of what gastropods, and invertebrates in general, are capable of with respect to priming of their defense systems.

OPPORTUNITIES TO EXPAND THE RELEVANCE OF GASTROPOD IMMUNITY

The primary focus of this chapter has been on basic gastropod immune mechanisms, often as revealed by exposure to digenetic trematodes. However, there are several additional more applied contexts in which an understanding of gastropod immunology is useful, and that provide additional opportunities to justify studies of this topic. Some of these are highlighted and summarized in Table 2.

CONCLUSION

The current state of gastropod immunology is marked by anticipation of the insights soon to be forthcoming from genome projects and by—thanks to the development of a number of new tools and approaches—the possibility of better understanding a backlog of interesting immune candidates that await rigorous testing with respect to their functional significance. The study of gastropod immunology currently is in need of approaches such as RNAi to enable more comprehensive functional testing. RNA knockdown can be achieved in *B. glabrata* by injection of dsRNA of target genes,[133] but the degree of knockdown needs to be higher so we can draw clearer conclusions from such experiments, and the consistency with which significant knockdown can be achieved improved. Using a different approach based on injection of multiple short dsRNAs, knockdown of FREP2 has again been achieved and is being further investigated (Hanington, personal communication). RNAi has been reported in *L. stagnalis*[134] and bivalves,[81,135] so progress can be anticipated on this front.

Except for recent studies documenting viruses in abalones (Table 1), we are almost completely ignorant with respect to the diversity and impacts of gastropod viruses and the nature of the immune response they engender. Availability of the *Biomphalaria glabrata* embryonic (Bge) cell line as a means to search for viruses from natural populations of planorbid and possibly other snails should be encouraged, both to find viruses that might be potential control agents for medically important snails and to harvest viruses which can then be used to probe basic gastropod defense responses. Another need is to develop more comprehensive overviews of gastropod transcriptional profiles, which should be enabled as genome sequences projects are completed. Representation of the

Table 2. The following topics represent important ways in which the scope and relevance of studies of gastropod immunobiology can be expanded

Maintaining Gastropods as Sources of Food
Learning how abalones respond to viruses[13,153] or bacteria,[63] to improve abalone culture
Control of Snail-Transmitted Human Parasites
Replacing snails in natural habitats susceptible to *S. mansoni* with snails that are resistant, thus achieving biological control of schistosomiasis[159,160]
To Identify Compounds with Potential Biomedical Relevance
Antimicrobial peptides, like abhisin from the disk abalone *Haliotis discus discus*,[148] may have activity against bacteria, fungi and cancer cells.
To Better Protect Endangered Gastropods, Including those Threatened by Disease
Perhaps an improved understanding of gastropod immunity could have been used to save the land snail *Partula turgida* from extinction at the hands of a microsporidian parasite[161]
Elimination of Invasive Gastropods
A thorough understanding of the immunobiology of invasive gastropods could lead to approaches to exploit co-evolved or new pathogens in their control.[162,163]
To Return Balance to Ecosystems
Improve abalone health and abundance to provide normal food sources for endangered sea otters[164]
As Models to Understand Commitment of Internal Defense Systems to Symbionts
The bacterial associates of gastropods are poorly known but range from near monocultures of specialized bacteria[9] to surprisingly diverse floras in the guts of diet generalists like *Biomphalaria* or *Helisoma*.[19] The nature of the accommodations made to symbionts in such diverse cases is unknown.
As Models to Study the Impact of Pollutants or Other Stressors on Immune Function, and to Develop Biomarkers for Stress
Snails with shells fouled by epibionts were also more likely to be infected with digeneans, and though multiple explanations might exist, interactions between stressors are likely.[165]
Snails stressed by fasting, exposure to cold or to detergent became more susceptible to trematode infection[166]
Toxin exposure may influence susceptibility to pathogens like trematodes[103,167] Exposure to herbicides such as fomesafen[168] or organic chemicals[169] may diminish defense capacities Use of selenium-dependent glutathione peroxidase as a possible biomarker for physical stress in abalones[170]

continued on next page

Table 2. Continued

To Investigate the Impact of Mounting Immune Responses on Life History Attributes
Repeated exposure to bacteria necessitating immune responses affected short-lived and long-lived land snails differently, with long-lived species checking growth and mating frequency, thus potentially favoring long-term survival[171]
B. glabrata snails resistant to *E. caproni* exhibit delayed maturity compared to susceptible snails[172] Rigby and Jokela (2000)[173] showed that immune response measured by increase in hemocytes trades off with reproduction and survival in snails, and predator avoidance may modify expression and costs of immune defense.
To Examine the Interface between Immunity and Behavior
Zbikowska (2004)[174] found that snails infected with digeneans tended to select lower temperatures than uninfected controls, and postulated this slowed down the rate of parasite production and gave the snail time to respond to the damage produced, No "behavioral fever" was noted.
To Monitor Climate Change and Record its Impacts
A 1°C temperature change increases abalone mortality due to *Vibrio harveyi*[18]
Different abalone species respond differently to withering syndrome with temperature change[175]

transcriptomic repertoires on current snail microarrays is modest. More comprehensive arrays can be used to assess responses to different pathogen exposures, and how such responses might be modulated in snails from natural habitats, including those under environmental stress or contending with preexisting infections. Such studies may very well also reveal novel immune responses undertaken by components that today we can only classify as "unknowns".

Thanks to the continuing availability of new genome sequences for gastropods and other invertebrates, in combination with further studies to assess the importance of various immune pathways to survival, we will begin to approach for the first time higher order questions pertaining to the evolution of immunity. For example, (1) does the presence of prominent lineage-specific pathogens—such as digeneans in snails—drive the development of unique immune capabilities not seen in other host groups?; (2) how similar are the immune systems of diverse gastropods, and do the patterns we discern among their defense systems more strongly reflect a phylogenetic signal, or do they diverge more as a consequence of their immediate ecological challenges?; (3) as previously argued,[136] are the internal defense systems of gastropods and bivalves fundamentally different, with the former showing narrower recognition diversity than the latter?; and finally, (4) do the immune capacities of major lineages, such as at the phylum level, evolve along similar trajectories, or has selection placed a premium on the development of novel immune solutions that prevent easy pathogen transfer from one group to the next? Exciting times lie ahead, and as noted at the beginning of this chapter, we must take care not to overgeneralize the gastropod case for no doubt, embraced within this ancient and diverse lineage, are fascinating alternative solutions to the problem of achieving effective internal defense.

ACKNOWLEDGMENTS

The author thanks Ms. Michelle Forys for assistance, and Dr. Coen Adema, Dr. Patrick Hanington and Dr. Si-Ming Zhang for critically reading the manuscript. The support of NIH grant AI24340 and CETI which is supported by NIH grant number 1P20RR18754 from the Institutional Development Award (IDeA) Program of the National Center for Research Resources, are gratefully acknowledged.

REFERENCES

1. Fryda J, Nutzel A, Wagner PJ. Paleozoic Gastropoda. In: Ponder WF, Lindberg DR, eds. Phylogeny and Evolution of the Mollusca. Berkeley: University of California Press, 2008:239-270.
2. Parkhaev PY. The early Cambrian radiation of Mollusca. In: Ponder WF, Lindberg DR, eds. Phylogeny and Evolution of the Mollusca. Berkeley: University of California Press, 2008:33-69.
3. Lindberg DR, Ponder WF, Haszprunar G. The Mollusca: relationships and patterns from their first half-billion years. In: Cracaft J, Donoghue MJ, eds. Assembling the Tree of Life. New York: Oxford University Press, 2004:252-278.
4. Aktipis SW, Giribet G, Lindberg DR et al. Gastropoda an overview and analysis. In: Ponder WF, Lindberg DR, eds. Phylogeny and Evolution of the Mollusca. Berkeley: University of California Press, 2008:201-237.
5. Bouchet P, Lozouet P, Maestrati P et al. Assessing the magnitude of species richness in tropical marine environments: exceptionally high numbers of molluscs at a New Caledonia site. Biol J Linn Soc 2002; 75:421-436.
6. Heller J. Longevity in mollusks. Malacologia 1990; 31:259-295.

7. Curtis LA. Spatial heterogeneity in size and parasitism: How it arises in an estuarine snail population. J Exp Mar Biol Ecol 2007; 352:317-330.
8. Storey KB. Life in the slow lane: molecular mechanisms of estivation. Comp Biochem Physiol A Mol Integr Physiol 2002; 133:733-754.
9. Bates AE. Feeding strategy, morphological specialisation and presence of bacterial episymbionts in lepetodrilid gastropods from hydrothermal vents. Mar Ecol-Progr Ser 2007; 347:87-99.
10. Burghardt I, Evertsen J, Johnsen G et al. Solar powered seaslugs - Mutualistic symbiosis of aeolid nudibranchia (Mollusca, Gastropoda, Opisthobranchia) with Symbiodinium. Symbiosis 2005; 38:227-250.
11. Suzuki Y, Kojima S, Watanabe H et al. Single host and symbiont lineages of hydrothermalvent gastropods Ifremeria nautilei (Provannidae): biogeography and evolution. Mar Ecol-Progr Ser 2006; 315:167-175.
12. Prince JS. Presumptive alphavirus in the gastropod mollusc, Aplysia californica. Bull Mar Sci 2003; 73:673-677.
13. Hooper C, Hardy-Smith P, Handlinger J. Ganglioneuritis causing high mortalities in farmed Australian abalone (Haliotis laevigata and Haliotis rubra). Aust Vet J 2007; 85:188-193.
14. Travers MA, Barbou A, Le Goic N et al. Construction of a stable GFP-tagged Vibrio harveyi strain for bacterial dynamics analysis of abalone infection. FEMS Microbiol Lett 2008; 289:34-40.
15. Travers MA, Mirella da Silva P, Le Goic N et al. Morphologic, cytometric and functional characterisation of abalone (Haliotis tuberculata) haemocytes. Fish Shellfish Immunol 2008; 24:400-411.
16. Cheng L, Huang J, Shi CY et al. Vibrio parahaemolyticus associated with mass mortality of postlarval abalone, Haliotis divericolor supertexta (L.), in Sanya, China. Journal of the World Aquaculture Society. 2008; 39:746-757.
17. Moore JD, Finley CA, Robbins TT et al. Withering syndrome and restoration of southern California abalone populations. Cal Coop Ocean Fish 2002; 43:112-117.
18. Travers MA, Basuyaux O, Le Goic N et al. Influence of temperature and spawning effort on Haliotis tuberculata mortalities caused by Vibrio harveyi: an example of emerging vibriosis linked to global warming. Global Change Biol 2009; 15:1365-1376.
19. Van Horn DJ, Hall JR, Loker ES et al. Gut bacterial community composition in three planorbid snail species. In review.
20. Cribb TH, Bray RA, Olson PD et al. Life cycle evolution in the Digenea: a new perspective from phylogeny. Advan Parasitol 2003; 54:197-254.
21. Lockyer AE, Jones CS, Noble LR et al. Trematodes and snails: an intimate association. Can J Zool. 2004; 82:251-269.
22. Gros O, Frenkiel L, Aranda DA. Structural analysis of the digestive gland of the queen conch Strombus gigas Linnaeus, 1758 and its intracellular parasites. J Mollus Stud. 2009; 75:59-68.
23. McClymont HE, Dunn AM, Terry RS et al. Molecular data suggest that microsporidian parasites in freshwater snails are diverse. Int J Parasitol 2005; 35:1071-1078.
24. Selman BJ, Jones AA, Barker GM. Microsporida (Microspora) parasitic in terrestrial gastropods. In: Barker GM, ed. Natural Enemies of Terrestrial Molluscs. Wallingford and Cambridge: CABI Press, 2004:579-597.
25. Hertel LA, Bayne CJ, Loker ES. The symbiont Capsaspora owczarzaki, nov. gen. nov. sp., isolated from three strains of the pulmonate snail Biomphalaria glabrata is related to members of the Mesomycetozoea. Int J Parasitol 2002; 32:1183-1191.
26. Morand, S, Wilson, MJ, Glen DM et al. Nematodes (Nematoda) parasitic in terrestrial gastropods. In: Barker GM, ed. Natural Enemies of Terrestrial Molluscs. Wallingford and Cambridge: CABI Press, 2004:525-557.
27. Schrodl M. Heavy infestation by endoparasitic copepod crustaceans (Poecilostomatoida: Splanchnotrophidae) in Chilean opisthobranch gastropods, with aspects of splanchnotrophid evolution. Organisms Diversity and Evolution 2002; 2:19-26.
28. Schupbach HU, Baur B. Parasitic mites influence fitness components of their host, the land snail Arianta arbustorum. Invertebr Biol 2008; 127:350-356.
29. Lemaitre B, Hoffmann J. The host defense of Drosophila melanogaster. Annu Rev Immunol. 2007; 25:697-743.
30. Ziegler K, Pujol N. C. elegans defence mechanisms. Med Sci (Paris) 2009; 25:497-503.
31. Hibino T, Loza-Coll M, Messier C et al. The immune gene repertoire encoded in the purple sea urchin genome. Dev Biol 2006; 300:349-365.
32. Vieira TC, Costa-Filho A, Salgado NC et al. Acharan sulfate, the new glycosaminoglycan from Achatina fulica Bowdich 1822. Structural heterogeneity, metabolic labeling and localization in the body, mucus and the organic shell matrix. Eur J Biochem 2004; 271:845-854.
33. Ehara T, Kitajima S, Kanzawa N et al. Antimicrobial action of achacin is mediated by L-amino acid oxidase activity. FEBS Letters 2002; 531:509-512.
34. Guo F, Huang ZB, Huang MQ et al. Effects of small abalone, Haliotis diversicolor, pedal mucus on bacterial growth, attachment, biofilm formation and community structure. Aquaculture 2009; 293:35-41.
35. Matricon-Gondran M, Letocart M. Internal defenses of the snail Biomphalaria glabrata. I. Characterization of hemocytes and fixed phagocytes. J Invertebr Pathol 1999; 74:224-234.

36. Albrecht U, Keller H, Gebauer W et al. Rhogocytes (pore cells) as the site of hemocyanin biosynthesis in the marine gastropod Haliotis tuberculata. Cell Tissue Res 2001; 304:455-462.
37. Sminia T. Structure and function of blood and connective tissue cells of fresh water pulmonate Lymnaea stagnalis studied by electron microscopy and enzyme histochemistry. Zeitschrift fur Zellforschung und Mikroskopische Anatomie 1972; 130:497-526.
38. Gorbushin AM, Iakovleva NV. Haemogram of Littorina littorea. J Mar Biol Assoc UK. 2006; 86:1175-1181.
39. Martin GG, Oakes CT, Tousignant HR et al. Structure and function of haemocytes in two marine gastropods, Megathura crenulata and Aplysia californica. J Mollus Stud 2007; 73:355-365.
40. Hermann PM, Nicol JJ, Bulloch AGM et al. RGD-dependent mechanisms in the endoneurial phagocyte response and axonal regeneration in the nervous system of the snail Lymnaea stagnalis. J Exp Biol 2008; 211:491-501.
41. Lie KJ, Heyneman D, Yau P. Origin of amebocytes in Biomphalaria glabrata. J. Parasitol 1975; 61:574-576.
42. Rondelaud D, Barthe D. The development of the amebocyte producing organ in Lymnaea truncatula Muller infected by Fasciola hepatica L. Z Parasitenkd 1981; 65:331-341.
43. Sullivan JT, Spence JV, Nunez JK. Killing of Schistosoma mansoni sporocysts in Biomphalaria glabrata implanted with amoebocyte-producing organ allografts from resistant snails. J Parasitol 1995; 81:829-833.
44. Vasquez RE, Sullivan JT. Hematopoietic tissue allografts in Biomphalaria glabrata (Mollusca: Pulmonata) induce humoral immunity to Schistosoma mansoni. Dev Comp Immunol 2001; 25:561-564.
45. Souza SD, Andrade ZA. On the origin of the Biomphalaria glabrata hemocytes. Mem Inst Oswaldo Cruz 2006; 101:213-218.
46. Sminia T. Haematopoiesis in the freshwater snail Lymnaea stagnalis studied by electron microscopy and autoradiography. Cell Tissue Res 1974; 150:443-454.
47. Hooper C, Day R, Slocombe R et al. Stress and immune responses in abalone: limitations in current knowledge and investigative methods based on other models. Fish Shellfish Immunol 2007;22:363-379.
48. Salamat Z, Sullivan JT. Involvement of protein kinase C signalling and mitogen-activated protein kinase in the amebocyte-producing organ of Biomphalaria glabrata (Mollusca). Dev Comp Immunol 2009; 33:725-727.
49. Noda S. Effects of excretory-secretory products of Echinostoma paraensei larvae on the hematopoietic organ of M line Biomphalaria glabrata snails. J Parasitol 1992; 78:512-517.
50. Hanington PC, Lun CM, Adema CM et al. Time series analysis of the transcriptional responses of Biomphalaria glabrata throughout the course of intramolluscan development of Schistosoma mansoni and Echinostoma paraensei. Int J Parasitol 2010; 40(7):819-31.
51. Adema CM, Hertel LA, Miller RD et al. A family of fibrinogen-related proteins that precipitates parasite-derived molecules is produced by an invertebrate after infection. Proc Natl Acad Sci USA 1997; 94:8691-8696.
52. Iijima R, Kisugi J, Yamazaki M. L-Amino acid oxidase activity of an antineoplastic factor of a marine mollusk and its relationship to cytotoxicity. Dev Comp Immunol 2003; 27:505-512.
53. Yoshino TP, Dinguirard N, Kunert J et al. Molecular and functional characterization of a tandem-repeat galectin from the freshwater snail Biomphalaria glabrata, intermediate host of the human blood fluke Schistosoma mansoni. Gene 2008; 411:46-58.
54. Davids BJ, Yoshino TP. Integrin-like RGD-dependent binding mechanism involved in the spreading response of circulating molluscan phagocytes. Dev Comp Immunol 1998; 22:39-53.
55. Bender RC, Broderick EJ, Goodall CP et al. Respiratory burst of Biomphalaria glabrata hemocytes: Schistosoma mansoni-resistant snails produce more extracellular H^2O^2 than susceptible snails. J Parasitol 2005; 91:275-279.
56. Gorbushin AM, Iakovleva NV. Functional characterization of Littorina littorea (Gastropoda: Prosobranchia) blood cells. J Mar Biol Assoc UK 2007; 87:741-746.
57. Wright B, Lacchini AH, Davies AJ et al. Regulation of nitric oxide production in snail (Lymnaea stagnalis) defence cells: a role for PKC and ERK signalling pathways. Biol Cell 2006; 98:265-278.
58. Zelck UE, Gege BE, Schmid S. Specific inhibitors of mitogen-activated protein kinase and PI3-K pathways impair immune responses by hemocytes of trematode intermediate host snails. Dev Comp Immunol 2007; 31:321-331.
59. Humphries JE, Yoshino TP. Regulation of hydrogen peroxide release in circulating hemocytes of the planorbid snail Biomphalaria glabrata. Dev Comp Immunol 2008; 32:554-562.
60. Lacchini AH, Davies AJ, Mackintosh D et al. Beta-1, 3-glucan modulates PKC signalling in Lymnaea stagnalis defence cells: a role for PKC in H^2O^2 production and downstream ERK activation. J Exp Biol 2006; 209:4829-4840.
61. Plows LD, Cook RT, Davies AJ et al. Phagocytosis by Lymnaea stagnalis haemocytes: a potential role for phosphatidylinositol 3-kinase but not protein kinase A. J Invertebr Pathol 2006; 91:74-77.
62. Gorbushin AM, Klimovich AV, Iakovleva NV. Himasthla elongata: effect of infection on expression of the LUSTR-like receptor mRNA in common periwinkle haemocytes. Exp Parasitol 2009; 123:24-30.

63. Travers MA, Le Bouffant R, Friedman CS et al. Pathogenic Vibrio harveyi, in contrast to non-pathogenic strains, intervenes with the p38 MAPK pathway to avoid an abalone haemocyte immune response. J Cell Biochem 2009; 106:152-160.
64. Goodson MS, Kojadinovic M, Troll JV et al. Identifying components of the NF-kappaB pathway in the beneficial Euprymna scolopes-Vibrio fischeri light organ symbiosis. Appl Environ Microbiol 2005; 71:6934-6946.
65. Qiu LM, Song LS, Xu W et al. Molecular cloning and expression of a Toll receptor gene homologue from Zhikong scallop, Chlamys farreri. Fish Shellfish Immunol 2007; 22:451-466.
66. Wang L, Song L, Zhao J et al. Expressed sequence tags from the Zhikong scallop (Chlamys farreri): discovery and annotation of host-defense genes. Fish Shellfish Immunol 2009; 26:744-750.
67. Jiang Y, Wu X. Characterization of a Rel\NF-kappaB homologue in a gastropod abalone, Haliotis diversicolor supertexta. Dev Comp Immunol 2007; 31:121-131.
68. Zhang SM, Zeng Y, Loker ES. Characterization of immune genes from the schistosome host snail Biomphalaria glabrata that encode peptidoglycan recognition proteins and gram-negative bacteria binding protein. Immunogenetics 2007; 59:883-898.
69. Coustau C, Gourbal B, Mitta G et al. Echinostomes and snails: exploring complex interactions. In: Fried B, Toledo R, eds. The Biology of Echinostomes. New York: Springer, 2008:35-60.
70. Bayne, CJ. Successful parasitism of vector snail Biomphalaria glabrata by the human blood fluke (trematode) Schistosoma mansoni: a 2009 assessment. Mol Biochem Parasitol 2009; 165:8-18.
71. Adema CM, Hanington PC, Lun CM et al. Differential transcriptomic responses of Biomphalaria glabrata (Gastropoda, Mollusca) to bacteria and metazoan parasites, Schistosoma mansoni and Echinostoma paraensei (Digenea, Platyhelminthes). Mol Immunol 2010; 47:849-860.
72. Matricon-Gondran M, Letocart M. Internal defenses of the snail Biomphalaria glabrata. III. Observations on tubular helical filaments induced in the hemolymph by foreign material. J Invertebr Pathol 1999; 74:248-254.
73. Bouchut A, Roger E, Coustau C et al. Compatibility in the Biomphalaria glabrata/Echinostoma caproni model: potential involvement of adhesion genes. Int J Parasitol 2006; 36:175-184.
74. Sanchez JF, Lescar J, Chazalet V et al. Biochemical and structural analysis of Helix pomatia agglutinin. A hexameric lectin with a novel fold. J Biol Chem 2006; 281:20171-20180.
75. Hathaway JJ, Adema CM, Stout BA et al. Identification of protein components of egg masses indicates parental investment in immunoprotection of offspring by Biomphalaria glabrata (Gastropoda, Mollusca). Dev Comp Immunol 2010; 34:425-435.
76. Vergote D, Bouchut A, Sautiere PE et al. Characterisation of proteins differentially present in the plasma of Biomphalaria glabrata susceptible or resistant to Echinostoma caproni. Int J Parasitol 2005; 35:215-224.
77. Guillou F, Mitta G, Galinier R et al. Identification and expression of gene transcripts generated during an anti-parasitic response in Biomphalaria glabrata. Dev Comp Immunol 2007; 31:657-671.
78. Smith VJ, Soderhall K. A comparison of phenoloxidase activity in the blood of marine invertebrates. Dev Comp Immunol 1991; 15:251-261.
79. Bahgat M, Doenhoff M, Kirschfink M et al. Serine protease and phenoloxidase activities in hemocytes of Biomphalaria glabrata snails with varying susceptibility to infection with the parasite Schistosoma mansoni. Parasitol Res 2002; 88:489-494.
80. Bai G, Brown JF, Watson C et al. Isolation and characterization of phenoloxidase from egg masses of the gastropod mollusc, Biomphalaria glabrata. Comp Biochem Physiol B Biochem Mol Biol 1997; 118:463-469.
81. Suzuki M, Saruwatari K, Kogure T et al. An acidic matrix protein, Pif, is a key macromolecule for nacre formation. Science 2009; 325:1388-1390.
82. Berriman M, Haas BJ, LoVerde PT et al. The genome of the blood fluke Schistosoma mansoni. Nature 2009; 460:352-358.
83. Zhou Y, Zheng H, Liu F et al. The Schistosoma japonicum genome reveals features of host-parasite interplay. Nature 2009; 460:345-351.
84. Guillou F, Roger E, Mone Y et al. Excretory-secretory proteome of larval Schistosoma mansoni and Echinostoma caproni, two parasites of Biomphalaria glabrata. Mol Biochem Parasitol 2007; 155:45-56.
85. Wu XJ, Sabat G, Brown JF et al. Proteomic analysis of Schistosoma mansoni proteins released during in vitro miracidium-to-sporocyst transformation. Mol Biochem Parasitol 2009; 164:32-44.
86. Roger E, Grunau C, Pierce RJ et al. Controlled chaos of polymorphic mucins in a metazoan parasite (Schistosoma mansoni) interacting with its invertebrate host (Biomphalaria glabrata). PLoS Negl Trop Dis 2008; 2:e330.
87. Roger E, Mitta G, Mone Y et al. Molecular determinants of compatibility polymorphism in the Biomphalaria glabrata/Schistosoma mansoni model: new candidates identified by a global comparative proteomics approach. Mol Biochem Parasitol 2008; 157:205-216.
88. Raghavan N, Knight M. The snail (Biomphalaria glabrata) genome project. Trends Parasitol 2006; 22:148-151.
89. Adema CM, Luo MZ, Hanelt B et al. A bacterial artificial chromosome library for Biomphalaria glabrata, intermediate snail host of Schistosoma mansoni. Mem Inst Oswaldo Cruz 2006; 101 Suppl 1:167-177.

90. Hanelt B, Lun CM, Adema CM. Comparative ORESTES-sampling of transcriptomes of immune-challenged Biomphalaria glabrata snails. J Invertebr Pathol 2008; 99:192-203.
91. Veriovski-Almeida S, Venancio TM, Oliveira K et al. Use of a 44k oligoarray to explore the transcriptome of Schistosoma mansoni adult worms. Expt Parasitol 2007; 117:236-245.
92. Lockyer AE, Spinks J, Kane RA et al. Biomphalaria glabrata transcriptome: cDNA microarray profiling identifies resistant- and susceptible-specific gene expression in haemocytes from snail strains exposed to Schistosoma mansoni. BMC Genomics 2008; 9:634.
93. Wang SH, Wang YL, Zhang ZX et al. Response of innate immune factors in abalone Haliotis diversicolor supertexta to pathogenic or nonpathogenic infection. J Shellfish Res 2004; 23:1173-1177.
94. Mitta G, Galinier R, Tisseyre P et al. Gene discovery and expression analysis of immune-relevant genes from Biomphalaria glabrata hemocytes. Dev Comp Immunol 2005; 29:393-407.
95. Zhang SM, Zeng Y, Loker ES. Expression profiling and binding properties of fibrinogen-related proteins (FREPs), plasma proteins from the schistosome snail host Biomphalaria glabrata. Innate Immun 2008; 14:175-189.
96. Bixler LM, Lerner JP, Ivanchenko M et al. Axenic culture of Schistosoma mansoni sporocysts in low O^2 environments. J Parasitol 2001; 87:1167-1168.
97. Peterson NA, Hokke CH, Deelder AM et al. Glycotype analysis in miracidia and primary sporocysts of Schistosoma mansoni: Differential expression during the miracidium-to-sporocyst transformation. Int J Parasitol 2009; 39:1331-1344.
98. Lehr T, Geyer H, Maass K et al. Structural characterization of N-glycans from the freshwater snail Biomphalaria glabrata cross-reacting with Schistosoma mansoni glycoconjugates. Glycobiology 2007; 17:82-103.
99. Lehr T, Beuerlein K, Doenhoff MJ et al. Localization of carbohydrate determinants common to Biomphalaria glabrata as well as to sporocysts and miracidia of Schistosoma mansoni. Parasitology 2008; 135:931-942.
100. Nowak TS, Woodards AC, Jung Y et al. Identification of transcripts generated during the response of resistant Biomphalaria glabrata to Schistosoma mansoni infection using suppression subtractive hybridization. J Parasitol 2004; 90:1034-1040.
101. Knight M, Raghavan N, Goodall C et al. Biomphalaria glabrata peroxiredoxin: Effect of Schistosoma mansoni infection on differential gene regulation. Mol Biochem Parasitol 2009; 167:20-31.
102. Ittiprasert W, Miller A, Myers J et al. Identification of immediate response genes dominantly expressed in juvenile resistant and susceptible Biomphalaria glabrata snails upon exposure to Schistosoma mansoni. Mol Biochem Parasitol 2010; 169:27-39.
103. Ittiprasert W, Nene R, Miller A et al. Schistosoma mansoni infection of juvenile Biomphalaria glabrata induces a differential stress response between resistant and susceptible snails. Exp Parasitol 2009; 123:203-211.
104. Lie KJ, Jeong KH, Heyneman D. Molluscan host reactions to helminthic infection. Immune Responses in Parasitic Infections: Immunology, Immunopathology, and Immunoprophylaxis. Boca Raton: CRC Press, 1987:211-270.
105. Lie KJ, Heyneman D, Richards CS. Studies on resistance in snails - interference by nonirradiated echinostome larvae with natural resistance to Schistosoma mansoni in Biomphalaria glabrata. J Invertebr Pathol 1977; 29:118-125.
106. Gorbushin AM, Iakovleva NV. The enigma of the haemogram "left-shift" in periwinkles infected with trematodes. Fish Shellfish Immunol 2008; 24:745-751.
107. Noda S, Loker ES. Phagocytic activity of hemocytes of M line Biomphalaria glabrata snails: effect of exposure to the trematode Echinostoma paraensei. J Parasitol 1989; 75:261-269.
108. Loker ES, Cimino DF, Hertel LA. Excretory-secretory products of Echinostoma paraensei sporocysts mediate interference with Biomphalaria glabrata hemocyte functions. J Parasitol 1992; 78:104-115.
109. Adema CM, Arguello DF, 2nd, Stricker SA et al. A time-lapse study of interactions between Echinostoma paraensei intramolluscan larval stages and adherent hemocytes from Biomphalaria glabrata and Helix aspersa. J Parasitol 1994; 80:719-727.
110. Humbert E, Coustau C. Refractoriness of host haemocytes to parasite immunosuppressive factors as a putative resistance mechanism in the Biomphalaria glabrata-Echinostoma caproni system. Parasitology 2001; 122:651-660.
111. Iakovleva NV, Shaposhnikova TG, Gorbushin AM. Rediae of echinostomatid and heterophyid trematodes suppress phagocytosis of haemocytes in Littorina littorea (Gastropoda: Prosobranchia). Exp Parasitol 2006; 113:24-29.
112. Lie KJ. Survival of Schistosoma mansoni and other trematode larvae in the snail Biomphalaria glabrata - a discussion of the interference theory. Trop Geogr Med 1982; 34:111-122.
113. Hertel LA, Stricker SA, Loker ES. Calcium dynamics of hemocytes of the gastropod Biomphalaria glabrata: effects of digenetic trematodes and selected bioactive compounds. Invertebr Biol. 2000; 119:27-37.
114. Walker AJ. Do trematode parasites disrupt defence-cell signalling in their snail hosts? Trends Parasitol 2006; 22:154-159.

115. Plows LD, Cook RT, Davies AJ et al. Carbohydrates that mimic schistosome surface coat components affect ERK and PKC signalling in Lymnaea stagnalis haemocytes. Int J Parasitol 2005; 35:293-302.
116. Zahoor Z, Davies AJ, Kirk RS et al. Disruption of ERK signalling in Biomphalaria glabrata defence cells by Schistosoma mansoni: Implications for parasite survival in the snail host. Dev Comp Immunol 2008; 32:1561-1571.
117. Stout BA, Adema CM, Zhang S et al. Biology of FREPs: diversified lectins with fibrinogen-related domains from the freshwater snail Biomphalaria glabrata. In: Vasta GR, Ahmed H, eds. Animal Lectins: A Functional View. Boca Raton: CRC Press, 2009:475-491.
118. Basch PF. An interpretation of snail-trematode infection rates: specificitity based on concordance of compatible phenotypes. Int J Parasitol 1975; 5:449-452.
119. Theron A, Coustau C. Are Biomphalaria snails resistant to Schistosoma mansoni? J Helminthol 2005;79:187-191.
120. Messier-Solek C, Buckley KM, Rast JP. Highly diversified innate receptor systems and new forms of animal immunity. Semin Immunol 2010; 22(1):39-47.
121. Loker ES, Adema CM, Zhang SM et al. Invertebrate immune systems-not homogeneous, not simple, not well understood. Immunol Rev 2004; 198:10-24.
122. Litman GW, Dishaw LJ, Cannon JP et al. Alternative mechanisms of immune receptor diversity. Curr Opin Immunol 2007; 19:526-534.
123. Horak P, VanderKnaap WPW. Lectins in snail-trematode immune interactions: a review. Folia Parasitologica 1997; 44:161-172.
124. Vasta GR, Ahmed H, Tasumi S et al. Biological roles of lectins in innate immunity: molecular and structural basis for diversity in self/non-self recognition. Adv Exp Med Biol 2007; 598:389-406.
125. Dong Y, Dimopoulos G. Anopheles fibrinogen-related proteins provide expanded pattern recognition capacity against bacteria and malaria parasites. J Biol Chem 2009; 284:9835-9844.
126. Zhang SM, Adema CM, Kepler TB et al. Diversification of Ig superfamily genes in an invertebrate. Science 2004; 305:251-254.
127. Gorbushin AM, Panchin YV, Iakovleva NV. In search of the origin of FREPs: characterization of Aplysia californica fibrinogen-related proteins. Dev Comp Immunol 2010; 34:465-473.
128. Costa MM, Prado-Alvarez M, Gestal C et al. Functional and molecular immune response of Mediterranean mussel (Mytilus galloprovincialis) haemocytes against pathogen-associated molecular patterns and bacteria. Fish Shellfish Immunol 2009; 26:515-523.
129. Raghavan N, Tettelin H, Miller A et al. Nimbus (BgI): an active non-LTR retrotransposon of the Schistosoma mansoni snail host Biomphalaria glabrata. Int J Parasitol 2007; 37:1307-1318.
130. Hauton C, Smith VJ. Adaptive immunity in invertebrates: a straw house without a mechanistic foundation. Bioessays 2007; 29:1138-1146.
131. Lie KJ, Heyneman D, Lim HK. Studies on resistance in snails: specific resistance induced by irradiated miracidia of Echinostoma lindoense in Biomphalaria glabrata snails. Int J Parasitol 1975; 5:627-631.
132. Lie KJ, Heyneman D. Studies on resistance in snails. 3. Tissue reactions to Echinostoma lindoense sporocysts in sensitized and resensitized Biomphalaria glabrata. J Parasitol 1976; 62:51-58.
133. Jiang Y, Loker ES, Zhang SM. In vivo and in vitro knockdown of FREP2 gene expression in the snail Biomphalaria glabrata using RNA interference. Dev Comp Immunol 2006; 30:855-866.
134. Korneev SA, Kemenes I, Straub V et al. Suppression of nitric oxide (NO)-dependent behavior by double-stranded RNA-mediated silencing of a neuronal NO synthase gene. J Neurosci 2002; 22:RC227.
135. Fabioux C, Corporeau C, Quillien V et al. In vivo RNA interference in oyster-vasa silencing inhibits germ cell development. FEBS J 2009; 276:2566-2573.
136. Yakovleva NV, Samoilovich MP, Gorbushin AM. The diversity of strategies of defense from pathogens in molluscs. J Evol Biochem Physl 2001; 37:358-367.
137. Klussmann-Kolb A, Dinapoli A, Kuhn K et al. From sea to land and beyond - New insights into the evolution of euthyneuran Gastropoda (Mollusca). BMC Evol Biol 2008; 8:57.
138. Adamowicz A, Bolaczek M. Blood cells morphology of the snail Helix aspersa maxima (Helicidae). Zoologica Poloniae 2003; 48:93-101.
139. Iakovleva NV, Gorbushin AM, Zelck UE. Partial characterization of mitogen-activated protein kinases (MAPK) from haemocytes of the common periwinkle, Littorina littorea (Gastropoda: Prosobranchia). Fish Shellfish Immunol 2006; 20:665-668.
140. Takamatsu N, Shiba T, Muramoto K et al. Molecular cloning of the defense factor in the albumen gland of the sea hare Aplysia kurodai. FEBS Lett 1995; 377:373-376.
141. Iijima R, Kisugi J, Yamazaki M. Antifungal activity of aplysianin-E, a cytotoxic protein of sea hare (Aplysia kurodai) eggs. Dev Comp Immunol 1995; 19:13-19.
142. Butzke D, Hurwitz R, Thiede B et al. Cloning and biochemical characterization of APIT, a new L-amino acid oxidase from Aplysia punctata. Toxicon 2005; 46:479-489.

143. Mah SA, Moy GW, Swanson WJ et al. A perforin-like protein from a marine mollusk. Biochem Bioph Res Co 2004; 316:468-475.
144. Wang GD, Zhang KF, Zhang ZP et al. Molecular cloning and responsive expression of macrophage expressed gene from small abalone Haliotis diversicolor supertexta. Fish Shellfish Immunol 2008; 24:346-359.
145. Nikapitiya C, De Zoysa M, Lee J. Molecular characterization and gene expression analysis of a pattern recognition protein from disk abalone, Haliotis discus discus. Fish Shellfish Immunol 2008; 25:638-647.
146. Wu LJ, Wu XZ. Molecular cloning and expression analysis of a beta-thymosin homologue from a gastropod abalone, Haliotis diversicolor supertexta. Fish Shellfish Immunol 2009; 27:379-382.
147. Wu LJ, Wu XZ, Zhu BJ et al. Identification and functional characterization of a novel cytidine deaminase in a gastropod abalone, Haliotis diversicolor supertexta. Dev Comp Immunol 2009; 33:709-717.
148. De Zoysa M, Nikapitiya C, Whang I et al. Abhisin: A potential antimicrobial peptide derived from histone H2A of disk abalone (Haliotis discus discus). Fish Shellfish Immunol 2009; 27:639-646.
149. De Zoysa M, Jung S, Lee J. First molluscan TNF-alpha homologue of the TNF superfamily in disk abalone: Molecular characterization and expression analysis. Fish Shellfish Immunol 2009; 26:625-631.
150. Ren HL, Xu DD, Gopalakrishnan S et al. Gene cloning of a sigma class glutathione S-transferase from abalone (Haliotis diversicolor) and expression analysis upon bacterial challenge. Dev Comp Immunol 2009; 33:980-990.
151. De Zoysa M, Nikapitiya C, Moon DO et al. A novel Fas ligand in mollusk abalone: Molecular characterization, immune responses and biological activity of the recombinant protein. Fish Shellfish Immunol 2009; 27:423-432.
152. Wang N, Whang I, Lee J. A novel C-type lectin from abalone, Haliotis discus discus, agglutinates Vibrio alginolyticus. Dev Comp Immunol 2008; 32:1034-1040.
153. De Zoysa M, Kang HS, Song YB et al. First report of invertebrate Mx: Cloning, characterization and expression analysis of Mx cDNA in disk abalone (Haliotis discus discus). Fish Shellfish Immunol 2007; 23:86-96.
154. Cheng W, Hsiao IS, Hsu CH et al. Change in water temperature on the immune response of Taiwan abalone Haliotis diversicolor supertexta and its susceptibility to Vibrio parahaemolyticus. Fish Shellfish Immunol 2004; 17:235-243.
155. Iakovleva NV, Gorbushin AM, Storey KB. Modulation of mitogen-activated protein kinases (MAPK) activity in response to different immune stimuli in haemocytes of the common periwinkle Littorina littorea. Fish Shellfish Immunol 2006; 21:315-324.
156. Iehata S, Inagaki T, Okunishi S et al. Colonization and probiotic effects of lactic acid bacteria in the gut of the abalone Haliotis gigantea. Fisheries Sci 2009; 75:1285-1293.
157. Tan J, Lancaster M, Hyatt A et al. Purification of a herpes-like virus from abalone (Haliotis spp.) with ganglioneuritis and detection by transmission electron microscopy. J Virol Methods 2008; 149:338-341.
158. Wang KJ, Ren HL, Xu DD et al. Identification of the up-regulated expression genes in hemocytes of variously colored abalone (Haliotis diversicolor Reeve, 1846) challenged with bacteria. Dev Comp Immunol 2008; 32:1326-1347.
159. Rosa FM, Godard ALB, Negrao-Correa D et al. Biomphalaria tenagophila: dynamics of populations of resistant and susceptible strains to Schistosoma mansoni, with or without pressure of the parasite. Mem Inst Oswaldo Cruz 2006; 101:247-251.
160. Negrao-Correa D, Pereira CAJ, Rosa FM et al. Molluscan response to parasite: Biomphalaria and Schistosoma mansoni interaction. Invertebrate Survival Journal 2007; 4:101-111.
161. Cunningham AA, Daszak P. Extinction of a species of land snail due to infection with a microsporidian parasite. Conserv Biol 1998; 12:1139-1141.
162. Adema CM, Lun CM, Hanelt B et al. Digenean trematode infections of native freshwater snails and invasive Potamopyrgus antipodarum in the Grand Teton National Park/John D. Rockefeller Memorial Parkway Area. J Parasitol 2009; 95:224-227.
163. Zbikowski J, Zbikowska E. Invaders of an invader - trematodes in Potamopyrgus antipodarum in Poland. J Invertebr Pathol 2009; 101:67-70.
164. Johnson CK, Tinker MT, Estes JA et al. Prey choice and habitat use drive sea otter pathogen exposure in a resource-limited coastal system. Proc Natl Acad Sci USA 2009; 106:2242-2247.
165. Mouritsen KN, Bay GM. Fouling of gastropods: a role for parasites? Hydrobiologia 2000; 418:243-246.
166. Abrous M, Rondelaud D, Dreyfuss G. The stress of Lymnaea truncatula just before miracidial exposure with Fasciola hepatica increased the prevalence of infection. Exp Parasitol 2001; 99:49-51.
167. Salice CJ, Roesijadi G. Resistance to cadmium and parasite infection are inversely related in two strains of a freshwater gastropod. Environ Toxicol Chem 2002; 21:1398-1403.
168. Russo J, Madec L, Brehelin M. Effect of a toxicant on phagocytosis pathways in the freshwater snail Lymnaea stagnalis. Cell Tissue Res 2008; 333:147-158.
169. Gopalakrishnan S, Thilagam H, Huang WB et al. Immunomodulation in the marine gastropod Haliotis diversicolor exposed to benzo(a)pyrene. Chemosphere 2009; 75:389-397.

170. De Zoysa M, Whang I, Lee Y et al. Transcriptional analysis of antioxidant and immune defense genes in disk abalone (Halitois discus discus) during thermal low-salinity and hypoxic stress. Comp Biochem Phys B 2009; 154:387-395.
171. Fevrier Y, Russo J, Madec L. Intraspecific variation in life history traits of a land snail after a bacterial challenge. J Zool 2009; 277:149-156.
172. Langand J, Jourdane J, Coustau C et al.Cost of resistance, expressed as a delayed maturity, detected in the host-parasite system Biomphalaria glabrata/Echinostoma caproni. Heredity 1998; 80:320-325.
173. Rigby MC, Jokela J. Predator avoidance and immune defence: costs and trade-offs in snails. Proc Roy Soc Lond B Bio 2000; 267:171-176.
174. Zbikowska E. Does behavioural fever occur in snails parasitised with trematode larvae? J Therm Biol 2004; 29:675-679.
175. Moore JD, Juhasz CI, Robbins TT et al. Green abalone, Haliotis fulgens infected with the agent of withering syndrome do not express disease signs under a temperature regime permissive for red abalone, Haliotis rufescens. Mar Biol 2009; 156:2325-2330.
176. Allegretti SM, Carvalho JF, Magalhaes LA et al. Behaviour of albino and melanic variants of Biomphalaria glabrata Say, 1818 (Mollusca: Planorbidae) following infection by Schistosoma mansoni Sambon, 1907. Braz J Biol 2009; 69:217-222.
177. Martins-Souza RL, Pereira CA, Martins Filho OA et al. Differential lectin labelling of circulating hemocytes from Biomphalaria glabrata and Biomphalaria tenagophila resistant or susceptible to Schistosoma mansoni infection. Mem Inst Oswaldo Cruz 2006; 101 Suppl 1:185-192.
178. Barbosa L, Caldeira RL, Carvalho OS et al. Resistance to Schistosoma mansoni by transplantation of APO Biomphalaria tenagophila. Parasite Immunol 2006; 28:209-212.
179. Hahn UK, Bender RC, Bayne CJ. Killing of Schistosoma mansoni sporocysts by hemocytes from resistant Biomphalaria glabrata: role of reactive oxygen species. J Parasitol 2001; 87:292-299.
180. Goodall CP, Bender RC, Broderick EJ et al. Constitutive differences in Cu/Zn superoxide dismutase mRNA levels and activity in hemocytes of Biomphalaria glabrata (Mollusca) that are either susceptible or resistant to Schistosoma mansoni (Trematoda). Mol Biochem Parasitol 2004; 137:321-328.
181. Goodall CP, Bender RC, Brooks JK et al. Biomphalaria glabrata cytosolic copper/zinc superoxide dismutase (SOD1) gene: association of SOD1 alleles with resistance/susceptibility to Schistosoma mansoni. Mol Biochem Parasitol 2006; 147:207-210.
182. Bender RC, Goodall CP, Blouin MS et al. Variation in expression of Biomphalaria glabrata SOD1: a potential controlling factor in susceptibility/resistance to Schistosoma mansoni. Dev Comp Immunol 2007; 31:874-878.
183. Coelho JR, Bezerra FS. Compatibility of Biomphalaria tenagophila with Schistosoma mansoni: a study of homologous plasma transference. Mem Inst Oswaldo Cruz 2006; 101:111-112.
184. Pereira CAJ, Martins-Souza RL, Correa A et al. Participation of cell-free haemolymph of Biomphalaria tenagophila in the defence mechanism against Schistosoma mansoni sporocysts. Parasite Immunol 2008; 30:610-619.
185. Myers J, Ittiprasert W, Raghavan N et al. Differences in cysteine protease activity in Schistosoma mansoni-resistant and -susceptible Biomphalaria glabrata and characterization of the hepatopancreas cathepsin B full-length cDNA. J Parasitol 2008; 94:659-668.
186. Guillou F, Mitta G, Dissous C et al. Use of individual polymorphism to validate potential functional markers: case of a candidate lectin (BgSel) differentially expressed in susceptible and resistant strains of Biomphalaria glabrata. Comp Biochem Physiol B Biochem Mol Biol 2004; 138:175-181.
187. Bouchut A, Sautiere PE, Coustau C et al. Compatibility in the Biomphalaria glabrata/Echinostoma caproni model: Potential involvement of proteins from hemocytes revealed by a proteomic approach. Acta Trop 2006; 98:234-246.
188. Jannotti-Passos LK, Andrade HM et al. Proteome analysis of the cardiac and pericardial tissue of Biomphalaria tenagophila populations susceptible and resistant to Schistosoma mansoni infection. Acta Trop 2008; 105:229-234.
189. Sullivan JT, Pikios SS, Alonzo AQ. Mitotic responses to extracts of miracidia and cercariae of Schistosoma mansoni in the amebocyte-producing organ of the snail intermediate host Biomphalaria glabrata. J Parasitol 2004; 90:92-96.
190. Raghavan N, Miller AN, Gardner M et al. Comparative gene analysis of Biomphalaria glabrata hemocytes pre- and post-exposure to miracidia of Schistosoma mansoni. Mol Biochem Parasitol 2003; 126:181-191.
191. Hertel LA, Adema CM, Loker ES. Differential expression of FREP genes in two strains of Biomphalaria glabrata following exposure to the digenetic trematodes Schistosoma mansoni and Echinostoma paraensei. Dev Comp Immunol 2005; 29:295-303.

CHAPTER 3

BIVALVE IMMUNITY

Linsheng Song,* Lingling Wang, Limei Qiu and Huan Zhang
Key Laboratory of Experimental Marine Biology, Institute of Oceanology, Chinese Academy of Sciences, Qingdao, China
**Corresponding Author: Linsheng Song—Email: lshsong@ms.qdio.ac.cn*

Abstract: Bivalves are comprised of animals unclosed in two shell valves, such as mussels, oysters, scallops and clams. There are about 7,500 bivalve species and some of them are of commercial importance. Recently, interest in bivalve immunity has increased due to the importance in worldwide aquaculture and their role in aquatic environmental science and their position in phylogenetic research. This chapter provides a short review of bivalve immunity, including cellular and humoral immunity and the key components and the interactions involved in humoral immunity.

INTRODUCTION

The phylum Mollusca is one of the most large, various and important groups in the animal kingdom. The monophyletic Mollusca is sister to the clade that unites annelids with nemerteans, phoronids and brachiopods.[1] Together with platyhelminthes, they constitute Lophotrochozoa and further constitute Protostomia with Ecdysozoa (molting animals).[1] Bivalvia has about 7,500 species and was the second most diverse class of molluscs after Gastropoda. Many of them are sources of seafood and are important for pearl production and therefore of great commercial importance. In addition, as sedentary filter feeders, bivalves may concentrate bacteria, viruses, pesticides, industrial wastes, toxic metals and petroleum derivatives, making them important markers for biomonitoring pollution in aquatic ecosystems and ideal species for investigating the effects of environmental contaminants.[2]

In the long course of evolution, bivalves have developed an array of effective strategies to protect themselves from the attacks of various pathogens and environmental stresses. Interest in bivalve immunity has increased continuously in recent years due to

Invertebrate Immunity, edited by Kenneth Söderhäll.

serious diseases and mortality problems threatening the healthy development **of bivalve** aquaculture. Although pathology information has been accumulating, research on bivalve immune systems and the underlying molecular mechanisms are still at an early stage, with only some investigations in a comparatively small number of species.[3,4]

As invertebrates, bivalves rely exclusively on an innate, nonlymphoid system of immune reactions.[5] The internal defense of bivalve is mediated by both cellular and humoral components. The former includes phagocytosis or encapsulation, with subsequent pathogen destruction via enzyme activity and oxygen metabolite release, while the latter includes various reactions mediated by series of molecules.[6] Here we review our current knowledge of bivalve immunity, mainly focusing on immune recognition, signal transduction and effector synthesis involved in cellular and humoral immunity.

HEMOCYTES AND PHAGOCYTOSIS

Bivalves have an 'open' circulatory system in which the hemolymph, passing out of the open ends of arteries, bathes all the organs before returning to the heart by the way of sinuses and respiratory structures (gills). Hemocytes, the circulating cells of bivalves, are primarily responsible for the defense against pathogens.[7] Phagocytosis and encapsulation are two major mechanisms for hemocytes to eliminate nonself substances and dead cells.[8] In addition, antibacterial effectors, opsonins, nonspecific hydrolysis and toxic oxygen intermediates found in bivalve hemolymph together coordinate immune response.

The assortation of bivalve hemocytes has been controversial for a long time. Based on morphology and histochemistry research results from mussels and clams, bivalve hemocytes are generally categorized into hyalinocytes and granulocytes,[9-11] and the latter can be further subdivided into eosinophilic granular hemocytes and basophilic granular hemocytes.[6,11] There are also other hemocyte types in some species. For instance, the Type 'III' eosinophil, morula-like cells and blast-like cells were found in *Cerastoderma edule*, *Tridacna derasa* and *Scrobicularia plana* hemolymph, respectively.[6,12] Among all these cell types, granular hemocytes are the most numerous, rich in a variety of hydrolytic enzymes and the major cell, executing the process of phagocytosis.[10]

Phagocytosis is a process to recognize and ingest nonself molecules and cell debris. Bivalve hemocytes can engulf a variety of particles including bacteria, algae, yeast, foreign blood cells and latex spheres.[7] The first step of phagocytosis is the attachment of phagocyte to the targeted particle.[13] Many studies have revealed that hemocytes of many bivalve species exhibit chemotactic as well as chemokinetic reactions, type of which is dependent upon the nature of the molecules presented.[14] In the mussel *Mytilus edulis*, lipopolysaccharides (LPS) from both *Serratia marcescens* and *Escherichia coli* stimulated the migration of cells. In European flat oyster *Ostrea edulis*, hemocytes migrated from circulatory system to connective tissues after *Bonamia ostreae* infection.[15] Hemocytes of *Mercenaria mercenaria* migrated toward not only peptides or small proteins secreted by both Gram-positive and Gram-negative bacteria. After chemotactism, the phagocytes adhere to pathogens, followed by cytoskeleton modification, internalization and destruction of the engulfed target within phagosomes.[13] In *Mytilus galloprovincialis*, both hyalinocytes and granulocytes could execute phagocytosis by formation of coated vesicles and uncoated endocytic vesicles,[10] and cells of different morphology presented different levels of phagocytosis towards zymosan, latex beads and bacteria.[16]

After the nonself is phagocytosed, the phagosomes and lysosomes fuse together and the engulfed target is destroyed within phagosomes by lysosomal enzymes, reactive oxygen species (ROS), nitric oxide (NO) as well as antimicrobial factors. During phagocytosis, the release of degradative enzymes for the destruction of foreign material is accomplished by a sudden release of ROS within hemocytes, which is referred to as respiratory burst. ROS act as killing agents, either alone or in combination with lysosomal enzymes and are important in the phagocyte-mediated killing of microorganisms.[17] Studies on oyster *Crassostrea ariakensis* revealed that granulocytes were most active in spontaneous ROS production[12]. In clam *M. mercenaria*, hemocyte oxidative burst was active in response to the stimulation of zymosan and bacterial extracellular products, accompanied by the increase of ROS production in hemocytes.[18] The internalization of different phagocytic targets and the production of ROS and NO in *M. galloprovincialis* were found to be blocked by inhibition of phosphatidylinositol 3-kinase, protein kinase C and extracellular signal-regulated kinase.[16] Furthermore, many other hemolymph factors also help phagocytosis, including agglutinins (e.g., lectins) and various antimicrobial peptides. Bivalve lectins have different carbohydrate-binding specificities and are involved in nonself-recognition.[19] Antimicrobial peptides are engaged in the destruction of bacteria inside the phagocyte before being released into hemolymph to participate in systemic responses.[7] These factors will be further discussed in the following parts.

Another mechanism of invertebrate cellular immune response is encapsulation.[13] In invertebrates, encapsulation is the common immune defense reaction for foreign bodies which are too large to be phagocytosed. In general, a capsule of hemocytes encloses the foreign body (e.g., multicellular parasites) and cytotoxic products (e.g., degradative enzymes and free radicals) are released by the hemocytes in an attempt to destroy the invader. In the encapsulation response of *Crassostrea gigas* against copepods *Myicola ostreae*, the copepods were observed on the gill surface of *C. gigas* engulfed by a massive agglomerate of hemocyte-like cells encircled by a thin layer of fibroblast-like cells.[20] The clam *Dreissena polymorpha* also employed such a defense mechanism when infected with trematode *Bucephalus polymorphus*.[21] Studies on *C. edule* revealed that positively charged targets stimulated the most vigorous response and that nonspecific electrostatic forces and humoral plasma factors have a synergistic role in hemocyte attachment and the encapsulation response. Phagocytosis and encapsulation are important and complicated processes, the involved molecules and the detailed mechanism should be addressed in future studies.

IMMUNE RECOGNITION

Immune recognition is the first step in activating immune response and occupies a very important position in the immune system to discriminate nonself from self substances. The immune responses begin when specialized, soluble or cell-bound Pattern Recognition Receptors (PRRs) recognize (and bind to) the major targets, called Pathogen-Associated Molecular Patterns (PAMPs).[22,23] PAMPs are common in microorganisms but rare or absent in host animals, such as LPS or peptidoglycan (PGN) in bacterial cell walls and β-1,3-glucan on fungal cell walls.[24] Invertebrates rely only on innate immunity and develop a sophisticated system of PRRs. Seven groups of distinct PRR are identified in bivalves, including peptidoglycan recognition proteins (PGRPs), Gram-negative binding proteins (GNBPs), C type lectins, galectins, thioester-containing proteins (TEPs), scavenger receptors (SRs) and Toll-like receptors (TLRs). In the following, we will briefly introduce the bivalve PRRs, focusing on their binding specificity and functions.

Peptidoglycan Recognition Protein

PGRP is a member of PRRs that specifically bind to PGN, a unique cell wall component of all virtual bacteria but not present in eukaryotic cells. The knowledge of invertebrate PGRPs comes mainly from insects. These PGRPs played a central and diverse role in activating immune reactions, such as melanization cascade, phagocytosis and activating the Toll or IMD signal transduction pathways for the production of antimicrobial products to hydrolyze peptidoglycan and protect host against infection. Some bivalve PGRPs have been identified from the Pacific oyster *C. gigas* (*CgPGRP-S1S*, *-S1L*, *-S2*, *-S3* and *CgPGRP-L*),[25] bay scallop *Argopecten irradians* (*AiPGRP*)[26] and Zhikong scallop *Chlamys farreri* (*CfPGRP-S1*).[27] The identified bivalve PGRPs were all short type with a conserved amidase PGRP domain in their C-terminus. Interestingly, there was an additional goose-type (g-type) lysozyme domain in CgPGRP-L, while a defensin-like domain was present in both CgPGRP-S1S and CgPGRP-S1L.[25] The recombinant protein rCfPGRP-S1 from scallop could bind not only with PGN but also with chitin and LPS moderately. More importantly, rCfPGRP-S1 exhibited strong activities to agglutinate Gram-positive bacteria *Micrococcus luteus* and *Bacillus subtilis*, while slightly with the Gram-negative bacterium *E. coli* (unpublished data). The short type PGRP is widely present in various bivalve species and seems to be a versatile PRR not limited to the function of recognizing and binding the PAMPs and thus an indispensable component in bivalve innate immunity.

Gram-Negative Binding Protein

The GNBP family includes members that bind Gram-negative bacteria, LPS and β-1,3-glucan.[28] Research on bivalve GNBP is still limited and they have only two related reports. The LPS and β-1,3-glucan binding protein (LGBP) is one of the GNBPs with various biological functions, including the activation of the prophenoloxidase (pro-PO) system, cytolysis, bacterial aggregation and opsonic reaction. The LGBP gene was cloned from scallop *C. farreri* (*CfLGBP*) with conserved domains of the LPS-binding site and glucan-binding site. The initial up-regulation after *Vibrio anguillarum* challenge indicated that *CfLGBP* was sensitive to bacterial infection.[27] The recombinant CfLGBP could bind not only LPS and β-glucan, but also PGN and exhibited obvious agglutination activity towards Gram-negative bacteria *E. coli*, Gram-positive bacteria *B. subtilis* and fungi *Pichia pastoris* in vitro. Additionally, a beta-1,3-glucan binding protein (βGBP) was purified from the plasma of marine mussel *Perna viridis*[29] with an inherent serine protease activity. It agglutinated bakers yeast, bacteria and erythrocytes and enhanced proPO activity of the plasma. Although its gene sequence and molecular structure are still unknown, βGBP is thought to be a multifunctional molecule and functions as a recognition molecule for beta-1,3-glucan on the surface of microbial cell walls.

C-Type Lectin

C-type lectins are a superfamily of diverse proteins with one or more carbohydrate-recognition domains (CRDs) of ~130 amino acid residues. They recognize and bind to terminal sugars on glycoproteins and glycolipids and function in nonself recognition and clearance of invaders.[30] In bivalves, lectins are undoubtedly involved in nonself recognition and also have agglutination or opsonic roles in hemocyte phagocytosis. We will quote those lectins in immune recognition of bivalve in the following and adduce their agglutination or opsonic roles later.

The forepart research on C-type lectins mainly focused on characterizing the features or detecting their activities. With the development of molecular biotechnology, much more attention was paid to their gene structure and recognition mechanism. Recent EST analysis in scallops, clams and oysters revealed a high content and variety of lectin gene homologues.[27] Some C-type lectins or proteins with similar activities have been characterized from bivalves.[31] The mRNA expression of these genes could be significantly up-regulated by bacteria or parasite, especially bacteria of *Vibrio* genus, suggesting that they were involved in the immune response against invading microbes.

Bivalve C-type lectins are diverse in their domain structure and organization. Besides most lectins with single CRD, there are also three and four CRDs in Cflec-3 and Cflec-4 from *C. farreri*, respectively, while most insect C-type lectins contain tandem CRDs.[32] The architecture and phylogenic analysis of these proteins together with those from *Drosophila* and *Caenorhabditis elegans* suggested that multidomain C-type lectins in different lineages did not arise from a common multidomain progenitor and these proteins served distinct functions in different animal lineages.

Bivalve C-type lectins have different carbohydrate-binding specificities and are believed to be a kind of antibodies in nonself-recognition. For instance, in clam *Ruditapes philippinarum*, MCL can bind to the surfaces of purified hypnospores and zoospores of *Perkinsus olseni* parasite by recognizing and binding the terminal GalNAc/Gal residues. Meanwhile, the hemo-agglutinating activity of MCL3 could be inhibited by GalNAc, Mannose, lactose, raffinose polysaccharides bovine mucin Type II and Candida mannan.[33] These two lectins, as well as MCL-4, contributed to the phagocytic ability to eliminate bacteria or parasite via recognition of terminal carbohydrated residues on the surface of microbes.[34] Chiletin from *Ostrea chilensis* could agglutinate sheep red blood cells through binding galactose and mannose.[35] The EPN motif in the canonical binding sites of Codakine from clam *Codakia orbicularis* proved to be important for calcium-dependence and mannose/GlcNAc-binding activity.[36] In some species, the lectins with similar carbohydrate-binding specificity may distinguish different invading microbes. For example, the recombinant C-type lectins Cflec-1, Cflec-2, Cflec-3 and Cflec-5 from *C. farreri* agglutinated *E. coli*, *Staphylococcus haemolyticus*, *Pseudomonas stutzeri* and *P. pastoris*, respectively, though they all possessed mannose-binding specificity.

The great number of C-type lectins identified from bivalves shared similar structural features and displayed binding and agglutinating activities towards a range of microbes. Since there is no antibody-mediated immunity in invertebrates, abundant lectins with diverse expression profiles and bioactivities might function as nonclonal effectors in the bivalve immune system.

Galectin

Galectins are a family of β-galactoside-binding lectins and they are probably the most conserved and ubiquitous lectin family found in multicellular organisms.[22,23,28] The bivalve galectin was purified first from oyster *Pinctada fucata martensii* in the 1980s, but the genes were cloned and identified recently, including *CvGal* from *Crassostrea virginica*[37], *CgGal* from *C. gigas*,[31] *AiGal1* from *A. irradians*,[38] *MCGal* from clam *R. philippinarum*[39] and *Pf-galectin* from *Pinctada fucata*. Except for CgGal, which contains a single CRD, the other four galectins possess multiple CRDs. AiGal1, CvGal and Pf-galectin are quadruple-CRD galectins, which are so unique in bivalves that it has never been reported in other species. From a phylogenetic point of view, all four CRDs of AiGal1, CvGal and Pf-galectin form a single clade, suggesting that the bivalve galectin CRDs share a common ancient and the four individual CRDs of each galectin are originated by repeated duplication of a single galectin gene.

CvGal and MCGal are two major galectins for functional study. CvGal could facilitate recognition of a variety of potential microbial pathogens, unicellular algae and preferentially *Perkinsus marinus* trophozoites. Attachment and spreading of hemocytes to foreign surfaces induced localization of CvGal to the cell periphery, its secretion and binding to the plasma membrane. CvGal subsequently promoted phagocytosis for both potential infectious challenges and phytoplankton components.[37] Moreover, MCGal had an affinity towards galactose and N-acetylgalactosamine and could bind to the surface of *Perkinsus olseni* and agglutinate *Vibrio tapetis* in vitro.[39]

Thioester-Containing Protein

TEPs are a family of proteins characterized by the unique intrachain β-cysteinyl-γ-glutamyl thioester bond and a propensity for multiple conformationally sensitive binding interactions.[40] This protein family consists of complement components C3, C4, C5, protease inhibitor alpha$_2$-macroglobulin (α_2M), CD109 and a set of insect TEPs. Among them, the invertebrate C3-like molecules and insect TEPs were thought to be involved in the innate immune defense as PRRs.[28]

The bivalve TEPs were recently identified from the clam *Ruditapes decussates*[41] and scallop *C. farreri*,[42] termed as *Rd-C3* and *CfTEP*, respectively. They both contained canonical thioester motif GCGEQ, proteolytic cleavage sites and catalytic histidine residues similar to C3 molecules. However, CfTEP possessed additional features distinguished it from C3 molecules, including: (1) the absence of anaphylatoxin-like and C345C domains, (2) a distinctive cysteine signature in the C-terminus which characterized the TEP subfamily apart from complement factors and α_2M subfamilies and (3) the highly variable central region.[42] Due to these structural differences, CfTEP was phylogenetically related to the insect TEPs, while Rd-C3 was related to the invertebrate C3-like molecules. The above results supported the view that TEP and complement factors shared a common ancestor but they separated from each other at a rather early lineage.[43]

The genomic organization of *CfTEP* was similar to human and mouse C3 rather than ciona C3-1 and *Drosophila dTEP2*, indicating a complicated evolutionary history of this gene family. It was of great interest that seven different *CfTEP* transcripts were produced by alternative splicing and they displayed different expression patterns in gonads in response to different bacterial challenges, which suggested an important role of diverse *CfTEP* transcripts in the innate immune defense of scallops.[44] These results provided new insights into the role of TEPs in bivalve immune responses, as well as the evolutionary origin of this important, widespread and functionally diversified family of proteins.

Scavenger Receptor

Scavenger receptors (SRs) are a main type of endocytic receptors with multifunctions to recognize and engulf various PAMPs.[45] In contrast with other PRRs, the information about invertebrate SR is extremely limited. There are only two invertebrate SRs identified from *Drosophila* (*dSR-CI*) and *C. farreri* (*CfSR*, GQ260639.1). CfSR is structurally different from all the characterized SRs. It contains six scavenger receptor cysteine-rich (SRCR) domains absent in dSR-CI and UPAR-like and ShK toxin-like domain do not exist in any other member of SRs. CfSR shares a similar attachment site with anchor protein Sgp-2 and it was mainly detected on the outer surface of hemocytes by immunofluorescence approach, indicating that CfSR was anchored on the outer-membrane of cells. The recombinant CfSR displayed a significantly strong activity to bind not only acetylated LDL, dextran sulfate but also with various PAMPs, including LPS, PGN, zymosan particle and mannan.

CfSR is one of the most primitive SR found so far in invertebrates. It displays unique structure and broader ligand binding ability. The existence of SR protein in bivalve will contribute not only to the origin and evolution of the diverse molecules, but also to the understanding of the complex mechanism of immune recognition.

Toll-Like Receptor

TLRs are an ancient family of proteins with the hallmark structure of extracellular leucine-rich repeats (LRRs), intracellular Toll/Interleukin-1 receptor (TIR) domains, which play key roles in detecting various nonself substances and then initiating and activating immune system. The reports of TLRs in mollusc are still rare although they were widely distributed in nearly all animal phyla. Only one TLR gene (*CfToll-1*) and three TLR EST fragments had been identified in *C. farreri*, *C. virginica*[46], *A. irradians*[47] and *M. mercenaria*.[48] *CfToll-1* shared the same domain architecture with *D. melanogaster* Tolls (*DmTolls*) and *Tachypleus tridentatus* Toll (*tToll*). The mRNA expression of *CfToll-1* was upregulated by LPS in a dose-dependent manner.[49] The finding suggested that *CfToll-1* might be involved in immune response against bacterial invasion. Considering their importance in innate immunity, the study of the number, assortation and the roles of bivalve TLR in immune recognition and signal transduction (discussed in other section) should be addressed in the future.

IMMUNITY SIGNALING PATHWAYS

Innate immune system is under the control of a complex network of evolutionary conserved signaling pathways, which are activated depending upon different invasions or stimuli. Several signaling pathways, such as TLR, the Janus kinase/signal transducer and activator of transcription (JAK-STAT), mitogen-activated protein kinase (MAPK) and NF-κB pathways have been extensively studied in recent years for their important roles in regulating the immune system in both vertebrates and invertebrates. In bivalves, effective immune defense systems have been developed during evolution which can protect them from infection successfully. When bivalves are challenged by a pathogen, different signaling pathways triggered by PRRs induce the systemic immune response and produce responding effectors. Although the studies are preliminary, some genes involved in conserved signaling pathways have been identified in bivalves. In this section, we will briefly introduce what is currently known about these signal transduction pathways, which may be a crucial step for our understanding about bivalve immune system.

The Canonical NF-κB Signaling Pathway

NF-κB signaling pathway is an evolutionarily conserved process to activate NF-κB,[50] a nuclear factor which has a central role in coordinating the expression of a wide variety of genes that control immune responses.[51] In resting cells, NF-κB proteins bind to the inhibitor of κB (IκB). The NF-κB signaling pathway is activated once IκB is degraded. This degradation of IκB is catalyzed by serine kinase I kinase (IKK) which leads to the translocation of released NF-κB dimers to nucleus. The activated NF-κB proteins then bind to DNA and activate gene transcription.[50]

In recent years, the key molecules involved in bivalve NF-κB signaling pathway, such as NF-κB, IκB, IKK genes, have been identified successively, offering clues for the existence of NF-κB signaling pathway in bivalves. Now three IκB genes have been cloned from scallops *C. farreri*[27] and *A. irradians*,[47] and oyster *C. gigas*,[52] and they showed a high level of identity with insect IκB-like proteins and vertebrate IκB isoforms. The *Cg-Rel*, a gene encoding the first NF-κB homologue in bivalve, shares the structural organization with Rel/NF-κB family members. The C-terminal transactivation domain and κB binding sites were indispensable for activation of the expression of genes controlled by NF-κB pathway. The NF-κB signaling pathways were well studied in pearl oyster *P. fucata*. Three genes, *Pf-Rel*,[53] *Pf-IKK*[54] and *poIκB*[55], were cloned and they were constitutively and ubiquitously expressed in tissues of pearl oyster. LPS could transiently stimulate IκB degradation, but couldn't influence the expression level of *Pf-IKK*. Transfection experimentation in NIH3T3 cells with Pf-IKK demonstrated that Pf-IKK triggered the gene expression by activating NF-κB in an IκB-dependent manner just as mammalian counterparts do.[54] The findings above favored the hypothesis that NF-κB signaling pathway was an ancient scheme of immune gene regulation pathway, which was conserved in bivalves. Although the complex members and their detailed information of NF-κB signaling pathway are not well understood in bivalve, the accumulating evidence indicates that it bears considerable similarity with mammalian NF-κB signaling pathway.

MAPK Pathway

MAPK cascades are one of the most important signaling pathways controlling a variety of physiological processes including cell proliferation, growth, differentiation, cell death, innate immunity and development.[56] Three subfamilies of MAPKs have been well-characterized in multicellular organism. They are extracellular signal-regulated kinases (ERKs), c-Jun amino-terminal kinases (JNKs) and p38 MAPKs. The pathway represents a characteristic phosphorylation system in which a series of three protein kinases phosphorylate and activate one another.[57] MAPK pathways were proven to exist in bivalve by immunoblot or ELISA techniques.[58] The pathways are not only sensitive to various environmental stressful stimuli, but also can be activated by growth factor insulin, cytokines, hormone substances and bacterial challenge. Several ESTs homologous to the MAPK pathway components were screened from cDNA library of oyster (*Cg-MAPKK1*,

Cg-MAPKK2, *Cg-c-jun* and *Cg-phosphatase*, *Cg-focal*, *Cg-FAK*) and Manila clam (c-jun).[59] Further investigations in molecular structure and their functions are needed for the precise defining of all molecules involved in MAPK pathway of bivalves.

JAK-STAT Pathway

JAK-STAT pathway is one of the important signaling pathways, downstream cytokine receptors.[60] Although JAK-STAT pathway seems to be present throughout evolution, there is no such report regarding the existence of JAK or STAT molecule in bivalves. Nevertheless, STAT activation has been observed recently in *M. galloprovincialis* under the stimulation of cytokine and bacteria. The microbicidal activity against *E. coli* in mussel hemocytes increased significantly after the pretreatment with human recombinant IFNγ, while no effect was observed with IFNα. IFNγ induced a rapid and time dependent increase of phosphorylated STAT1-like protein, evaluated by polyclonal antibodies specific for the tyrosine phosphorylated sites of STAT1.[58] It was suspected that JAK-STAT pathway possibly existed in bivalves. More and more cytokines and their receptors found in bivalves (details in other section) provided the clue for further investigation of bivalve JAK-STAT pathway. The roles of JAK-STAT pathway and their relationship with the newly indentified cytokines from these animals may be an interesting research field which is helpful to our comprehensive understanding of bivalve immunity.

The Toll-Like Receptors Signal Pathway

The characterization of TLR and its signaling pathway is one of the greatest propellers to acquaint with the immune system in the last decades. The receptors, adaptors and transducers in this pathway, such as TLR, myeloid differentiation factor88 (Myd88), tumor necrosis factor receptor-associated factor 6 (TRAF6) and IL-1 receptor-associated kinase (IRAK) displayed striking similarity from *Drosophila* to mammals, while the function of this pathway is different between human and fly. TLR pathway in mammals is a skillful system that detects invasion of various pathogens and plays a key role in bridging innate and adaptive immunity. Toll pathway in *Drosophila* is indispensable for both development and antifungal/anti Gram-positive bacteria immunity. It is worth noting that the absence of Myd88 orthologue and their very low similarity of TRAF and IRAK homologues strongly imply the evolutionary 'loss' of Toll signaling pathway in nematodes.[61] Moreover, *tol-1* in *C. elegans* is not involved in immune response but only in development. More information about TLR pathway in other invertebrates is required for better understanding of its function and evolution.

Fortunately, the key components involved in TLR pathway, including TLR,[49] Myd88,[62,63] TRAF6,[64] TRAF3,[65] and IRAK[46] have also been identified in bivalves. And the expressions of *CfToll-1*,[46] *CfMyd88*,[62] *CfTRAF6*[64] in scallops were up-regulated by stimulation of LPS or PGN. Furthermore, when the *CfToll-1* was "knock down" by RNAi technique, the scallop was more susceptable to pathogen and the expression of downstream genes was also down-regulated. Undoubtedly, the canonical TLR signaling pathway existed in bivalve. Considering scallop is a relatively primitive animal, the presence of TLR signaling pathway is epochmaking to understanding its origin, evolution and crucial roles in innate immunity.

Complement Pathway

The complement system is a major component of vertebrate innate immunity and also an essential bridge between innate and adaptive immunity. It can be activated mainly through four separate pathways, referred to as the classical, alternative, lectin[66] and coagulation[67] pathways. Recent investigation on the evolution of complement system has demonstrated the origins of complement system could be traced to near to the beginnings of multi-cellular animal life and the components of complement system should have some evolutionary traces in inferior species. However, there are very few reports on complement system in invertebrates and even less in bivalves. To our knowledge, only two complementary factors, C3 (*Rd-C3*) and B factor-like (*Rd-Bf-like*) molecules have been identified as representatives of the alternative pathway from carpet-shell clam (*Ruditapes decussatus*). The Rd-C3 shares distinctive structural characteristics with complement proteins (C3/C4/C5). Rd-Bf-like is composed of two complement control protein modules (CCP domains) and shares about 30% similarity with other known Bf proteins.[68] Additionally, three complement-like factors, including *CfTEP*[42] (details in other section), *CfC1qDC*[69] and *AiFREP*[70] were identified from scallops. TEP reported in scallop suggested that it might be an ancient equivalent component of the key factor, C3, and there was likely to be a primitive, simple complement-like system in bivalves. CfC1qDC is a novel C1q-domain-containing protein with LPS binding activity and is suspected to be a candidate of classical pathway in scallop. AiFREP, a member of protein with fibrinogen-like (FBG) domain at the C-terminus, exhibited a very high similarity with FBG domain of mammalian ficolin (at 61% approximately with *Rattus norvegicus*) and other invertebrate FREP proteins. *AiFREP*, the EST fragment of MBL and ficolin from oyster and kinds of lectin in bivalves indicated the high existent probability of lectin pathway. The results place us in a puzzle about the bivalve complement system and its activation.

IMMUNE EFFECTOR

Innate immunity encompasses a complex array of defense reactions, in which immune effectors are fundamental molecules and utilized as executor for the incapacitation and elimination of invaders. The immunity of bivalves relies upon the production of immune effectors that are active against a large range of pathogens or sensitive for the environmental stress. The importance of those molecules is underlined by increasing outbreak of a variety of diseases and summer mortalities in bivalve aquaculture. The following details recent knowledge of immune effectors, such as antimicrobial peptides (AMPs), cytokines, complement components, antioxidant enzymes, acute phase proteins and draws their repertoire in bivalve immune response.

Antimicrobial Peptides

AMPs have been characterized as one of the key immune effectors in innate immunity and are widespread in plants and animals. All the known AMPs are classified into four structure groups according to their amino acid sequences, secondary structures and functional similarities: (1) the linear basic peptides forming amphipathic α-helices conformation and deprived of cysteine residues; (2) peptides containing cysteine residues with one to six intra-molecular disulphide bonds; (3) peptides rich in regular amino acids like proline with a variable structure; (4) the peptides produced by the hydrolysis of large inactive or proteins with little activity. Approximately 20 AMPs have been identified from bivalves, with 10 from *M. edulis* and *M. galloprovincialis*, which were organized into four distinct groups according to their antimicrobial action based on their hydrophobic and cationic properties and the amphipathic structure: MGD (defensin), mytilin, myticin and mytimycin.[71,72]

MGDs (MGD1 and MGD2) characterized from *M. galloprovincialis* with eight cysteines showed high similarities to the arthropod defensin family, which contains 6 cysteins. They were believed to be the original members of the arthropod defensin family because of the presence of two extra cysteines.[73] MGD1 and MGD2 share significant homology with the high conserved cysteines, hydrophilic and hydrophobic residues,[72] and they are essentially active against Gram-positive bacteria, including some pathogens for marine invertebrates. There are five kinds of mytilin isoforms in the bivalve mussels. Mytilin A and B were isolated from *M. edulis* plasma and mytilin B, C, D and G_1 were isolated from *M. galloprovincialia* hemocytes. Mytilin A, B, C, D exhibited significant activities against both Gram-positive and Gram-negative strains, while mytilin G1 was only active on Gram-positive bacteria. Mytilin B and D were also active against the filamentous fungus *Fusarium oxysporum*. Mytilin C was tested against the protozoan parasite, *P. marinus*.

Myticins were identified as novel cysteine-rich peptides. At present, three classes of myticin, A, B and C have been reported. Only one isoform has been described for myticin A and B and a total of 74 different isoforms have been reported for myticin C so far.[74] Myticin A and B were purified from the hemocytes and the former was also from the plasma of *M. galloprovincialis*. The mature peptides of myticin A and B comprise 40 residues with four intra-molecular disulfide bridges and a cysteine array in the primary structure different to that of the previously characterized cysteine-rich AMPs. The two myticins had marked activity against the gram-positive strains, *M. luteus, Bacillus megaterium* and *Aerococcus viridans*. Myticin B was also active against the filamentous fungus, *F. oxysporum* and moderately against the Gram-negative, *E. coli D31*. Myticin C had a high variability on the nucleotide sequence. It was a ubiquitous peptide expressed early in the development and associated with mussel survival. Different myticin C isoforms were observed along their life when mussel touched with new potential pathogens or nonself molecules in general. Therefore, myticin C could deal with a huge range of potential pathogens present in the marine ecosystem. However, the mechanisms to generate diverse isoforms and its implication in the fight against pathogens should be further studied.[74]

There is only one mytimycin identified from *M. edulis*[75] with molecular weight of 6.2 kD. The search in the peptide sequence data bases did not yield any homology with known peptides. There are twelve cysteines engaged in the formation of six intra-molecular disulfide bridges. Its antimicrobial activity was strictly antifungal.

In addition to the AMPs found in *M. galloprovincialis*, defensins have been identified from other bivalves, including *Cg-Def*, *Cg-Defh1* and *Cg-Defh2* from *C. gigas*,[76] and a big defensin (*AiBD*) from *A. irradians*.[77] Cg-Defh1 and Cg-Defh2 identified from hemocyte were almost same and shared around 80% identity with mantle Cg-Def.[76] Recombinant Cg-Def was active in vitro against Gram-positive bacteria but displayed no or limited activities against Gram-negative bacteria and fungi. *AiBD* was the first bivalve big defensin gene cloned from bay scallop *A. irradians*. It consisted of 531 nucleotides with a canonical polyadenylation signal sequence AATAAA and a poly(A) tail, encoding a polypeptide of 122 amino acids. Recombinant AiBD showed activities against both Gram-positive and Gram-negative bacteria and some fungi.

As the most important effector in bivalve immunity, more AMPs will be found and well characterized, which will provide a better understanding of the immune defense mechanisms of bivalve and new insights into health management and disease control in aquaculture.

Lysozymes

Lysozyme is a ubiquitous enzyme existing in numerous phylogenetically diverse organisms such as bacteria, bacteriophages, fungi, plants and animals, which catalyzes the hydrolysis of β-1,4-glycosidic linkage between *N*-acetylmuramic acid and *N*-acetylglucosamine of PGN and causes bacterial cell lysis.[78] It has been widely accepted that lysozyme functions as a crucial effector molecule in innate immunity. In general, the lysozymes are classified into six types according to the organisms where they are first identified, chicken-type (c-type), goose-type (g-type), invertebrate-type (i-type), phage, bacterial and plant lysozyme.

Several lysozymes and their activities have been characterized in bivalves. All previous reported bivalve lysozymes are i-type.[79] But recently, two g-type lysozymes (*CFLysG*, *AILysG*) were identified in scallops and no such thing has been reported before. It demonstrated that g-type lysozyme was not vertebrate-specific and its origin should precede the divergence of invertebrate and vertebrate.[80] Chlamysin from *Chlamys icelandic*,[81] lysozymes from *Tapes japonica*[82] and *C. virginica*[83] have been verified as i-type lysozymes with remarkable antibacterial activity against both Gram-positive and Gram-negative bacteria. The g-type lysozyme CFLysG possessed all conserved features critical for the fundamental structure and function of g-type lysozymes, such as three catalytic residues (Glu 82, Asp 97, Asp 108) and showed more potent inhibitive activities against Gram-positive bacteria. The self-defense activity of lysozyme have been detected widely in the body fluid and various tissues of bivalves, including the genus of *Mytilus*, *Bathymodiolus*, *Calyptogena* and *Chlamys*. A recent study has demonstrated that lysozyme could augment the activity of AMPs through a synergistic mechanism.[84] The lysozyme can also serve as a digestive enzyme in the digestive organs,[85] which has been further verified by lysozyme activities in hepatopancreas of several bivalves.

Lectins

The function of lectins includes not only self/nonself recognition but also engaging associated effector mechanisms, such as complement-mediated opsonization and killing of potential pathogens. Diverse lectins have been reported in oysters,[31,35] scallops and clams,[86,87] and they are involved in the immune response against pathogens. The recombinant C-type lectins Cflec-1, Cflec-2, Cflec-3 and Cflec-5 from *C. farreri* agglutinated *E. coli*, *S. haemolyticus*, *P. stutzeri* and *P. pastoris*, respectively. The lectin from *C. virginica* agglutinates a wide variety of bacteria.[88] The sialic acid-binding lectin of horse mussel, *Modiolus modiolus*, has strong antibacterial activity against *Vibrio* strains.[89] A calcium independent lectin isolated from the foot muscle of marine bivalve *Macoma birmanica,* named MBA, could interact with both Gram-positive and Gram-negative bacteria.[90] There is increasing evidence to support the suggestion that the presence of isoforms and diverse roles of lectins provide bivalves with a functional diversity to the innate immune response.

Cytokines

Cytokines comprise a large number of regulatory molecules, interleukins (IL), IFN, TNF and chemokines and many of them function vitally in the vertebrate immune system.[91] In view of the significant functions, the studies of cytokine in bivalves have been pursued since the early 1990s. In the first publication, oyster cells were found to respond to IL-1 and TNF in a manner similar to that of human granulocytes.[92] Subsequently, the information about the effect of cytokines on immunity such as cell motility, chemotaxis, phagocytosis and cytotoxicity has been accumulating in bivalves.[93] However, the previous knowledge about bivalve cytokine is from immunological experiments with human antibodies, which lacks molecular evidence.[94] Recently, cytokine homologs and their receptors have been validated at molecular level.

The first identified cytokine is oyster IL-17 homologue (*CgIL-17*). Bacteria challenge induced a large and rapid elevation of *CgIL-17* transcript in oyster.[91] This is the solely IL identified in bivalve. Although other cytokines are not available in bivalves, two TNF receptor genes were cloned from Zhikong scallop, which was homologous respectively with p75 neurotrophin receptor (*p75NTR*) and osteoprotegerin.[95] Another receptor for the transforming growth factor beta (*TGFβ*) has been identified from oyster *C. gigas*.[96] The phylogenetic and structural analysis as well as the expression pattern during early development suggested that Cg-TGFβRI belonged to the TGFβs.s./activin type I receptor. The existence of different receptors suggested the corresponding cytokines should be present in bivalves.

It is exciting that a bivalve TNF (*RpTNF*) gene was cloned from clam *R. philippinarum*. The typical structure indicated that RpTNF was a true Type II (i.e., intracellular N terminus and extracellular C terminus) transmembrane protein. The recombinant RpTNF induced the death of tumour cells just like its mammalian homologs (unpublished data). Considering the broad immune response of cytokines, a multiplex cytokine-receptor system must exist and mediate bivalve immunity in a specific "ligand-receptor" manner, which may take important roles in regulating the bivalve immunity.

Complement Components

The complement system is one of the major effector arms of immune response in vertebrates and a necessary complement for antibodies to play a role in cytolytic. Its effector functions include opsonization leading to enhanced phagocytosis and lysis of microbes. In the evolutionary progress, it appears earlier than the acquired immunity. The previous research on sea urchins, tunicates and horseshoe crabs also revealed a simple opsonic complement defense system in invertebrates.[97] However, there are few reports on its role in bivalves.

The recent identification of C3 and B factor-like (details in other section) from carpet-shell clam provides insight into their conserved characters critical for function. Three other molecules with the hallmark domain structure of complement factors have also been cloned from scallops (details in other section). A C1qDC protein has been identified from *C. farreri* to have LPS binding activity.[69] The *CfTEP* transcripts were mainly detected in the tissues of hepatopancreas and gonad and remarkably up-regulated by microbial challenge.[42] Its expression was complicatedly mediated by alternative splicing mechanism. Recombinent AiFREP agglutinated chicken and human A, B, O-type erythrocytes. The agglutinating activities were calcium-dependent and could be inhibited by acetyl group-containing carbohydrates. It also agglutinated Gram-negative bacteria *E. coli* JM109, *V. anguillarum* and Gram-positive bacteria *M. luteus* in the presence of calcium ions. The attractive studies collectively favored that those complement factors play significant role in bivalve immune responses and complement system should be an intriguing driver for understanding the characteristic of innate immune system in bivalves.

The Antioxidant Enzymes

ROS are free radicals that contain the oxygen atom, constantly generated when the organism is attacked by invaders or contaminant exposures.[98] Low concentrations of ROS, such as superoxide anion (O_2^-), hydrogen peroxide (H_2O_2) and hydroxyl radical (OH^-) may be beneficial or even indispensable in processes including intracellular signaling and defense against micro-organisms. Excessive production of ROS may, however, lead to oxidative stress, loss of cell function and ultimately apoptosis or necrosis.[99] As with chemical antioxidants, cells are protected against oxidative stress by an interacting network of antioxidant enzymes. In bivalves, a large number of antioxidant enzymes have been identified over the last decades and their response against microbe challenge and environmental stress have been studied. The knowledge of superoxide dismutase (SOD), catalase (CAT), glutathione peroxidase (GPx) and glutathione-S-transferase (GST) is summarized briefly in this section.

Superoxide Dismutase

SODs are a class of antioxidant enzymes that catalyze the dismutation of superoxide into oxygen and hydrogen peroxide. According to their metal content, SODs are classified into four distinct groups: iron SOD (FeSOD), manganese SOD (MnSOD), copper/zinc SOD (Cu/ZnSOD) and nickel SOD (NiSOD). However, there were only Cu/ZnSOD and MnSOD identified presently in bivalves.[100] Most of SODs were constitutive proteins that could play a crucial role in bivalve defense. Many studies have indicated the importance of SODs to the immune response as well as the role in protecting cells against various challenges.[100] For example, the SOD genes have been identified from scallop *C. farreri* and *Nodipecten subnodosus* and a rapid elevation of SOD activities were observed after microbe infection.[101,102] The SOD activity in gills and mantle of *M. galloprovincialis* increased obviously at the toxic chemical pollutants, especially heavy metals.[103]While in *Chamelea gallina*, the SOD activity decreased in haemocytes with increasing temperature or exposure to benzo[a]pyrene.[104,105] A higher-level of mRNA expression of MnSOD was detected in gill and mantle of *A. irradians* after being challenged with *V. anguillarum* and the expression level in gill was even higher, indicating that MnSOD was necessary in the immune responses against *V. anguillarum* infection. The responses of bivalve SOD to environmental stress were quite variable depending upon isoforms and tissues. When exposed to heavy metals, SOD is predominantly detected in the cytosolic fractions of gill and digestive gland, with the highest amount in the gill. The mitochondrial Mn-SOD activity in the gill and digestive gland is lower than cytosolic SOD.[106] The whole set of results make them potential target biomarkers aiming at the given environment factors in monitoring strategy.

Catalase

CAT is one of the central enzymes involved in scavenging the high level of ROS. It catalyzes the decomposition of hydrogen peroxide to gaseous oxygen and water molecules.[107] This enzyme is ubiquitous and present in archaea, prokaryotes and eukaryotes. In mammals, the functional catalase is a tetramer of four identical subunits with a molecular weight of approximate 240 kD.[108] Recently, more and more proteins with catalase activities and the genes have been reported in bivalves, such as *M. edulis*,[109] *D. polymorpha*,[110] *C. gigas*[111] and *C. farreri*.[112] The expression of CAT from *C. farreri* increased gradually after-*Vibrio* infection. *M. galloprovincialis* CAT activity was increased 2-3 times at the polluted coastal areas, with high activity in winter and spring.[103] This rise of *M. galloprovincialis* catalase activity was significant with temperature, salinity and light duration. The importance of these enzymes in regulating oxidative stress is recognized. CAT is considered as an important and sensitive biomarker of environmental stress, used to reveal the biological effect on the redox status of bivalve organisms.

Glutathione Peroxidase

GPx is the general name of an enzyme family with peroxidase activity whose main biological role is to protect the organism from oxidative damage. It reduces lipid hydroperoxides to their corresponding alcohols and reduces free hydrogen peroxide to water. Two isoforms of GPx had been identified, selenium-dependent GPx (Se-GPx) and selenium-independent GPx. Se-GPx catalyses the reduction of both organic and inorganic peroxides like hydrogen peroxide (H_2O_2) while selenium-independent GPx reduces only organic peroxide. The detection of GPx activity and identification of GPx molecules collectively suggest that GPx may respond to bacterial infection and hydrogen peroxide exposure and be involved in protection against oxidative stress and immune defense in bivalves.[2,113-115] To date, the activity of GPx has been detected in bivalves *Corbicula fluminea*,[113] *Pinna nobilis*,[114] *M. galloprovincialis*[115] and *Ruditapes decussatus*.[2] There have also been several reports on the GPx genes cloned from bivalves, such as *D. polymorpha*, *Unio tumidus* and *C. gigas*.

Glutathione S-Transferase

GSTs are comprised of classes of dimeric enzymatic proteins that catalyse the conjugation of glutathione to a wide variety of hydrophobic compounds through the formation of a thioether bond with their electrophilic centre. Most mammalian GSTs are cytosolic enzymes with molecular masses of 23 to 28 kD as homodimers or heterodimers. The cytosolic GSTs have been classified into at least 12 different classes based on their N-terminal amino acid sequence, substrate specificity, antibody cross-reactivity and sensitivity to inhibitors. To date, many novel classes of GST sequences have been identified and classified from nonmammalian organisms. A few GSTs have been studied in bivalves, including *Atactodea striata*, *R. decussates*, *M. edulis*, *C. fluminea,* etc.

GSTs are major Phase II detoxication enzymes found mainly in cytosol and function as a substrate of antioxidant enzymes to eliminate the reactive oxygen induced by xenobiotic compounds,[116] providing protection against electrophiles and products of oxidative stress. Most GST studies in bivalves focused on purification and/or biochemical measurement of total GST or different GST isoforms. GST has been used as a biomarker in environmental assess in bivalves for a number of years. The level of GST in bivalves was responsive, at some extent, to the potential environmental contamination exposure. Mussel population affected by nontreated wastewaters exhibited significantly higher GST activities as a result of an increase in conjugating activities.[117] A rapid increase of GST activities was observed in primary cultured digestive gland acini of *Pecten maximus* treated with Tributyltin, ethylmethane sulfonate and the water-soluble fraction of crude oil, in a time and dose dependent manner.[118]

Acute Phase Proteins

In addition to the discription above, there are other effectors, such as acute phase proteins, to protect bivalve against the toxic effects of contaminants or bacteria challenge. Although nonspecific, acute phase response serves as a core of the innate immune response involving physical and molecular barriers and responses that serve to prevent infection, clear potential pathogens, initiate inflammatory processes and contribute to resolution and the healing process. Acute phase proteins, an integral part of the acute phase response, have been identified in many bivalve species. Among these, heat shock protein and metallothioneins (MT) are two families of acute phase proteins recently studied in bivalves.

Heat shock proteins (HSPs) are ubiquitous and highly conserved stress proteins, not only playing important roles in response to potentially deleterious stress conditions, but also preventing cell toxicity and cell death to protect cells and tissues against damage. More recently, it has also been suggested that HSPs could function as potent activators of the innate immune system.[119] The principal HSPs range in molecular mass from 15 to 110 kD. According to apparent molecular mass, they are classified into several families. Members of HSP22, HSP60, HSP70 and HSP90 have been identified from invertebrate and most of them were found in bivalves. Recent studies in different species of bivalves have confirmed the relevant physiological roles of HSP expression in thermal tolerance and multiple stress response.[120] The different forms of HSPs might be highly affected by temperature and salinity, as well as by a great variety of chemical stressors that might be often found in seawater or in the sediment.[120,121] Meanwhile, the response of some HSPs against bacteria challenge has also been reported in bivalves. After *Vibro* stimulation, a clearly time-dependent expression pattern of HSP22, HSP70 and HSP90 was observed in scallops. HSPs might serve as powerful biomarkers of marine pollution and be helpful for health management of bivalve aquaculture.

Metallothionein (MT) is a superfamily of cysteine-rich proteins with low molecular weight, no aromatic amino acids and high metal binding affinity and widely found in a large variety of organisms.[122] In the past years, multiple functions have been attributed to MT proteins, such as homeostasis of essential metals, detoxification of toxic metals, protecting against ionizing radiation and oxidative stress, the scavenging of free radicals and response to estrogenic compounds. Bivalves are known to accumulate high concentrations of heavy metals in their tissue and are widely used as bioindicators for pollution in marine and freshwater environments. There has been considerable research dedicated to the diversity of metal-inducibility and expression in different MT isoforms from bivalves *C. virginica*, *M. galloprovincialis* and *D. polymorpha*.[123,124] The mRNA expression of MT from scallop was increased drastically to hundred fold post*Vibrio* challenge respectively.[125] The sensitivity of MT from scallop to bacteria challenge offers us a hint of its regulation in scallop immune defense. The progress opens research perspectives for the use of this marker to assess the effect of various pollutants in the aquatic environment and a better knowledge of its functional multiplicity in the bivalve immune system.

CONCLUSION

Bivalves are always challenged by their environment with high bacterial and viral loads, pollutants and they have evolved a multiplicity of efficient defense strategies to defend against microbial attack. In recent years considerable progress has been made in our understanding of the bivalve immunity. As described in this chapter, bivalve immune system is effective, multifaceted and incorporates cellular and humoral components. It is well accepted that the immune system in bivalves lacks the components of adaptive immunity present in the vertebrates, but there is also a complex mechanism involving an array of different molecules and various multi-step cascade processes. Many aspects of bivalve immunity are not well understood. Even some of the molecules and immune response mechanisms are found to be structurally and functionally similar to that in vertebrate animals. The research progress about the bivalve immunity has been hampered by absence of genome, tools for genetic manipulation and mutants and stable long-term cell lines for in vitro studies. Thus, there is no doubt that the collaborative efforts among immunologists, cell biologists, physiologists and geneticists are necessary for the study of bivalve immunity. It is still an essential task for us to fully characterize the molecules, responses, cascade pathways involved in bivalve immunity. The analysis of the precise underlying molecular mechanism of bivalve immunity will assist in understanding the nature and evolution of immune system and the connections between immune defense in invertebrates and vertebrates.

ACKNOWLEDGMENTS

The authors thank Mr. S.S. Vinu, Mr. Xiaowei Shi, Miss Chuanyan Yang and Dr. Fengmei Li for helpful discussion. L. Song and L. Wang acknowledge NSFC (Grant number No. 30730070, 30925028 and 40806064) and the Chinese Ministry of Science and Technology (Grant number 2010CB126404) for supporting their research on scallop immunity.

REFERENCES

1. Dunn CW, Hejnol A, Matus DQ et al. Broad phylogenomic sampling improves resolution of the animal tree of life. Nature 2008; 452:745-749.
2. Bebianno MJ, Geret F, Hoarau P et al. Biomarkers in Ruditapes decussatus: a potential bioindicator species. Biomarkers 2004; 9:305-330.
3. Tanguy A, Bierne N, Saavedra C et al. Increasing genomic information in bivalves through new EST collections in four species: development of new genetic markers for environmental studies and genome evolution. Gene 2008; 408:27-36.
4. Gueguen Y, Cadoret JP, Flament D et al. Immune gene discovery by expressed sequence tags generated from hemocytes of the bacteria-challenged oyster, Crassostrea gigas. Gene 2003; 303:139-145.
5. Loker ES, Adema CM, Zhang SM et al. Invertebrate immune systems—not homogeneous, not simple, not well understood. Immunol Rev 2004; 198:10-24.
6. Wootton EC, Dyrynda EA, Ratcliffe NA. Bivalve immunity: comparisons between the marine mussel (Mytilus edulis), the edible cockle (Cerastoderma edule) and the razor-shell (Ensis siliqua). Fish Shellfish Immunol 2003; 15:195-210.
7. Canesi L, Gallo G, Gavioli M et al. Bacteria-hemocyte interactions and phagocytosis in marine bivalves. Microsc Res Tech 2002; 57:469-476.
8. Takahashi KG, Muroga K. Cellular Defense Mechanisms in Bivalve Molluscs. Fish Pathology 2008; 43:1-17.
9. Cheng TC, ed. Bivalves. London: Acdemic Press; 1981. Ratcliffe NA, Rowley AF, eds. Invertebrate blood cell.
10. Cajaraville MP, Pal SG. Morphofunctional study of the haemocytes of the bivalve mollusc Mytilus galloprovincialis with emphasis on the endolysosomal compartment. Cell Struct Funct 1995; 20:355-367.
11. Lopez C, Carballal MJ, Azevedo C et al. Morphological characterization of the hemocytes of the clam, Ruditapes decussatus (Mollusca: Bivalvia). J Invertebr Pathol 1997; 69:51-57.
12. Donaghy L, Kim BK, Hong HK et al. Flow cytometry studies on the populations and immune parameters of the hemocytes of the Suminoe oyster, Crassostrea ariakensis. Fish Shellfish Immunol 2009; 27:296-301.
13. Lemaitre B, Hoffmann J. The host defense of Drosophila melanogaster. Annu Rev Immunol 2007; 25:697-743.
14. Schneeweiss H, Renwrantz L. Analysis of the attraction of haemocytes from Mytilus edulis by molecules of bacterial origin. Dev Comp Immunol 1993; 17:377-387.
15. Cochennec-Laureau N, Auffret M, Renault T et al. Changes in circulating and tissue-infiltrating hemocyte parameters of European flat oysters, Ostrea edulis, naturally infected with Bonamia ostreae. J Invertebr Pathol 2003; 83:23-30.
16. Garcia-Garcia E, Prado-Alvarez M, Novoa B et al. Immune responses of mussel hemocyte subpopulations are differentially regulated by enzymes of the PI 3-K, PKC and ERK kinase families. Dev Comp Immunol 2008; 32:637-653.
17. Chu FE, ed. Defense mechanisms of marine bivalves. Enfield Science Publishers Inc; 2000. Fingerman M, Nagabhushanam R, eds. Recent advances in marine biotechnology No. 5.
18. Bugge DM, Hegaret H, Wikfors GH et al. Oxidative burst in hard clam (Mercenaria mercenaria) haemocytes. Fish Shellfish Immunol 2007; 23:188-196.
19. Olafsen JA. Role of lectins (C-reactive protein) in defense of marine bivalves against bacteria. Adv Exp Med Biol 1995; 371A:343-348.
20. Batista FM, Boudry P, Dos Santos A et al. Infestation of the cupped oysters Crassostrea angulata, C. gigas and their first-generation hybrids by the copepod Myicola ostreae: differences in susceptibility and host response. Parasitology 2009; 136:537-543.
21. Laruelle F, Molloy DP, Roitman VA. Histological analysis of trematodes in Dreissena polymorpha: their location, pathogenicity and distinguishing morphological characteristics. J Parasitol 2002; 88:856-863.
22. Janeway CA, Jr., Medzhitov R. Innate immune recognition. Annu Rev Immunol 2002; 20:197-216.
23. Medzhitov R, Janeway CA, Jr. Decoding the patterns of self and nonself by the innate immune system. Science 2002; 296:298-300.
24. Medzhitov R, Janeway C, Jr. Innate immune recognition: mechanisms and pathways. Immunol Rev 2000; 173:89-97.
25. Itoh N, Takahashi KG. A novel peptidoglycan recognition protein containing a goose-type lysozyme domain from the Pacific oyster, Crassostrea gigas. Mol Immunol 2009; 46:1768-1774.
26. Ni D, Song L, Wu L et al. Molecular cloning and mRNA expression of peptidoglycan recognition protein (PGRP) gene in bay scallop (Argopecten irradians, Lamarck 1819). Dev Comp Immunol 2007; 31:548-558.

27. Wang L, Song L, Zhao J et al. Expressed sequence tags from the zhikong scallop (Chlamys farreri): discovery and annotation of host-defense genes. Fish Shellfish Immunol 2009; 26:744-750.
28. Christophides GK, Zdobnov E, Barillas-Mury C et al. Immunity-related genes and gene families in Anopheles gambiae. Science 2002; 298:159-165.
29. Jayaraj SS, Thiagarajan R, Arumugam M et al. Isolation, purification and characterization of beta-1,3-glucan binding protein from the plasma of marine mussel Perna viridis. Fish Shellfish Immunol 2008; 24:715-725.
30. Zelensky AN, Gready JE. The C-type lectin-like domain superfamily. FEBS J 2005; 272:6179-6217.
31. Yamaura K, Takahashi KG, Suzuki T. Identification and tissue expression analysis of C-type lectin and galectin in the Pacific oyster, Crassostrea gigas. Comp Biochem Physiol B Biochem Mol Biol 2008; 149:168-175.
32. Watanabe A, Miyazawa S, Kitami M et al. Characterization of a novel C-type lectin, Bombyx mori multibinding protein, from the B. mori hemolymph: mechanism of wide-range microorganism recognition and role in immunity. J Immunol 2006; 177:4594-4604.
33. Kim JY, Adhya M, Cho SK et al. Characterization, tissue expression and immunohistochemical localization of MCL3, a C-type lectin produced by Perkinsus olseni-infected Manila clams (Ruditapes philippinarum). Fish Shellfish Immunol 2008; 25:598-603.
34. Takahashi KG, Kuroda T, Muroga K. Purification and antibacterial characterization of a novel isoform of the Manila clam lectin (MCL-4) from the plasma of the Manila clam, Ruditapes philippinarum. Comp Biochem Physiol B Biochem Mol Biol 2008; 150:45-52.
35. Minamikawa M, Hine M, Russell S et al. Isolation and partial characterization of a calcium-dependent lectin (chiletin) from the haemolymph of the flat oyster, Ostrea chilensis. Fish Shellfish Immunol 2004; 17:463-476.
36. Gourdine JP, Cioci G, Miguet L et al. High affinity interaction between a bivalve C-type lectin and a biantennary complex-type N-glycan revealed by crystallography and microcalorimetry. J Biol Chem 2008; 283:30112-30120.
37. Tasumi S, Vasta GR. A galectin of unique domain organization from hemocytes of the Eastern oyster (Crassostrea virginica) is a receptor for the protistan parasite Perkinsus marinus. J Immunol 2007; 179:3086-3098.
38. Song X, Zhang H, Zhao J et al. An immune responsive multidomain galectin from bay scallop Argopectens irradians. Fish and Shellfish Immunology 2010;In Press, Corrected Proof.
39. Kim JY, Kim YM, Cho SK et al. Noble tandem-repeat galectin of Manila clam Ruditapes philippinarum is induced upon infection with the protozoan parasite Perkinsus olseni. Dev Comp Immunol 2008; 32:1131-1141.
40. Blandin S, Levashina EA. Thioester-containing proteins and insect immunity. Mol Immunol 2004; 40:903-908.
41. Prado-Alvarez M, Rotllant J, Gestal C et al. Characterization of a C3 and a factor B-like in the carpet-shell clam, Ruditapes decussatus. Fish Shellfish Immunol 2009; 26:305-315.
42. Zhang H, Song L, Li C et al. Molecular cloning and characterization of a thioester-containing protein from Zhikong scallop Chlamys farreri. Mol Immunol 2007; 44:3492-3500.
43. Dishaw LJ, Smith SL, Bigger CH. Characterization of a C3-like cDNA in a coral: phylogenetic implications. Immunogenetics 2005; 57:535-548.
44. Zhang H, Wang L, Song L et al. The genomic structure, alternative splicing and immune response of Chlamys farreri thioester-containing protein. Dev Comp Immunol 2009.
45. Mukhopadhyay S, Gordon S. The role of scavenger receptors in pathogen recognition and innate immunity. Immunobiology 2004; 209:39-49.
46. Tanguy A, Guo X, Ford SE. Discovery of genes expressed in response to Perkinsus marinus challenge in Eastern (Crassostrea virginica) and Pacific (C. gigas) oysters. Gene 2004; 338:121-131.
47. Song L, Xu W, Li C et al. Development of expressed sequence tags from the bay scallop, Argopecten irradians irradians. Mar Biotechnol (NY) 2006; 8:161-169.
48. Mickael Perrigault ATaBA. Identification and expression of differentially expressed genes in the hard clam, Mercenaria mercenaria, in response to quahog parasite unknown (QPX). BMC Genomics 2009; 10.
49. Qiu L, Song L, Xu W et al. Molecular cloning and expression of a Toll receptor gene homologue from Zhikong Scallop, Chlamys farreri. Fish Shellfish Immunol 2007; 22:451-466.
50. Moynagh PN. The NF-kappaB pathway. J Cell Sci 2005; 118:4589-4592.
51. Li Q, Verma IM. NF-kappaB regulation in the immune system. Nat Rev Immunol 2002; 2:725-734.
52. Montagnani C, Labreuche Y, Escoubas J. Cg-I¦ÊB, a new member of the I¦ÊB protein family characterized in the pacific oyster Crassostrea gigas. Developmental and Comparative Immunology 2008; 32:182-190.
53. Wu X, Xiong X, Xie L et al. Pf-Rel, a Rel/nuclear factor-kappaB homolog identified from the pearl oyster, Pinctada fucata. Acta Biochim Biophys Sin (Shanghai) 2007; 39:533-539.

54. Xiong X, Feng Q, Chen L et al. Cloning and characterization of an IKK homologue from pearl oyster, Pinctada fucata. Dev Comp Immunol 2008; 32:15-25.
55. Zhang D, Jiang S, Qiu L et al. Molecular characterization and expression analysis of the IkappaB gene from pearl oyster Pinctada fucata. Fish Shellfish Immunol 2009; 26:84-90.
56. Lamprou I, Mamali I, Dallas K et al. Distinct signalling pathways promote phagocytosis of bacteria, latex beads and lipopolysaccharide in medfly haemocytes. Immunology 2007; 121:314-327.
57. Johnson G, Lapadat R. Mitogen-activated protein kinase pathways mediated by ERK, JNK and p38 protein kinases. Science's STKE 2002; 298:1911.
58. Canesi L, Betti M, Ciacci C et al. Tyrosine kinase-mediated cell signalling in the activation of Mytilus hemocytes: possible role of STAT-like proteins. Biol Cell 2003; 95:603-613.
59. Kang YS, Kim YM, Park KI et al. Analysis of EST and lectin expressions in hemocytes of Manila clams (Ruditapes philippinarum) (Bivalvia: Mollusca) infected with Perkinsus olseni. Dev Comp Immunol 2006; 30:1119-1131.
60. Huang S, Yuan S, Guo L et al. Genomic analysis of the immune gene repertoire of amphioxus reveals extraordinary innate complexity and diversity. Genome Res 2008; 18:1112-1126.
61. Pujol N, Link EM, Liu LX et al. A reverse genetic analysis of components of the Toll signaling pathway in Caenorhabditis elegans. Current Biology 2001; 11:809-821.
62. Qiu L, Song L, Yu Y et al. Identification and characterization of a myeloid differentiation factor 88 (MyD88) cDNA from Zhikong scallop Chlamys farreri. Fish Shellfish Immunol 2007; 23:614-623.
63. Gueguen Y, Cadoret J-P, Flament D et al. Immune gene discovery by expressed sequence tags generated from hemocytes of the bacteria-challenged oyster, Crassostrea gigas. Gene 2003; 303:139-145.
64. Qiu L, Song L, Yu Y et al. Identification and expression of TRAF6 (TNF receptor-associated factor 6) gene in Zhikong scallop Chlamys farreri. Fish Shellfish Immunol 2009; 26:359-367.
65. Goetz SRGGSWF. Analysis of Genes Isolated from Plated Hemocytes of the Pacific Oyster, Crassostreas gigas. Mar Biotechnol (NY) 2009; 11:24-44.
66. Morgan BP. The Human Complement System in Health and Disease. Annals of the Rheumatic Diseases 1998; 57:581-581.
67. Huber-Lang M, Sarma JV, Zetoune FS et al. Generation of C5a in the absence of C3: a new complement activation pathway. Nat Med 2006; 12:682-687.
68. Prado-Alvarez M, Rotllant J, Gestal C et al. Characterization of a C3 and a factor B-like in the carpet-shell clam, Ruditapes decussatus. Fish and Shellfish Immunology 2009; 26:305-315.
69. Zhang H, Song L, Li C et al. A novel C1q-domain-containing protein from Zhikong scallop Chlamys farreri with lipopolysaccharide binding activity. Fish and Shellfish Immunology 2008.
70. Zhang H, Wang L, Song L et al. A fibrinogen-related protein from bay scallop Argopecten irradians involved in innate immunity as pattern recognition receptor. Fish Shellfish Immunol 2009; 26:56-64.
71. Bulet P, Stocklin R, Menin L. Anti-microbial peptides: from invertebrates to vertebrates. Immunol Rev 2004; 198:169-184.
72. Li C, Zhao J, Song L. A review of advances in research on marine molluscan antimicrobial peptides and their potential application in aquaculture. Molluscan Research 2009; 29:17-26.
73. Mitta G, Vandenbulcke F, Roch P. Original involvement of antimicrobial peptides in mussel innate immunity. FEBS Lett 2000; 486:185-190.
74. Costa MM, Dios S, Alonso-Gutierrez J et al. Evidence of high individual diversity on myticin C in mussel (Mytilus galloprovincialis). Dev Comp Immunol 2009; 33:162-170.
75. Charlet M, Chernysh S, Philippe H et al. Innate immunity. Isolation of several cysteine-rich antimicrobial peptides from the blood of a mollusc, Mytilus edulis. J Biol Chem 1996; 271:21808-21813.
76. Gonzalez M, Gueguen Y, Desserre G et al. Molecular characterization of two isoforms of defensin from hemocytes of the oyster Crassostrea gigas. Dev Comp Immunol 2007; 31:332-339.
77. Zhao J, Song L, Li C et al. Molecular cloning, expression of a big defensin gene from bay scallop Argopecten irradians and the antimicrobial activity of its recombinant protein. Mol Immunol 2007; 44:360-368.
78. Prager EM, Jolles P. Animal lysozymes c and g: an overview. EXS 1996; 75:9-31.
79. Bachali S, Jager M, Hassanin A et al. Phylogenetic analysis of invertebrate lysozymes and the evolution of lysozyme function. J Mol Evol 2002; 54:652-664.
80. Zhao J, Song L, Li C et al. Molecular cloning of an invertebrate goose-type lysozyme gene from Chlamys farreri and lytic activity of the recombinant protein. Mol Immunol 2007; 44:1198-1208.
81. Nilsen IW, Overbo K, Sandsdalen E et al. Protein purification and gene isolation of chlamysin, a cold-active lysozyme-like enzyme with antibacterial activity. FEBS Lett 1999; 464:153-158.
82. Takeshita K, Hashimoto Y, Thujihata Y et al. Determination of the complete cDNA sequence, construction of expression systems and elucidation of fibrinolytic activity for Tapes japonica lysozyme. Protein Expr Purif 2004; 36:254-262.
83. Xue QG, Schey KL, Volety AK et al. Purification and characterization of lysozyme from plasma of the eastern oyster (Crassostrea virginica). Comp Biochem Physiol B Biochem Mol Biol 2004; 139:11-25.

84. Haug T, Stensvag K, Olsen MO et al. Antibacterial activities in various tissues of the horse mussel, Modiolus modiolus. J Invertebr Pathol 2004; 85:112-119.
85. McHenery JG, Birkbeck TH. Characterization of the lysozyme of Mytilus edulis (L). Comp Biochem Physiol B 1982; 71:583-589.
86. Bulgakov AA, Park KI, Choi KS et al. Purification and characterisation of a lectin isolated from the Manila clam Ruditapes philippinarum in Korea. Fish Shellfish Immunol 2004; 16:487-499.
87. Gourdine JP, Smith-Ravin EJ. Analysis of a cDNA-derived sequence of a novel mannose-binding lectin, codakine, from the tropical clam Codakia orbicularis. Fish Shellfish Immunol 2007; 22:498-509.
88. Fisher WS, DiNuzzo AR. Agglutination of bacteria and erythrocytes by serum from six species of marine molluscs. J Invertebr Pathol 1991; 57:380-394.
89. Olafsen JA, Fletcher TC, Grant PT. Agglutinin activity in Pacific oyster (Crassostrea gigas) hemolymph following in vivo Vibrio anguillarum challenge. Dev Comp Immunol 1992; 16:123-138.
90. Adhya M, Singha B, Chatterjee BP. Purification and characterization of an N-acetylglucosamine specific lectin from marine bivalve Macoma birmanica. Fish Shellfish Immunol 2009; 27:1-8.
91. Roberts S, Gueguen Y, de Lorgeril J et al. Rapid accumulation of an interleukin 17 homolog transcript in Crassostrea gigas hemocytes following bacterial exposure. Dev Comp Immunol 2008; 32:1099-1104.
92. Hughes TK, Jr., Smith EM, Chin R et al. Interaction of immunoactive monokines (interleukin 1 and tumor necrosis factor) in the bivalve mollusc Mytilus edulis. Proc Natl Acad Sci USA 1990; 87:4426-4429.
93. Ottaviani E, Malagoli D, Franchini A. Invertebrate humoral factors: cytokines as mediators of cell survival. Prog Mol Subcell Biol 2004; 34:1-25.
94. Beschin A, Bilej M, Torreele E et al. On the existence of cytokines in invertebrates. Cell Mol Life Sci 2001; 58:801-814.
95. Li L, Qiu L, Song L et al. First molluscan TNFR homologue in Zhikong scallop: molecular characterization and expression analysis. Fish Shellfish Immunol 2009; 27:625-632.
96. Herpin CLA, Becker T, Rosa FM et al. Structural and functional evidences for a type 1 TGF-β sensu stricto receptor in the lophotrochozoan Crassostrea gigas suggest conserved molecular mechanisms controlling mesodermal patterning across bilateria. Mechanisms of Development 2005; 122:695-705.
97. Smith LC, Clow LA, Terwilliger DP. The ancestral complement system in sea urchins. Immunol Rev 2001; 180:16-34.
98. Aguirre J MM, Hewitt D, Hansberg W. Reactive oxygen species and development in microbial eukaryotes. Trends Microbiol 2005; 13:111-118.
99. Winston GW. Oxidants and antioxidants in aquatic animals. Comp Biochem Physiol C 1991; 100:173-176.
100. Bao Y, Li L, Zhang G. The manganese superoxide dismutase gene in bay scallop Argopecten irradians: cloning, 3D modelling and mRNA expression. Fish Shellfish Immunol 2008; 25:425-432.
101. Estrada N, de Jesus Romero M, Campa-Cordova A et al. Effects of the toxic dinoflagellate, Gymnodinium catenatum on hydrolytic and antioxidant enzymes, in tissues of the giant lions-paw scallop Nodipecten subnodosus. Comp Biochem Physiol C Toxicol Pharmacol 2007; 146:502-510.
102. Xing J, Lin T, Zhan W. Variations of enzyme activities in the haemocytes of scallop Chlamys farreri after infection with the acute virus necrobiotic virus (AVNV). Fish Shellfish Immunol 2008; 25:847-852.
103. Vlahogianni T, Dassenakis M, Scoullos MJ et al. Integrated use of biomarkers (superoxide dismutase, catalase and lipid peroxidation) in mussels Mytilus galloprovincialis for assessing heavy metals' pollution in coastal areas from the Saronikos Gulf of Greece. Mar Pollut Bull 2007; 54:1361-1371.
104. Monari M, Foschi J, Matozzo V et al. Investigation of EROD, CYP1A immunopositive proteins and SOD in haemocytes of Chamelea gallina and their role in response to B[a]P. Comp Biochem Physiol C Toxicol Pharmacol 2009; 149:382-392.
105. Monari M, Matozzo V, Foschi J et al. Effects of high temperatures on functional responses of haemocytes in the clam Chamelea gallina. Fish Shellfish Immunol 2007; 22:98-114.
106. Geret F, Manduzio H, Company R et al. Molecular cloning of superoxide dismutase (Cu/Zn-SOD) from aquatic molluscs. Mar Environ Res 2004; 58:619-623.
107. Chelikani P, Fita I, Loewen PC. Diversity of structures and properties among catalases. Cell Mol Life Sci 2004; 61:192-208.
108. Fita I, Rossmann MG. The active center of catalase. J Mol Biol 1985; 185:21-37.
109. Letendre J, Chouquet B, Manduzio H et al. Tidal height influences the levels of enzymatic antioxidant defences in Mytilus edulis. Mar Environ Res 2009; 67:69-74.
110. Contardo-Jara V, Krueger A, Exner HJ et al. Biotransformation and antioxidant enzymes of Dreissena polymorpha for detection of site impact in watercourses of Berlin. J Environ Monit 2009; 11:1147-1156.
111. Damiens G, His E, Gnassia-Barelli M et al. Evaluation of biomarkers in oyster larvae in natural and polluted conditions. Comp Biochem Physiol C Toxicol Pharmacol 2004; 138:121-128.

112. Li C, Ni D, Song L et al. Molecular cloning and characterization of a catalase gene from Zhikong scallop Chlamys farreri. Fish Shellfish Immunol 2008; 24:26-34.
113. Bigot A, Doyen P, Vasseur P et al. Metallothionein coding sequence identification and seasonal mRNA expression of detoxification genes in the bivalve Corbicula fluminea. Ecotoxicol Environ Saf 2009; 72:382-387.
114. Box A, Sureda A, Deudero S. Antioxidant response of the bivalve Pinna nobilis colonised by invasive red macroalgae Lophocladia lallemandii. Comp Biochem Physiol C Toxicol Pharmacol 2009; 149:456-460.
115. Soldatov AA, Gostiukhina OL, Golovina IV. [Antioxidant enzyme complex of tissues of the bivalve Mytilus galloprovincialis Lam. under normal and oxidative-stress conditions: a review]. Prikl Biokhim Mikrobiol 2007; 43:621-628.
116. Hoarau P, Garello G, Gnassia-Barelli M et al. Purification and partial characterization of seven glutathione S-transferase isoforms from the clam Ruditapes decussatus. Eur J Biochem 2002; 269:4359-4366.
117. Moreira SM, Guilhermino L. The use of Mytilus galloprovincialis acetylcholinesterase and glutathione S-transferases activities as biomarkers of environmental contamination along the northwest Portuguese coast. Environ Monit Assess 2005; 105:309-325.
118. Le Pennec G, Le Pennec M. Induction of glutathione-S-transferases in primary cultured digestive gland acini from the mollusk bivalve Pecten maximus (L.): application of a new cellular model in biomonitoring studies. Aquat Toxicol 2003; 64:131-142.
119. Gao Q, Zhao J, Song L et al. Molecular cloning, characterization and expression of heat shock protein 90 gene in the haemocytes of bay scallop Argopecten irradians. Fish Shellfish Immunol 2008; 24:379-385.
120. Ivanina AV, Taylor C, Sokolova IM. Effects of elevated temperature and cadmium exposure on stress protein response in eastern oysters Crassostrea virginica (Gmelin). Aquat Toxicol 2009; 91:245-254.
121. Pantzartzi CN, Kourtidis A, Drosopoulou E et al. Isolation and characterization of two cytoplasmic hsp90s from Mytilus galloprovincialis (Mollusca: Bivalvia) that contain a complex promoter with a p53 binding site. Gene 2009; 431:47-54.
122. Dabrio M, Rodriguez AR, Bordin G et al. Recent developments in quantification methods for metallothionein. J Inorg Biochem 2002; 88:123-134.
123. Bourdineaud JP, Baudrimont M, Gonzalez P et al. Challenging the model for induction of metallothionein gene expression. Biochimie 2006; 88:1787-1792.
124. Fasulo S, Mauceri A, Giannetto A et al. Expression of metallothionein mRNAs by in situ hybridization in the gills of Mytilus galloprovincialis, from natural polluted environments. Aquat Toxicol 2008; 88:62-68.
125. Wang L, Song L, Ni D et al. Alteration of metallothionein mRNA in bay scallop Argopecten irradians under cadmium exposure and bacteria challenge. Comp Biochem Physiol C Toxicol Pharmacol 2009; 149:50-57.

CHAPTER 4

EARTHWORM IMMUNITY

Martin Bilej,* Petra Procházková, Marcela Šilerová and Radka Josková
Institute of Microbiology, Academy of Sciences of the Czech Republic, Prague, Czech Republic
**Corresponding Author: Martin Bilej—Email: mbilej@biomed.cas.cz*

Abstract: Earthworms belonging to oligochaete annelids became a model for comparative immunologists in the early sixties with the publication of results from transplantation experiments that proved the existence of self/nonself recognition in earthworms. This initiated extensive studies on the earthworm immune mechanisms that evolved to prevent the invasion of pathogens. In the last four decades important cellular and humoral pathways were described and numerous biologically active compounds were characterized and often cloned.

INTRODUCTION

Considering the fact that the majority of the immunologists all over the world are focused on the mammalian or particularly human immunology, it may sound surprising that invertebrates—and among others earthworms—have been an important experimental model since the very beginning of immunology. For example, phagocytosis, an important and evolutionarily conserved defense mechanism of innate immunity, was discovered in the late 1800s by a Nobel Prize winner Elie Mechnikoff while studying the origin of the digestive organs in the floating larvae of starfish.[1]

From the total number of extant animal species, certainly surpassing 2 millions, 95% are included in the invertebrate taxa. Invertebrates have evolved for hundreds of millions of years, often surviving in very hostile environments. Their successful survival strategies are likely based on short life span combined with numerous offspring. More importantly, all invertebrate species have developed a variety of defense mechanisms efficiently recognizing and responding to nonself substances.[2,3]

In contrast to adaptive immunity, which is a highly sophisticated system based on antigen-specific T and B cells and antibodies and which is observed in vertebrates only,

Invertebrate Immunity, edited by Kenneth Söderhäll.

many innate immunity mechanisms are conserved from invertebrates to vertebrates. Cellular mechanisms of invertebrate innate immunity include wound repair, clotting and coagulation responses, phagocytosis of invading microorganisms and encapsulation reactions. Apart from these cellular mechanisms, invertebrates possess a broad range of antimicrobial factors such as lysozyme-like proteins, proteases, cytolytic proteins, antimicrobial peptides and enzyme activation-base cascades; humoral defense also includes lectin-like and pattern recognition molecules that are designed to recognize a few highly conserved structures present in many different microorganisms. The majority of the above immune responses of invertebrates are nonadaptive with no or very limited ability either to "remember" or to respond more vigorously and effectively to repeated exposures to the same pathogens.[4]

BASIC INFORMATION ON EARTHWORM ANATOMY

The earthworms are protostomian animals possessing true coelom of mesenchymal origin. The coelomic cavity is filled with coelomic fluid containing free wandering cells, named coelomocytes, originating in the mesenchymal lining of the cavity. The coelomic cavity is metameric and the segments are separated by transversal septa. Transport of the coelomic fluid and coelomocytes between the segments is regulated by channels. Each segment of the coelomic cavity is opened to the outer environment by a pair of nephridia and by a dorsal pore (Fig. 1).

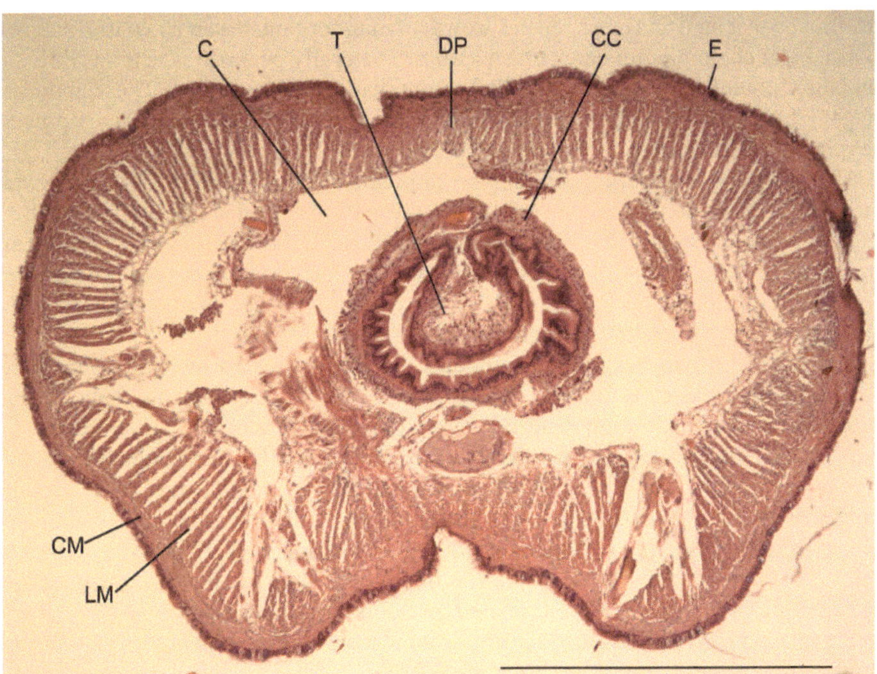

Figure 1. Transverse section of *Eisenia fetida* stained with hematoxylin-eosin. C—coelom, T—typhlosole, DP—dorsal pore, CC—chloragogen cells, E—epidermis, CM—circular muscles, LM—longitudinal muscles. Bar 1 mm. Reproduced by courtesy of Dr. Pavel Rossmann, Prague.

In terms of systematic biology, the earthworm family—Lumbricidae—is the largest member of the class Oligochaeta, phylum Annelida. The earthworms are found in leaf litter, manure, under stones and logs as well as some arid areas, but most species prefer wetter, more heavily vegetated regions. Earthworms range in size from two centimeters to over one meter. There are over 3000 described earthworm species known worldwide and they have adapted to a wide range of soil habitats as well as freshwater lakes and streams (for a review see ref. 5). Immunological research is performed mainly in two genera—*Lumbricus* and *Eisenia.*

CELLULAR DEFENSE MECHANISMS

The first nonspecific barrier of earthworms is the skin covering their entire body. The skin consists of the epidermis and a thin cuticle, which contains mucopolysaccharides acting as an antimicrobial barrier.[6,7] The epidermis is formed by a single layer epithelium of supporting cells, basal cells and secretory cells. The basal cells play an important role in wound healing and graft rejection, often exerting phagocytic activity.[8,9] Thus, these basal cells are sometimes considered not to be of epidermal origin, but rather homologous to coelomocytes.[10-12]

Each segment of the coelomic cavity communicates with outer environment by a dorsal pore; the skin therefore cannot sufficiently prevent the microorganisms from entering the coelomic cavity. Consequently, the coelomic cavity is not aseptic and always contains bacteria, protozoans and fungi from the outer environment. Nevertheless, there are efficient mechanisms that keep the growth of microorganisms under control.[13] It was reported that coelomic fluid contains 6×10^5/ml naturally occurring bacteria while the number of potentially phagocytic cells is more than ten times higher. These abundant phagocytes combined with the presence of various humoral factors can easily prevent the microorganisms from outgrowth.

The invading microorganisms can be eliminated by a number of ways. First, they can be excreted by nephridia[14] or engulfed by the cells of nephrostome or middle tube.[15] Second, as mentioned above, the microorganisms can be phagocytosed by certain coelomocytes and phagocytic cells that, when they become exhausted, are expelled through dorsal pores. The pores are equipped with muscular sphincters controlling intracoelomic pressure and the exchange of material between the outer and inner environments.[14] Third, large foreign bodies, e.g., agglutinated bacteria or parasites are eliminated by encapsulation.[16,17] This process begins, similarly to phagocytosis, by the recognition of foreign material which, however, cannot be engulfed due to its size. Within the first day, the foreign body is surrounded by free coelomocytes and after several days a dense capsule (often called a brown body because of its melanin content as a consequence of the prophenoloxidase cascade) composed of flattened cells is formed. When the capsule is about 1-2 mm in diameter, its external cells lose their adhesiveness so that the capsule can migrate towards the posterior segments of the coelomic cavity where it is eliminated by autonomy followed by wound.[18-21] It was documented in *Eisenia fetida* earthworms that most brown bodies contain tissue wastes, agglutinated bacteria, gregarines or nematodes.[17]

As mentioned above, coelomic fluid contains different types of coelomocytes. Their nomenclature is based mainly on morphological and cytochemical criteria (for a review see refs. 22,23) though more recent studies attempt to determine superficial and functional markers for cell classification.[24] In general, there are three main coelomocyte types—eleocytes, free chloragogen cells with nutritive and accessory functions and either hyaline or granular amoebocytes, both representing effectory immunocytes involved in a broad range of defense functions including phagocytosis.

Although both types of amoebocytes have phagocytic properties, their activity differs. In contrast to granular amoebocytes, the cytoplasm of hyaline amoebocytes is occasionally full of engulfed material.[22] It should be mentioned that amoebocytes engulf all kinds of material including inert particles, microbial cell wall components as well as foreign cells. However, the phagocytosis of eukaryotic cells depends on the source of the cells. Unlike allologous cells, xenologous cells (both from different earthworm species and from noninvertebrate species) are rapidly phagocytosed.[14,25]

Phagocytosis by coelomocytes, similarly to that of vertebrates, can be modulated by humoral components, opsonins, which coat the engulfed particle and thus promote its phagocytosis. It was proven that preincubation of both yeast and synthetic copolymer particles with the coelomic fluid significantly increased their phagocytosis.[26,27] It is noteworthy that also mammalian opsonins, IgG immunoglobulin and C3b complement fragment, were described to enhance coelomocyte phagocytic activity, in contrast to IgM and C3d fragment, which did not affect phagocytosis.[28]

The earthworms are regarded as an important model organism of comparative immunology since 1960s when transplantation experiments were performed (for a review see ref. 29) and cell-mediated short-term memory was observed.[30] All these experiments proved the existence of self and nonself recognition in earthworms and initiated extensive studies of earthworm immune mechanisms. The ability to recognize and respond to allografts as well as xenografts and, on the other hand, the ability to accept or not to destroy autografts was observed in many annelid species.[29] This process begins like the reaction to injury. The first major change, which occurs after the healing of wounds is, regardless of the graft origin, the accumulation of coelomocytes near the graft sites and their infiltration into the matrix. The response to the xenografts results in complete walling off of the graft and its destruction by encapsulation reaction.[31] The number of invading coelomocytes during the autograft transplantation is markedly lower[32,33] but the reaction seems to be more rapid. The maximum number of coelomocytes surrounding the graft was detected within 24 hours, returning to the normal level by 72 hours. In contrast, the peak response to xenografts is on day 3 or 4 and normal levels are not reached before day 7. The destruction of xenografts is completed approximately by day 17 after the transplantation. If a second graft is transplanted at this time, an accelerated rejection within 6 or 7 days occurs. Moreover, the number of the invading coelomocytes is 20-30 % higher. The increased number of coelomocytes during the retransplantation is probably caused by an increased proliferating activity of mesenchymal lining of the coelomic cavity and the septa. These data suggest the existence of short-term and very limited memory, which is based solely on cells as the transfer of either the coelomic fluid or other substances does not induce any accelerated reaction.[30,33]

Allo- and xenorecognition mediated by earthworm coelomocytes was evidenced in experiments showing cell-mediated cytotoxicity to allogeneic coelomocytes[34] and, more interestingly, capability of earthworm non/phagocytic coelomocytes to kill efficiently tumor cell targets such as K562.[35,36]

HUMORAL DEFENSE MECHANISMS

The coelomic fluid of annelids exerts numerous biological activities that are involved in effective defense mechanisms against invaders. It was documented that it contains various antimicrobial factors like lysozyme[37,38] and antimicrobial peptides.[39-41] Moreover, coelomic fluid was found to cause the lysis of vertebrate erythrocytes and, subsequently, several hemolytic factors were isolated and described. The majority of proteins with hemolytic properties have hemagglutination activity as well and, more interestingly, a spectrum of antibacterial and bacteriostatic activities against pathogenic soil bacteria.[42-44] Furthermore, it was observed that coelomic fluid lyses eukaryotic cells other than erythrocytes, namely fibroblasts and insect hemocytes[45] and various tumor cell lines.[46,47] Hereinafter, each activity of the coelomic fluid is thoroughly described.

Lysozyme is a bacteriolytic enzyme which catalyzes the hydrolysis of 1,4-β-D-links between N-acetylmuramic acid and N-acetyl-D-glucosamine residues in the peptidoglycan of bacterial cell walls and thus efficiently protects against infections caused particularly by Gram-positive bacteria. Lysozyme activity was observed in coelomocyte extracts as well as in the coelomic fluid.[37] Later, the active protein was isolated and partially sequenced.[48] Based on the N-terminal sequence, a novel class of lysozymes including those of molluscs, echinoderms, nematodes and earthworms was proposed.[48] Recently, cDNA coding for lysozyme-like molecule of *E. andrei* earthworms was characterized and cloned.[38] Earthworm lysozyme exhibited both lysozyme and isopeptidase activity and shared homology with other invertebrate lysozymes, with the highest similarity (72 % identity) to destabilase I from medicinal leech. Moreover, lysozyme expression can be up-regulated after a challenge with Gram-positive as well as Gram-negative bacteria.

Antimicrobial peptides are an abundant and diverse group of molecules that are produced by many cell types in invertebrates, vertebrates and plants. To date, only a limited number of bioactive peptides have been described in annelids. An antimicrobial peptide named Lumbricin I was identified in *Lumbricus rubellus*.[39] Lumbricin I is a proline-rich antimicrobial peptide which is constitutively expressed in adult animals and is not induced by bacterial infection. A Lumbricin I analog named PP-1 was found in the Asian earthworm *Pheretima tschiliensis*, is synthesized in the body wall only and its localization in the mucus of the epidermis suggests its role in the mucosal defense.[40] Furthermore, an antimicrobial short peptide OEP3121 of only 5 amino acids was found in *Eisenia fetida* earthworms.[41]

The coelomic fluid of *Eisenia* earthworms was described to exhibit strong *hemolytic activity* that is tightly connected with *bacteriostatic and antibacterial properties* against pathogenic soil bacteria. The first hemolytic proteins were described by Du Pasquier and Duprat[49] and later on they were named *EFAF* (*Eisenia fetida andrei* factors) and characterized as two glycoproteins secreted by chloragocytes and eleocytes.[42,50,51] The 45-kDa protein is encoded by a single nonpolymorphic gene and has a pI of 6.0, while the 40-kDa protein is encoded by a gene having four alleles, each representing one of isoforms with pI of 6.3, 6.2, 5.95 and 5.9. Each individual earthworm possesses the 45-kDa protein and 1 or 2 isoforms of the 40-kDa protein.[42,52] In addition to EFAFs hemolytic activity, these proteins were found to agglutinate red blood cells[53] and to participate in the cytotoxic activity of the coelomic fluid.[45] Moreover, they exhibit antibacterial activity against both Gram-positive and Gram-negative bacteria,[43,54,55] particularly against strains that are pathogenic for earthworms.[44,52,56] In addition to their bacteriolytic activity, they may also mediate opsonization[57] and participate in the clotting of the coelomic fluid.[58]

It was documented, that upon binding to sphingomyelin, a major lipid constituent of plasma membranes of most mammalian cells, these proteins polymerize and form 10-nm channels through the lipid bilayer.[50,59]

Later, EFAFs were characterized at the molecular level and were named **fetidins**. A gene for the 40-kDa protein was cloned[60,61] and it was found that its putative amino acid sequence comprises an N-glycosylation site and a peroxidase motif. This is in accordance with the finding that both fetidins have peroxidase activity.

Independently, a 41-kDa hemolytic protein, which is produced by coelomocytes and causes contraction of rat vascular smooth muscles, was characterized and named **lysenin**.[62] Simultaneously, two 42-kDa lysenin-related proteins with weak contractive activity were identified.[63] More recently, a new member of this lysenin-like multi-gene family has been cloned and provisionally called lysenin-related protein 3.[64] Lysenin has a high amino acid sequence homology with fetidin (89% identity, 95% positivity), with lysenin-related protein 1 (76% identity, 89% positivity) and lysenin-related protein 3 (81% identity, 90% positivity). Amino acid sequence of lysenin-related protein 2 corresponds to that of fetidin. All these data suggest a close relationship between these lytic molecules. Individual sequence analyses have revealed that fetidin and lysenin are encoded by two distinct highly homologous genes but their expression level differs in individual earthworms.[65]

The hemolytic activity of lysenin is dependent on the presence of sphingolipids in the membrane.[66] Moreover, the presence of cholesterol in the membrane facilitates hemolysis. Upon binding to the sphingomyelin, lysenin forms oligomers and subsequently pores 3 nm in diameter in the target membranes.[66,67] Oligomerization does not occur on bacterial membranes since they are devoid of sphingomyelin. Therefore the mechanism of antibacterial activity must be different from its cytolytic activity.[64]

As sphingomyelin is crucial for the cytolytic activity of lysenin, it has been proposed to use lysenin as a valuable probe for sphingomyelin detection in sphingomyelin storage diseases, particularly in the cells of Niemann-Pick A patients,[66] although the multiplicity of hemolysins in the natural source and the cytolytic activity appeared to be a major obstacle.

Independently on fetidin and lysenin, eiseniapore and hemolysins H_1, H_2, H_3, CL_{39} and CL_{41} were described.[68-71] Nevertheless, their more detailed analyses revealed close relationships with fetidin, lysenin or lysenin-related proteins.

Coelomic Cytolytic Factor as a Pattern Recognition Molecule

The coelomic fluid does not cause either the lysis of the coelomocytes of other earthworm species or of the hemocytes of mollusks, nematodes and protozoans. However, it was documented that coelomic fluid of *E. fetida* lyses a broad spectrum of various cell types including chicken fibroblasts, guinea-pig polymorphonuclear leukocytes and insect hemocytes.[45] A proteinase-independent cytolytic effect of the coelomic fluid was observed in experiments with TNF-sensitive tumor L929 cell line. Subsequent isolation of lytic proteins led to the identification of a 42-kDa protein, which was named coelomic cytolytic factor—CCF.[46]

CCF acts in earthworm defense as a **pattern-recognition molecule**. Upon binding microbial pathogen-associated molecular patterns, namely O-antigen of LPS of Gram-negative bacteria, muramyl dipeptide and muramic acid of peptidoglycan from the cell walls of Gram-positive bacteria and β-1,3-glucans and *N, N′*-diacetylchitobiose of yeast, CCF triggers the activation of the **prophenoloxidase cascade**, which results in the formation of cytotoxic and antimicrobial compounds and thus represents an important invertebrate defense mechanism.[72-75] The broad specificity of CCF for pathogen-associated molecular patterns results from the presence of two spatially distinct pattern recognition lectin-like domains. One domain, which shows homology with the polysaccharide and glucanase motifs of β-1,3-glucanases and invertebrate defense molecules, is located in the central part of CCF molecule and interacts with LPS and β-1,3-glucans. The C-terminal tryptophan-rich domain mediates interactions of CCF with *N, N′*-diacetylchitobiose, muramyl dipeptide and muramic acid (Fig. 2).[74]

The binding activity of C-terminal domain of CCF is rather unique for *Eisenia fetida* earthworms. Comparative analysis of CCF-like pattern-recognition proteins in seven other lumbricid species (*Aporrectodea caliginosa, A. icterica, A. longa, A. rosea, Dendrobaena veneta, Lumbricus rubellus* and *L. terrestris*) revealed high homology in polysaccharide-binding and glucanase motifs while C-terminal part was more heterogeneous.[76] This is in a good agreement with the absence of cytolytic activity and binding capacity for *N, N′*-diacetylchitobiose and peptidoglycan components. *E. fetida* is an epigeic earthworm living in decaying organic matter, in compost and mold, where

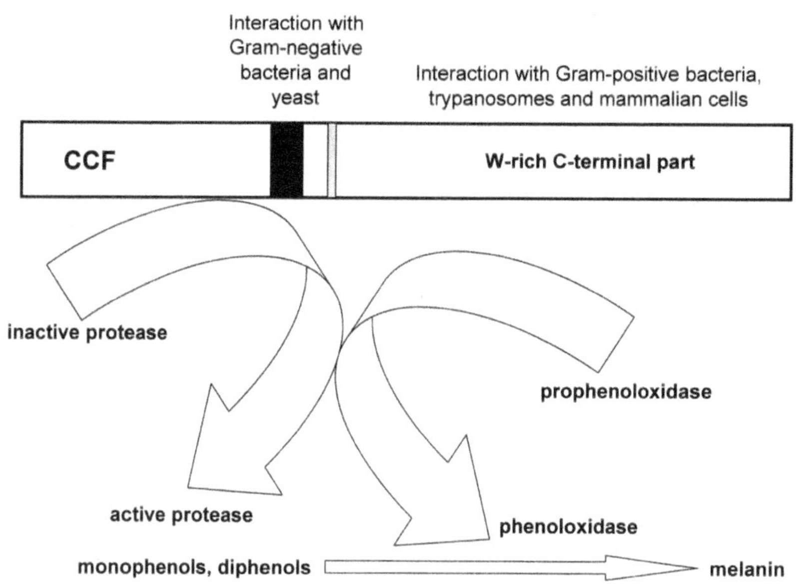

Figure 2. The role of CCF as a pattern-recognition molecule. CCF encompasses two distinct domains: the central part with polysaccharide recognition (black box) and glucanase (grey box) motifs interacts with LPS of Gram-negative bacteria and β-1.3-glucan of yeast while C-terminal tryptophan-rich domain interacts with peptidoglycan constituents of Gram-positive bacteria and N-acetylglucosamine-linked polysaccharides on trypanosomes or mammalian cells. Upon binding of pathogen-associated molecular patterns CCF activates proteolytic enzymes that cleave inactive prophenoloxidase to active phenoloxidase. Phenoloxidase then catalyzes hydroxylation and oxidation of monophenols and diphenols that leads to melanin production. Melanin exhibits antimicrobial and cytotoxic activities and potentiates other defense functions (phagocytosis, encapsulation).

the diversity as well as quantity of microorganisms is substantially higher as compared to other soil layers. It is therefore obvious that *E. fetida* appears to be best equipped to resist microbial load as reflected by the broader CCF pattern-recognition repertoire.

As indicated above, CCF displays amino acid sequence homology with bacterial and animal β-1,3-glucanases but it does not exhibit their enzymatic activity.[77-79] Moreover, CCF shows homology with the α subunit of the β-1,3-glucan sensitive factor G from the horseshoe crab *Tachypleus tridentatus*,[80] with the Gram-negative bacteria-binding proteins of various insects[81-84] and β-1,3-glucan recognition protein of arthropods.[85,86] All these invertebrate homologs have been suggested to play a role in invertebrate innate immunity by acting as pattern recognition molecules.

Further, it was shown that CCF agglutinates both Gram-positive and Gram-negative bacteria[72] and contributes to the opsonizing properties of the coelomic fluid, thereby providing an efficient mechanism for phagocytosis in earthworm defense reactions.[46] CCF is also involved in the cell-mediated cytotoxic reactions and potentiates the lytic activity of coelomic fluid against red blood cells from various species.[34]

More interestingly, CCF shares functional analogies with mammalian tumor necrosis factor. The ability of the coelomic fluid to lyse TNF-sensitive tumor cell line L929 is caused by CCF. This activity is not inhibited by anti-TNF neutralizing monoclonal antibodies, suggesting that the structure of TNF and CCF as well as the mechanism of TNF and CCF mediated lysis differ. In addition to this TNF-like lytic activity, CCF exhibits other similarities with this cytokine. CCF is secreted by phagocytic coelomocytes upon LPS stimulation (Fig. 3), while TNF is produced by LPS-activated macrophages.[34,87] TNF and CCF have opsonizing properties[46,88] bind β-1,3-glucans and *N, N′*-diacetylchitobiose via lectin-like interactions.[72,89] In addition, monoclonal antibodies elicited against the lectin-like TIP domain of TNF cross-react with CCF and, conversely, monoclonal antibody against CCF reacts with TNF without impairing the interaction of TNF with its specific receptor.[90,91] However, the activity of CCF is not inhibited by anti-TNF antibody suggesting different mechanisms of TNF- and CCF-mediated lysis.

The lectin-like domain of TNF was shown to be involved in the killing of African and American trypanosomes.[90-92] Hence, in view of the similar lectin-like activity of CCF and TNF, the possible trypanolytic activity of CCF was investigated.[73] The coelomic fluid of *E. fetida* as well as purified CCF were described to have a potent trypanolytic activity that can be inhibited not only by anti-CCF monoclonal antibodies but also by *N, N′*-diacetylchitobiose and anti-TNF antibodies. The possible target for both CCF and TNF on the trypanosome surface is the N-linked *N, N′*-diacetylchitobiose core of the variant-specific glycoprotein (VSG) that acts as a protective coat. This idea is strongly supported by the fact that CCF and TNF are able to lyse only bloodstream forms of parasites expressing VSG but not insect-stage procyclic forms expressing procyclin as a surface protein.

It was documented that TNF increases the membrane conductance in mammalian cells, interacting with ion-channels or ion-channel-coupled molecules through a lectin-like domain.[93,94] Similarly, when endothelial cells or macrophages were activated with CCF, an increase in membrane conductance occurred.[95] As observed with TNF, the ion-gating effect of CCF appeared when cells from TNF-receptor I and TNF-receptor II knockout mice were used. Moreover, this effect is blocked by *N, N′*-diacetylchitobiose and amiloride—an epithelial sodium channel inhibitor—suggesting that the effect is mediated by the lectin-like domain of CCF. In macrophages, CCF-induced depolarization results in the release of TNF, IL-6 and nitric oxide via NF-κB signaling. This pathway based

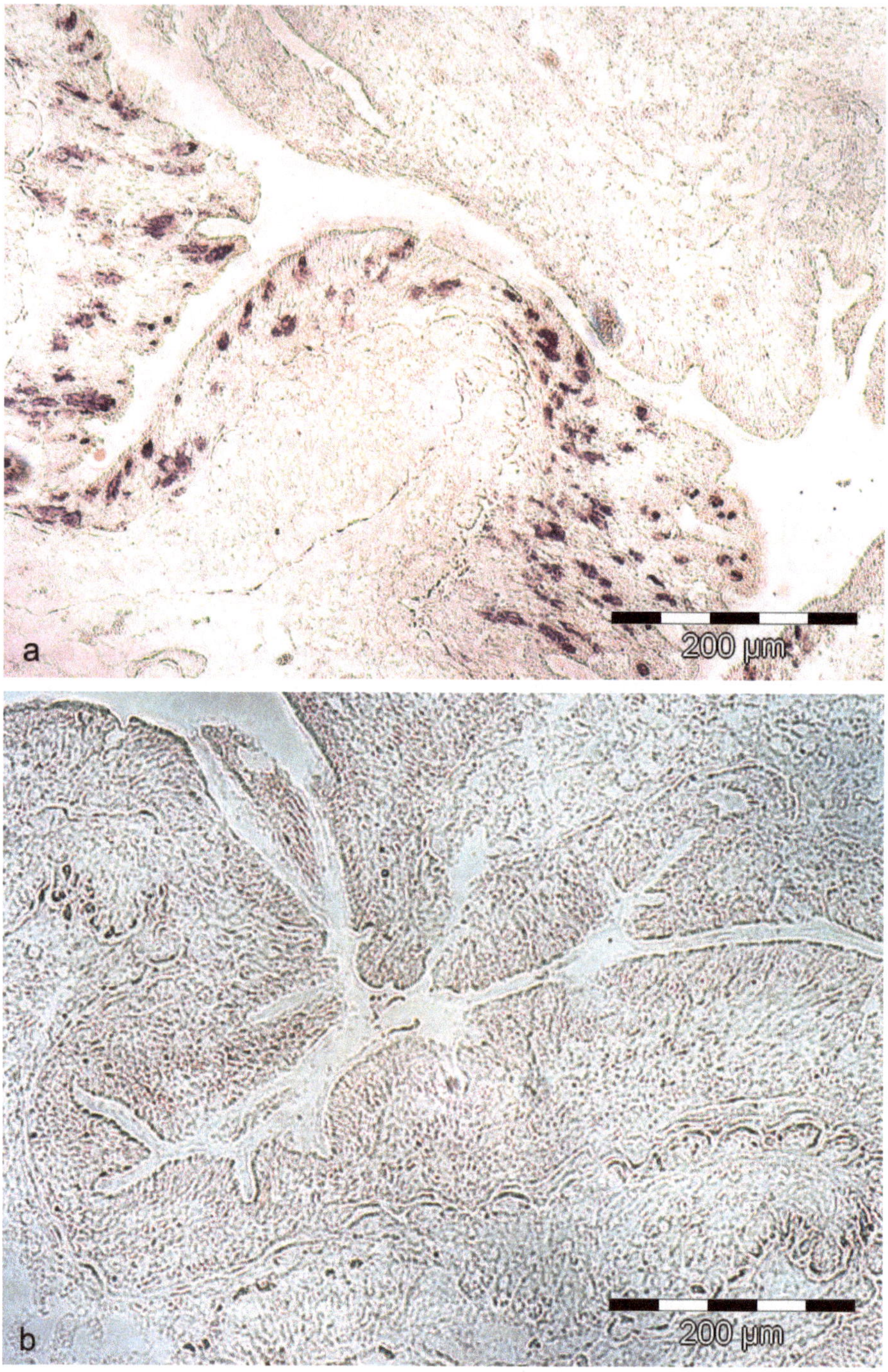

Figure 3. Up-regulated tissue expression of CCF in chloragogen tissue after LPS challenge (a) as compared to nonstimulated controls (b) followed by in situ hybridization. Reproduced by courtesy of Dr. Ellen Kauschke, Giessen.

on an interaction of lectin domain with saccharide moiety of ion channel may represent an evolutionary ancient mechanism of cell activation.[96]

Surprisingly, despite the functional analogies of CCF and TNF and cross-reactivity of anti-CCF and anti-TNF antibodies, these molecules do not show any gene or amino acid sequence homology, indicating a lack of common evolutionary origin.[73]

CONCLUSION

Earthworms rely on innate defense mechanisms that are sufficient for survival in often hostile environment. The choice of earthworms for comparative immunology studies was pertinent since they represent an inexpensive, appropriate and noncontroversial model for experimentation. Described defense mechanisms and molecules help to better understand more sophisticated immunity in vertebrates. Moreover, earthworms were found useful in monitoring environmental pollution. The Organization for Economic Cooperation and Development (OECD Guidelines for testing of Chemicals 1984) and the American Environmental Protection Agency accepted official protocols involving screening of earthworm immunological parameters as markers/indicators of impaired environmental conditions.[97,98] Furthermore, earthworms might be considered as a source of biologically active compounds with potential industrial or medical use. Actually, earthworm powder has been used as a traditional medicine in some South Asia countries for years to treat various diseases. Currently, the therapeutic effect of earthworm active factors is being evaluated by a modern scientific approach. Some therapeutics containing fibrinolytic enzymes from *Lumbricus rubellus* and *Eisenia fetida* earthworms[99-102] are already commercially available to support coagulation and fibrinolysis balance in the body and thus prevent or treat cardiac and cerebrovascular diseases (Boluoke® (lumbrokinase), Canada RNA Biochemical Inc.).

ACKNOWLEDGEMENTS

The authors acknowledge the support by the Czech Science Foundation (206/07/0378), Grant Agency of the Academy of Sciences of the Czech Republic (A600200704), the Ministry of Education, Youth and Sports (2B06155) and an Institutional Research Concept (AV0Z50200510).

REFERENCES

1. Mechnikoff EE. Sur la lutte des cellules de l'organisme contre de l'invasion des microbes. Ann Inst Pasteur 1887; 1:322-340.
2. Janeway CA. Approaching the asymptote? Evolution and revolution in immunology. Cold Spring Harb Symp Quant Biol 1989; 54:1-13.
3. Little TJ, Hultmark D, Read AF. Invertebrate immunity and the limits of mechanistic immunology. Nat Immunol 2005; 6:651-654.
4. Schulenburg H, Boehnisch C, Michiels NK. How do invertebrates generate a highly specific innate immune response? Mol Immunol 2007; 44:3338-3344.
5. Mill PJ. Physiology of Annelids. New York: Academic Press, 1978:1-683.
6. Rahemtulla F, Lovtrup S. The comparative biochemistry of invertebrate mucopolysaccharides II. Nematoda; Annelida. Comp Biochem Physiol B 1974; 49:639-646.

7. Rahemtulla F, Lovtrup S. The comparative biochemistry of invertebrate mucopolysaccharides III. Oligocheta and Hirudinea. Comp Biochem Physiol B 1975; 50:627-629.
8. Chapron C. Etude histologique, infrastructurale et experimentale de la regeneration cephalique chez les lombriciens, Eisenia foetida. Ann Embr Morphol 1970; 3:235-239.
9. Valembois P. Degenerescence et regeneration de l'epiderme a la suite d'une xenogreffe de paroi du corps entre lombriciens. C R Acad Sci Paris 1971; 96:59-64.
10. Burke JM. Wound healing in Eisenia foetida (Oligochaeta). I. Histology and ^{3}H-thymidine radioautography of the epidermis. J Exp Zool 1974; 188:49-63.
11. Burke JM. Wound healing in Eisenia foetida (Oligochaeta). II. A fine structural study of the role of the epidermis. Cell Tissue Res 1974; 154:61-82.
12. Burke JM. Wound healing of Eisenia foetida (Oligochaeta). III. A fine structural study of the role of non-epidermal tissues. Cell Tissue Res 1974; 154:83-112.
13. Dales RP, Kalac Y. Phagocytic defence by the earthworm Eisenia foetida against certain pathogenic bacteria. Comp Biochem Physiol 1992; 101A:487-490.
14. Cameron GR. Inflammation in earthworms. J Pathol 1932; 35:933-972.
15. Villaro AC, Sesma P, Alegría D et al. Relationship of symbiotic microorganisms to metanephridium: phagocytic activity in the metanephridial epithelium of two species of Oligochaeta. J Morphol 1985; 186:307-314.
16. Ratcliffe NA, Rowley AF, Fitzgerald SW et al. Invertebrate immunity: basic concepts and recent advances. Internat Rev Cytol 1985; 97:183-349.
17. Valembois P, Lassegues M, Roch P. Formation of brown bodies in the coelomic cavity of the earthworm Eisenia fetida andrei and attendant changes in shape and adhesive capacity of constitutive cells. Dev Comp Immunol 1992; 16:95-101.
18. Keilin ND. Parasitic autotomy of the host as a mode of liberation of coelomic parasites from the body of the earthworm. Parasitology 1925; 17:170-172.
19. Herlan-Meewis H. Regeneration in annelids. Advances in Morphogenesis, Vol. 4. M. Abercrombie and J. Brachet. New York:Academic Press, 1965:155-215.
20. Alonso-Bedate M, Sequeros E. Neorosecretory phenomena in the cerebral ganglia of clitellated Allolobophora caliginosa. Acta Embryol Morphol Exp 1983; 4:93-103.
21. Alonso-Bedate M, Sequeros E. Suggested regulatory mechanisms for caudal regeneration in Allolobophora molleri (Annelida: Oligochaeta). Comp Biochem Physiol 1985; 81A:225-228.
22. Stein E, Avtalion RR, Cooper EL. The coelomocytes of the earthworm Lumbricus terrestris: morphology and phagocytic properties. J Morphol 1977; 153:467-477.
23. Šíma P. Annelid coelomocytes and haemocytes: roles in cellular immune reactions. In: Větvička V, Šíma P, Cooper EL, Bilej M, Roch P, eds. Immunology of Annelids. Boca Raton:CRC Press, 1994:115-165.
24. Engelmann P, Pal J, Berki T et al. Earthworm leukocytes react with different mammalian antigen-specific monoclonal antibodies. Zoology 2002; 105:257-265.
25. Stein E, Cooper EL. The role of opsonins in phagocytosis by coelomocytes of the earthworm, Lumbricus terrestris. Dev Comp Immunol 1981; 5:415-425.
26. Bilej M, Scheerlinck JP, Van Den Driessche T et al. The flow cytometric analysis of in vitro phagocytic activity of earthworm coelomocytes (Eisenia foetida, Annelida). Cell Biol Internat Rep 1990; 14:831-837.
27. Bilej M, Vetvicka V, Tuckova L et al. Phagocytosis of synthetic particles in earthworms. Effect of antigenic stimulation and opsonization. Folia Biol (Praha) 1990; 36:273-280.
28. Laulan A, Lestage J, Bouc AM et al. The phagocytic activity of Lumbricus terrestris leukocytes is enhanced by the vertebrate opsonins: IgG and complement C3b fragment. Dev Comp Immunol 1988; 12:269-277.
29. Cooper EL, Roch P. Immunological profile of annelids: transplantation immunity In: Větvička V, Šíma P, Cooper EL, Bilej M, Roch P, eds. Immunology of Annelids. Boca Raton: CRC Press, 1994:201-243.
30. Bailey S, Miller BJ, Cooper EL. Transplantation immunity in annelids. II. Adoptive transfer of the xenograft reaction. Immunology 1971; 21:81-86.
31. Parry MJ. Survival of body wall autografts, allografts and xenografts in the earthworm Eisenia foetida. J Invert Pathol 1978; 31:383-388.
32. Cooper EL. Transplantation immunity in helminths and annelids. Transplantation Proc 1970; 2:216-221.
33. Hostetter R, Cooper EL. Cellular anamnesis in earthworms. Cell Immunol 1973; 9:384-392.
34. Bilej M, Rossmann P, Sinkora M et al. Cellular expression of the cytolytic factor in earthworms Eisenia foetida. Immunol Lett 1998; 60:23-29.
35. Cotuk A, Dales RP. Lysozyme activity in the coelomic fluid and coelomocytes of the earthworm Eisenia foetida SAV in relation to bacterial infection. Comp Biochem Physiol 1984; 78A:469-474.
36. Cossarizza A, Cooper EL, Suzuki MM et al. Earthworm leukocytes that are not phagocytic and cross-react with several human epitopes can kill human tumor cell lines. Exp Cell Res 1996; 224:174-182.

37. Quaglino D, Cooper EL, Salvioli S et al. Earthworm coelomocytes in vitro: cellular features and "granuloma" formation during cytotoxic activity against the mammalian tumor cell target K562. Eur J Cell Biol 1996; 70:278-288.
38. Josková R, Šilerová M, Procházková P et al. Identification and cloning of an invertebrate-type lysozyme from Eisenia andrei. Dev Comp Immunol 2009; 33:932-938.
39. Cho JH, Park CB, Yoon YG et al. Lumbricin I, a novel proline-rich antimicrobial peptide from the earthworm: purification, cDNA cloning and molecular characterization. Biochim Biophys Acta 1998; 1408:67-76.
40. Wang X, Wang X, Zhang Y et al. An antimicrobial peptide of the earthworm Pheretima tschiliensis: cDNA cloning, expression and immunolocalization. Biotechnol Lett 2003; 25:1317-1323.
41. Liu YQ, Sun ZJ, Wang C et al. Purification of a novel antibacterial short peptide in earthworm Eisenia foetida. Acta Biochim Biophys Sin (Shanghai) 2004; 36:297-302.
42. Roch P. Protein analysis of earthworm coelomic fluid: 1) polymorphic system of the natural hemolysin of Eisenia fetida andrei. Dev Comp Immunol 1979; 3:599-608.
43. Valembois P, Roch P, Lassegues M et al. Antibacterial activity of the haemolytic system from the earthworm Eisenia fetida andrei. J Invert Pathol 1982; 40:21-27.
44. Roch P, Lassegues M, Valembois P. Antibacterial activity of Eisenia fetida andrei coelomic fluid: III. Relationship within the polymorphic hemolysins. Dev Comp Immunol 1991; 15:27-32.
45. Kauschke E, Mohrig W. Cytotoxic activity in the coelomic fluid of the annelid Eisenia foetida Sav. J Comp Physiol [B] 1987; 157:77-83.
46. Bilej M, Brys L, Beschin A et al. Identification of a cytolytic protein in the coelomic fluid of Eisenia foetida earthworms. Immunol Lett 1995; 45:123-128.
47. Cooper EL, Cossarizza A, Suzuki MM et al. Autogeneic but not allogeneic earthworm effector coelomocytes kill the mammalian tumor cell target K562. Cell Immunol 1995; 166:113-122.
48. Ito Y, Yoshikawa A, Hotani T et al. Amino acid sequences of lysozymes newly purified from invertebrates imply wide distribution of a novel class in the lysozyme family. Eur J Biochem 1999; 259:456-461.
49. Du Pasquier L, Duprat P. Humoral and cellular aspects of a nonspecific natural immunity in the oligochetee Eisenia foetida Sav. (Lumbricinae), C R Acad Sci Hebd Seances Acad Sci D 1968; 266:538-541.
50. Roch P, Valembois P, Davant N et al. Protein analysis of earthworm coelomic fluid. II. Isolation and biochemical characterization of the Eisenia foetida andrei factor (EFAF). Comp Biochem Physiol 1981; 69B:829-836.
51. Roch P. Isolation of agglutinins and lysins from earthworm coelomic fluid by gel filtration followed by chromatofocusing. J Chromat 1984; 290:231-235.
52. Roch P, Valembois P, Lassegues M. Genetic and biochemical polymorphism of earthworm humoral defenses. Prog Clin Biol Res 1987; 233:91-102.
53. Valembois P, Roch P, Lassegues M. Simultaneous existence of haemolysins and hemagglutinins in the coelomic fluid and in the cocoon albumen of the earthworm Eisenia fetida andrei. Comp Biochem Physiol 1984; 78A:141-145.
54. Lassegues M, Roch P, Valembois P. Antibacterial activity of Eisenia fetida andrei coelomic fluid: evidence, induction and animal protection. J Invert Pathol 1989; 53:1-6.
55. Hirigoyenberry F, Lassegues M, Roch P. Antibacterial activity of Eisenia foetida andrei coelomic fluid: immunological study of the two major antibacterial proteins. J Invert Pathol 1992; 59:69-74.
56. Valembois P, Roch P, Lassegues M. Antibacterial molecules in annelids. In: Brehelin M, ed. Immunity in Invertebrates. Berlin:Springer-Verlag, 1986:74-93.
57. Sinkora M, Bilej M, Tuckova L et al. Hemolytic function of opsonizing proteins of earthworm's coelomic fluid. Cell Biol Int 1993; 17:935-939.
58. Valembois P, Roch P, Lassegues M. Evidence of plasma clotting system in earthworms. J Invertebr Pathol 1988; 51:221-228.
59. Roch P, Canicatti C, Valembois P. Interactions between earthworm hemolysins and sheep red blood cell membranes. Biochim Biophys Acta 1989; 983:193-198.
60. Lassegues M, Milochau A, Doignon F et al. Sequence and expression of an Eisenia-fetida-derived cDNA clone that encodes the 40-kDa fetidin antibacterial protein. Eur J Biochem 1997; 246:756-762.
61. Milochau A, Lassegues M, Valembois P. Purification, characterization and activities of two hemolytic and antibacterial proteins from coelomic fluid of the annelid Eisenia fetida andrei. Biochim Biophys Acta 1997; 1337:123-132.
62. Sekizawa Y, Hagiwara K, Nakajima T et al. A novel protein, lysenin, that causes contraction of the isolated rat aorta: its purification from the coelomic fluid of the earthworm, Eisenia foetida. Biomed Res 1996; 17:197-203.
63. Sekizawa Y, Kubo T, Kobayashi H et al. Molecular cloning of cDNA for lysenin, a novel protein in the earthworm Eisenia foetida that causes contraction of rat vascular smooth muscle. Gene 1997; 191:97-102.

64. Bruhn H, Winkelmann J, Andersen C et al. Dissection of the mechanisms of cytolytic and antibacterial activity of lysenin, a defence protein of the annelid Eisenia fetida. Dev Comp Immunol 2006; 30:597-606.
65. Procházková P, Šilerová M, Felsberg J et al. Relationship between hemolytic molecules in Eisenia fetida earthworms. Dev Comp Immunol 2006; 30:381-392.
66. Yamaji A, Sekizawa Y, Emoto K et al. Lysenin, a novel sphingomyelin-specific binding protein. J Biol Chem 1998; 273:5300-5306.
67. Yamaji-Hasegawa A, Makino A, Baba T et al. Oligomerization and pore formation of a sphingomyelin-specific toxin, lysenin. J Biol Chem 2003; 278:22762-22770.
68. Lange S, Nussler F, Kauschke E et al. Interaction of earthworm hemolysin with lipid membranes requires sphingolipids. J Biol Chem 1997; 272:20884-20892.
69. Lange S, Kauschke E, Mohrig W et al. Biochemical characteristics of Eiseniapore, a pore-forming protein in the coelomic fluid of earthworms. Eur J Biochem 1999; 262:547-556.
70. Eue I, Kauschke E, Mohrig W et al. Isolation and characterization of earthworm hemolysins and agglutinins. Dev Comp Immunol 1998; 22:13-25.
71. Koenig S, Wagner F, Kauschke E et al. Mass spectrometric analyses of CL(39), CL(41) and H(1), H(2), H(3) confirm identity with fetidin and lysenin produced by earthworm leukocytes. Dev Comp Immunol 2003; 27:513-520.
72. Beschin A, Bilej M, Hanssens F et al. Identification and cloning of a glucan- and lipopolysaccharide-binding protein from Eisenia foetida earthworm involved in the activation of prophenoloxidase cascade. J Biol Chem 1998; 273:24948-24954.
73. Beschin A, Bilej M, Brys L et al. Convergent evolution of cytokines. Nature 1999; 400:627-628.
74. Bilej M, De Baetselier P, Van Dijck E et al. Distinct carbohydrate recognition domains of an invertebrate defense molecule recognize Gram-negative and Gram-positive bacteria. J Biol Chem 2001; 276:45840-45847.
75. Procházková P, Šilerová M, Stijlemans B et al. Evidence for proteins involved in prophenoloxidase cascade in Eisenia fetida earthworms. J Comp Physiol B 2006; 176:581-587.
76. Šilerová M, Procházková P, Josková R et al. Comparative study of the CCF-like pattern recognition protein in different lumbricid species. Dev Comp Immunol 2006; 30:765-771.
77. Yamamoto M, Aono R, Horikoshi K. Structure of the 87-kDa beta-1,3-glucanase gene of Bacillus circulans IAM1165 and properties of the enzyme accumulated in the periplasm of Escherichia coli carrying the gene. Biosci Biotechnol Biochem 1993; 57:1518-1525.
78. Bachman ES, McClay DR. Molecular cloning of the first metazoan beta-1,3-glucanase from eggs of the sea urchin Strongylocentrotus purpuratus. Proc Natl Acad Sci USA 1996; 93:6808-6813.
79. Kozhemyako VB, Rebrikov DV, Lukyanov SA et al. Molecular cloning and characterization of an endo-1,3-beta-D-glucanase from the mollusk Spisula sachalinensis. Comp Biochem Physiol B Biochem Mol Biol 2004; 137:169-178.
80. Seki N, Muta T, Oda T et al. Horseshoe crab (1,3)-beta-D-glucan-sensitive coagulation factor G. A serine protease zymogen heterodimer with similarities to beta-glucan-binding proteins. J Biol Chem 1994; 269:1370-1374.
81. Lee WJ, Lee JD, Kravchenko VV et al. Purification and molecular cloning of an inducible gram-negative bacteria-binding protein from the silkworm, Bombyx mori. Proc Natl Acad Sci USA 1996; 93:7888-7893.
82. Dimopoulos G, Richman A, Muller HM et al. Molecular immune responses of the mosquito Anopheles gambiae to bacteria and malaria parasites. Proc Natl Acad Sci USA 1997; 94:11508-11513.
83. Shin SW, Park SS, Park DS et al. Isolation and characterization of immune-related genes from the fall webworm, Hyphantria cunea, using PCR-based differential display and subtractive cloning. Insect Biochem Mol Biol 1998; 28:827-837.
84. Kim YS, Han SJ, Ryu JH et al. Lipopolysaccharide-activated kinase, an essential component for the induction of the antimicrobial peptide genes in Drosophila melanogaster cells. J Biol Chem 2000; 275:2071-2079.
85. Ma C, Kanost MR. A beta1,3-glucan recognition protein from an insect, Manduca sexta, agglutinates microorganisms and activates the phenoloxidase cascade. J Biol Chem 2000; 275:7505-7514.
86. Ochiai M, Ashida M. A pattern-recognition protein for beta-1,3-glucan. The binding domain and the cDNA cloning of beta-1,3-glucan recognition protein from the silkworm, Bombyx mori. J Biol Chem 2000; 275:4995-5002.
87. Aggarwal BB, Kohr WJ, Hass PE et al. Human tumor necrosis factor. Production, purification and characterization. J Biol Chem 1985; 260:2345-2354.
88. Luo G, Niesel DW, Shaban RA et al. Tumor necrosis factor alpha binding to bacteria: evidence for a high-affinity receptor and alteration of bacterial virulence properties. Infect Immun 1993; 61:830-835.
89. Olson EJ, Standing JE, Griego-Harper N et al. Fungal beta-glucan interacts with vitronectin and stimulates tumor necrosis factor alpha release from macrophages. Infect Immun 1996; 64:3548-3554.

90. Lucas R, Magez S, De Leys R et al. Mapping the lectin-like activity of tumor necrosis factor. Science 1994; 263:814-817.
91. Magez S, Geuskens M, Beschin A et al. Specific uptake of tumor necrosis factor-alpha is involved in growth control of Trypanosoma brucei. J Cell Biol 1997; 137:715-727.
92. Olivares Fontt EO, De Baetselier P, Heirman C et al. Effects of granulocyte-macrophage colony-stimulating factor and tumor necrosis factor alpha on Trypanosoma cruzi trypomastigotes. Infect Immun 1998; 66:2722-2727.
93. Hribar M, Bloc A, van der Goot FG et al. The lectin-like domain of tumor necrosis factor-alpha increases membrane conductance in microvascular endothelial cells and peritoneal macrophages. Eur J Immunol 1999; 29:3105-3111.
94. van der Goot FG, Pugin J, Hribar M et al. Membrane interaction of TNF is not sufficient to trigger increase in membrane conductance in mammalian cells. FEBS Lett 1999; 460:107-111.
95. Bloc A, Lucas R, Van Dijck E et al. An invertebrate defense molecule activates membrane conductance in mammalian cells by means of its lectin-like domain. Dev Comp Immunol 2002; 26:35-43.
96. Bilej M, Josková R, Van den Bergh R et al. An invertebrate TNF functional analogue activates macrophages via lectin-saccharide interaction with ion channels. Int Immunol 2006; 18:1663-1670.
97. Green JC, Bartels CL, Warren-Hicks WJ et al. Protocols of Short-Term Toxicity Screening of Hazardous Waste Sites, US Environmental Protection Agency, EPA/600/3-88/029, ERLC, Corvallis (OR); 1989.
98. Goven AJ, Kennedy J. Environmental pollution and toxicity in invertebrates: an earthworm model for immunotoxicology. Adv Comp Environ Physiol 1996; 24:169-211.
99. Mihara H, Maruyama M, Sumi H. Novel thrombolytic therapy discovered from traditional oriental medicine using the earthworm. Southeast Asian J Trop Med Public Health 1992; 23 (Suppl 2):131-140.
100. Mihara H, Sumi H, Yoneta T et al. A novel fibrinolytic enzyme extracted from the earthworm, Lumbricus rubellus. Jpn J Physiol 41:461-472.
101. Nakajima N, Mihara H, Sumi H. Characterization of potent fibrinolytic enzymes in earthworm, Lumbricus rubellus. Biosci Biotechnol Biochem 1993; 57:1726-1730.
102. Nakajima N, Ishihara K, Sugimoto M et al. Chemical modification of earthworm fibrinolytic enzyme with human serum albumin fragment and characterization of the protease as a therapeutic enzyme. Biosci Biotechnol Biochem 1996; 60:293-300.

CHAPTER 5

LEECH IMMUNITY:
From Brain to Peripheral Responses

Aurélie Tasiemski and Michel Salzet
Université Lille Nord De France, CNRS, Laboratoire de Neuroimmunologie et Neurochimie Evolutives, Université Lille 1, France
Emails: aurelie.tasiemski@univ-lille1.fr and michel.salzet@univ-lille1.fr

Abstract: In the present chapter, we will emphasize the immune response in two compartments (Central nervous system and peripheral system) in two blood sucking leeches i.e., the medicinal leech and the bird leech *Theromyzon tessulatum*. In the medicinal leech, the neuroimmune response has been described in the context of septic trauma at the cellular and humoral levels through microglia, Toll-like, cannabinoids and chemoattractant factors activation and modulation. In the bird leech, the antimicrobial responses have been dissected at the cellular and molecular levels. Altogether, this chapter presents a complete integrate immune response from the brain and the systemic compartments with high similarity to the vertebrates one. These points that the neuroimmune and immune responses evolved sooner than can be expected.

INTRODUCTION

Leeches are derived from their cousin's earthworm in the class of annelids called hirudinea. Leeches evolved during the Cambrian Explosion, a time of rapid biological development 540 million years ago.[1-6] There are now approximately 700 species of leeches distributed throughout freshwater, marine and terrestrial ecosystems worldwide. Phylogenetic studies assess that the common leech ancestor was probably a bloodsucking leech with a proboscis rather than an unspecialized ectocommensal.[7] During the course of leech evolution, a reduction of the proboscis could have taken place in predatory arhynchobdellid ancestors to enable swallowing of larger prey. A second gain of sanguivory by the jawed Hirudiniforms could have been facilitated

Invertebrate Immunity, edited by Kenneth Söderhäll.

by pre-adaptations to ectoparasitic blood feeding.[7] Recently, leech EST data from the jaw leech *hirudo medicinalis* have been obtained[8] and it can be seen high sequence homologies between the medicinal leech and mammals[8] suggesting a co-evolution between the parasite and the host. Thus, taking into account such evolutive data, we will present in this manuscript, data obtained from two haematophageous leeches, the jawed leech *Hirudo medicinalis* and the gut leech *Theromyzon tessulatum* focused on the immune responses angle. Due to their anatomic differences, *Hirudo medicinalis* appears as a good model for studying the immune response of the nervous system although the easy access to the body fluids of *Theromyzon tessulatum* makes this animal interesting for understanding the systemic response. Taken together, we will try to show that immune response from these blood sucking animals evolve closely to their vertebrate host, pointing the fact that their use currently in hospital never trigger patient immune responses. Host immune tolerance is due to co-evolution and molecular mimicry.[9]

THE MEDICINAL LEECH AS A MODEL FOR STUDYING THE IMMUNE RESPONSE OF THE CNS

The central nervous system (CNS) of the leech has a fixed number of bilateral neuromeres, 32. The 4 anterior-most neuromeres fuse to form the sub-esophageal ganglion and the 7 posterior-most fuse to form the tail ganglion; single bilateral neuromeres comprise the individual ganglia found in each of the corresponding body segments. A supra-esophageal ring of nonsegmental origin, together with the sub-esophageal ganglion, comprises the head ganglion. The central ganglia are connected to each other by a bilateral pair of nerves (the lateral "connectives") and a single small medial nerve (Faivre's) and to the periphery by two bilateral pairs of nerves (the "roots") that branch in a stereotypic pattern that allows the identification of branches up to fourth-order and even to fifth-order in some cases (Fig. 1). In hirudinid leeches, each segmental ganglionic primordium gives rise to about 400 neurons.[10] Most of these are bilateral pairs (~180-190 pairs), but perhaps 5-8% are unpaired, with at least some becoming unpaired through cell death.[10,11] Thus, understanding how a leech segmental ganglion functions requires, in principle, detailed knowledge of the function and connectivity of only ~200-220 individual neurons. Moreover, since each segmental ganglion is a variation on a theme (with the exception of the "sex" ganglia of body segments 5 and 6, which have additional complements of neurosecretory cells), the leech has one of the most accessible nervous systems from a systems analysis point of view.

An important property of leeches is their capacity to regenerate neurites and synaptic connections in the adult CNS. Neurites that have been damaged or severed can sprout, establish de novo growth cones and extend and reconnect specifically with normal targets.[12] Early stages of leech CNS regeneration following a mechanical lesion are characterized by two events that appear to be crucial for successful repair: one is the increased activity of epithelial nitric oxide synthase (NOS) in the area of the lesion and the generation and diffusion of nitric oxide (NO) and the second is the induced migration of microglia towards and their accumulation at, the injury site.[13,14] Microglial cells are considered as the brain immune cells.

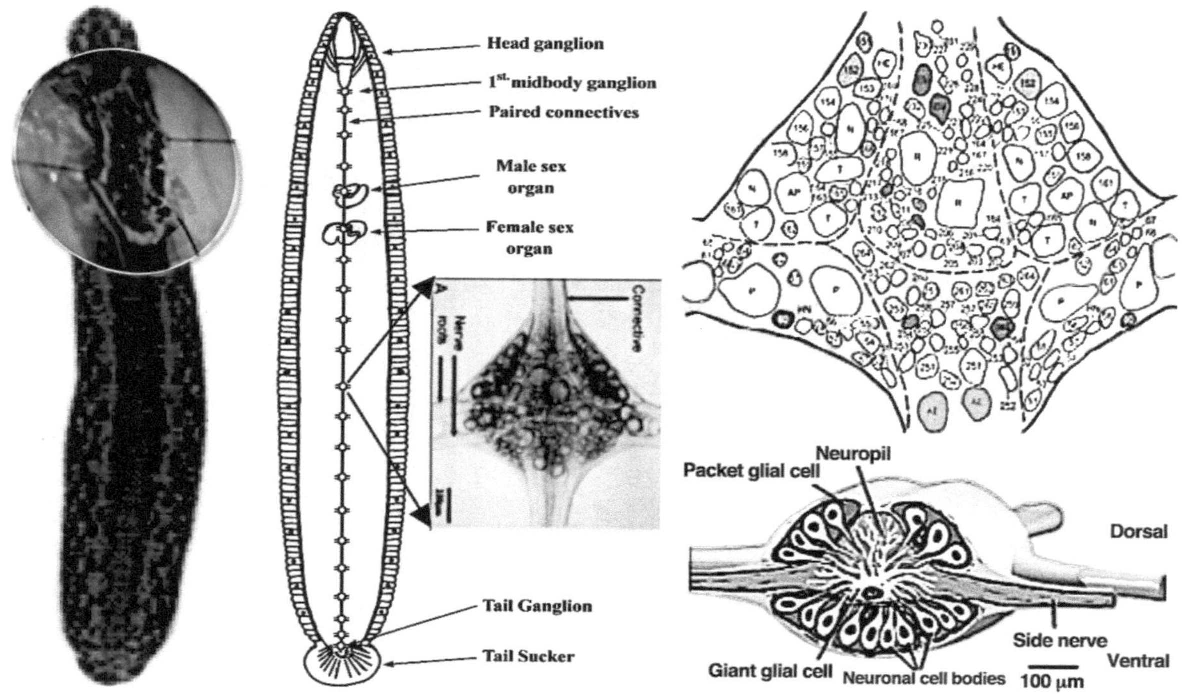

Figure 1. The central nervous system (CNS) of the leech has a fixed number of 32 bilateral neuromeres. The four anterior-most neuromeres fuse to form the sub-esophageal ganglion and the seven posterior-most fuse to form the tail ganglion; single bilateral neuromeres comprise individual ganglia found in each of the corresponding mid-body segments. The central ganglia (insert, inset photography) are connected to each other by a bilateral pair of nerves (lateral "connectives") and a single small medial nerve (Faivre's) and to the periphery by two bilateral pairs of nerves ("roots") that branch in the body wall and carry motor activity centrifugally and sensory signals centrally. B) (Top) Diagram of a leech ganglion with some well-known neurons labeled. One of our goals is to create protein profiles for all of these neurons in order to understand the molecular mechanisms that give rise to their different functions. (Bottom) Drawing showing the typical organization of a leech ganglion: a rind of cell bodies and a central neuropil containing processes and synapses.

NO and Cannabinoids in Leech Microglia Chemotaxis Involvement

To assay directly for a role of NO on microglial accumulation at the injury site, Chen et al[15] modulated NO levels in several ways. As demonstrated by NOS immunoreactivity, a large increase in NOS occurs at the crush site within 5 min of injury and this high level persists for at least 24 hrs. Microglial accumulation at the lesion, however, is not detectable at 5 min but is quite strong after a few hours and peaks at ~24 hrs. Inhibition of NO synthesis by the prior application of the NOS inhibitor L-NAME effectively blocks microglial accumulation, while the presence of its inactive enantiomer D-NAME has little or no effect. Interestingly, increasing NO levels with the NO donor spermine NOate (SPNO) also inhibits accumulation of microglia at the crush, but not in the presence of the NO scavenger cPTIO (2-(4-Carboxyphenyl)-4,4,5,5-tetramethylimidazoline-l-oxyl-3-oxide).Examination of microglial kinetics in living nerve cords shows that the effect of SPNO application occurs by the reduction of average microglial migratory speeds, even to no movement. Thus, NO is clearly implicated as a modulator of microglial movement and indeed appears to function as a stop signal at high levels, leading to the higher density of these cells at the injury site. Moreover, we recently found that the NO time-course profile in the injured leech brain is partially under the control of endocannabinoids, namely, anandamide (AEA) and 2-arachidonylglycerol (2AG), which affect the NO time-course through their modulation of cannabinoid-like receptors.[16] AEA blocked microglial cell accumulation before their arrival to the lesion site in a concentration dependent-manner and this effect of AEA on microglial cell recruitment was also demonstrated in vitro. Moreover, stimulation with a concentration of AEA determined elsewhere by (Matrix assisted laser desorption ionization time of flight/ time of flight (MALDI TOF/TOF)mass spectrometry[17] enhanced the release of NO at the lesion site of the harmed connective. This NO release may be related to the AEA-activated CB1-like receptor carried by neurons of the leech present in the injury site. This same report suggested that this pathway leads to a microglia stop signal. In a complementary mode of action, the second most described endocannabinoid 2AG has been shown to enhance and drive the chemotaxis of microglial cells in a dose-dependent manner. Moreover, a better result in chemotaxis was obtained using a concentration of 2AG detected 30 min after injury of the leech nerve cords.[17] This result is reinforced by Cabral et al who suggested that 2AG could act through the autocrine/paracrine system to chemoattract microglial cells after brain insult.[18] In leech brain, AEA level declines[17] and in contrast the 2AG climbs during the first 4 h following a lesion of the leech CNS.[16]

Knowing that these two endocannabinoids are able to act as chemoattractant factors on leech microglia in a dose-dependent manner in combination with NO release, the role of their receptors in the control of NO release and chemotaxis towards the lesion site has been investigated.[16] Blocking the CB1-like receptor in the leech with a specific antagonist of the mammalian CB1 receptor (AM-251) failed to block the accumulation of microglial cells at a distance from the lesion site. This latter result underscores the role of the CB1-like receptor in microglial chemotaxis. On the other hand, blocking the CB2-like receptor in the leech with the specific antagonist AM-630 completely abolished the recruitment of microglial cells at the lesion site and costimulation of a crush with 2AG and AM-630 failed to reverse the accumulation of microglia as compared to the 2AG treatment alone.[16] This latter result suggests that the CB2-like receptor in the leech is triggered by 2AG in order to promote chemotaxis and the direct recruitment of microglial cells to the site of injury. In parallel, the treatment of injured connectives with a physiological concentration of 2AG released upon CNS injury in the leech provoked a specific response in NO production at the lesion site just after the stimulation.[16] Blockade of the CB2-like receptor by AM-630 revealed an inhibition of NO secretion and the time-course of NO production was significantly delayed in comparison with the control condition. In this way, the whole of these data demonstrated that at least part of the NO produced by the injured leech was related to the 2AG-activated CB2-like receptor, as was described in experiments using Guinea pig mast cells.[19]

Even if 2AG and AEA are both able to produce NO during a lesion of the leech brain, only 2AG seems to play a major role in the chemotaxis of microglia and its capacity to do so is closely related to a putative functional CB2-like receptor expressed by activated microglia. Previous pharmacological studies have named the CB2 receptor as playing a crucial role in the early inflammatory process, thereby implicating microglia in this process as well in mammals. The CB2 receptor is expressed very early in the different activation steps of microglia, thus describing a "window" of functional relevance for the expression of the CB2 receptor in microglial cells.[20] The delayed time of the initiation of microglial cell activation is linked to changes of their morphology from resting to responsive and allows them to acquire CB2 receptors in correlation with chemotaxis and phagocytosis. The next steps of activation, named "primed" and "responsive," were discarded from the initial step of CB2 receptor expression. As a diffusible molecule, NO has been demonstrated to take part in the migration of microglia and their accumulation at lesions of the leech CNS in a dose-dependent manner. The mechanisms by which the resting microglia becomes responsive are not yet fully understood. However, we hypothesize that the NO immediately released by the damaged neurons might stimulate the resting microglial cells present at the lesion site and allow them to be responsive for CB2-like receptor expression.

Thus, the cannabinoid system becomes activated at the lesion site and produces the two major endocannabinoids (AEA and 2AG) from membrane precursors in an opposite concentration time-course of,[17] consistent with our previous work where the 2AG concentration was found to progressively increase in opposition with the AEA concentration after the lesion. It has been hypothesized that when the diffusible lipid 2AG increases, the microglia become responsive and start to accumulate at the lesion site via changes in cell morphology related to the activation of the CB2-like receptor expressed in the lamellipodia of responsive microglia.[21] This hypothesis is reinforced by our ex vivo results regarding the stimulation of crushed connectives with 2AG (30 μM) 1 h after the lesion.[16] In comparison with the connectives simultaneously crushed and treated with 30 μM 2AG, the same stimulation 1h after the crush is stimulated NO production more rapidly. This observation suggests that microglial cells are more responsive to 2AG 1h after the lesion than at the instant of the lesion, at which time microglia might be resting. During the responsive step, the leech microglia might express functional CB2 receptors at the cell surface and when these receptors are activated by the endocannabinoid 2AG, NO might be produced and participate in cell recruitment as a chemical gradient from the lesion site towards the periphery. On the other hand, the AEA released at the lesion site decreased in concentration after the lesion and it can activate NO release by targeting the CB1-like receptor and inhibiting the accumulation of microglia at the lesion site. This pathway has been described as leading to a microglia stop signal.[22] The dual activity of these two endocannabinoids working in apparent opposition might be a means of controlling microglial cell recruitment to the lesion site. However, both molecules control the long-term release of NO, as shown in our time-course measurements of NO after stimulation of injured connectives with cannabinoids.[16] This can be explained by the immunosuppressive response of microglial cells described both in vertebrate[23] and invertebrate models.[22]

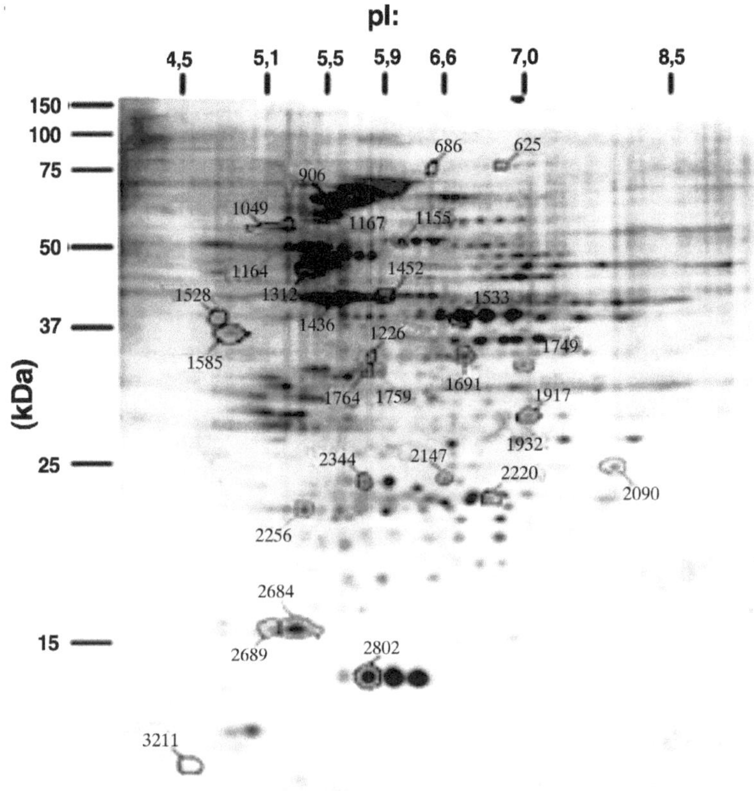

Figure 2. 2D gel between control and immune challenged leech nervous systems. Differentially represented spots are highlighted by arrows and depends on the variation times (variations at 1 h, 6 h, 12 h and 24 h) with permission from Vergote D et al. Proteomics 2006;6(17): 4817-25.[26]

Bacterial Infection and Leech Brain Regeneration

Differential display proteomic analyses using 2D gel electrophoresis (Fig. 2), coupled to mass spectrometry, yielded evidence that the leech CNS responds to bacterial infection by modulating the expression of at least sixteen proteins. These proteins appear between 1 and 24h after bacterial challenge and have been assigned to the immune response because they are not induced by exposure to control sterile medium. These immune-response induced proteins include cytoskeletal and metabolic proteins, foldases, calcium sensors, kinases and neurohemerythrin, reflecting specific cytoskeletal rearrangements linked to cell migration, vesicular trafficking and/or phagocytosis, as well as the modulation of synaptic activity. Interestingly, several of these up-regulated proteins, such as gliarin[24] and neurohemerythrin[25] are expressed specifically in glial and microglial cells, suggesting a key role for these cells in the immune response of the leech nervous system, similar to what has been observed in vertebrates.[26] Gliarin up-regulation, in particular, could thus serve as a new marker of proliferation and maturation of leech glial cells, it's up-regulation reflecting glial activation in response to the immune challenge. The accompanying cytoskeletal rearrangements might result from morphological changes associated with phagocytosis or with the migration of cells. Indeed, it has been shown that leech microglial cells are able to migrate within the ganglionic chain to the site of a lesion and that they can play a phagocytic role.[27] By contrast, the expected properties of neurohemerythrin have led to the proposal of several putative functions for this protein in the responses of leech nervous tissue: (i) a role as an oxygen supplier for metabolism, (ii) a role as a trap for reactive oxygen species and NO, protecting cells from cell death and (iii) a role as an antibacterial factor depriving bacteria of iron.[28]

The data from these initial studies of the effects of bacterial toxins show that the leech CNS is able to respond in an intrinsic manner to a septic stimulus, mounting a "neuroimmune" response. Careful study of potential roles of the identified proteins will be essential to fully understand the mechanisms involved, but some candidates can be proposed on the basis of the proteins identified in this study (ref. 26): (i) cytoskeletal rearrangements potentially responsible for morphological changes, cell migration, vesicular trafficking and/or phagocytosis, (ii) modulation of synaptic activity, (iii) calcium signalling and (iv) unfolded protein response controlling the functionality of proteins affected by the stress generated by the sepsis. The involvement in innate immunity of some proteins or protein families we identified here as previously been described by transcriptomic studies, but at the peripheral level and not within the nervous system.[29,30] Moreover, the protein families involved in the immune response of the medicinal leech nervous system appear to also be involved in nerve regeneration, as shown by Blackshaw *et al*, also in the medicinal leech[31] (discussed above) and by Perlson *et al* in *Lymnaea stagnalis*[32] (Table 1).

These observations suggest certain parallelism between CNS defence mechanisms and CNS regeneration leading the possibility of the involvement of neuroinflammation in such phenomena in leeches. Modulation of neuroinflammation after a crush seems to be necessary but highly regulated in the CNS during regeneration through cannabinoids as we previously demonstrated.[22,33-36] Interestingly, evidences that microbial infection triggers the leech brain regeneration (Fig. 3) has led us to perform some biochemical studies focused on antimicrobial peptides present in the leech CNS and expressed in course of infection or trauma. Two novel antimicrobial peptides, *Hm*-lumbricin and neuromacin have been fully characterized.[37] Neuromacin and *Hm*-lumbricin exert bactericidal activities against Gram positive bacteria without any haemolytic properties.[37] We have observed that in addition to exert antimicrobial activities, *Hm*-lumbricin and neuromacin have regenerative effects on the leech CNS.[37] The capacity of both peptides to promote the regeneration of the leech nerve cord was tested ex vivo by adding the neuromacin and/or *Hm*-lumbricin antibody(ies) to axotomized nerve cords in presence of killed bacteria.[37] Due to the presence of bacteria, the reconnection process should have started two days post-axotomy. It appeared that the presence of antibodies in the culture medium blocked the regeneration process since no reconnection was observed even seven days post-axotomy.[37] These observations were corroborated by the data obtained by adding native neuromacin to axotomized nerve cords under aseptic conditions. Nerve repair was evident sooner in the presence of neuromacin, reconnection starting in less than one day instead of four without an exogenous contribution in neuromacin.The participation of endogenous neuromacin and *Hm*-lumbricin in the neural repair is sustained by the accumulation of both peptides at the lesion site upon bacterial challenge of injured nerve cords.[37] Further investigations based on single cell RT-PCR analysis and on immunohistochemical analysis of a model, developed by our group, of leech CNS almost completely devoid of microglial cells allowed

Table 1. Similarities between proteins involved in innate immunity within invertebrates and those involved in central nervous system regeneration within Lophotrochozoaria, all identified by proteomic or transcriptomic approaches

Immunity		Leech CNS	Leech Periphery	Insect Periphery
Cytosk.	Microfilaments IF	Tropomyosin gliarin	Tropomyosin-2 Actin-2	Myosin II reg. light-chain Actin 5C
Calcium	Calcium sensor	NCS-2/Neurocalcin		Calmodulin, sarc. CaBP1
Metabolism	AA/nt metabolism Energy Others	AA dehydrogenase ATP synthase β subunit Acetyl transferase	Aldehyde dehydr.	Aldehyde dehydr. ATP synth. β subunit
Hsp and chaperones	Cyclophilin/PPI PDI	Cyclophilin PDI	PPI	FK506-BP-PPI, Cyclophylin PDI (ERp60), CaBP1-PDI
Metal-oxidation	Resp. molecule Others	Neurohemerythrin		Hemocyanin Thioredox., Transf., ferritin
Regeneration		**Leech CNS**	**Mollusc CNS**	
Cytosk.	Microfilaments IF Microtubules	Protein 4.1 Synapsin α- and β-tubulin	Tropomyosin Intermediate filament Tubulin	
Calcium	Calcium sensor Others	Calmodulin-like	Calmodulin, Calbindin Calpain	
Metabolism	AA/nt metabolism Energy	ATPase inhibitor	Glutamine Synthase ATP Synthase	
Hsp and chaperones	Cyclophilin/PPI PDI Others	Hsp90	Cyclophilin PDI Hsp60, 14-3-3	
Metal-oxidation	Resp. molecule Others	Myohemerythrin COX I	Peroxiredoxin, Ferritin	

AA, aminoacyl; ATP synth., ATP synthase; calcin., calcineurin; CNS, central nervous system; COX I, Cyclooxygenase I; Cycloph., cyclophilin; dehydr., dehydrogenase; IF, intermediate filaments; nt, neurotransmitters; phosph., phosphatase; PK, protein kinase; PPI, peptidyl-prolyl cis-trans isomerase; sarc., sarcoplasmic; reg, regulatory; resp. molecule, respiratory molecule; Transf, transferring. Used with permission from Vergote D et al. Proteomics 2006;6(17): 4817-25.

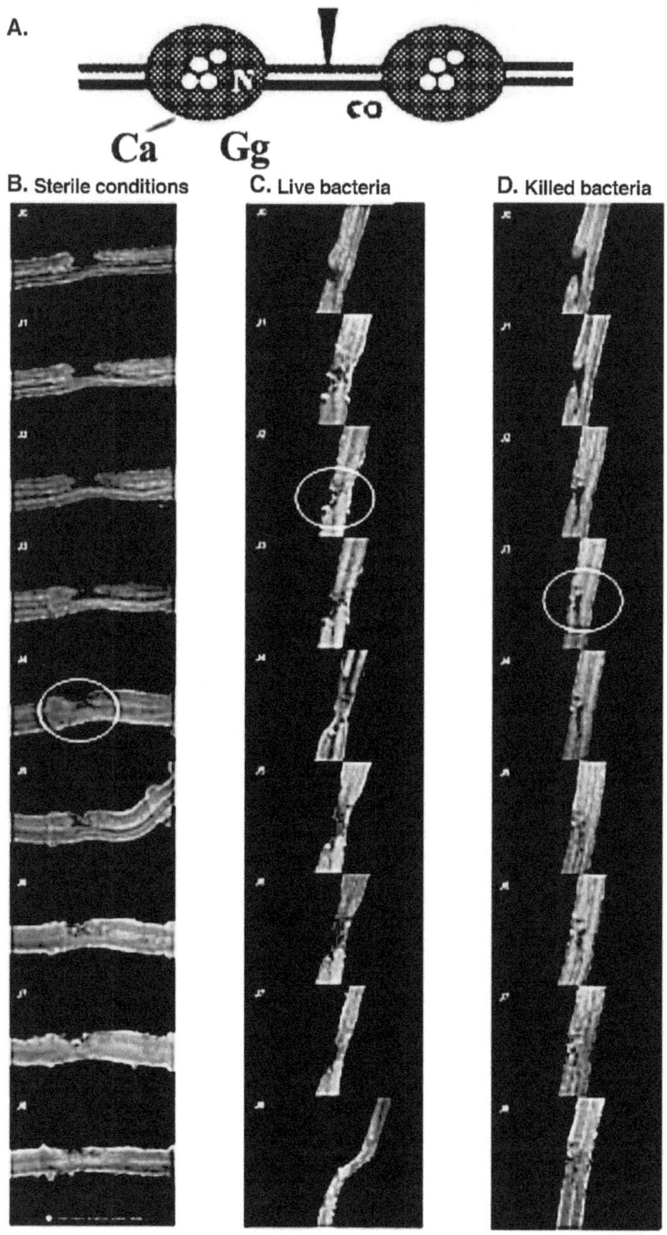

Figure 3. Effects on nerve regeneration of exposure of excised leech CNS to live or heat-killed bacteria (A) Diagram of the leech CNS in culture preparation. Neuron cell bodies (N) within ganglia (Gg) project axons into connectives (co) towards adjacent ganglia. V indicates the location of the cut of one of the two connectives linking two segmental ganglia. Microglial cells, evenly distributed in the nerve cord, are represented by dots. The nervous system is protected by a fibrous capsule (Ca). B-D) Sequential micrographs, taken 24 hr apart, from one (J1) to eight days (J8) post-axotomy, documenting the regeneration of the severed connective nerve. B) Preparation in sterile culture medium, (C) incubated with live bacteria and (D) incubated with killed bacteria. Used with permission from Schiskorski et al. J. Immol 2008;181(2): 1083 95. ©2008. The American Association of Immunologists, Inc.

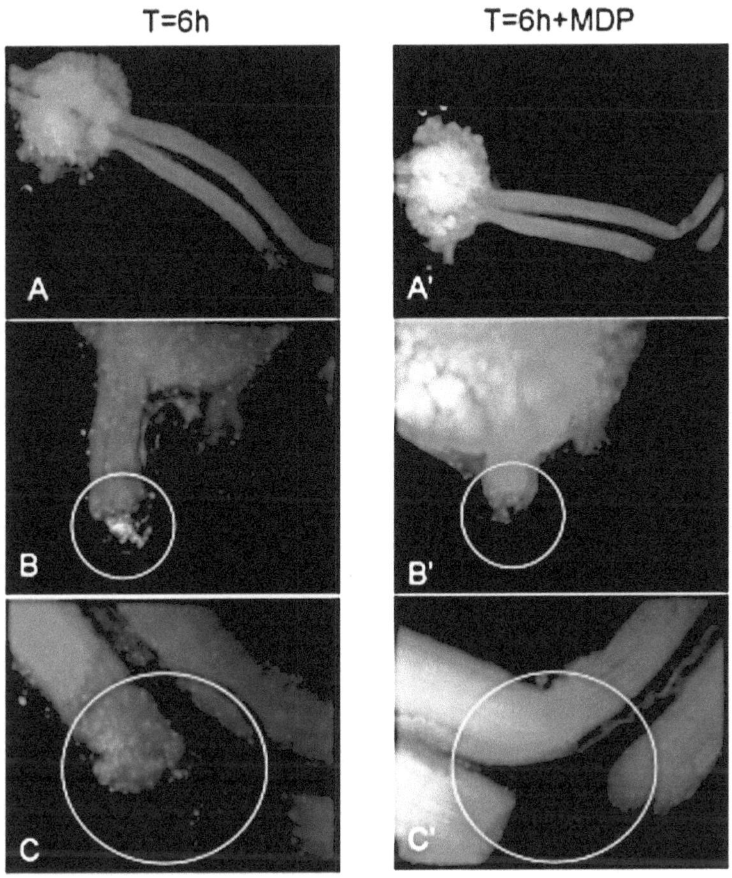

Figure 4. Effects on the recruitment of microglial cells of exposure of injured leech CNS to Muramyl DiPeptide (MDP). A) without addition of MDP to the medium culture, an accumulation of microglial cells is observed by nuclear staining at the injured site of both lesioned lateral (B) and interganglionary connective (C). (A′, B′, C′) This cell accumulation is not visible anymore post treatment with MDP.

us to conclude that the presence of both *Hm*-lumbricin and neuromacin at the axotomized site implicates peptide production by neurons and by the microglial cells recruited at the lesion site.[37] However silencing studies have not allowed connecting such antimicrobial peptides with *Hm*TLR1 receptor. Moreover, although bacteria cocktail enhance leech brain regeneration, Muramyl DiPeptide (MDP) blocks microglia migration (Fig. 4) which is line with the data obtained on macrophages migration, also inhibited by MDP.[38,39]

Taken together, these data confirm the immune role of the leech microglia and the presence of specific sensing receptor in these cells. The next step is the characterization of these receptors and effectors contained in leech brain. For this purpose, we started by *in silico* analyses through medicinal leech adult EST obtained by a consortium between Professsor E. Macagno, Professor Terry Gaasterland and Professor Michel Salzet. 91,233 transcripts were obtained from Genoscope (France) and JGI (USA) before annotated and 31,232 sequences were obtained.[8]

Leech Brain Immune Receptors and Effectors

Medicinal Leech Toll-Like Receptors (HmTLRs)

Five medicinal leech Toll-like receptors (TLRs) have been detected, one has already been fully characterized and data from the other ones are due soon. The first *Hm*TLR characterized is presented in Figure 5.[40] TLRs in general, share similarities in their extracellular Leucine Rich Repeat (LRR) and their intracellular Toll/IL-1 Receptor (TIR) domains. The TIR domain plays a central role in TLR signaling. All TLRs contain a cytoplasmic TIR domain, which, upon activation, acts as a scaffold to recruit adaptor proteins. It is well established that the differential recruitment of adaptors to TLRs provides a significant amount of specificity

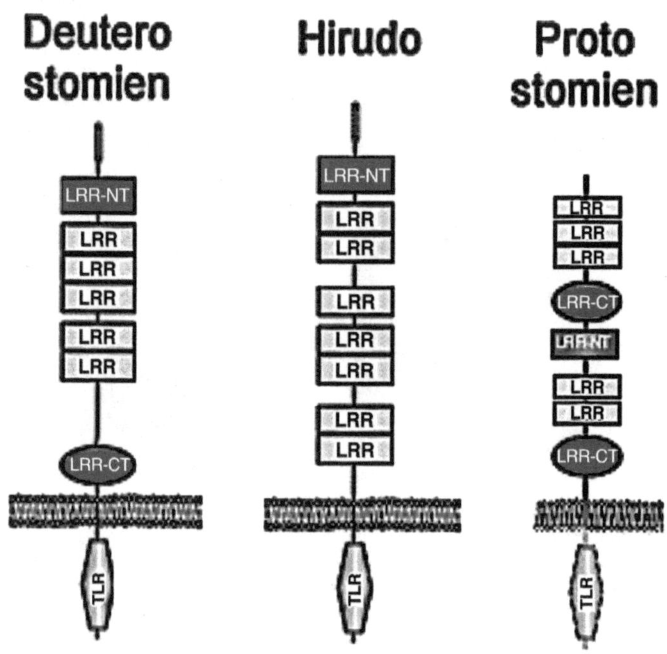

Figure 5. Structure comparison of *Hm*TLR1 and Protostomian and Deurterostomian TLRs.[40] SMART™ sequence analysis of the Nterminal part of *Hm*TLR1 revealed the presence of one LRRNT followed by six LRRs. *Hm*TLR1 presents the originality to exhibit an array of LRRs capped by one LRRNT only. By contrast to most TLRs described in invertebrates and vertebrates species, no LRRCT domains were identified from the analysis of the ectodomain of *Hm*TLR1. Blastp analysis was realized on the entire amino acid sequence of *Hm*TLR1. Data reveal great homologies mainly with TLR13s characterized in vertebrate species such as the mouse *Mus musculus* and the opossum *Monodelphis domestica* and in a lesser extent with TLRs of animals living in freshwater such as the zebra fish *Danio rerio*, the goldfish *Carassius auratus*, the salmon *Salmo salar* or again the rainbow trout *Oncorhynchus mykiss*. No homology with molecules identified in other lophotrochozoa was noticed. A second step, Blastp analysis of the LRR and the TIR domains of *Hm*TLR1 were performed separately, in order to get information on the function and the signalling pathway associated with this receptor respectively. The LRR domain of *Hm*TLR1 significantly (e-values < e-12) matches with the sequences of LRRs implicated in (i) pathogen recognition such as the LRR domain of the TLR3 and those of some Variable Lymphocyte Receptors (VLRs) and (ii) with the LRRs of vasorin, decorin and netrin known to participate in tissue remodelling and/or axonal guidance in vertebrates. Thus, both regenerative and immune functions could be attributable to this receptor. Concerning the TIR domain of *Hm*TLR1, Blastp analyses evidence a great percentage of homology with the TLR13.

to the TLR-signalling pathways. Among these adapter proteins MyD88 and TRIF are now considered as the signalling ones and hence the TLR pathways can be categorized as MyD88-dependent and TRIF-dependent. The LRR domain is an extracellular domain implicated in the detection of pathogens. Based on the organization of the extracellular LRR array, two types of TLRs have been described. Vertebrate TLRs have an array of LRRs capped by cysteine-rich domains located at the N- and C-terminal LRR domains (LRRNT and LRRCT, respectively). By contrast, most of invertebrate TLRs also contain LRRNT and LRRCT domains, but instead of capping the LRR array, these are located within the array in a tandem orientation. Interestingly, leech *HmTLR*1 presents the originality to exhibit an array of LRRs capped by one LRRNT only sharing sequence similarity with mouse TLR3. Based on EST medicinal leech sequence and Helobdella genome, four other HmTLR has been detected and their complete characterization is now in progress.

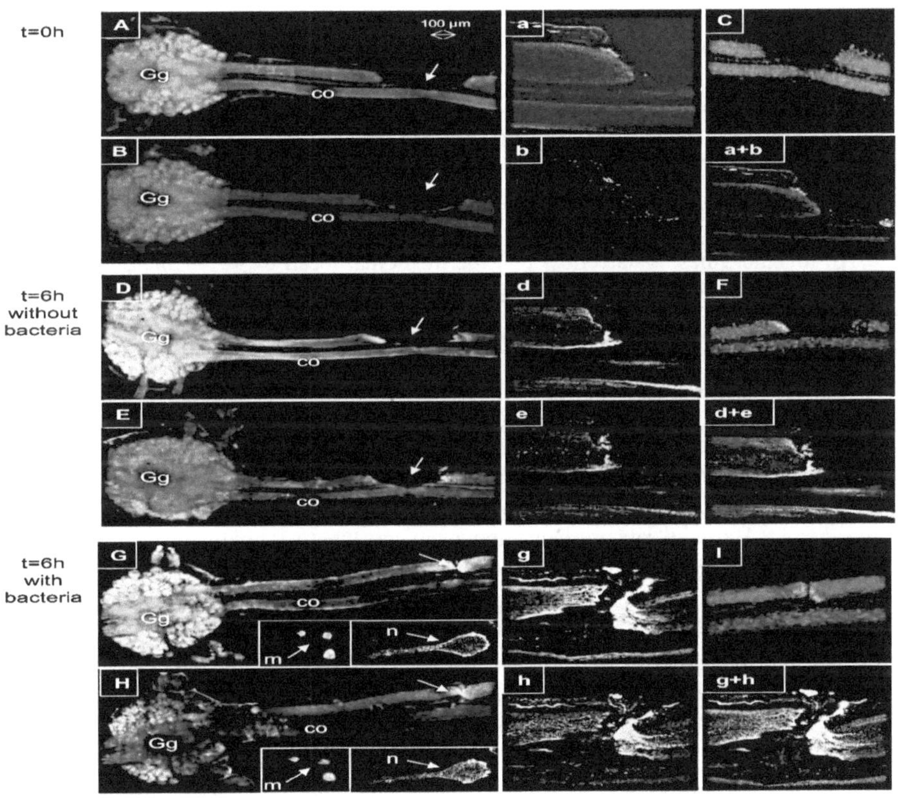

Figure 6. Co-appearance of *Hm*p43/EMAPII (Green) and *HmTLR*1 (Red) in the injured CNS incubated or not for 6 h with killed bacteria. Double staining of injured nerve cords was performed at t = 0 and t = 6 h post-axotomy using the fluorescent nuclear dye Hoechst 33258 (C, F and I) with both the anti-p43/EMAPII (A, D and G) and the anti-*HmTLR*1 (B, E and H) polyclonal Ab. Immunodetection was performed using green-labelled anti-EMAPII and red labeled anti-*HmTLR*1 secondary Ab. The results demonstrate an accumulation of *Hm*-EMAPII (G, g) and *HmTLR*1 (H, h) at the lesion site 6 h after axotomy in the presence of bacteria from reference 40. Used with permission from Schiskorski et al. J. Immol 2009;183(11): 7119-28. ©2009. The American Association of Immunologists, Inc. A color version of this figure is available at www.landesbioscience.com/curie.

We recently demonstrate that *Hm*TLR1[40] is localized in both neurons and microglia and expressed upon septic trauma (Fig. 6). *Hm*TLR1 is co-expressed with a cytokine related to the Endothelial Monocyte Activating Polypeptide (*Hm*EMAPII) sharing chemoattractive activity (Fig. 6). The complete intracellular trafficking upon pathogen challenges has to be undertaken to confirm the functionnal similarity between *Hm*TLR1 and mammalian TLR3.

*Hm*TLR1 and the Leech Cytokine (EMAPII)

Functional studies using silencing studies have not allowed connecting such antimicrobial peptides with *Hm*TLR1 receptor. By contrast, a cytokine sharing microglia chemoattactant activity recently characterized by our group in the medecinal leech e.g., *Hm*EMAPII,[40] shown based on RNAi silencing qPCR, western blot experiments (Fig. 7) and biological tests an association to *Hm*TLR1.[40] *Hm*EMAPII is processed from *Hm*p43 like mammals EMAP II. We hypothesized that *Hm*EMAPII could exert a chemoattractant effect on microglial cells as mammalian EMAPII does on monocytes. The chemoattractive effect of *Hm*EMAPII is blocked when an antihuman EMAPII antibody underscoring for the first time the ability of EMAPII to exert chemotactic effect toward microglial cells through CXCR3.[40] Leech CXCR3-related receptor cloning is on the way. These data points out that *Hm*TLR1 is linked to leech EMAPII and gives for the first time an immune function to a TLR in a nonecdysozoan model *i.e.,* in an invertebrate model different from *Caenorhabditis elegans* and *Drosophila melanogaster*.

All together, these data reflect that the medicinal leech express *Hm*TLR related to mammalian TLR. The first receptor characterized showed an intracellular localization and *Hm*TLR1 is linked to the cytokine related to EMAPII which exerts chemottractive effect after brain trauma or upon pathogen challenge. All these data confirm for the first time the presence of a complete TLR-signalling-Cytokine pathway implicated in immune response in medicinal leech nervous system. Such complex is also present in mammals reflecting again a co-evolution between the medicinal leech and its host mammals. It has also to be noted that such mechanism conservation is in line with a common origin of nervous system centralization between annelids and vertebrates as shown in the polychaete, *Platylnereis dumeriliis*.[41-43] Thus, the data presented above indicate that major players in innate immune response like danger sensing receptors coupling to cytokines or antimicrobial peptides and microglia are present in leech and strongly resemble that in vertebrates.

THEROMYZON TESSULATUM AS A MODEL FOR STUDYING THE PERIPHERAL IMMUNE RESPONSE

T. tessulatum is an ectoparasite of aquatic birds. Its life cycle was arbitrarily subdivided in stages (these are not larval stages) defined by taking, as indicators, the three blood meals. The third stage which corresponds to the gametogenesis phase is characterized by an important water uptake making the collection of the body fluid easy (Fig. 8). For this reason, *T. tessulatum* constitutes a convenient model for studying the antimicrobial response which takes place at the systemic level in coelomic fluid.[44] As a comparison, the medicinal leech has a parenchymatous body, coelomic cavities are reduced and the botryoidally tissue is immersed in a connective tissue. This makes the collection of the body fluid impossible. The antimicrobial response of *T. tessulatum* was investigated at the molecular level by

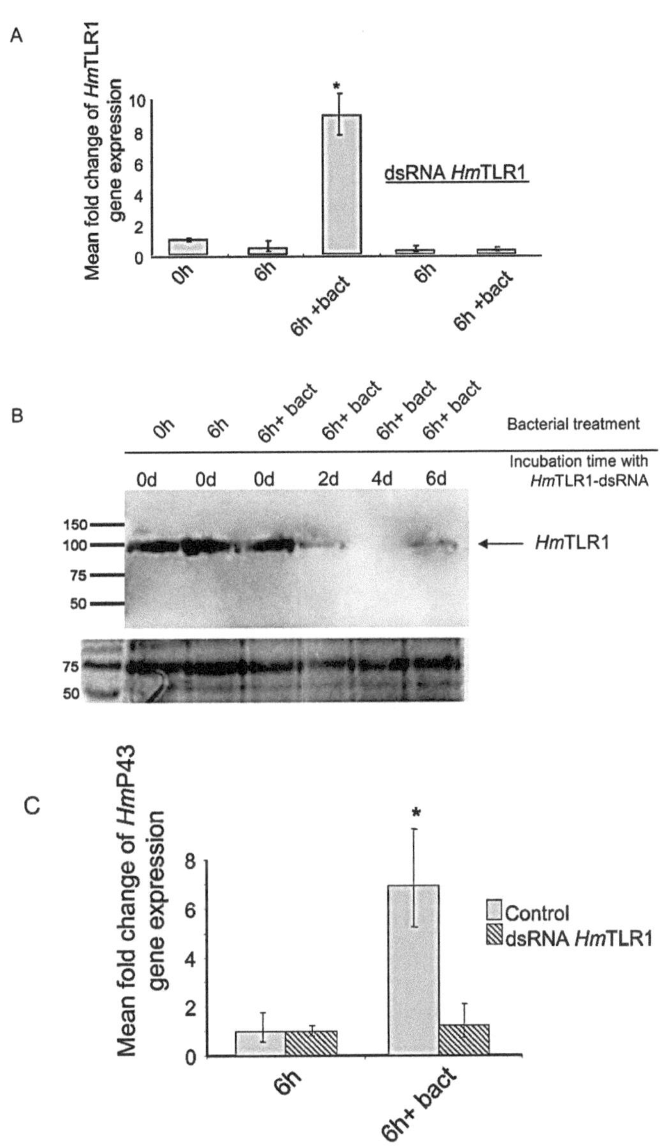

Figure 7. Impact of the *Hm*TLR1 gene silencing on induction of the HmP43/EMAPII gene in the leech CNS incubated with bacteria. A) The efficiency of the knock-down was quantified by measuring the level of *HmTLR*1 expression in nerve cords incubated with or without dsRNA. Data are expressed as relative levels comparatively to the basal level of expression measured in nerve cords processed immediately after sampling (0 h). *HmTLR*1 expression was quantified after 6 h of culture without bacteria (6 h) or with a mix of heat-killed Gram+ and Gram- bacteria (6 h + bact), revealing an induction of *HmTLR*1 gene under septic conditions. The bacterial induction of *HmTLR*1 is significantly reduced when the CNS is incubated for 4 days with *HmTLR*1-dsRNA (B) Western blot analyses of *HmTLR*1 protein level in the same conditions as for Figure 1. Best protein extinction is observed at 4 days (4d) of incubation with specific dsRNA confirming the efficiency of the knock down and the specificity of the anti *HmTLR*1 antibody. d, day (C) *HmTLR*1 gene silencing abolished the bacterial gene induction of *Hm*EMAPII observed in the control (without dsRNA *HmTLR*1), indicating a role of the *HmTLR*1 in the gene regulation of this cytokine in the leech CNS under septic conditions.[40]

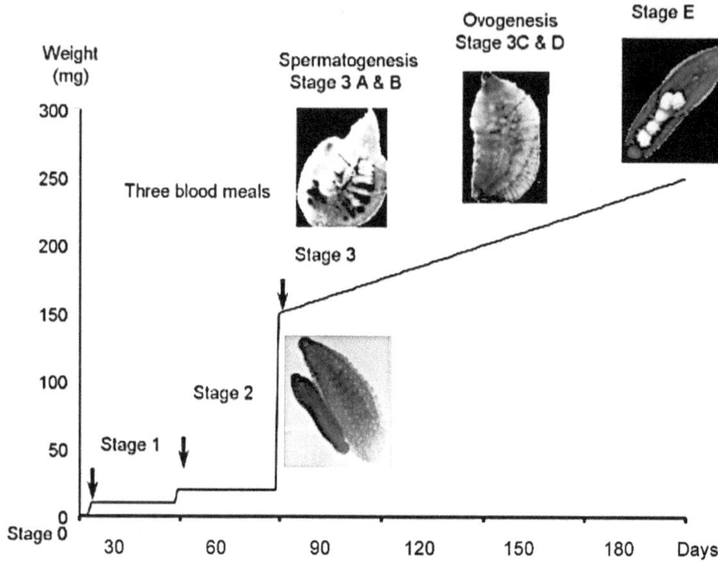

Figure 8. Life cycle of *Theromyzon tessulatum* from reference 61.

focusing on the antimicrobial peptides (AMPs) released in the coelomic fluid and at the cellular level by determining the immune functions of the coelomocytes.

AMPs of *T. tessulatum*

Three antimicrobial peptides (AMPs) were isolated and fully characterized from the body fluid of *T. tessulatum*. These are theromacin, a cysteine rich AMP exhibiting bactericidal activities, theromyzin an anionic peptide with bacteriostatic properties[44] and peptide B an anionic peptide matured from a neuropeptide precursor, proenkephalin A (PEA).[45] They all present an activity directed against Gram positive bacteria. Recently, a cDNA encoding a peptide presenting high percentage homologies with lumbricin-1, an AMP firstly characterized from the earthworm *Lumbricus rubellus* was cloned in *T. tessulatum* (Fig. 9).

Theromacin belongs to the cysteine rich AMP family. In invertebrates, most of them share the disulfide array of the insect/arthropod defensin.[46] In addition to having ten cysteine residues instead of six, theromacin does not harbor this consensus sequence. Theromacin by contrast with neuromacin (see before) has never been evidenced in other lophotrochozoan models or in ecdysozoan and thus seems to be restricted to leeches.

Lumbricin is a linear peptide without any posttranslational modifications. As the majority of AMPs described in the literature, theromacin and lumbricin possess a global positive charge presumably allowing their interaction with the negatively charged bacterial membrane (Fig. 10).

Theromyzin and peptide B, in contrast to theromacin and lumbricin, are anionic molecules. The mode of action of anionic AMPs is still unknown even if several hypotheses have been advanced. In vertebrates, AMPs with anionic properties were evidenced in the human and the sheep lung[47] AMPs are anionic because of homopolymeric regions of aspartate and require zinc as a cofactor for bactericidal activity.[48] Histatins,

A

```
atgttcagcaagtacgaagaaggcaaaaggacaaaagaagttacggtgaaagggttcagc
 M  F  S  K  Y  E  E  G  K  R  T  K  E  V  T  V  K  G  F  S
atggttcacaggccctcagttcatttcgcccccgagaggatcaaacccaacaaaattct
 M  V  H  R  P  S  V  H  F  A  P  R  E  D  Q  T  Q  Q  N  S
ccagtggggatggggaggggcatgcccatttacgccacgtcgggggctgcagccgagtaa
 P  V  G  M  G  R  G  M  P  I  Y  A  T  S  G  A  A  A  E  -
tgactcgagccccaatcactagtgaattaacaaaaaaaaaaaaaaa
```

B

```
lumb_H.m.    MFSKYERQKDKRSYGER-FSMFTGPQFISPP--ERIKPNKILQWDGEGMPIYATSGAAE 55
lumb_L.r.     FSKYERQKDKRPYSER-KNQYTGPQFLYPP--ERIPPQKVIKWNEEGLPIYEIPGAAE 55
lumb_T.t.    MFSKYEEGKRTKEVTVKGFSMVHRPSVHFAPREDQTQQNSPVGMG-RGMPIYATSGAAE 57
              *****. * .:    :  .    *..  .*  ::    :. :   . .*:***  .****
```

Figure 9. A) Nucleotide sequence of *Theromyzon tessulatum* lumbricin cDNA. The deduced amino acid sequence of the open reading frame is presented under the nucleotide sequence. B) Alignment of the *Tt* lumbricin (Lum T.t.) with the lumbricins characterized from the earthworm *Lumbricus rubellus* (Lum L.r.) and from the medicinal leech *Hirudo medicinalis* (Lum H.m.).

a family of histidin rich AMPs found in human saliva, also need the presence of zinc ions for bactericidal activities. Circular dichroism studies showed that the antimicrobial activities of histatin-5 require a conformational change that results from the interaction of the peptide with both zinc ions and negatively charged membranes.[49] The abundance of histidine residues at the N-terminal part of theromyzin could argue in favor of some common structures between the leech antibacterial peptide and histatins.

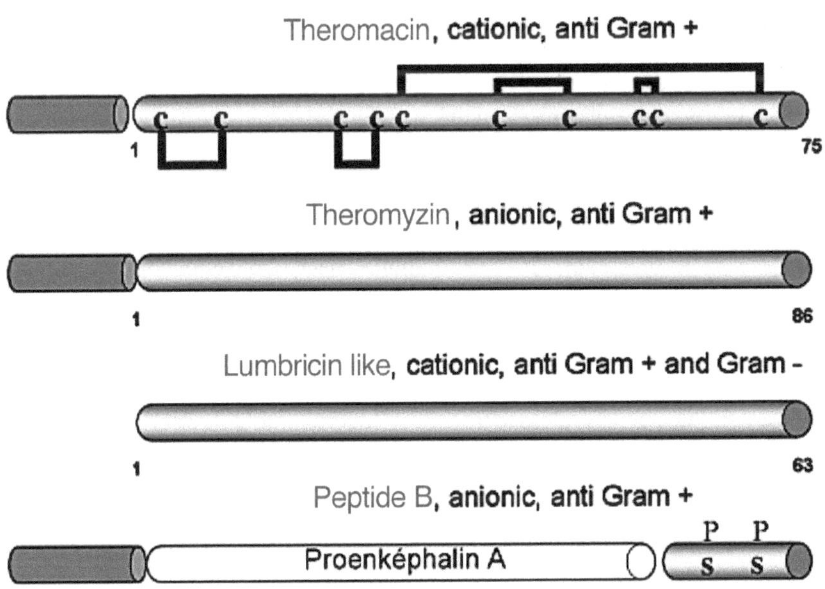

Figure 10. AMPs characterized in *T. tessulatum* from the coelomic liquid.[57]

Consequently, it clearly appears that leeches present a relatively large variety of AMPs. It is interesting to remark that at present no defensins have been isolated from leeches and from annelids in general. Defensins which are considered as the most widespread family of invertebrate AMPs have not been found neither in the genomes of the leech *Helobdella robusta* ant the polychaeta *Capitella* nor in the EST libraries of the earthworm *Lumbricus terrestris*, *Eiseinia fetida* and *Hirudo medicinalis*. Reciprocally most AMPs described in annelids have not been found in the genomes of ecdysozoan invertebrate such as *C. elegans* and *Drosophila melanogaster*.

Expression Site and Regulation of the AMPs Synthesis in *T. tessulatum*

Theromacin, *theromyzin* and *Tt lumbricin* genes are preferentially expressed in large fat cells (LFC) evenly distributed in the leech and in contact with the coelomic liquid from which the AMPs were purified. Their transcriptional level is enhanced after bacteria challenge evidencing a regulation of the leech AMPs similar to that of the insect antimicrobial peptides genes. Indeed, in the fruit fly, genes encoding antibiotic peptides are rapidly induced following a septic injury.[50] The similarity between the antibacterial response of the leech and those of holometabol insects is also supported by the functional resemblance between the leech LFC and the insect fat body which possess the common capacity to produce egg-yolk proteins.[51]

Moreover, based on our first investigations, no difference in gene expression was observed after Gram positive or Gram negative injection suggesting that the antibacterial response of *Theromyzon* is aspecific. This nonspecificity has also been assumed in *Drosophila* until the work of Lemaitre *et al* demonstrated that the humoral antimicrobial response of the fruit fly discriminates between various classes of microorganisms and mounts a response that is adapted to the infection.[52] That suggested that in a more natural mode of infection the leech could also adapt its antimicrobial response, what was recently confirmed in our annelid models by using bacteria living in the environment of the leech.[37]

The peptide sequences deduced from the *theromacin* and *theromyzin* genes contain putative signal peptides, indicating that mature peptides correspond to secreted molecules. As for the lumbricin like characterized in the medicinal leech, the *Tt* lumbricin precursor lacks the typical signal peptide. We assume that *Tt* lumbricin as demonstrated in *Hirudo* could be secreted through a nonconventional mechanism already observed but still unexplained for several molecules in mammals also.

Peptide B is not produced by the LFC although it was also isolated from the body fluid of the leech. Its precursor, PEA, was immunodetected into circulating coelomocytes suggesting that peptide B could be released from these cells. In contrast to the other leech AMPs, the production of peptide B seems to be more regulated at the translational level by the enzymes implicated in the PEA processing than at the transcriptional level. Of equal importance is the finding that enkephalin such as methionin enkephalins (ME) and peptide B are simultaneously released from PEA. Invertebrate and vertebrate immunocytes contain delta 2 opioid receptors that appear to mediate activation of these cells. In this regard, ME can be envisioned to activate immunocytes and provide a chemotactic signal to further stimulate cell recruitment.

However, since this process may take many minutes to accomplish, the bactericidal peptide B may cover this activation latency period. In this scenario, peptide B is broken down with time, it could release during this time, the heptapeptide MERF. Since we demonstrated that this peptide was able to interact with delta 2 opioid receptors, MERF

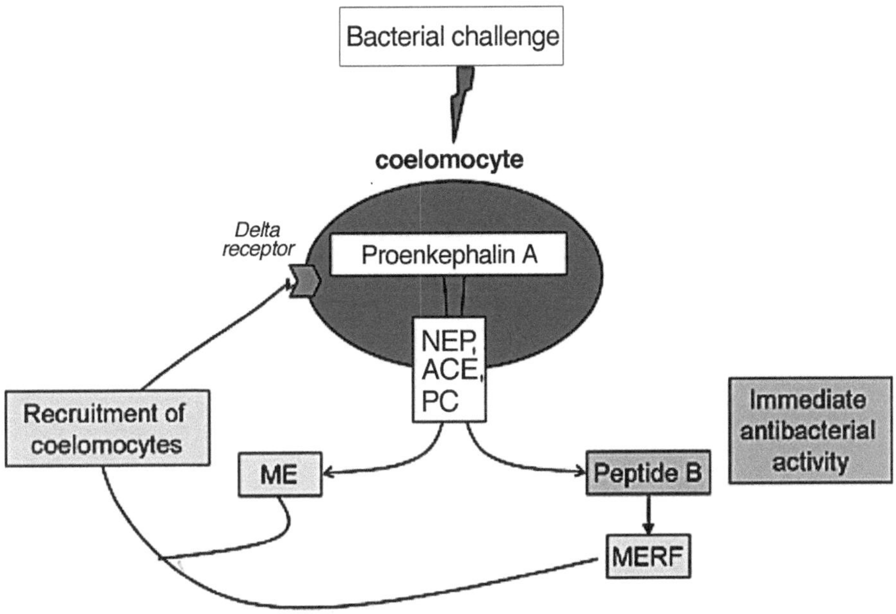

Figure 11. Illustration for the involvement of PEA derived peptides in the immune response of *T. tessulatum*. Upon an initial stimulus, i.e., bacterial challenge, smaller bioactive peptides Met enkephalin (ME) and peptide B processed by different enzymes _PC, NEP, ACE are released into the coelomic fluid of the leech. Peptide B plays its antibacterial role and is cleaved in a second time into Met enkephalin arg phe (MERF) which as ME does, can recruit coelomocytes.

could keep the rate of immunocyte activity.[45] This hypothesis is supported by previous studies demonstrating an immune activating role for MERF in human and invertebrate immunocytes. MERF as well as ME was shown to induce rounded invertebrate immunocytes to become mobile and amoeboid as well as to initiate chemotaxis. (Fig. 11).

Whatever the mode of regulation is, the accumulation of AMPs into the coelomic fluid after septic injury suggests that these peptides play their antimicrobial activities through a systemic action. Moreover the presence of theromacin and theromyzin in the intestinal epithelial cells and at the epidermis level also evokes participation in epithelial defense and/or in the control of the symbionts. The localization of antibiotic molecules in gastrointestinal tract has also been reported in insects and in vertebrates where they provide a rapid local immune response against exogenous pathogens brought in during feeding.[53] Theromacin, lumbricin and theromyzin were detected in the mucous covering the animal. That reminds the local defensive response reported in frogs in which antibacterial peptides secreted in the mucous prevent bacteria colonization and/or subsequent infection.[54]

As for lumbricin-1 in *L. rubellus*, physiological events occurring during gametogenesis phase appeared to be inducers of the AMPs gene expression in *T. tessulatum*. These data suggest that several hormonal factors implicated in sexual maturation may participate in the induction of genes encoded AMPs in annelids as described in *Drosophila* by Meister et al.[55] Interestingly, leech AMPs were detected

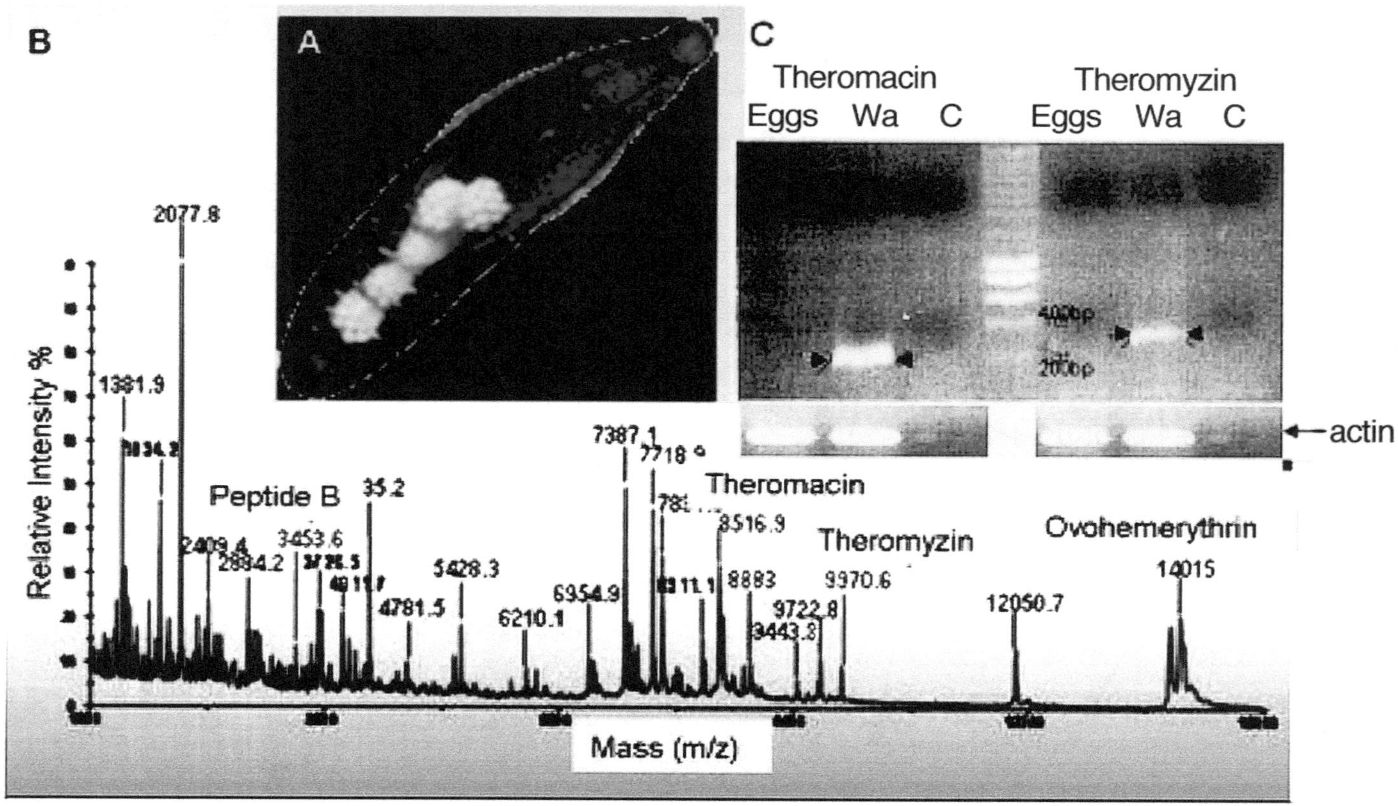

Figure 12. A) Ventral view of a Stage 3 *Theromyzon tessulatum* showing eggs (yellow balls) regrouped within five cocoons. B) Detection by mass spectrometry of molecular masses corresponding to theromyzin and theromacin in eggs extracts. C) Transcripts encoding theromacin and theromyzin are amplified by RT-PCR in whole animals (Wa) but not in eggs. C corresponds to the water control. A color version of this figure is available at www.landesbioscience.com/curie.

A

	Theromacin		Theromyzin	
	Unchallenged	Challenged	Unchallenged	Challenged
Stage 0	0	0	0	0
Stage 1	0	+	0	+
Stage 2	0	+	0	+
Stage 3	+	++	+	++

B

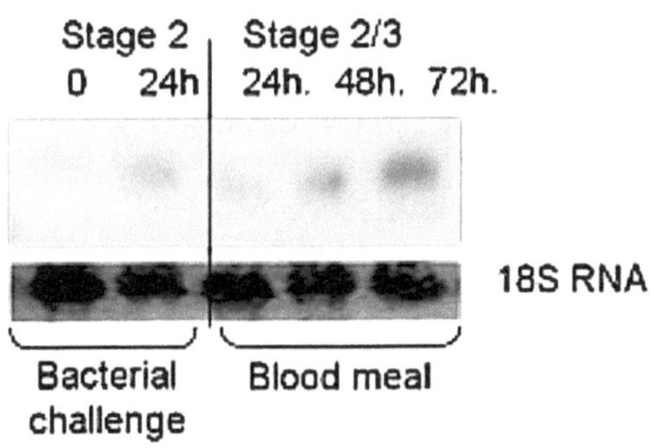

Figure 13. A) Table presenting the detection of transcripts coding for theromacin and theromyzin in leeches pricked or not with a mix of bacteria, at different stages of their life cycle. B) Northern blot showing that the theromacin gene expression is inducible upon bacteria challenge in Stage 2 leeches and after the last blood meal which marks the transition between the Stage 2 and 3.

in the eggs by mass spectrometry analysis whereas the transcripts were not amplified by RT-PCR (Fig. 12).

These results suggest a vertical transmission of the PAMs which could exert a protective role against bacteria during eggs development. AMPs may accumulate into the eggs by being captured from the body fluid of the mother though a pinocytosis mechanism before laying as described for egg yolks proteins and/or from the mucous covering the eggs after laying. A mass corresponding to ovohemerythrin, an egg yolk protein abundantly presents in the leech body fluid was detected by mass spectrometry analysis of eggs extracts supporting the first hypothesis without excluding the second one since AMPs were also detected in the mucous covering the leech. The study of the gene expression in course of the post-embryonic development demonstrated that *T. tessulatum* starts to synthesize its own AMPs from the Stage 1 of its life cycle and that a bacterial challenge is necessary for observing this synthesis. Indeed, a basal level of transcripts encoding theromacin and theromyzin was detected in unchallenged Stage 3 leeches only. We presume that AMPs are expressed at a basal level in adults for providing an immune protection to the eggs by a vertical transmission of the antibiotic molecules. By contrast, a bacterially inducible response is observable from the Stage 1 to the Stage 3 suggesting that leeches acquire the ability to establish an immune response after their first blood meal (Fig. 13).

Consequently, *T. tessulatum* is an original invertebrate model which has developed two modes of fighting infections by AMPs: (i) storage of antibacterial peptide derived from PEA and release of the peptide into the coelomic fluid after immune challenge (ii) induction after septic injury of gene coding for more classical AMPs, mainly in LFC and rapid release into the body fluid of the antibiotic peptides. Data collected from our group (unpublished) suggest that the same AMPs participate to the systemic antimicrobial response of the medicinal leech, *Hirudo medicinalis*. Interestingly, the PEA processing appeared to be very well conserved in course of evolution since we have demonstrated that the same mechanism is observable during surgery in human patients undergoing cardiopulmonary bypass.[56,45]

The Cellular Immune Response

Annelids are primitive coelomates known to possess specially developed cellular immunity against microorganisms including phagocytosis, encapsulation and spontaneous cytotoxicity against allogenic or xenogenic cells.[57,58] In leeches, our group has characterized three distinct populations of coelomocytes in *T. tessulatum* (Fig. 14).

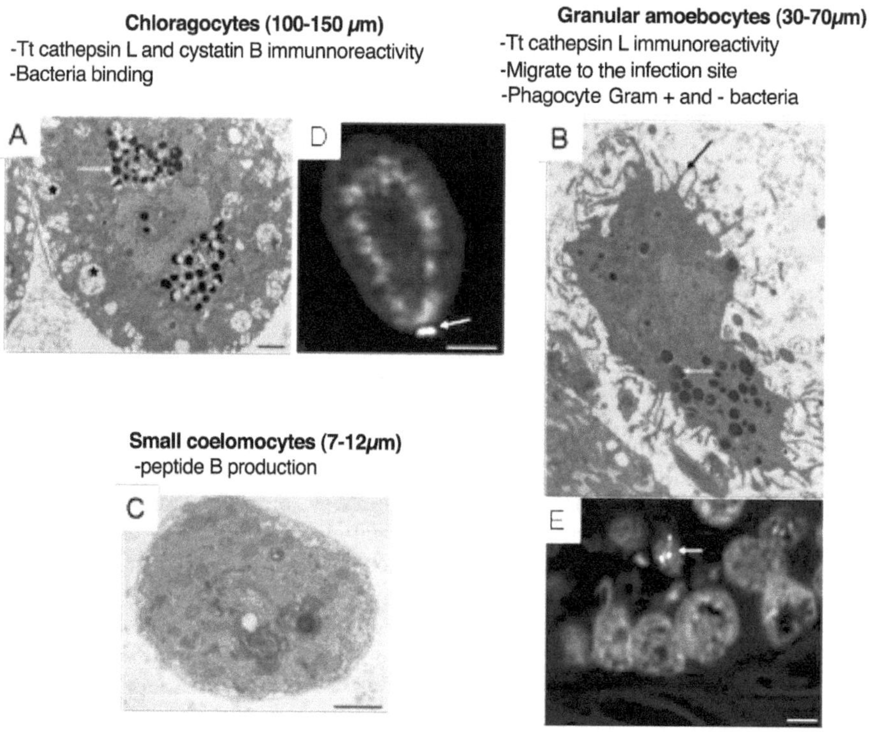

Figure 14. Coelomocytes of *T. tessulatum*. A) Large coelomocytes present large electron-dense granules (white arrow) and electron-lucent vesicles (black stars). B) Intermediate size coelomocytes show long cytoplasmic pseudopods (black arrow) and large electron-dense granules (white arrow). C) Small circulating cells are rich in endoplasmic reticulum and small granules. D-E) Animals were injected with FITC-labelled bacteria (killed *M.luteus* or *E. coli*). After 24 hours of incubation, bacteria are still observed near chloragocytes (D, white arrow) but phagocytosed bacteria are observed in granular amoebocytes only (E, white arrow).

These are constituted by chloragocytes, granular amoebocytes and small coelomocytes.[30] Leech chloragocytes are the only cells expressing both the *Tt* cathepsin L and the *Tt* cystatin B. Granular amoebocytes present an immunoreactivity to the anti *Tt* cathepsin L antibody (Ab) although the small coelomocytes are not recognized by neither the anti *Tt* cathepsin L Ab nor the anti *Tt* cystatin B Ab. The immune functions were investigated by essentially focusing on the phagocytic activity and the migrating property of these circulating cells. As resumed in the Figure 14, the granular amoebocytes are able to migrate to the injection site of microorganisms and to phagocyte without any apparent distinction both killed Gram positive and Gram negative bacteria. The molecular mechanism of recruitment may imply enkephalin peptides derived from the PEA maturation process, such as ME or MERF. By contrast, leech chloragocytes are not able to phagocyte bacteria. However, while no phagocytosis was detected, confocal microscopy analysis evidenced chloragocytes-bacteria interaction suggesting the presence of recognition molecule expressed at the surface of these cells (Fig. 14D). Leech chloragocytes may also be implicated in encapsulation reactions as described in oligochaeta annelids. The third type represented by the small coelomocytes presents the morphology of invertebrate hyaline cells i.e., a cytoplasm deprived of granules. However the leech cells have been considered not to be hyaline cells due to their incapacity to phagocytose killed bacteria. Interestingly, our group has very recently observed that the use of live bacteria was a prerogative for inducing the phagocytosis process in a population of leech blood cells. Thus, the small coelomocytes should be incubated with live bacteria to determinate whether they could be assimilated to hyaline cells or not.

This heterogeneous population of coelomocytes reminds the population described in other annelids. Numerous studies performed in oligochætes as *Lumbricus sp.* and *Eisenia sp*[59] used optical and electron microscopy. Although monoclonal antibodies were carried out against various invertebrate taxa, the majority of the studies related to insect hæmocytes. De Eguileor *et al* identified three coelomic cell populations, i.e., macrophage-like, NK-like and granular cells, using human monoclonal antibodies in the hirudinea *Glossiphonia complanata*.[58] Engelmann *et al* produced monoclonal antibodies against coelomic cells in *Eisenia fetida* earthworm.[60] While anti-EFCC1 antibody (*Eisenia fetida* coelomocyte differentiation cluster) is able to recognize antigenic motifs on various tissues, three other antibodies named anti-EFCC2, anti-EFCC3 and anti-EFCC4 allowed to respectively discriminate chloragocytes, hyaline amoebocytes and granular amoebocytes.[60] Because oligochætes and leeches are closed relatives, it should be interesting to test these anti-EFCC antibodies for discriminating our coelomic cells and possibly identify the presence of phagocytic hyaline cells in *T. tessulatum*.

Unlike other invertebrates, one of the particularities of annelids is to possess a closed circulatory system separated from the coelomic cavity. Our group is investigating the immune function of the circulating blood cells of the medicinal leech. Most of the reported data being focused on coelomic cells, the obtained results will constitute the first description of the function of the blood cells in an annelid. Moreover since the leech central nervous system permanently baths into the blood, these data may open up new avenues for discovering the impact of the immune response on the neural repair of the medicinal leech.

CONCLUSION AND PERSPECTIVES

The observations we have reviewed in this chapter attest to the broad range of inquiry, from structure to function, and to the breadth of the techniques currently employed to study the defence, repair and maintenance of the leech nervous and immune systems. Clearly, the increasing application of biochemical and molecular genetic tools is beginning to yield insights into the nature of the molecular mechanisms responsible for these phenomena. But, while progress is being made, it is also clear that there is a strong need to accelerate the implementation and application of genomic, transcriptomic and proteomic tools to the leech. The leech model has many important advantages, perhaps the most important being the ability to bridge from the immune response to neural repair, in the context of having, or the possibility of having, detailed knowledge about all the neurons in the CNS and thus an unparalleled level of completeness. Most other invertebrate systems, as well as all vertebrate systems, under current study, can only afford partial, because many or most of their neurons are inaccessible and only population properties are attainable. However, other systems, including *C. elegans*, *Drosophila*, zebrafish, and some rodents, greatly benefit from the possibility of doing standard genetics, their shorter reproductive cycles (particularly in comparison to Hirudo, not as much with respect to small leeches like Helobdella), and the availability of the complete sequence of their genomes. Nonetheless, the ability to conduct systems level functional studies, and the possibility of relating the physiological programs to genetic programs that encode the underlying circuitry and its properties, the capacity of the system to defend itself, regenerate and repair, and the availability of reverse genetic tools to study gene expression and regulation, all strongly justify continuing and enhancing our efforts to understand neuroimmune responses in the leech.

ACKNOWLEDGMENTS

This work was supported by the Centre National de la Recherche Scientifique (CNRS), the Ministère de l'Enseignement, de la Recherche et des Technologies (MERT) and the Genoscope.

REFERENCES

1. Bely AE, Weisblat DA. Lessons from leeches: a call for DNA barcoding in the lab. Evol Dev 2006; 8(6):491-501.
2. Borda E, Siddall ME. Arhynchobdellida (Annelida: Oligochaeta: Hirudinida): phylogenetic relationships and evolution. Mol Phylogenet Evol 2004; 30(1):213-225.
3. Kaigorodova IA, Shcherbakov D. [Molecular phylogenetic study of the systematic position of Baikalian oligochaetes in the system Clitellata]. Genetika 2006; 42(12):1647-1655.
4. McDougall C, Hui JH, Monteiro A et al. Annelids in evolutionary developmental biology and comparative genomics. Parasite 2008; 15(3):321-328.
5. Pfeiffer I, Brenig B, Kutschera U. Molecular phylogeny of selected predaceous leeches with reference to the evolution of body size and terrestrialism. Theory Biosci 2005; 124(1):55-64.
6. Utevsky SY, Trontelj P. A new species of the medicinal leech (Oligochaeta, Hirudinida, Hirudo) from Transcaucasia and an identification key for the genus Hirudo. Parasitol Res 2005; 98(1):61-66.
7. Trontelj P, Utevsky SY. Celebrity with a neglected taxonomy: molecular systematics of the medicinal leech (genus Hirudo). Mol Phylogenet Evol 2005; 34(3):616-624.

8. Macagno ERGTE L, Bafna V, Soares MB et al. Construction of a medicinal leech transcriptome database and its application to the identification of leech homologs of neural and innate immune genes. BMC Genomics 2010; In press.
9. Salzet M, Capron A, Stefano GB. Molecular crosstalk in host-parasite relationships: schistosome- and leech-host interactions. Parasitol Today 2000; 16(12):536-540.
10. Macagno ER. Number and distribution of neurons in leech segmental ganglia. J Comp Neurol 1980; 190(2):283-302.
11. Sawyer RT. Leech biology and behaviour. Clarendon Press, Oxford, England 1986; I, II, III.
12. Nicholls JG, Hernandez UG. Growth and synapse formation by identified leech neurones in culture: a review. Q J Exp Physiol 1989; 74(7):965-973.
13. McGlade-McCulloh E, Morrissey AM, Norona F et al. Individual microglia move rapidly and directly to nerve lesions in the leech central nervous system. Proc Natl Acad Sci USA 1989; 86(3):1093-1097.
14. Shafer OT, Chen A, Kumar SM et al. Injury-induced expression of endothelial nitric oxide synthase by glial and microglial cells in the leech central nervous system within minutes after injury. Proc Biol Sci 1998; 265(1411):2171-2175.
15. Chen A, Kumar SM, Sahley CL et al. Nitric oxide influences injury-induced microglial migration and accumulation in the leech CNS. J Neurosci 2000; 20(3):1036-1043.
16. Arafah K, Desmons A, Vizioli J et al. Crosstalk between the CB2-like receptor and nitric oxide systems in the injured leech brain: involvement of the endocannabinoid 2AG in microglia recruitment at the injured site. J Neurosci 2010; Submitted.
17. Lipitz JB, Arafah K, Zhou X et al. Cannabinoid control of microglia migration to lesions in the leech CNS. J Neurosci 2010; Submitted.
18. Raborn ES, Marciano-Cabral F, Buckley NE et al. The cannabinoid delta-9-tetrahydrocannabinol mediates inhibition of macrophage chemotaxis to RANTES/CCL5: linkage to the CB2 receptor. J Neuroimmune Pharmacol 2008; 3(2):117-129.
19. Vannacci A, Giannini L, Passani MB et al. The endocannabinoid 2-arachidonylglycerol decreases the immunological activation of Guinea pig mast cells: involvement of nitric oxide and eicosanoids. J Pharmacol Exp Ther 2004; 311(1):256-264.
20. Cabral GA, Marciano-Cabral F. Cannabinoid receptors in microglia of the central nervous system: immune functional relevance. J Leukoc Biol 2005; 78(6):1192-1197.
21. Walter L, Neumann H. Role of microglia in neuronal degeneration and regeneration. Semin Immunopathol 2009; 31(4):513-525.
22. Salzet M, Breton C, Bisogno T et al. Comparative biology of the endocannabinoid system possible role in the immune response. Eur J Biochem 2000; 267(16):4917-4927.
23. Ehrhart J, Obregon D, Mori T et al. Stimulation of cannabinoid receptor 2 (CB2) suppresses microglial activation. J Neuroinflammation 2005; 2:29.
24. Xu Y, Bolton B, Zipser B et al. Gliarin and macrolin, two novel intermediate filament proteins specifically expressed in sets and subsets of glial cells in leech central nervous system. J Neurobiol 1999; 40(2):244-253.
25. Vergote D, Sautiere PE, Vandenbulcke F et al. Up-regulation of neurohemerythrin expression in the central nervous system of the medicinal leech, Hirudo medicinalis, following septic injury. J Biol Chem 2004; 279(42):43828-43837.
26. Vergote D, Macagno ER, Salzet M et al. Proteome modifications of the medicinal leech nervous system under bacterial challenge. Proteomics 2006; 6(17):4817-4825.
27. Morgese VJ, Elliott EJ, Muller KJ. Microglial movement to sites of nerve lesion in the leech CNS. Brain Res 1983; 272(1):166-170.
28. Deloffre L, Salzet B, Vieau D et al. Antibacterial properties of hemerythrin of the sand worm Nereis diversicolor. Neuro Endocrinol Lett 2003; 24(1-2):39-45.
29. Lefebvre C, Salzet M. Annelid neuroimmune system. Curr Pharm Des 2003; 9(2):149-158.
30. Lefebvre C, Vandenbulcke F, Bocquet B et al. Cathepsin L and cystatin B gene expression discriminates immune coelomic cells in the leech Theromyzon tessulatum. Dev Comp Immunol 2008; 32(7):795-807.
31. Blackshaw SE, Babington EJ, Emes RD et al. Identifying genes for neuron survival and axon outgrowth in Hirudo medicinalis. J Anat 2004; 204(1):13-24.
32. Perlson E, Medzihradszky KF, Darula Z et al. Differential proteomics reveals multiple components in retrogradely transported axoplasm after nerve injury. Mol Cell Proteomics 2004; 3(5):510-520.
33. Matias I, Bisogno T, Melck D et al. Evidence for an endocannabinoid system in the central nervous system of the leech Hirudo medicinalis. Brain Res Mol Brain Res 2001; 87(2):145-159.
34. Stefano GB, Rialas CM, Deutsch DG et al. Anandamide amidase inhibition enhances anandamide-stimulated nitric oxide release in invertebrate neural tissues. Brain Res 1998; 793(1-2):341-345.
35. Stefano GB, Salzet B, Rialas CM et al. Morphine- and anandamide-stimulated nitric oxide production inhibits presynaptic dopamine release. Brain Res 1997; 763(1):63-68.

36. Stefano GB, Salzet B, Salzet M. Identification and characterization of the leech CNS cannabinoid receptor: coupling to nitric oxide release. Brain Res 1997; 753(2):219-224.
37. Schikorski D, Cuvillier-Hot V, Leippe M et al. Microbial challenge promotes the regenerative process of the injured central nervous system of the medicinal leech by inducing the synthesis of antimicrobial peptides in neurons and microglia. J Immunol 2008; 181(2):1083-1095.
38. Nagao S, Tanaka A, Onozaki K et al. Differences between macrophage migration inhibitions by lymphokines and muramyl dipeptide (MDP) or lipopolysaccharide (LPS): migration enhancement by lymphokines. Cell Immunol 1982; 71(1):1-11.
39. Fukuyama R, Takeda H, Fushiki S et al. Muramyl dipeptide injected into crushed sciatic nerve, activates macrophages and promotes recovery of walking locomotion in rats. Restor Neurol Neurosci 1998; 13(3-4):213-219.
40. Schikorski D, Cuvillier-Hot V, Boidin-Wichlacz C et al. Deciphering the immune function and regulation by a TLR of the cytokine EMAPII in the lesioned central nervous system using a leech model. J Immunol 2009; 183(11):7119-7128.
41. Denes AS, Jekely G, Steinmetz PR et al. Molecular architecture of annelid nerve cord supports common origin of nervous system centralization in bilateria. Cell 2007; 129(2):277-288.
42. Hui JH, Raible F, Korchagina N et al. Features of the ancestral bilaterian inferred from Platynereis dumerilii ParaHox genes. BMC Biol 2009; 7:43.
43. Raible F, Tessmar-Raible K, Osoegawa K et al. Vertebrate-type intron-rich genes in the marine annelid Platynereis dumerilii. Science 2005; 310(5752):1325-1326.
44. Tasiemski A, Vandenbulcke F, Mitta G et al. Molecular characterization of two novel antibacterial peptides inducible upon bacterial challenge in an annelid, the leech Theromyzon tessulatum. J Biol Chem 2004; 279(30):30973-30982.
45. Tasiemski A, Verger-Bocquet M, Cadet M et al. Proenkephalin A-derived peptides in invertebrate innate immune processes. Brain Res Mol Brain Res 2000; 76(2):237-252.
46. Levy F, Bulet P, Ehret-Sabatier L. Proteomic analysis of the systemic immune response of Drosophila. Mol Cell Proteomics 2004; 3(2):156-166.
47. Brogden KA, Ackermann M, Huttner KM. Detection of anionic antimicrobial peptides in ovine bronchoalveolar lavage fluid and respiratory epithelium. Infect Immun 1998; 66(12):5948-5954.
48. Melino S, Rufini S, Sette M et al. Zn(2+) ions selectively induce antimicrobial salivary peptide histatin-5 to fuse negatively charged vesicles. Identification and characterization of a zinc-binding motif present in the functional domain. Biochemistry 1999; 38(30):9626-9633.
49. Brewer D, Lajoie G. Evaluation of the metal binding properties of the histidine-rich antimicrobial peptides histatin 3 and 5 by electrospray ionization mass spectrometry. Rapid Commun Mass Spectrom 2000; 14(19):1736-1745.
50. Hoffmann JA. The immune response of Drosophila. Nature 2003; 426(6962):33-38.
51. Baert JL, Britel M, Slomianny MC et al. Yolk protein in leech. Identification, purification and characterization of vitellin and vitellogenin. Eur J Biochem 1991; 201(1):191-198.
52. Lemaitre B, Reichhart JM, Hoffmann JA. Drosophila host defense: differential induction of antimicrobial peptide genes after infection by various classes of microorganisms. Proc Natl Acad Sci USA 1997; 94(26):14614-14619.
53. Zasloff M. Antimicrobial peptides of multicellular organisms. Nature 2002; 415(6870):389-395.
54. Zasloff M. Antibiotic peptides as mediators of innate immunity. Curr Opin Immunol 1992; 4(1):3-7.
55. Meister M, Richards G. Ecdysone and insect immunity: the maturation of the inducibility of the diptericin gene in Drosophila larvae. Insect Biochem Mol Biol 1996; 26(2):155-160.
56. Tasiemski A, Salzet M, Benson H et al. The presence of antibacterial and opioid peptides in human plasma during coronary artery bypass surgery. J Neuroimmunol 2000; 109(2):228-235.
57. Salzet M, Tasiemski A, Cooper E. Innate immunity in lophotrochozoans: the annelids. Curr Pharm Des 2006; 12(24):3043-3050.
58. Tettamanti G, Malagoli D, Benelli R et al. Growth factors and chemokines: a comparative functional approach between invertebrates and vertebrates. Curr Med Chem 2006; 13(23):2737-2850
59. Linthicum DS, Stein EA, Marks DH et al. Electron-microscopic observations of normal coelomocytes from the earthworm, Lumbricus terrestris. Cell Tissue Res 1977; 185(3):315-330.
60. Engelmann P, Pal J, Berki T et al. Earthworm leukocytes react with different mammalian antigen-specific monoclonal antibodies. Zoology (Jena) 2002; 105(3):257-265.
61. Verger-Bocquet M, Wattez C, Salzet M et al. Immunohistochemical identification of peptidergic neurons in compartment 4 of the supraœsophageal ganglion of the leech Theromyzon tessulatum (O.F.M). Can J Zool 1993; 70:856-865.

CHAPTER 6

INNATE IMMUNITY IN *C. ELEGANS*

Ilka Engelmann and Nathalie Pujol*

Centre d'Immunologie de Marseille-Luminy, Université de la Méditerranée, Marseille, France
**Corresponding Author: Nathalie Pujol—Email: pujol@ciml.univ-mrs.fr*

Abstract: The nematode *Caenorhabditis elegans* is proving to be a powerful invertebrate model to study host-pathogen interactions. In common with other invertebrates, *C. elegans* relies solely on its innate immune system to defend itself against pathogens. Studies of the nematode response to infection with various fungal and bacterial pathogens have revealed that the innate immune system of *C. elegans* employs evolutionary conserved signalling pathways. They regulate the expression of various effectors molecules, some of which are also conserved. Here, we summarize the current knowledge of the pathways and effector molecules involved in the nematode immune response, with a particular focus on the antifungal immune response of the *C. elegans* epidermis.

INTRODUCTION

C. elegans is a free-living soil nematode that feeds on bacteria and is therefore constantly exposed to potential pathogens.[1] Like other invertebrates, *C. elegans* lacks an adaptive immune system. In contrast to many invertebrate species, however, *C. elegans* does not appear to have specialized immune cells. For example, while Drosophila has macrophage-like hemocytes, which engulf invading microbes, the only cells in the nematode body cavity, the 6 coelomocytes, do not seem to be capable of phagocytosis but function as scavenger cells with a high endocytic capacity.[2]

C. elegans possesses three major mechanisms of defences against microbial attacks:[1] Avoidance behaviour: It has been demonstrated that worms are able to distinguish between different bacteria. Whereas most bacteria attract *C. elegans*, some repel the nematode and cause an avoidance behaviour. Such an aversive response can be specifically directed against the pathogenic strains of a bacterial species (reviewed in ref. 3). Olfactory neurons, G protein coupled receptors and the only Toll-like receptor (TLR) in *C. elegans*, TOL-1,

Invertebrate Immunity, edited by Kenneth Söderhäll.

are involved in triggering the avoidance behaviour to pathogenic *Serratia marcescens*.[4,5] Worms can "remember" odours[6] and can even learn to avoid bacteria that are recognized as noxious.[7] This discrimination relies in part on pairs of asymmetric chemosensory neurons.[8] Their correct development requires a signalling cassette that includes an intracellular TIR-domain adapter protein (TIR-1) acting upstream of a p38 MAPK cascade.[9] This cassette, which will be described in more detail below, appears also to play a direct behavioural role as it has been found to be involved in the neuroendocrine regulation of serotonin-dependent aversion to *Pseudomonas aeruginosa*.[10,2] The second axis of protection against pathogen invasion is a strong cuticle, made of collagen and chitin and constituting the exoskeleton of the worm. It acts as a physical barrier that is relatively resistant to puncturing. As a complement, the pharyngeal grinder destroys pathogens that are taken up during feeding. It prevents live pathogens from reaching the intestine and establishing an infection. Indeed, mutants with defective grinder function are more susceptible to infection.[11,12,3] The third line of defence involves inducible mechanisms. These will be the main focus of this chapter. *C. elegans* possesses a complex inducible defence system involving multiple signalling cascades that regulate the production of antimicrobial peptides (AMP) and proteins in a pathogen- and tissue-specific way.

ROUTES OF INFECTION

Most of the known pathogens of *C. elegans* use two main routes of infection, through the pharynx or the epidermis (Fig. 1). Many Gram-positive and Gram-negative bacteria as well as yeast, infect worms upon oral up-take during feeding and establish an intestinal infection. They must survive the passage through the grinder to reach the intestine, proliferate and establish an infection. In some cases, it has been shown that the pathogen destroys the grinder,[13] in others it appears that the infectious particles, such as the spores of *Bacillus thuringensis* are resistant to the mechanical action of the grinder.[14] Almost all characterised intestinal pathogens of *C. elegans* remain extracellular, apart from *Salmonella typhimurium* and the microsporidium *Nematocida parisii*, that have been shown to establish intracellular infection in the intestinal cells.[15,16]

Some pathogenic bacteria and fungi can adhere to the cuticle and infect the *C. elegans* epidermis. For example, *Microbacterium nematophilum* adheres to the anal region of the nematode and induces hindgut swelling[17] and *Leucobacter chromiireducens* is capable of causing lethal uterine infections[18] (Fig. 1). Different fungi that are pathogenic for nematodes, including *Drechmeria coniospora* and species of *Haptocillium*, produce spores that adhere and then penetrate the cuticle and grow into the epidermis (Figs. 1 and 2).[19,20] Although some pathogens, such as certain strains of *P. aeruginosa*, produce fast-acting toxins,[21] against which *C. elegans* appears defenceless, in many cases, infection provokes an immune response.

PATHOGEN RECOGNITION

The first step of an inducible defence is the recognition of the pathogen. Conserved structures on pathogens that are not present in the host and thus recognized as foreign, so called microbe-associated molecular patterns (MAMPs), bind to pattern recognition receptors (PRRs) in many organisms.[22] PRRs include peptidoglycan recognition proteins (PGRP), Gram negative binding proteins (GNBP), nucleotide-binding oligomerization domain (NOD) and NACHT domain proteins.[23] Genes encoding proteins of these families are absent from the *C. elegans* genome.

One prominent class of PRRs, in vertebrates the TLRs, can sense outer membrane components of the bacteria, RNA or DNA.[22] As mentioned above, the single worm TLR, TOL-1, is involved in behavioural avoidance of some pathogenic bacteria,[4,5] but does not seem to play a role in the resistance to several pathogens,[5] nor in the regulation of certain immune effectors.[24] One study showed that *tol-1* mutants are more susceptible to *S. typhimurium* infection,[25] but it is unclear whether this is due to an involvement of *tol-1* in a protective immune response or rather due to a defect in cell adherence in the pharynx of the *tol-1* mutant leading to a defect in a physical barrier thus favouring pathogen invasion.

TLRs, as well as a number of other PRR families, in both plants and animals, share a common domain, the leucine rich repeat (LRR) domain. In a recent study, the role in host defences of each of the 14 predicted transmembrane proteins with LRR domains encoded in the *C. elegans* genome, was assayed. Loss-of-function mutants in one gene, *fshr-1*, which encodes a glycopeptide hormone receptor homologue, were found to be more susceptible to infection by Gram positive and Gram negative bacteria. It has yet to be determined if FSHR-1, which is expressed in the intestine, acts as a pathogen receptor or rather functions as a positive modulator of the nematode immune response.[26]

C-type lectins are carbohydrate-binding proteins that can exhibit very narrow ligand specificity. In mammals, a number of C-type lectins have established roles in innate immunity. For example, Dectin-1 is highly expressed on macrophages and recognizes beta-glucan, a component of the fungal cell wall and thereby acts as a PRR.[23] *C. elegans* possesses 278 genes encoding C-type lectins, but it is currently unclear as to whether any of them function as PRRs or rather as effector molecules (see below).

While there is no clear Dectin-1 orthologue in *C. elegans*, there are a number of potential scavenger receptors (SR), another class of protein known to be involved in pathogen recognition in other species.[27] Indeed, there are six proteins homologous to CD36 and Croquemort, members of the SR-B family and one well-characterised SCARF orthologue CED-1. Because of its expression in the intestine throughout development, one of these, C03F11.3, was suggested a number of years ago to be potentially involved in the recognition of microbial molecules.[28] A study published last year supports such an idea, as CED-1/SCARF and C03F11.3/CD36 appear to function in host resistance to *Candida albicans* and *Cryptococcus neoformans* in *C. elegans*.[29] Whether in the nematode these proteins in fact recognize yeast cell wall beta-glucans and act as PRRs has not been formally demonstrated. Alternatively, given CED-1's known function in recognizing dying cells during programmed cell death, it might instead recognize damaged host material and then induce the unfolded protein response (UPR, see below) in an attempt to contain this damage. So we still do not know whether the worm responds through the detection of specific MAMPs or more generally to the cellular damage and stress caused by the pathogen (so-called danger theory[30]) or both. Nevertheless, the finding that *C. elegans* shows distinct immune responses to different pathogens that infect via the same route and have similar levels of virulence,[31,32] clearly supports a model of *C. elegans* specifically recognizing pathogens.

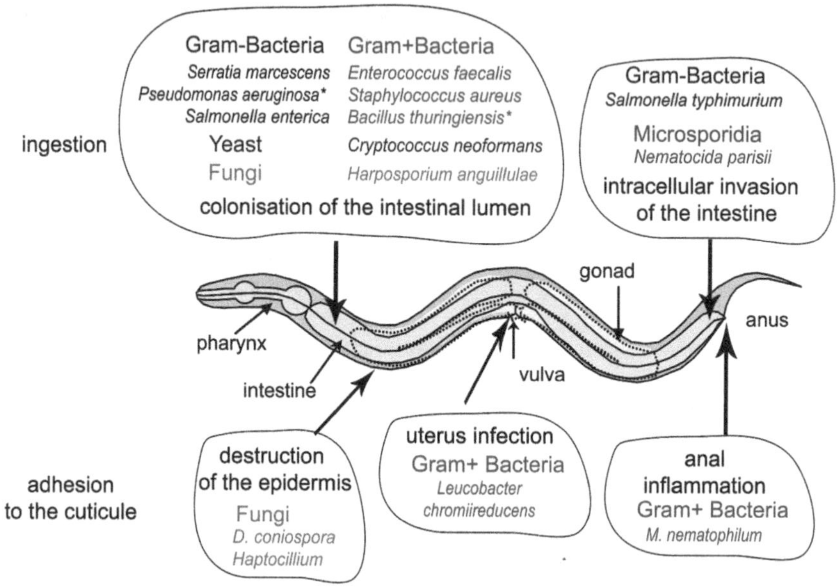

Figure 1. Pathogens of *C. elegans* and their route of infection. Most known pathogens of *C. elegans* are ingested and establish an infection in the intestinal lumen. Certain bacteria produce toxins (*) that can kill the nematode. The fungus *D. coniospora* and the bacteria *M. nematophilum* adhere to the cuticle and infect the nematode via the epidermis. Not all known pathogens of *C. elegans* are shown.

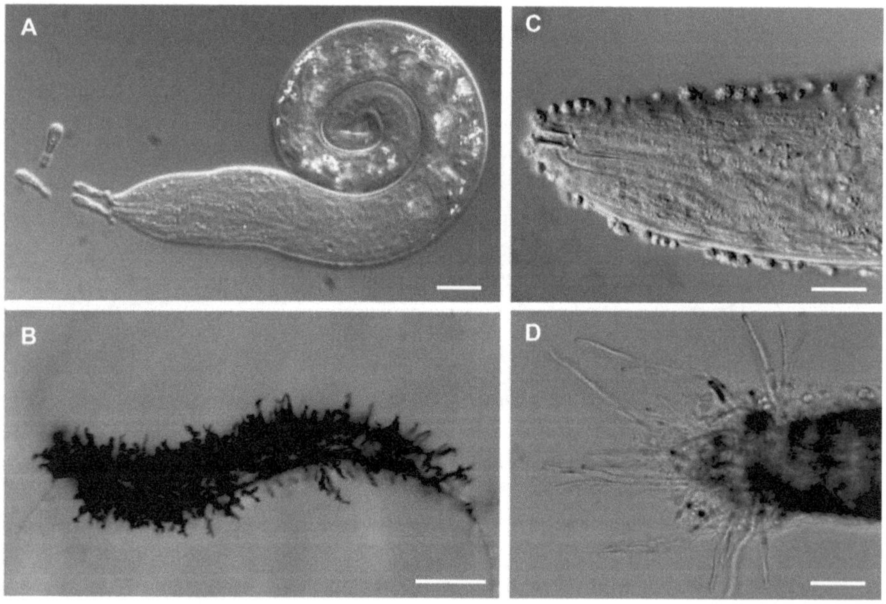

Figure 2. Fungal infection of *C. elegans*. (A and B) *D. coniospora*, (C and D) *Haptocillium*. (A and C) adhesion of the spores to the cuticle after few hours, (B and D) after 2 days fungal hyphae grow out of the worm. Scale bars are 10 μm (A), 100 μm (B) and 50 μm (C and D).

Table 1. Summary of the major signalling pathways in the *C. elegans* Immune System (updated from ref. 86)

Pathway	Tissue	Components	Homologues	References
p38 MAPK	Epidermis	GPA-12, RACK-1	G protein subunits	82
		EGL-8, PLC-3	Phospholipase C	82
		NIPI-3	Tribbles kinase	35
	Epidermis and intestine	TPA-1	Protein kinase C	82,87
		TIR-1	SARM	24,34,88
		NSY-1, SEK-1, PMK-1	MAP kinases	12,35
FSHR-1	Intestine	FSHR-1	G protein coupled receptor	26
ZIP-2	Intestine	ZIP-2	b-zip transcription factor	65
Insulin signalling	Nervous system	INS-7	Insulin-like peptide	76
	Intestine	DAF-2	Insulin receptor	42
		AGE-1	PI3 kinase	42
		AKT-1, AKT-2	Akt kinase	43
		DAF-16	FOXO transcription factor	42
TGF-β	Nervous system	DBL-1	TGF-β	54,55
	epidermis	SMA-6	TGF-βreceptor	55
		SMA-3	SMAD protein	55
Wnt/Hox	Intestine/ Hindgut	BAR-1	β-catenin	61
		EGL-5	Hox transcription factor	61,64
ERK MAPK	Hindgut	LIN-45, MEK-2, MPK-1	ERK MAP kinase	39
		EGL-8	Phospholipase C	89
		SUR-2	Mediator component	39
UPR[1]	Intestine	XBP-1	X box protein	50,52
		HSP-4	Heat shock protein	
	Pharynx	CED-1, C03F11.3	Scavenger receptor	51
Autophagy	Intestine	BEC-1, LGG-1	ATG proteins	16

[1]The recent results of Richardson et al suggest that the primary function of the UPR is to protect against ER stress arising from the increase secretory response. Whether this is the case for Bt toxin[50] and for the noncanonical UPR[51] remains to be seen.

SIGNALLING PATHWAYS INVOLVED IN THE IMMUNE RESPONSE

Even if the manner in which the immune response in *C. elegans* is initiated has not been fully elucidated, several signalling cascades have been described that are activated specifically by certain pathogens (Table 1) and lead to the production of effector molecules which have the potential to destroy pathogens.

Mitogen-Activated Protein Kinase (MAPK) Pathways

Mitogen-activated protein kinase (MAPK) pathways are considered to be the most ancient signal transduction cascades in immunity, found in both animals and plants. Three MAPK cascades are implicated in *C. elegans* immunity. A role for the p38 MAPK pathway in *C. elegans* defence was first revealed in a genetic screen for mutants hypersensitive to infection by *Pseudomonas aeruginosa*.[12] Since then, the p38 MAPK cascade has been shown to protect the worm against other Gram negative and positive bacteria and also fungi and seems to be one of the main signal transduction cascades in the worm's innate immune response.[33-38]

A second MAPK cascade implicated in *C. elegans* immunity is the extracellular signal-regulated kinase (ERK) pathway that is involved in the resistance of *C. elegans* to infection by the Gram positive bacterium *M. nematophilum*.[39] Thirdly, the MAPK kinase protein MEK-1 of the c-Jun N-terminal kinase (JNK) pathway is required for full activation of the p38 MAPK PMK-1, revealing an interaction between the different MAPK pathways.[40]

DAF-2/Insulin-Like Receptor (ILR) Pathway

The DAF-2/insulin-like receptor (ILR) pathway, which involves the Foxo family transcription factor DAF-16, is also clearly important for the immune response of *C. elegans*, but its precise role is less clear. Active DAF-2 retains DAF-16/FOXO in the cytoplasm. In *daf-2* mutants, DAF-16 is predominantly in the nucleus. This results in an increase in DAF-16-dependent gene expression.[41] DAF-2 is well known to be important for the control of lifespan. In addition to being long-lived, *daf-2* mutants show increased resistance to infection by several bacteria.[42] Genetic evidence suggests, however, that the role of DAF-2 in immune signalling is distinct from its role in ageing. Downstream of DAF-2, four known serine threonine kinases, PDK-1, SGK-1, AKT-1 and AKT-2, regulate lifespan in at least 3 independent pathways. Mutants in these four kinases are long-lived, but only *akt-1* and *akt-2* mutants are more resistant to infection by *P. aeruginosa*.[43]

It has been postulated that increased resistance of *daf-2* mutants may be linked to changes in expression for multiple antimicrobial genes.[44] But a direct comparison of the genes transcriptionally regulated by DAF-16/FOXO and the genes regulated after infection reveals a surprisingly limited overlap. Indeed most of the pathogen-induced immunity genes downstream of the PMK-1/p38 pathway are repressed by DAF-16/FOXO.[45,46] Further, in contrast to what is seen upon exposure of *C. elegans* to several different abiotic stresses, nuclear translocation of DAF-16/FOXO has never been detected after infection. Additionally, it has been shown that DAF-16 transcriptionally regulates many genes involved in stress responses.[44,47] It is therefore more probable that the DAF-2/DAF-16 pathway is part of a general stress response rather than a specific immune response. It should, however, be mentioned that the DAF-2/DAF-16 pathway controls multiple aspects of *C. elegans* physiology. Among other things, the DAF-2/DAF-16 pathway appears to influence the worm's pathogen avoidance behaviour, by an as yet undetermined mechanism.[48]

The Unfolded Protein Response

In vertebrates, the endoplasmic reticulum (ER) unfolded protein response (UPR) is particularly important for the development and survival of highly secretory cells such as plasma cells and exocrine gland acinar cells, which secrete immunoglobulins and digestive enzymes, respectively, as well as in dendritic cells and other antigen presenting cells.[49] In *C. elegans*, the UPR has been shown to be protective against *B. thuringiensis*. As detailed more fully below, it is activated by the poreforming toxins through the p38 MAPK pathway.[50] The UPR is also involved in the immune response to *S. typhimurium* and it appears that the scavenger receptor CED-1 is required for the activation of the UPR pathway.[51]

Very recently, the IRE-1-XBP-1 branch of the UPR was shown to be involved in defences against *P. aeruginosa*. Abrogation of *xbp-1* blocks part of the UPR and leads to a disruption of ER morphology. This has no major detrimental effect when worms are cultured under normal conditions, but if they are raised on *P. aeruginosa*, they are unable to complete their development and arrest as larvae. The developmental requirement for XBP-1 is bypassed in mutants of the p38/PMK-1 pathway, such that *xbp-1;pmk-1* double mutants can grow on *P. aeruginosa*. This led the authors to suggest that the production of antimicrobial proteins and peptides places a stress on the ER, which needs to be balanced by the activation of the UPR. In other words, the UPR may protect the host from the potentially damaging effect of its own innate immune against microbes.[52]

TGF-β

A comparison of known targets of the developmentally important transforming growth factor β (TGF-β)/DBL-1 pathway[53] with those upregulated in adults upon infection with *S. marcescens* revealed a number of genes in common, including some encoding lectins and lysozymes.[54] More recently, TGF-β has been shown to be necessary for the regulation of AMP expression after a fungal infection (see below, ref. 55).

Autophagy, Apoptosis and Necrosis

A transcriptome analysis comparing the host genes affected by different bacterial infections revealed that among the genes induced by multiple bacteria were ones required for necrotic cell death. One might interpret this as indicating that necrosis could be a protective host defence mechanism. But when necrosis-defective mutants were tested, they were found to be more resistant to a bacterial infection than wild type.[32] Similarly, a recent study has shown that a loss-of-function mutation in *ced-3* that encodes a caspase involved in apoptosis, also protects the worm against infection with *S. typhimurium*.[16] This could be consistent with a deliberate triggering of necrotic cell death or apoptosis by pathogenic bacteria, as a strategy to increase their effective virulence.

Conversely, autophagy appears to be protective against the intracellular pathogen *S. typhimurium*. Thus *bec-1* or *lgg-1* mutants that are autophagy-defective show an increased susceptibility to infection, with an accumulation of *Salmonella* containing vacuoles (SCV) in the intestinal cell compared to wild type worms. Interestingly, these autophagy-defective mutants suppress the enhanced resistance to *S. typhimurium* infection of *daf-2/Insulin receptor* mutants and of a strain overexpressing DAF-16/FOXO,[16] while at the same time increasing normal life span.[56] This suggests that increased intestinal epithelial cell autophagic activity may partially underlie the resistance of *daf-2* mutants to intracellular pathogens.[16] It will be interesting to establish whether intestinal cell autophagy also contributes to the resistance of *C. elegans* to extracellular pathogens.

TRANSCRIPTION FACTORS INVOLVED IN THE IMMUNE RESPONSE

The transcription factor NF-kB links the reception and transmission of an infection signal to the expression of effector proteins in vertebrates and insects. Therefore, its absence from the nematode genome is remarkable and opens the possibility of studying alternative mechanisms of transcriptional regulation potentially conserved in other species. For example, having shown that most of the effectors induced by *P. aeruginosa* infection in the intestine of *C. elegans* are under the control of the GATA transcription factor ELT-2, Tan and colleagues were able to show that ELT-2 increases host resistance to intestinal infection with bacterial[45] and this was subsequently also shown to be the case for intestinal fungal pathogens.[57] Another GATA transcription factor ELT-3 contributes to the proper expression in the epidermis of AMP genes. But it was also shown to be required for the expression in the epidermis of genes important for osmoregulation, not directly related to innate immunity. This led to the suggestion that this GATA TF acts as a more generic transcription factor in the epidermis.[58] This latter conclusion is in line with a study published this year showing that ELT-3 in the epidermis, but also ELT-2 in the intestine, are essential for tissue-specific activation of osmosensitive gene expression and promote survival under osmotically stressful conditions.[59]

Just as the response to infection and osmotic adaptation may be controlled via regulation of common tissue-specific GATA transcription factors, so too is there a link between innate immunity and temperature adaptation. A mild heat shock has been shown to increase the resistance of *C. elegans* to infection with Gram positive and Gram negative bacteria. This resistance is independent of the p38 MAPK/PMK-1 pathway and requires the heat shock factor HSF-1 and heat shock proteins. The forkhead transcription factor DAF-16 is positively regulated by heat shock and is required for the induction of HSF-1 thus linking the heat shock pathway to the DAF-2/ILR pathway.[60]

The transcriptional cofactor BAR-1/β-catenin and the homeobox gene *egl-5* have been shown to play a role in *C. elegans* intestinal epithelial immunity and resistance to *S. aureus*,[61] in addition to its established role in cell fate decision during development.[62] EGL-5 is also necessary in the hindgut to induce swelling upon *M. nematophilum* infection.[63,64] Interestingly, the human homologues of EGL-5, HOXA9 and HOXA10 dampen NFkB-dependent TLR2 signalling, suggesting a conserved role in innate immune defence.[61]

Further insights into the complexity of innate immune signalling were obtained in a study that used a gene specifically induced by virulent *P. aeruginosa* strains called "infection response gene 1" (*irg-1*). It was chosen as its expression is independent of the PMK-1/p38 pathway. Several candidates required for the full induction of *irg-1* were identified from a screen of more than 300 transcription factors, based on RNA interference. Among them, most interest was focused on the bZIP transcription factor *zip-2*. It was shown to be required for the induction not only of *irg-1* but of several putative effector genes, in all cases independent of the PMK-1/p38 pathway and also of FSHR-1. Certain target genes, such as *irg-3*, were demonstrated to be regulated by yet another pathway, involving neither *zip*-2, nor p38, nor FSHR-1, suggesting that at least 4 independent pathways contribute to pathogen resistance in the *C. elegans* intestine upon *P. aeruginosa* infection.[65]

EFFECTOR MOLECULES INVOLVED IN THE IMMUNE RESPONSE

Antimicrobial Peptides

C. elegans possesses different types of antimicrobial proteins and several classes of AMP. Among them are the mollusc defensin/mycitin-like peptides (ABF-1 to ABF-6). One of them, ABF-2 has demonstrated antimicrobial activity.[66] It is strongly upregulated upon prolonged exposure to *S. typhimurium*.[67] Additionally, there are the neuropeptide-like proteins (NLPs) and the caenacins (CNC),[24,55,58] that are rapidly and strongly induced by fungal infection and that will be discussed in detail below.

Caenopores

Caenopores is the name given to a number of *C. elegans* proteins that contain the saposin domain, common to mammalian NK-lysin and granulysin and the protozoan amoebapores.[68] Members of this family were first identified more than a decade ago, when two among them, SPP-1 and SPP-5, were shown to have a bactericidal function.[69] SPP-5 is constitutively expressed and kills bacteria by permeabilising their membrane. Interestingly, another member of this family, SPP-3, is expressed both upon starvation and contact with certain bacteria, thus suggesting a potential link between nutrition and immunity.[68]

Lysozymes

Lysozymes are another class of molecules known to be involved in immune defence in many species. In contrast to arthropods, *C. elegans* does not have C-type lysozymes, but possesses a repertoire of 15 genes, falling into 3 classes, two related to protist lysozymes and one specific to invertebrates.[70] Certain lysozymes, including *lys-7*, are induced upon bacterial challenge and their inactivation has been shown to render worms more susceptible to *M. nematophilum* and *P. aeruginosa*.[54,71,72] The expression of other lysozymes, mainly from the invertebrate class, has been reported to be repressed upon infection. Although the exact function of these latter genes still remains to be determined, the amplification of the lysozyme family by gene duplication in the nematode is a clear example of evolutionary function diversification.[70]

Lectins

Lectins are also involved in innate defences in many species and can be involved in pathogen recognition but also in immune effector functions. In *C. elegans*, there are a very large number of lectin genes, including 11 galectins, in the *lec* gene class and 265 C-type lectins in the *clec* gene class. The expression of some *lec* and *clec* genes is up-regulated by several pathogens; for others their induction appears to be relatively pathogen-specific. This differential upregulation has led to the suggestion that they might be an element conferring specificity to the immune response of *C. elegans*.[31,32,54,72] In some cases, they have demonstrable role in host defence. Inactivation of some lectins, for example, renders worms more susceptible to *M. nematophilum*.[72] Unfortunately, there is currently little direct functional information about most of the large number of lectins. One exception is the glycolipid-binding galectin LEC-8 that has been shown recently to play a role in host defence against *B. thuringiensis* infection by competitively inhibiting the binding of the toxin Cry5B to its host glycolipid receptor.[73]

Reactive Oxygen Species

In addition to its arsenal of antimicrobial proteins, *C. elegans* also has the capacity to produce bactericidal reactive oxygen species (ROS) in response to exposure to pathogens. This has been best characterised in the case of infection with the Gram-positive pathogen *Enterococcus faecalis*, which provokes ROS production via the action of the dual oxidase BLI-3. ROS are relatively unspecific in their capacity to kill cells. They affect invading micro-organisms, but can also damage host tissues. As a result, increased levels of ROS triggers a protective stress response in the host. This involves up-regulation of the superoxide dismutase SOD-3 and the catalase CTL-2, which sequential detoxify the ROS. Both enzymes are targets of DAF-16. Indeed, their combined action is part of the mechanism underlying the increased resistance to infection of *daf-2* mutants. Consistent with a protective role for oxidative stress, the addition of compounds that scavenge ROS increase the sensitivity of *C. elegans* to infection with *E. faecalis*.[74,75]

MODULATION OF THE IMMUNE RESPONSE BY THE NERVOUS SYSTEM

ROS can act in a relatively unspecific manner over a distance to affect cells not in direct contact with a pathogen. Several recent papers attempt to provide evidence for more refined noncell autonomous mechanisms involved in controlling the response of *C. elegans* to infection. These examples involve the nervous system. Kawli and Tan demonstrated that the release of dense core vesicles (DCVs) from neurons suppresses the intestinal immune response of *C. elegans* to *P. aeruginosa* and that this neuronal control mechanisms is mediated in the intestine by the DAF-2/ILR pathway. The insulin-like peptide INS-7 has been proposed to provide the link between DCV release in neurons and the DAF-2/ILR pathway in the intestine.[76] In a second study, Stryer et al reported that NPR-1, a G-protein-coupled receptor related to mammalian neuropeptide Y receptors, functions to suppress innate immunity to *P. aeruginosa*, by acting upstream of the p38 MAPK signalling cascade.[77] It should be noted, however, that the reported changes in gene expression seen in the *npr-1* mutant are minimal compared to those seen upon infection by *P. aeruginosa*. Further, the results of Stryer et al have been contradicted by a more recent study showing that the difference in susceptibility in the *npr-1* mutant strain is due to its well-characterised behaviour of clumping, a behaviour which is linked to sensing oxygen concentration.[78] In a final example, as mentioned above and described more fully below, the expression of *cnc* AMP genes appears to involve regulation by neuronally-derived TGF-β.[55]

IMMUNE RESPONSE TO PORE-FORMING TOXINS

Some bacteria, such as *B. thuringiensis*, are able to produce multiple toxins that target host cells. These pore-forming toxins (PFTs) make holes in membranes and alone can cause the death of *C. elegans*. Not surprisingly, the nematode has evolved mechanisms to protect itself from the nefarious effects of PFTs. It has been demonstrated that the toxin directly binds glycolipids and that the major mechanism for PFT resistance in *C. elegans* entails a loss of glycolipid carbohydrates.[79] Moreover, in the case of the *B. thuringensis* Cry5B toxin, this response involves the PMK-1/p38 and cJunN-terminal kinase-like pathways. The mechanism has an inbuilt component of amplification, since these kinases are also transcriptionally upregulated by Cry5B.[38] Activation of the p38 MAPK pathway by Cry5B activates the IRE-1 UPR pathway. IRE-1 induces an alternative splicing of the transcription factor *xbp-1*. The resultant infection-specific transcript then drives the expression of a number of target genes which protect against the effects of the PFT.[50] There is some cross-talk between the mechanisms involved in the response to PFTs and those required to tolerate conditions of low oxygen. Indeed, activation of the hypoxia pathway also increases resistance against PFTs. Resistance to hypoxia also involves the UPR and in common with the response to PFTs, it is mediated by the transcription factor HIF-1.[80]

EPIDERMAL IMMUNE RESPONSE TO THE FUNGUS *DRECHMERIA CONIOSPORA*

Most of the bacterial or fungal pathogens described in the previous sections infect the worm through the intestinal lumen, which is primarily programmed for destroying microbes as part of normal feeding and digestion. In some instances, this can blur the distinction between an immune response and the consequence of a change of diet. Other pathogens infect worms via the cuticle. Among them, *Drechmeria coniospora* is a natural fungal pathogen of nematodes, including *C. elegans* (M.A Felix, personal communication). *D. coniospora* pierces the worm's cuticle and its hyphae proliferate first in the epidermis, then throughout the organism. This provokes a complex transcriptional response involving, among others, the upregulation of AMP genes.[24,58] These include members of two phylogenetically-related families, the *nlp* and *cnc* genes, which are present in clusters in the genome. Phylogenetic analysis shows that these AMP genes, arose through recent duplication and diversification, have been under selective pressure during evolution and are thus likely to be important in nature for the survival of *C. elegans*.[58]

As this fungal infection involves breaching the cuticle and epidermis, the question of whether *C. elegans* epidermis responds to physical injury was addressed. Needle pricking or laser wounding not only provokes an up-regulation of AMP genes, but also triggers a cellular wound-healing and scarring mechanism.[35] These two processes appears to be independent, but are normally kept in check by a common negative regulator, the nematode Death-associated protein kinase (DAPK).[81] While infection and injury induces the expression of both *nlp* and *cnc* family genes, the molecular mechanisms that regulate each class appear to be strikingly different.

Cell-Autonomous Regulation of *nlp* Gene Expression

Through direct genetic screens, proteomics and a candidate gene approach, two signalling pathways required for the regulation of *nlp* gene expression in the epidermis have been described. One is specific for infection, the second is also activated by wounding. Both pathways converge on a protein kinase C, TPA-1, which, in turn, acts upstream of the PMK-1/p38 pathway (Fig. 3). The proximal elements of the pathways are, however, distinct. The former requires the conserved protein kinase Tribbles, NIPI-3, while the latter involves heterotrimeric G proteins acting upstream of a phospholipase C. All of the characterised components act in a cell-autonomous manner to control *nlp* gene expression in the epidermis.[35,82]

For the time being, the identity of the putative G-protein coupled receptor (GPCR) that activates the heterotrimeric G proteins is unknown, nor is it known how NIPI-3 is activated. There are, nonetheless, marked similarities between the molecular architecture underlying these pathways and the organisation of the signalling pathways that regulate the innate immune response both in Drosophila and in vertebrates. These led to the speculation that the innate immune response to *D. coniospora* arose from a GPCR-dependent mechanism

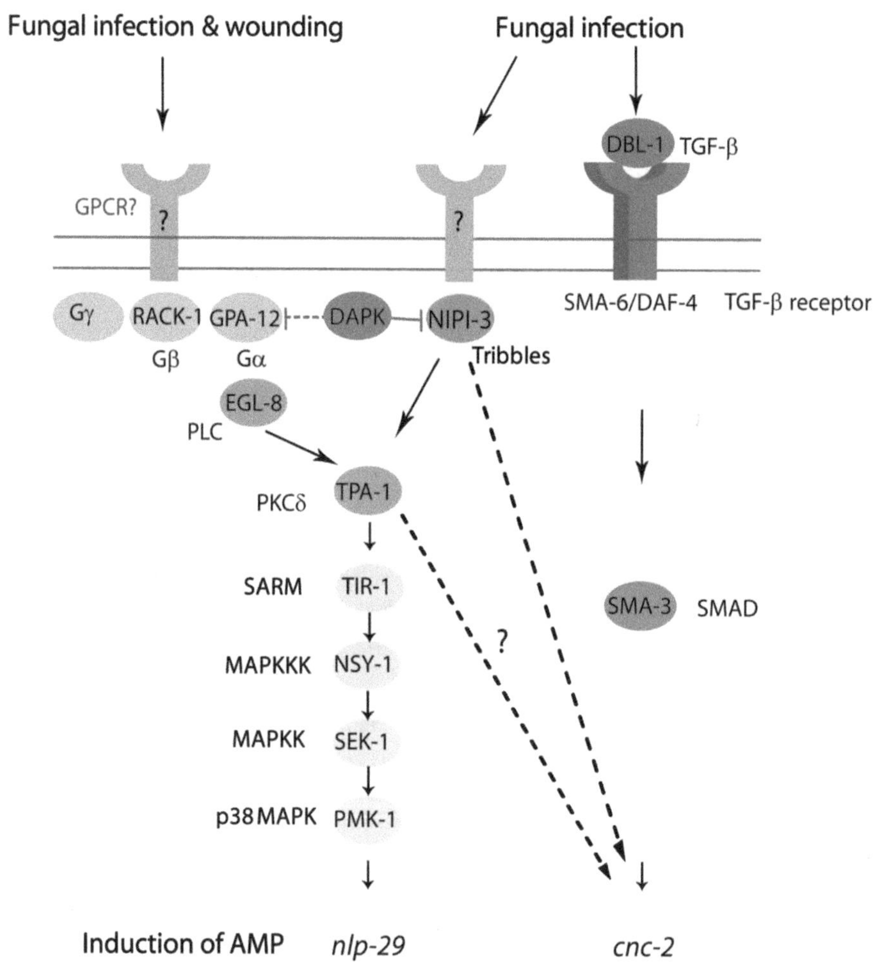

Figure 3. Schematic representation of the different signalling pathways and their components involved in the induction of antimicrobial peptides expression upon *D. coniospora* infection. Expression of the *nlp* genes is controlled by a PKC/SARM/p38 MAPK pathway and expression of *cnc* genes is controlled by a TGF-β pathway.

used to detect cellular damage, which was subsequently ameliorated by the addition of a pathogen-specific detection mechanism.

Paracrine Regulation of *cnc* Gene Expression

Surprisingly, the *cnc* gene family is regulated in an entirely different fashion, as it is largely independent of the PMK-1/p38 pathway. Induction of the *cnc* genes after infection requires the ligand DBL-1/TGF-β, produced by certain neuronal cells, which acts via its normal receptor SMA-6/DAF-4 expressed on epidermal cells (Fig. 3). The resultant signal is transduced by a noncanonical TGF-β pathway that does not involve all three SMAD proteins, hitherto considered to be indispensable for TGF-β signalling in *C. elegans*. Artificially altering the level of expression of the TGF-β modulated the strength of *cnc* induction after infection but did not affect the basal level of *cnc* gene expression, suggesting that infection triggers the conversion of an inactive precursor into an active TGF-β. This is reminiscent of the proteolytic activation of Spaetzle required for triggering the Toll pathway during the immune response in Drosophila. As the canonical TGF-β pathway undoubtedly existed before the appearance of the *cnc* genes, this is another clear example of the co-option and adaptation of a pre-existing signalling pathway for use by the innate immune system.[55]

CONCLUSION

In the last 10 years, our knowledge of the way *C. elegans* defends itself against microbes has greatly expanded. The worm clearly shows specific immune responses to pathogens and their toxins. The signalling processes and effector molecules involved in this response have been elucidated to some extent. Interestingly, the main signalling pathways involved in the *C. elegans* immune response are conserved in other species, in part because they also play important developmental roles.

On the other hand, much remains to be learnt about the way the nematode recognizes pathogens and the receptors involved in this process, as well as the temporal and spatial dynamics of the downstream signalling processes. For the latter, techniques to visualise proteins and transcripts at the single molecule level within living cells must be improved.

It is important to note how deeply the defence mechanisms are embedded in the physiology of the organism. We described how important are several aspects of behaviour, digestion or stress resistance to defence. Other studies have shown connections with reproduction, where sterile mutants are more resistant to bacterial infection in a DAF-16/FOXO dependent manner.[83] The same is true for osmotic stress since several osmotic mutants are more resistant to fungal infection and some immune effectors genes are induced upon osmotic stress although apparently regulated by a dedicated pathway.[58,84] Lastly, lipid synthesis has been shown to be required for the basal activity of the PMK-1/p38 pathway that influences resistance against *P. aeruginosa*.[71] Unbiased forward genetics screens combined with global functional genomic approaches will help to unravel the complex and intricate biology that underlies successful host defences in *C. elegans*.

Another powerful approach to understand more fully how the innate immune system works is to take advantage of the fact that pathogens can specifically interfere with the defence mechanisms of the host. For the moment there are only a few examples in

C. elegans, such as the down-regulation of several intestinal immune effectors through activation of DAF-2 by *P. aeruginosa* which requires bacterial virulence factors controlled by the signalling molecule GacA.[85] The study of these interactions will also teach us more about innate immune defence and the virulence strategies that pathogens have developed to escape host immunity.

ACKNOWLEDGEMENT

We thank J. Ewbank for insightful comments on the manuscript. Work in our laboratory is supported by institutional grants from INSERM and the CNRS and grants from the ANR and FRM. IE is supported by a Marie Curie fellowship.

REFERENCES

1. Barrière A, Felix MA. Isolation of C. elegans and related nematodes. The C. elegans Research Community: WormBook; 2006.
2. Fares H, Greenwald I. Genetic analysis of endocytosis in Caenorhabditis elegans: coelomocyte uptake defective mutants. Genetics 2001; 159(1):133-45.
3. Schulenburg H, Ewbank JJ. The genetics of pathogen avoidance in Caenorhabditis elegans. Mol Microbiol 2007; 66(3):563-70.
4. Pradel E, Zhang Y, Pujol N et al. Detection and avoidance of a natural product from the pathogenic bacterium Serratia marcescens by Caenorhabditis elegans. Proc Natl Acad Sci USA 104(7):2295-300.
5. Pujol N, Link EM, Liu LX et al. A reverse genetic analysis of components of the Toll signalling pathway in Caenorhabditis elegans. Curr Biol 2001; 11(11):809-21.
6. Remy JJ, Hobert O. An interneuronal chemoreceptor required for olfactory imprinting in C. elegans. Science (New York, NY) 2005; 309(5735):787-90.
7. Zhang Y, Lu H, Bargmann CI. Pathogenic bacteria induce aversive olfactory learning in Caenorhabditis elegans. Nature 2005; 438(7065):179-84.
8. Wes PD, Bargmann CI. C. elegans odour discrimination requires asymmetric diversity in olfactory neurons. Nature 2001; 410(6829):698-701.
9. Chuang CF, Bargmann CI. A Toll-interleukin 1 repeat protein at the synapse specifies asymmetric odorant receptor expression via ASK1 MAPKKK signaling. Genes Dev 2005; 19(2):270-81.
10. Shivers RP, Kooistra T, Chu SW et al. Tissue-specific activities of an immune signaling module regulate physiological responses to pathogenic and nutritional bacteria in C. elegans. Cell Host Microbe 2009; 6(4):321-30.
11. Labrousse A, Chauvet S, Couillault C et al. Caenorhabditis elegans is a model host for Salmonella typhimurium. Curr Biol 2000; 10(23):1543-5.
12. Kim DH, Feinbaum R, Alloing G et al. A conserved p38 MAP kinase pathway in Caenorhabditis elegans innate immunity. Science (New York, NY) 2002; 297(5581):623-6.
13. Kurz CL, Chauvet S, Andres E et al. Virulence factors of the human opportunistic pathogen Serratia marcescens identified by in vivo screening. EMBO J 2003; 22(7):1451-60.
14. Borgonie G, Claeys M, Leyns F et al. Effect of nematicidal Bacillus thuringiensis strains on free-living nematodes. 1. Light microscopic observations, species and biological stage specificity and identification of resistant mutants of Caenorhabditis elegans. Fundam appl Nematol 1996; 19(4):391-8.
15. Troemel ER, Felix MA, Whiteman NK et al. Microsporidia are natural intracellular parasites of the nematode Caenorhabditis elegans. PLoS Biol 2008; 6(12):2736-52.
16. Jia K, Thomas C, Akbar M et al. Autophagy genes protect against Salmonella typhimurium infection and mediate insulin signaling-regulated pathogen resistance. Proc Natl Acad Sci USA 2009; 106(34):14564-9.
17. Hodgkin J, Kuwabara PE, Corneliussen B. A novel bacterial pathogen, Microbacterium nematophilum, induces morphological change in the nematode C. elegans. Curr Biol 2000; 10(24):1615-8.
18. Muir RE, Tan MW. Leucobacter chromiireducens subsp. solipictus Exerts Virulence on Caenorhabditis elegans, Characterization of a Novel Host-Pathogen Interaction. Appl Environ Microbiol 2008.

19. Barron GL. Nematophagous destroying fungi. Topics in Mycobiology [serial on the Internet] 1977; 1.
20. Jansson HB. Adhesion of conidia of Drechmeria coniospora to Caenorhabditis elegans wild type and mutants. J Nematol 1994; 26:430-5.
21. Gallagher LA, Manoil C. Pseudomonas aeruginosa PAO1 kills Caenorhabditis elegans by cyanide poisoning. J Bacteriol 2001; 183(21):6207-14.
22. Akira S, Uematsu S, Takeuchi O. Pathogen recognition and innate immunity. Cell 2006; 124(4):783-801.
23. Palm NW, Medzhitov R. Pattern recognition receptors and control of adaptive immunity. Immunol Rev 2009; 227(1):221-33.
24. Couillault C, Pujol N, Reboul J et al. TLR-independent control of innate immunity in Caenorhabditis elegans by the TIR domain adaptor protein TIR-1, an ortholog of human SARM. Nature immunology 2004; 5:488-94.
25. Tenor JL, Aballay A. A conserved Toll-like receptor is required for Caenorhabditis elegans innate immunity. EMBO Rep 2008; 9(1):103-9.
26. Powell JR, Kim DH, Ausubel FM. The G protein-coupled receptor FSHR-1 is required for the Caenorhabditis elegans innate immune response. Proc Natl Acad Sci USA 2009; 106(8):2782-7.
27. Gordon S. Pattern recognition receptors. Doubling up for the innate immune response. Cell 2002; 111(7):927-30.
28. Nicholas HR, Hodgkin J. Responses to infection and possible recognition strategies in the innate immune system of Caenorhabditis elegans. Mol Immunol 2004; 41(5):479-93.
29. Means TK, Mylonakis E, Tampakakis E et al. Evolutionarily conserved recognition and innate immunity to fungal pathogens by the scavenger receptors SCARF1 and CD36. J Exp Med 2009.
30. Matzinger P. The danger model: a renewed sense of self. Science (New York, NY) 2002; 296(5566):301-5.
31. Schulenburg H, Hoeppner MP, Weiner J 3rd et al. Specificity of the innate immune system and diversity of C-type lectin domain (CTLD) proteins in the nematode Caenorhabditis elegans. Immunobiology 2008; 213(3-4):237-50.
32. Wong D, Bazopoulou D, Pujol N et al. Genome-wide investigation reveals pathogen-specific and shared signatures in the response of Caenorhabditis elegans to infection. Genome Biol 2007; 8(9):R194.
33. Sifri CD, Begun J, Ausubel FM et al. Caenorhabditis elegans as a model host for Staphylococcus aureus pathogenesis. Infect Immun 2003; 71(4):2208-17.
34. Liberati NT, Fitzgerald KA, Kim DH et al. Requirement for a conserved Toll/interleukin-1 resistance domain protein in the Caenorhabditis elegans immune response. Proc Natl Acad Sci USA 2004; 101(17):6593-8.
35. Pujol N, Cypowyj S, Ziegler K et al. Distinct innate immune responses to infection and wounding in the C. elegans epidermis. Curr Biol 2008; 18(7):481-9.
36. Begun J, Gaiani JM, Rohde H et al. Staphylococcal biofilm exopolysaccharide protects against Caenorhabditis elegans immune defenses. PLoS Pathog 2007; 3(4):e57.
37. Aballay A, Drenkard E, Hilbun LR et al. Caenorhabditis elegans innate immune response triggered by Salmonella enterica requires intact LPS and is mediated by a MAPK signaling pathway. Curr Biol 2003; 13(1):47-52.
38. Huffman DL, Abrami L, Sasik R et al. Mitogen-activated protein kinase pathways defend against bacterial pore-forming toxins. Proc Natl Acad Sci USA 2004; 101(30):10995-1000.
39. Nicholas HR, Hodgkin J. The ERK MAP kinase cascade mediates tail swelling and a protective response to rectal infection in C. elegans. Curr Biol 2004; 14(14):1256-61.
40. Kim DH, Liberati NT, Mizuno T et al. Integration of Caenorhabditis elegans MAPK pathways mediating immunity and stress resistance by MEK-1 MAPK kinase and VHP-1 MAPK phosphatase. Proc Natl Acad Sci USA 2004; 101(30):10990-4.
41. Lin K, Hsin H, Libina N et al. Regulation of the Caenorhabditis elegans longevity protein DAF-16 by insulin/IGF-1 and germline signaling. Nat Genet 2001; 28(2):139-45.
42. Garsin DA, Villanueva JM, Begun J et al. Long-lived C. elegans daf-2 mutants are resistant to bacterial pathogens. Science (New York, NY) 2003; 300(5627):1921.
43. Evans EA, Chen WC, Tan MW. The DAF-2 Insulin-like signaling pathway independently regulates aging and immunity in C. elegans. Aging Cell 2008; 7(6):879-93.
44. Murphy CT, McCarroll SA, Bargmann CI et al. Genes that act downstream of DAF-16 to influence the lifespan of Caenorhabditis elegans. Nature 2003; 424(6946):277-83.
45. Shapira M, Hamlin BJ, Rong J et al. A conserved role for a GATA transcription factor in regulating epithelial innate immune responses. Proc Natl Acad Sci USA 2006; 103(38):14086-91.
46. Troemel ER, Chu SW, Reinke V et al. p38 MAPK regulates expression of immune response genes and contributes to longevity in C. elegans. PLoS Genetics 2006; 2(11):e183.

47. Alper S, McBride SJ, Lackford B et al. Specificity and complexity of the Caenorhabditis elegans innate immune response. Mol Cell Biol 2007; 27(15):5544-53.
48. Hasshoff M, Bohnisch C, Tonn D et al. The role of Caenorhabditis elegans insulin-like signaling in the behavioral avoidance of pathogenic Bacillus thuringiensis. FASEB J 2007; 21(8):1801-12.
49. Todd DJ, Lee AH, Glimcher LH. The endoplasmic reticulum stress response in immunity and autoimmunity. Nat Rev Immunol 2008; 8(9):663-74.
50. Bischof LJ, Kao CY, Los FC et al. Activation of the unfolded protein response is required for defenses against bacterial pore-forming toxin in vivo. PLoS Pathog 2008; 4(10):e1000176.
51. Haskins KA, Russell JF, Gaddis N et al. Unfolded protein response genes regulated by CED-1 are required for Caenorhabditis elegans innate immunity. Dev Cell 2008; 15(1):87-97.
52. Richardson CE, Kooistra T, Kim DH. An essential role for XBP-1 in host protection against immune activation in C. elegans. Nature 2010; 463(6784):1092-5.
53. Mochii M, Yoshida S, Morita K et al. Identification of transforming growth factor-beta- regulated genes in Caenorhabditis elegans by differential hybridization of arrayed cDNAs. Proc Natl Acad Sci USA 1999; 96(26):15020-5.
54. Mallo GV, Kurz CL, Couillault C et al. Inducible antibacterial defense system in C. elegans. Curr Biol 2002; 12(14):1209-14.
55. Zugasti O, Ewbank JJ. Neuroimmune regulation of antimicrobial peptide expression by a noncanonical TGF-beta signaling pathway in Caenorhabditis elegans epidermis. Nature immunology 2009; 10(3):249-56.
56. Hashimoto Y, Ookuma S, Nishida E. Lifespan extension by suppression of autophagy genes in Caenorhabditis elegans. Genes Cells 2009; 14(6):717-26.
57. Kerry S, Tekippe M, Gaddis NC et al. GATA transcription factor required for immunity to bacterial and fungal pathogens. PLoS ONE 2006; 1:e77.
58. Pujol N, Zugasti O, Wong D et al. Anti-fungal innate immunity in C. elegans is enhanced by evolutionary diversification of antimicrobial peptides. PLoS Pathog 2008; 4(7):e1000105.
59. Rohlfing AK, Miteva Y, Hannenhalli S et al. Genetic and physiological activation of osmosensitive gene expression mimics transcriptional signatures of pathogen infection in C. elegans. PLoS One 2010; 5(2):e9010.
60. Singh V, Aballay A. Heat-shock transcription factor (HSF)-1 pathway required for Caenorhabditis elegans immunity. Proc Natl Acad Sci USA 2006; 103(35):13092-7.
61. Irazoqui JE, Ng A, Xavier RJ et al. Role for beta-catenin and HOX transcription factors in Caenorhabditis elegans and mammalian host epithelial-pathogen interactions. Proc Natl Acad Sci USA 2008; 105(45):17469-74.
62. Chisholm A. Control of cell fate in the tail region of C. elegans by the gene egl-5. Development 1991; 111(4):921-32.
63. Nicholas HR, Hodgkin J. The C. elegans Hox gene egl-5 is required for correct development of the hermaphrodite hindgut and for the response to rectal infection by Microbacterium nematophilum. Dev Biol 2009; 329(1):16-24.
64. Gravato-Nobre MJ, Nicholas HR, Nijland R et al. Multiple genes affect sensitivity of Caenorhabditis elegans to the bacterial pathogen Microbacterium nematophilum. Genetics 2005; 171(3):1033-45.
65. Estes KA, Dunbar TL, Powell JR et al. bZIP transcription factor zip-2 mediates an early response to Pseudomonas aeruginosa infection in Caenorhabditis elegans. Proc Natl Acad Sci USA 2010; 107(5):2153-8.
66. Kato Y, Aizawa T, Hoshino H et al. abf-1 and abf-2, ASABF-type antimicrobial peptide genes in Caenorhabditis elegans. Biochem J 2002; 361(Pt 2):221-30.
67. Alegado RA, Tan MW. Resistance to antimicrobial peptides contributes to persistence of Salmonella typhimurium in the C. elegans intestine. Cell Microbiol 2008; 10(6):1259-73.
68. Roeder T, Stanisak M, Gelhaus C et al. Caenopores are antimicrobial peptides in the nematode Caenorhabditis elegans instrumental in nutrition and immunity. Developmental and comparative immunology 2009.
69. Banyai L, Patthy L. Amoebapore homologs of Caenorhabditis elegans. Biochim Biophys Acta 1998; 1429(1):259-64.
70. Schulenburg H, Boehnisch C. Diversification and adaptive sequence evolution of Caenorhabditis lysozymes (Nematoda: Rhabditidae). BMC Evol Biol 2008; 8:114.
71. Nandakumar M, Tan MW. Gamma-linolenic and stearidonic acids are required for basal immunity in Caenorhabditis elegans through their effects on p38 MAP kinase activity. PLoS Genet 2008; 4(11):e1000273.
72. O'Rourke D, Baban D, Demidova M et al. Genomic clusters, putative pathogen recognition molecules and antimicrobial genes are induced by infection of C. elegans with M. nematophilum. Genome Res 2006; 16(8):1005-16.

73. Ideo H, Fukushima K, Gengyo-Ando K et al. A Caenorhabditis elegans glycolipid-binding galectin functions in host defense against bacterial infection. J Biol Chem 2009; 284(39):26493-501.
74. Chavez V, Mohri-Shiomi A, Maadani A et al. Oxidative stress enzymes are required for DAF-16-mediated immunity due to generation of reactive oxygen species by Caenorhabditis elegans. Genetics 2007; 176(3):1567-77.
75. Chavez V, Mohri-Shiomi A, Garsin DA. Ce-Duox1/BLI-3 generates reactive oxygen species as a protective innate immune mechanism in Caenorhabditis elegans. Infect Immun 2009; 77(11):4983-9.
76. Kawli T, Tan MW. Neuroendocrine signals modulate the innate immunity of Caenorhabditis elegans through insulin signaling. Nature immunology 2008; 9(12):1415-24.
77. Styer KL, Singh V, Macosko E et al. Innate immunity in Caenorhabditis elegans is regulated by neurons expressing NPR-1/GPCR. Science (New York, NY) 2008; 322(5900):460-4.
78. Reddy KC, Andersen EC, Kruglyak L et al. A polymorphism in npr-1 is a behavioral determinant of pathogen susceptibility in C. elegans. Science (New York, NY) 2009; 323(5912):382-4.
79. Griffitts JS, Haslam SM, Yang T et al. Glycolipids as receptors for Bacillus thuringiensis crystal toxin. Science (New York, NY.) 2005; 307(5711):922-5.
80. Bellier A, Chen CS, Kao CY et al. Hypoxia and the hypoxic response pathway protect against pore-forming toxins in C. elegans. PLoS Pathog 2009; 5(12):e1000689.
81. Tong A, Lynn G, Ngo V et al. Negative regulation of Caenorhabditis elegans epidermal damage responses by death-associated protein kinase. Proc Natl Acad Sci USA 2009; 106(5):1457-61.
82. Ziegler K, Kurz CL, Cypowyj S et al. Antifungal innate immunity in C. elegans: PKCdelta links G protein signaling and a conserved p38 MAPK cascade. Cell Host Microbe 2009; 5(4):341-52.
83. Miyata S, Begun J, Troemel ER et al. DAF-16-dependent suppression of immunity during reproduction in Caenorhabditis elegans. Genetics 2008; 178(2):903-18.
84. Lee KZ, Kniazeva M, Han M et al. The fatty acid synthase fasn-1 acts upstream of WNK and Ste20/GCK-VI kinases to modulate antimicrobial peptide expression in C. elegans epidermis. Virulence 2010; 1(3).
85. Evans EA, Kawli T, Tan MW. Pseudomonas aeruginosa suppresses host immunity by activating the DAF-2 insulin-like signaling pathway in Caenorhabditis elegans. PLoS Pathog 2008; 4(10):e1000175.
86. Partridge FA, Gravato-Nobre MJ, Hodgkin J. Signal transduction pathways that function in both development and innate immunity. Dev Dyn 2010.
87. Ren M, Feng H, Fu Y et al. Protein kinase D (DKF-2), a diacylglycerol effector, is an essential regulator of C. elegans innate immunity. Immunity 2009; 30(4):521-32.
88. Kurz CL, Shapira M, Chen K et al. Caenorhabditis elegans pgp-5 is involved in resistance to bacterial infection and heavy metal and its regulation requires TIR-1 and a p38 map kinase cascade. Biochem Biophys Res Commun 2007; 363(2):438-43.
89. Yook K, Hodgkin J. Mos1 Mutagenesis Reveals a Diversity of Mechanisms Affecting Response of Caenorhabditis elegans to the Bacterial Pathogen Microbacterium nematophilum. Genetics 2007; 175(2):681-97.

CHAPTER 7

IMMUNOCOMPETENT MOLECULES AND THEIR RESPONSE NETWORK IN HORSESHOE CRABS

Shun-ichiro Kawabata

Department of Biology, Faculty of Sciences, Kyushu University, Fukuoka, Japan
Email: skawascb@kyudai.jp

Abstract: Horseshoe crab hemocyte selectively responds to bacterial lipopolysaccharides (LPS), which depends critically on the proteolytic activity of the LPS-responsive serine protease zymogen factor C. In response to stimulation by LPS, the hemocyte secretes several kinds of immunocompetent proteins. The coagulation cascade triggered by LPS or β-1,3-D-glucans (BDG) results in the formation of coagulin fibrils that are subsequently stabilized by transglutaminase (TGase)-dependent cross-linking. Invading pathogens are recognized and agglutinated by lectins and then killed by antimicrobial peptides. Moreover, LPS-triggered hemocyte exocytosis is enhanced by a feedback mechanism in which the antimicrobial peptides serve as endogenous mediators. Factor C also acts as an LPS-sensitive complement C3 convertase. In addition, a sub-cuticular epidermis-derived protein forms a TGase-stabilized mesh at sites of injury. Horseshoe crabs have a sophisticated innate immune response network that coordinately effects pathogen recognition and killing, prophenoloxidase activation, complement activation and TGase-dependent wound healing.

INTRODUCTION AND HISTORICAL BACKGROUND

Extant horseshoe crabs comprise four species—*Limulus polyphemus*, *Tachypleus tridentatus*, *T. gigas* and *Carcinoscorpius rotundicauda*—each having a distinct geographic distribution. *L. polyphemus* is distributed along the east coast of North America and the other three species are distributed throughout Southeast Asia. In Japan, *T. tridentatus* inhabits coastal areas of the northern part of Kyushu Island as well as the Inland Sea. *T. tridentatus* is relatively long-lived; the embryo molts four

Invertebrate Immunity, edited by Kenneth Söderhäll.

times within the fertilized eggs and after hatching it molts every year for 15 years to become a mature adult.[1]

The first report of hemolymph coagulation involved *L. polyphemus* and was made by Howell in 1885[2] The early studies on hemolymph coagulation and granular hemocytes were performed by Loeb and granular hemocytes were named as amebocytes after their amoeboid ovement.[3,4] In 1903, Hideyo Noguchi—who later gained fame for his studies on snake venoms and syphilology—injected horse red blood cells into *L. polyphemus* and reported strong hemagglutinating activity in hemolymph plasma.[2] In 1956, Bang published a landmark study on hemolymph coagulation and reported that intravascular clotting is induced by an injection of a ubiquitous marine pathogen, *Vibrio*.[5] In 1964, Levin and Bang made the important observation that coagulation occurs when exposed to endotoxin (LPS).[6,7] Moreover, they found that procoagulant activity and a clottable protein are located in granular hemocytes, not in hemolymph plasma.[8] They immediately recognized the high sensitivity of LPS-mediated hemolymph coagulation and its applicability to hemocyte lysate, *Limulus* amoebocyte lysate (LAL), for assaying LPS.

In 1972, evidence for the proteolytic mediation of hemolymph coagulation was presented; LAL was fractionated, with one fraction containing the clottable protein and the other containing a coagulation accelerator with properties of proteolytic enzyme properties.[9] The clottable protein was designated coagulogen by virtue of its functional similarity to mammalian fibrinogen.[10] In 1977, Tai et al reported the mechanism underlying proteolytic conversion of coagulogen by the proclotting enzyme and they presented preliminary evidence for the activity of TGase in LAL as an unpublished observation.[11] Tai and Liu further demonstrated that the activation of the proclotting enzyme depends on the presence of Ca^{2+} and LPS and that the proclotting enzyme contains γ-carboxyglutamic acids, as does bovine prothrombin.[12] Later, Muta et al clearly indicated that the proclotting enzyme requires no Ca^{2+} in the proteolytic cascade and that it contains no γ-carboxyglutamic acid in the sequence.[13] Iwanaga et al developed chromogenic or fluorogenic substrates for proteases involved in hemolymph, resulting in the first discovery of a proteolytic cascade mediated by LPS and BDG.[14-16]

In 1985, Copeland and Levin examined the morphology and the fine structure of the granular hemocyte of *L. polyphemus*, showing that the hemocyte contains two types of granules: large ones and dense ones.[17] The immunocytochemical analysis of hemocytes showed that the large granules contain coagulogen and factor C, whereas the dense (or small) granules contain only an antimicrobial peptide, tachyplesin.[18] Moreover, the two types of granules were separated by ultracentrifugation and their protein components were purified by reverse-phase high-performance liquid chromatography.[19]

Here we review the current knowledge of horseshoe crab innate immunity at the molecular level with an emphasis on the importance of immunocompetent proteins derived from hemocytes, hemolymph plasma and sub-cuticular epidermis.

THE MOLECULAR MECHANISM UNDERLYING LPS-TRIGGERED HEMOCYTE EXOCYTOSIS

In horseshoe crabs, granular hemocytes constitute 99% of all hemocytes.[18] The hemocyte responds selectively to LPS but not to other pathogen-associated molecular patterns (PAMPs), such as BDG and peptidoglycans.[20] A variety of defense molecules are stored in the secretory granules of the hemocyte; large granules contain serine protease zymogens including factor C, factor G, factor B and the proclotting enzyme, the clottable protein coagulogen, serine protease inhibitors (serpins), lectins and substrates for TGase, whereas dense granules contain antimicrobial peptides.[21] In response to stimulation by LPS, these immunocompetent molecules are rapidly secreted by the hemocyte. Factor C is a unique LPS-responsive serine protease zymogen. A part of factor C is located on the hemocyte surface and acts as an LPS sensor to potentiate hemocyte exocytosis, which is localized in a punctate distribution on the hemocyte surface.[22,23] Factor C interacts with acidic phospholipids, and it also shows tight interaction with cholesterol.[20] The interaction between factor C and LPS is competitively inhibited by the addition of acidic phospholipids.[20] In contrast, cholesterol does not inhibit this interaction, suggesting that factor C interacts with cholesterol through a binding site that is distinct from that for LPS and raising the possibility that factor C may be localized on cholesterol-rich microdomains or lipid rafts on the hemocyte membrane.[20] Upon activation by LPS, activated factor C initiates hemocyte exocytosis via a G-protein-dependent exocytic pathway that is also dependent on the proteolytic activity of activated factor C. In this respect, LPS-triggered hemocyte exocytosis is analogous to the thrombin-thrombin receptor (the protease-activated G protein-coupled receptor, PAR) signaling axis in mammalian platelets. Hemocyte exocytosis requires 50 mM Mg^{2+} and 10 mM Ca^{2+} ions, equivalent to the concentrations of these cations in hemolymph plasma. Exclusion of these cations from the assay buffer inhibits hemocyte exocytosis even at high concentrations of LPS. Moreover, in the absence of LPS, hemocyte exocytosis can be triggered by synthetic hexapeptides corresponding to the tethered ligands of mammalian PARs, supporting the notion of a PAR-like receptor on the hemocyte surface (Fig. 1).

Factor C is a mosaic protein that contains a Cys-rich region, an epidermal growth factor-like domain, five Sushi domains (also designated as complement control protein modules or short consensus repeats) and a C-type lectin domain, in addition to a typical serine protease domain at the COOH-terminus[21]. Factor C interacts with LPS via the N-terminal Cys-rich domain ($K_d = 7.6 \times 10^{-10}$ M).[20,23] This contrasts with the findings of studies of the factor C ortholog from *C. rotundicauda*, which defined a region with the tandem sushi domains as important for LPS binding, using the truncated recombinant fragments, namely, sushi 123, sushi 1 and sushi 3 domains or their synthetic peptides.[24] However, these recombinant fragments or the peptides may not be properly folded by mixed disulfide bridges. The intermolecular disulfide bonding of the sushi 3-derived synthetic peptide, resulting in an unphysiological dimer, is practically indispensable for its interaction with LPS.[25]

ENDOGENOUS AMPLIFICATION SYSTEM FOR HEMOCYTE EXOCYTOSIS

The binding parameter between factor C and LPS could not explain the extraordinarily high sensitivity of horseshoe crab hemolymph for LPS as low as $\sim 10^{-13}$ g/ml ($\sim 10^{-14}$ M). We found that the hemocyte has an endogenous feedback mechanism for LPS-triggered hemocyte exocytosis.[26] The hemolymph normally contains hemocytes at $\sim 10^6$ cells/ml. LPS-triggered hemocyte exocytosis is highly dependent on cell density, namely, an increase in cell density (cells/ml) from 0.05×10^6 to 0.8×10^6 results in a 10^6-fold change in the apparent LPS sensitivity (from 10^{-7} to 10^{-13} g/ml of LPS). Interestingly, an antimicrobial tachyplesin in the exocytosed fluid acts as a secondary secretagogue, thereby dramatically enhancing the sensitivity of the hemocyte to LPS. Tachyplesin has structural properties in common with mastoparan, a basic tetradecapeptide from wasp venom. Mastoparan interacts directly with G proteins without direct stimulation of the upstream receptor and

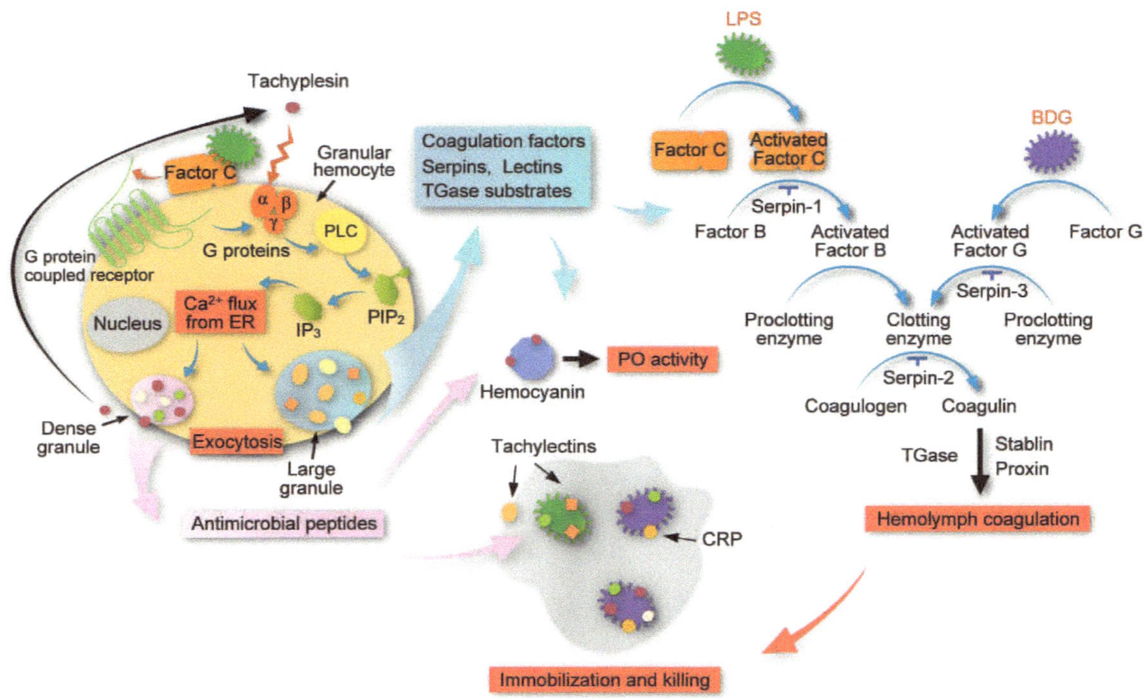

Figure 1. LPS-triggered hemocyte exocytosis and the resulting innate immune responses. In response to stimulation by LPS, granular components are rapidly secreted by the hemocyte via G protein-dependent exocytic pathway. The hemocyte stores a variety of defense molecules including coagulation factors, lectins, antimicrobial peptides and transglutaminase substrates. A serine protease zymogen, factor C, functions as an LPS sensor on the hemocyte surface and an initiator of hemolymph coagulation. Alternatively, activated factor G in the presence of β-1,3-D-glucans also triggers the coagulation cascade. Factor C and factor G independently serve to couple the recognition of pathogen-associated molecular patterns to form a physical barrier at the site of microbial invasion. LPS-induced hemocyte exocytosis leads not only to coagulation but also activates a sophisticated innate immune response network that culminates in immobilization and killing of pathogens and conversion of hemocyanin to phenoloxidase. PLC, phospholipase C; PIP_2, phosphatidylinositol-4,5-biphosphate; IP_3, inositol-1,4,5-triphosphate; PO, phenoloxidase.

induces exocytosis in the mast cell.[27] Consistent with these findings, mastoparan is able to induce hemocyte exocytosis in *T. tridentatus*.[20] Moreover, because tachyplesin binds to bovine G protein ($K_d = 8.8 \times 10^{-7}$ M), tachyplesin may interact with G-proteins in the hemocyte in a manner similar to that of mastoparan.[26] In addition, tachyplesin binds to hemocyanin ($K_d = 3.4 \times 10^{-6}$ M), the major protein in the hemolymph plasma; this suggests that hemocyanin may serve as a sink for tachyplesin, thereby spatially restricting its hemocyte-stimulating effect to the site of infection (Fig. 1).[28]

FUNCTION OF A TOLL-LIKE RECEPTOR ON THE HEMOCYTE

Insects and mammals conserve a signaling pathway of the innate immune system through cell-surface receptors called Tolls and Toll-like receptors.[29,30] The toll-like receptor of *T. tridentatus* is closely related to the *Drosophila* Toll in both domain architecture and overall length.[31,32] Horseshoe crab Toll is nonspecifically expressed in all tissues examined, suggesting that it does not act as an LPS receptor on granular hemocytes.[32] On the other hand, NF-κB and IκB homologues (CrNF-κB and CrIκB) have been identified in *C. rotundicauda*.[33] Gram-negative bacteria infection causes degradation of CrIκB and nuclear translocation of CrNF-κB, leading to upregulation of immune-related gene expression, including nitric oxide synthase and factor C, indicating that the NF-κB/IκB signaling cascade remains well conserved from horseshoe crabs to mammals. Although they must play a fundamental role in regulating the expression of critical immune defense molecules, further investigations are essential to reveal the physiological function of horseshoe crab Toll and its signaling.

THE COAGULATION CASCADE TRIGGERED BY LPS AND BDG

Factor C is autocatalytically activated in the presence of LPS and the resulting activated factor C activates coagulation factor B, which in turn converts the proclotting enzyme into the clotting enzyme. The clotting enzyme then promotes the proteolytic conversion of coagulogen to coagulin, which spontaneously forms an insoluble polymer (Fig. 1). Alternatively, activated factor G in the presence of BDG triggers the activation of the proclotting enzyme to the clotting enzyme. The mammalian coagulation system acts locally on the phospholipid surface in cooperation with Ca^{2+} ions at the site of vascular injury. In an analogous fashion, the horseshoe crab coagulation system is restricted to the surfaces of invading pathogens, such as Gram-negative bacteria and fungi. However, the coagulation cascade of horseshoe crabs does not have a common evolutionary origin with that of mammals. In fact, fibrinogen homologues of the horseshoe crab, tachylectins-5A and -5B, act as nonself-recognizing proteins rather than as target proteins of the coagulation cascade.[34] Also, coagulogen has no structural similarity or evolutionary relatedness to fibrinogen.[35]

A protease cascade in *Drosophila* has been well characterized as the morphogenetic cascade for determining embryonic dorsal-ventral polarity, leading to the production of the Toll ligand spätzle.[36] The *Drosophila* Toll pathway also controls resistance to fungal and Gram-positive bacteria infections.[37] Spätzle belongs to the nerve growth factor family and is homologous to coagulogen.[38,39] In addition, a clip-like domain (originally identified in the proclotting enzyme as a disulfide-knotted domain), located in the N-terminal region of coagulation factor B and the proclotting enzyme, has been identified in the proteins snake and easter of the *Drosophila* Toll pathway.[13,40] Therefore, the two functionally distinct cascades may have a common evolutionary origin.[41,42]

The coagulation cascade is regulated by three types of serpins that form stable 1:1 covalent complexes with target coagulation proteases: serpins-1, -2 and -3 inhibit activated factor C, the clotting enzyme and activated factor G, respectively.[43-45] These serpins are more closely related to mammalian serpins than they are to insect serpins: serpin-1 shows higher sequence identities to human plasminogen activator inhibitor (40%) than to an elastase inhibitor from *Manduca sexta* (29%) or silkworm antichymotrypsin (27%).[43] All three serpins are secreted upon hemocyte exocytosis in response to stimulation by LPS. These serpins appear to prevent diffusion of the activated forms of coagulation factors that escape into the hemolymph from the surfaces of microbes at the site of injury and thereby prevent unnecessary clot formation.

BDG RECOGNITION BY FACTOR G

In crustaceans and insects, the recognition of BDG triggers a serine protease cascade, leading to the activation of prophenoloxidase, a key enzyme in the melanization of pathogens and damaged tissues.[46,47] In mammals, the recognition of BDG by dectin-1, a C-type lectin family member, potentiates the production of cytokines and antifungal reactive oxygen species by dendritic cells and macrophages.[48] Factor G is another pattern-recognition protein in the coagulation cascade that acts as a sensitive sensor for BDG. Factor G is a heterodimeric serine protease zymogen composed of two noncovalently associated subunits, α and β.[49] The β subunit contains a serine protease domain, whereas the α subunit acts as a pattern-recognition subunit: it comprises three types of noncatalytic glycosidase-like modules, including a single β-1,3-D-glucanase A1-like module, three tandem xylanase A-like modules and two tandem xylanase Z-like modules, Z1 and Z2. Of these three types, the C-terminal Z1 and Z2 modules act as independent binding sites for BDG.[50] This observation, taken together with the high degree of sequence identity between Z1 and Z2 (91%), suggests that duplicated binding sites for BDG may increase avidity to allow stable and specific recognition of fungi.

The Z1 or Z2 module shows the highest sequence similarity to a carbohydrate-binding module of endoglucanase 5A from the aerobic soil bacterium *Cellvibrio mixtus* (45% sequence identity). Endoglucanase 5A contains an N-terminal catalytic domain and two tandem repeats of noncatalytic family 6 carbohydrate-binding modules, *Cm*CBM6-1 and *Cm*CBM6-2.[51] Our recent structural studies of the recombinant Z2 module by NMR spectroscopy clearly indicate that the ligand-binding site in the Z2 module is located in a cleft on a β-sheet in a predicted β-sandwich structure, which is superimposed onto cleft B in *Cm*CBM6-2.[52] Pattern recognition for BDG by factor G may be accomplished by a carbohydrate-binding cleft that is evolutionarily conserved between the horseshoe crab and the bacterium. In contrast, a crystal structure of the extracellular domain of mouse dectin-1 exhibits no structural similarity to the structure of the Z2 module.[53]

PATHOGEN RECOGNITION BY HEMOCYTE- AND PLASMA-DERIVED LECTINS

Four types of lectins have been identified in hemocytes: tachylectin-1, tachylectin-2, tachylectin-3 and tachylectin-4.[54-57] Tachylectin-1 interacts with 2-keto-3-deoxyoctonate on Gram-negative bacteria, whereas tachylectin-2 binds to GlcNAc or GalNAc and recognized lipoteichoic acids of Gram-positive bacteria. In contrast, tachylectin-3 specifically recognizes a certain sugar moiety on O-antigens of S-type LPS. Tachylectin-4 also recognizes the O-antigen of *E. coli* O111:B4 and shows ligand specificity for colitose (3-deoxy-L-fucose). In addition, fibrinogen homologues, tachylectins-5A and -5B, have been identified in hemolymph plasma as acetyl-group-recognizing lectins.[34] Tachylectin-5A possesses extraordinarily strong hemagglutinating activity against all types of human erythrocytes (minimum agglutinating concentration ~0.004 μg/ml). The concentration of tachylectin-5A in the plasma was ~10 μg/ml, suggesting that it plays an important role in the recognition of invading pathogens.

Hemolymph plasma also contains several lectins, such as isoforms of tachylectin-1[58,59] and three types of C-reactive proteins (CRPs) with functional and structural diversity.[60] Horseshoe crab CRP is a predominant LPS-binding protein and is upregulated at transcript levels by *Pseudomonas* infection, suggesting the importance of horseshoe crab CRP as a conserved molecule for pathogen recognition.[61] CRP from *L. polyphemus* forms extended fibrilar structures that encapsulate liposomes in the presence of Ca^{2+}· ions.[62] Although tachylectin-P, an isoprotein of tachylectin-1, is present in the perivitelline space of the egg, its physiological function in innate immunity remains unknown.[63]

The crystal structural analyses of tachylectin-2 and tachylectin-5A point out the importance of multivalency for achieving high specificity and high affinity. Tachylectin-2 contains five tandem WD repeats and adopts a five-bladed β-propeller structure with five equivalent GlcNAc/GalNAc-binding sites.[64] Each propeller blade has an independent binding site for the ligand. The specific recognition by tachylectin-2 is reinforced by the short distance between the individual binding sites, according to the pentagonal geometry. The crystal structure of tachylectin-5A is readily superimposed onto that of the C-terminal polymerization domain of the γ-chain of fibrinogen.[65] The polymerization pocket within the γ-chain structurally corresponds to the acetyl group binding site of tachylectin-5A, a finding that highlights the evolutionary connection between hemostasis and nonself recognition. Tachylectin-5A is present in oligomer in hemolymph plasma and its propeller like arrangement is evident by electron microscopy.[34] Given that tachylectins -2 and -5 exhibit virtually no side- or main-chain conformational changes upon ligand binding, it is likely that the polyvalent nature of these lectins underlies their avidity by allowing them to recognize specific densities or clustering ligands on pathogen surfaces.

STERILIZATION BY ANTIMICROBIAL PEPTIDES

Antimicrobial peptides are widely recognized to be important for innate immunity against invasive microbes. Several kinds of cysteine-rich peptides with antimicrobial activity are stored in the dense granules, including big defensin, tachycitin and tachystatins.[66-69] The NMR structural analyses have determined the solution structures of these antimicrobial peptides.[70-73] For example, the N-terminal domain of big defensin possesses a more potent antimicrobial activity against Gram-positive bacteria than the C-terminal domain.[66] In contrast, the C-terminal domain homologous to mammalian β-defensins displays more potent antimicrobial activity than the N-terminal domain against Gram-negative bacteria. The structure of big defensin reveals a new class within the defensin family; the C-terminal domain adopts a β-defensin structure, whereas the N-terminal domain forms a unique globular conformation. The hydrophobic N-terminal domain, but not the C-terminal domain, undergoes a conformational change in micelle solution, possibly associated with the antimicrobial activity against Gram-positive bacteria. Interestingly, the antimicrobial activity of an antimicrobial peptide (tachycitin) is synergistically enhanced when another antimicrobial peptide (big defensin) is present.[67] Horseshoe crab antimicrobial peptides show specific chitin-binding activity, which may play an important role in innate immunity.[66-69]

CONVERSION OF HEMOCYANIN TO PHENOLOXIDASE BY ANTIMICROBIAL PEPTIDES AND COAGULATION FACTORS

In crustaceans and insects, the prophenoloxidase activation system is an important part of innate immunity, where it acts to detect and kill invading pathogens as well as to synthesize melanin for wound healing and encapsulation of pathogens.[46,47] However, prophenoloxidase has not been identified in horseshoe crabs. Both prophenoloxidase and hemocyanin contain two functional copper-binding sites capable of reversibly binding an oxygen molecule.[74] Horseshoe crab hemocyanin is present at high concentrations in hemolymph plasma (~70 mg/ml) and acts as an oxygen carrier under physiological conditions. Arthropod prophenoloxidases are known to be nonenzymatically activated by treatment with detergents, lipids, or organic solvents.[75] In horseshoe crab as well, the induction of phenoloxidase activity is evident upon similar treatment.[76] Amphiphilic substances such as SDS and phosphatidylethanolamine convert horseshoe crab hemocyanin to phenoloxidase. Moreover, consistent with the amphiphilic nature of antimicrobial peptides, tachyplesin interacts with hemocyanin and induces its intrinsic phenoloxidase activity.[28] This tachyplesin-induced activity is inhibited by phenylthiourea, a typical inhibitor of phenoloxidase. Chemical modification of tachyplesin at Trp-2 or Tyr-8 and Try-13 on its hydrophobic face, but not modification of basic residues on its cationic face, significantly impairs its interaction with hemocyanin, implicating the hydrophobic face of tachyplesin in the functional conversion of hemocyanin to phenoloxidase. Although tachyplesin is the most effective activator of hemocyanin, other antimicrobial peptides, such as tachystatins and tachycitin, significantly induce its phenoloxidase activity.

On the other hand, the proclotting enzyme or coagulation factor B also converts hemocyanin to phenoloxidase and the resulting phenoloxidase activity reaches a plateau at a 1:1 molar ratio, whereas factor C, factor G, or trypsin does not convert hemocyanin.[77] The proteolytic cleavage of hemocyanin is not required for the functional conversion and the zymogen forms are effective activators. The clip domain of the proclotting enzyme or coagulation factor B may promote the interaction of these factors with hemocyanin to effect its functional conversion to an active phenoloxidase.

The origin of arthropod hemocyanins appears to be an ancient prophenoloxidase-like protein.[78] Under physiological conditions, arthropod prophenoloxidases require proteolytic cleavage for activation by a specific protease.[79] Tarantula hemocyanin expresses phenoloxidase activity after limited proteolysis with trypsin or chymotrypsin.[80] In the crayfish *Pacifastacus leniusculus*, the antibacterial peptide astacidin 1 is released from the C-terminal part of hemocyanin by a cysteine-like protease and is upregulated by LPS or BDG injection.[81] Horseshoe crab hemocyanin and human hemoglobin are also activated by microbial proteases, resulting in the production of reactive oxygen species as an antimicrobial strategy.[82]

TGase-DEPENDENT CROSS-LINKING IN THE IMMUNE SYSTEM

In crustaceans, hemolymph coagulation depends directly on intermolecular ε-(γ-glutamyl)lysine cross-linking of a vitellogenin-related protein by TGase without prior proteolytic cleavage.[83,84] Recently, in *Drosophila*, TGase and Fondue have been identified as being involved in larval hemolymph coagulation.[85] In horseshoe crabs, the coagulation cascade triggered by LPS or BDG promotes the conversion of coagulogen to coagulin, resulting in noncovalent polymerization of coagulins in a head-to-tail manner.[86] TGase of *T. tridentatus* is not present in hemolymph plasma and is restricted to cytoplasm of the hemocyte.[87,88] Coagulogen is not a substrate for TGase and coagulin is cross-linked to other hemocyte-derived proteins, such as the cysteine-rich protein stablin and the proline-rich protein proxin, resulting in coagulin fibrils with enhanced stability.[89,90] Stablin interacts with LPS and lipoteichoic acids and exhibits bacterial agglutinating activity against both Gram-negative and Gram-positive bacteria. Consequently, stablin colocalizes with coagulin fibrils that are cross-linked with proxin, effectively trapping bacteria. In addition, stablin binds to chitin, a major component of the arthropod cuticle. Proxin and stablin may promote not only the formation of the stable clotting fibrils but also the immobilization of invading microbes at sites of injury. Horseshoe crab cuticular proteins with chitin-binding activity have been identified as substrates for TGase.[91] One of these, caraxin-1 (carapace-derived chitin-binding protein for protein cross-linking) is specifically localized to the sub-cuticular epidermis.[92] Recombinant caraxin-1 exists as ~20-mer in solution and these oligomers are cross-linked by TGase to form an elaborate mesh of honeycomb structures that is distinguishable, by electron microscopy, from the clotting mesh triggered by LPS.

Horseshoe crab hemocytes are actively motile and one of the principal functions of these hemocytes is to seal scars in the cuticle.[93] According to observations on the wound repair process of *L. polyphemus*, a coagulation plug is formed within 10 min and the coagulum is then infiltrated by hemocytes to form a cellular plug within 24 h.[94] The sub-cuticular epithelial cells begin to migrate into the wound after 15 days; the epithelial cells span the wound between the cut ends of the exoskeleton by day 30 and then probably secrete cuticular components to complete the wound repair process. At the initial stage of this wound repair process, sufficient quantities of TGase may be secreted from hemocytes recruited to the site of injury and may then be immediately activated by Ca^{2+} ions in hemolymph plasma. At the same time, caraxin-1 may be secreted from the sub-cuticular epidermis. The resulting cross-linked clotting fibrils and caraxin mesh may function to seal the wound to stop bleeding, serve as a barrier to the entry of pathogens into the interior of the animal via the wound and operate as a transient extracellular matrix for the migration of epithelial cells that facilitate wound healing (Fig. 2).

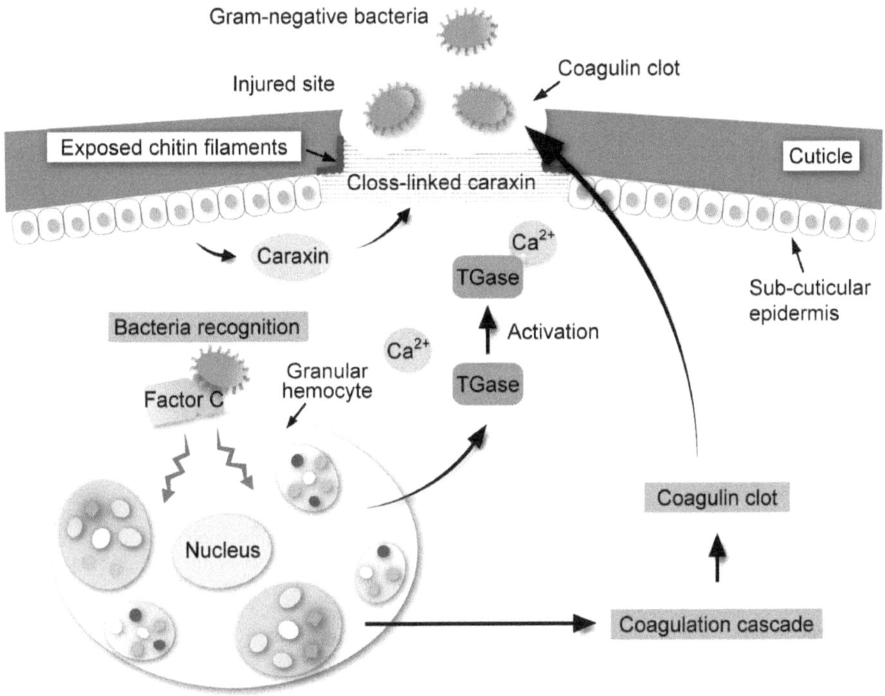

Figure 2. A hypothetical scheme for TGase-dependent cross-linking of caraxins at injured sites. In response to stimulation by LPS, TGase is secreted from the recruited hemocytes at injured sites and immediately activated by Ca^{2+} in plasma, which leads to the crosslinking of caraxins localized in the sub-cuticular epithelial cells. Caraxins may serve to provide an effective mesh to fix invading pathogens at injured sites in cooperation with the clotting mesh.

PATHOGEN RECOGNITION BY THE COMPLEMENT SYSTEM

A complement-related protein, α_2-macroglobulin, has been previously identified in horseshoe crabs,[95] and a homolog of complement component C3 has been identified in *C. rotundicauda* (CrC3)[96] and *T. tridentatus* (TtC3),[97] indicating the presence of a complement system capable of promoting the phagocytosis of invading microbes in protostomes. CrC3 or TtC3 consists of 1,716 residues, with an overall domain structure that is identical to that of mammalian C3, including α_2-macroglobulin domains, complement-urchin-bone domains, a thioester-containing domain, an anaphylatoxin domain and a C345C domain. The sequence identity between TtC3 and CrC3 (98%) is considerably higher than that between coagulogens from the two species (90%).[98] Horseshoe crab C3 is present as a disulfide-linked three-chain structure in hemolymph plasma, unlike mammalian C3, which is present in a two-chain form.[97]

The horseshoe crab complement system promotes the deposition of C3b on the surface of Gram-negative or Gram-positive bacteria. In TtC3, LPS, but not zymosan, peptidoglycan, or laminarin, strongly induces the proteolytic conversion of TtC3 to TtC3b, highlighting the selective response of the complement system to LPS stimulation. An antifactor C antibody inhibits both the proteolytic conversion of TtC3 and the deposition

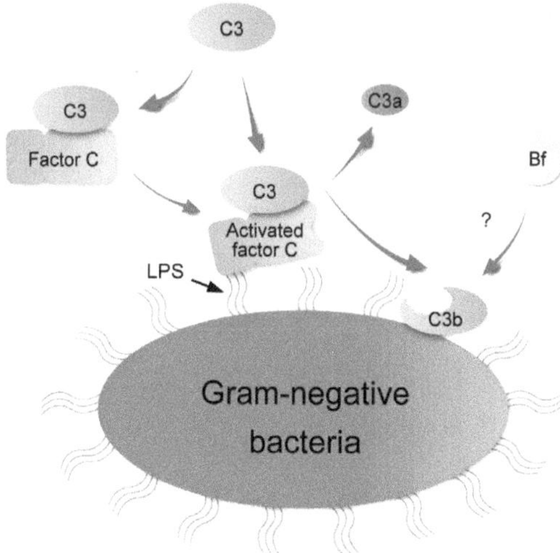

Figure 3. A proposed mechanism for the activation of horseshoe crab complement C3 by factor C localized on the surface of Gram-negative bacteria in the initial phase of the complement system. Factor C recognizes LPS on Gram-negative bacteria and recruits TtC3 by the affinity between the two proteins. The resulting activated factor C converts TtC3 to TtC3b, which is deposited on the surface of bacteria. Whether the activated factor C can convert horseshoe crab Bf leading to the formation of a vertebrate like C3 convertase (C3bBb) remains to be examined.

of TtC3b on the surface of Gram-negative bacteria. Moreover, activated factor C present on the surface of Gram-negative bacteria directly catalyzes the proteolytic conversion of TtC3, thereby promoting TtC3b deposition (Fig. 3). TtC3 interacts with factor C ($K_d = 4.9 \times 10^{-8}$M) and TtC3 is present at a concentration of at least 300 μg/ml in hemolymph plasma, whereas the amount of factor C in hemolymph plasma is very low (~10 μg/ml).[97] The relatively high concentration of TtC3 and its high affinity to factor C suggest that factor C exists in a complex with TtC3 in hemolymph plasma and that the formation of this complex is a prerequisite for the immediate activation of TtC3 by factor C on the surface of Gram-negative bacteria. In the alternative pathway of the mammalian complement system, the interaction between C3b and Bb is essential to form C3 convertase (C3bBb). A homolog of complement Bf has been identified in *C. rotundicauda*.[96] Although the physiological function of the Bf homolog in the horseshoe crab complement system remains unknown, it is likely that it may be responsible for the formation of the second C3 convertase (Fig. 3).

CONCLUSION AND FUTURE PROSPECTS IN THE HORSESHOE CRAB INNATE IMMUNE SYSTEM

In response to stimulation by LPS, hemocyte secretes several kinds of immunocompetent molecules, such as coagulation factors, lectins, antimicrobial peptides, protein substrates for TGase and also cytoplasm-derived TGase. The hemocyte possesses a feedback mechanism in which the antimicrobial peptides serve as endogenous mediators. The coagulation cascade results in the formation of coagulin fibrils that are subsequently stabilized by TGase-dependent cross-linking of proxin and stablin. A sub-cuticular epidermis-derived caraxin forms a TGase-stabilized mesh at sites of injury. Invading pathogens are agglutinated by both hemocyte- and plasma-derived lectins. In addition, coagulation factors and tachyplesin convert hemocyanin to phenoloxidase. In the plasma, factor C acts as an LPS-sensitive complement C3 convertase. In this manner, horseshoe crabs have a sophisticated innate immune response network that coordinately effects pathogen recognition and killing, prophenoloxidase activation, complement activation and TGase-dependent wound healing.

Interestingly, the antifactor C antibody exhibits no effect on the deposition of TtC3b on *Staphylococcus aureus*, suggesting the presence of a factor C-independent pathway to initiate the opsonization of Gram-positive bacteria.[97] Complement factors required for the deposition of TtC3b on Gram-positive bacteria remain to be examined. Factor C and complement Bf from *C. rotundicauda* interact with plasma-derived lectins, such as galactose-binding protein, carcinolectin-5 and CRP.[99] Therefore, the protease-lectin complexes on the surface of Gram-positive bacteria may enhance the deposition of C3b. Although the complement-dependent clearance system of invading pathogens in horseshoe crabs remains to be examined, phagocytosis of Gram-positive bacteria by hemocytes both in vivo and in vitro is inhibited by protease inhibitors, raising the possibility that the proteolytic dependence of opsonization by C3b may underlie phagocytosis by hemocytes.[96]

ACKNOWLEDGMENTS

We wish to express our deepest thanks to Toshio Shibata, a graduate student at Kyushu University, for the preparation of figures in this manuscript.

REFERENCES

1. Sekiguchi K, Yamamichi Y, Seshimo H et al. Normal development. In: Sekiguchi K, ed. Biology of Horseshoe Crabs, Tokyo: Science House 1988; 133-224.
2. Levin J, Hochstein HD, Novitsky TJ. Clotting cells and Limulus Amebocyte Lysate: An amazing analytical tool. In: Shuster CN Jr, Barlow RB, Brockmann HJ, eds. The American Horseshoe Crab, Cambridge, MA: Harvard University Press, 2003; 310-340.
3. Loeb L. On the blood lymph cells and inflammatory processes of Limulus. J Med Res 1902; 7:145-158.
4. Loeb L. Amoebocyte tissue and amoeboid movement. Protoplasma 1928; 4:596-625.
5. Bang FB. A bacterial disease of Limulus polyphemus. Bull Johns Hopkins Hosp 1956; 98:325-351
6. Levin J, Bang FB. The role of endotoxin in the extracellular coagulation of Limulus blood. Bull Johns Hopkins Hosp 1964; 115:265-274.
7. Levin J, Bang FB. A description of cellular coagulation in the Limulus. Bull Johns Hopkins Hosp 1964; 115:337-345.
8. Levin J, Bang FB. Clottable protein in Limulus: Its localization and kinetics of its coagulation by endotoxin. Thromb Diath Haemorrh 1968; 19:186-197.
9. Young NS, Levin J, Prendergast RA. An invertebrate coagulation system activated by endotoxin: Evidence for enzymatic mediation. J Clin Invest 1972; 51:1790-1797.
10. Solum NO. The coagulation of Limulus polyphemus hemocytes: A comparison of the clotted and nonclotted forms of the molecule. Thromb Res 1973; 2:55-70.
11. Tai JY, Seid RC Jr, Huhn RD et al. Studies on Limulus Amoebocyte Lysate II. J Biol Chem 1977; 252:4773-4776.
12. Tai JY, Liu T-Y. Studies on Limulus Amoebocyte Lysate. J Biol Chem 1977; 252:2178-2181.
13. Muta T, Hashimoto R, Miyata T et al. Proclotting enzyme from horseshoe crab hemocytes. J Biol Chem 1990; 265:22426-22433.

14. Ohki M, Nakamura T, Morita T et al. A new endotoxin sensitive factor associated with hemolymph coagulation system of horseshoe crab (Limulidae). FEBS Lett 1980; 120:217-220.
15. Nakamura T, Horiuchi T, Morita T et al. Purification and properties of intracellular clotting factor, factor B, from Horseshoe crab (Tachypleus tridentatus) hemocytes. J Biochem 1986; 99:847-857.
16. Kawabata S, Miura T, Morita T et al. Highly sensitive peptide-4-methylcoumaryl-7-amide substrates for blood-clotting proteases and trypsin. Eur J Biochem 1988; 172:17-25.
17. Copeland DE, Levin J. The fine structure of the Amebocyte in the blood of Limulus polyphemus: I. Morphology of the normal cell. Biol Bull 1985; 169:449-457.
18. Toh Y, Mizutani A, Tokunaga F et al. Morphology of the granular hemocytes of the Japanese horseshoe crab Tachypleus tridentatus and immunocytochemical localization of clotting factors and antimicrobial substances. Cell Tissue Res 1991; 266:137-147.
19. Shigenaga T, Takayenoki Y, Kawasaki S et al. Separation of large and small granules from horseshoe crab (Tachypleus tridentatus) hemocytes and characterization of their components. J Biochem 1993; 114:307-316.
20. Ariki S, Koori K, Osaki T et al. A serine protease zymogen functions as a pattern-recognition receptor for lipopolysaccharides. Proc Natl Acad Sci USA 2004; 101:953-958.
21. Iwanaga S, Kawabata S, Muta T. New type of clotting factors and defense molecules found in horseshoe crab hemolymph: their structures and functions. J Biochem 1998; 123:1-15.
22. Kurata S, Ariki S, Kawabata S. Recognition of pathogens and activation of immune response in Drosophila and horseshoe crab innate immunity. Immunobiol 2006; 211:237-249.
23. Koshiba T, Hashii T, Kawabata S. A structural perspective on the interaction between lipopolysaccharide and Factor C, a receptor involved in recognition of Gram-negative bacteria. J Biol Chem 2007; 282:3962-3967.
24. Tan NS, Patricia MLNG, Yau UH et al. Definition of endotoxin binding sites in horseshoe crab factor C recombinant sushi proteins and neutralization of endotoxin by sushi peptides. FASEB J 2000; 14:1801-1813.
25. Li P, Wohland T, Ho B et al. Perturbation of lipopolysaccharide (LPS) micelles by sushi 3 (S3) antimicrobial peptide. J Biol Chem 2004; 279:50150-50156.
26. Ozaki A, Ariki S, Kawabata S. An antimicrobial peptide tachyplesin acts as a secondary secretagogue and amplifies lipopolysaccharide-induced hemocyte exocytosis. FEBS J 2005; 272:3863-3871.
27. Higashijima T, Uzu S, Nakajima T et al. Mastoparan, a peptide toxin from wasp venom, mimics receptors by activating GTP-binding regulatory proteins (G proteins). J Biol Chem 1988; 263:6491-6494.
28. Nagai T, Osaki T, Kawabata S. Functional conversion of hemocyanin to phenoloxidase by horseshoe crab antimicrobial peptides. J Biol Chem 2001; 276:27166-27170.
29. Hoffmann JA. The immune response of Drosophila. Nature 2003; 426:33-38.
30. Akira S, Uematsu S, Takeuchi O. Pathogen recognition and innate immunity. Cell 2006; 124:783-801.
31. Inamori K, Koori, K, Mishima C et al. A horseshoe crab receptor structurally related to Drosophila Toll. J Endotoxin Res 2000; 6:397-399.
32. Inamori K, Ariki S, Kawabata S. A toll-like receptor in horseshoe crabs. Immunol Rev 2004; 198:106-115.
33. Wang XW, Tan NS, Ho B et al. Evidence for the ancient origin of the NF-B/IB cascade: Its archaic role in pathogen infection and immunity. Proc Natl Acad Sci USA 2006; 103:4204-4209.
34. Gokudan S, Muta T, Tsuda R et al. Horseshoe crab acetyl group-recognizing lectins involved in innate immunity are structurally related to fibrinogen. Proc Natl Acad Sci USA 1999; 96:10086-10091.
35. Bergner A, Oganessyan V, Muta T et al. Crystal structure of coagulogen, the clotting protein form horseshoe crab: a structural homologue of nerve growth factor. EMBO J 1996; 15:6789-6797.
36. Belvin MP, Anderson KV. A conserved signaling pathway-the Drosophila toll-dorsal pathway. Annu Rev Cell Dev Biol 1996; 12:393-416.
37. Ferrandon D, Imler J-L, Hetru C et al. The Drosophila systemic immune response: sensing and signaling during bacterial and fungal infections. Nature Rev Immunol 2007; 7:862-874.
38. Smith CL, DeLotto R. A common domain within the proenzyme regions of the Drosophila snake and easter proteins and Tachypleus proclotting enzyme defines a new subfamily of serine proteases. Protein Sci 1992; 1:1225-1226.
39. Bergner A, Muta T, Iwanaga S et al. Horseshoe crab coagulation is an invertebrate protein with a nerve growth factor-like domain. Biol Chem 1997; 378:283-287.
40. Muta T, Oda T, Iwanaga S. Horseshoe crab coagulation factor B. J Biol Chem 1993; 268:21384-21388.
41. Krem MM, Di Cera E. Evolution of enzyme cascades from embryonic development to blood coagulation. Trends Biochem Sci 2002; 27:67-74.
42. Kawabata S, Osaki T, Iwanaga S. Innate immunity in the horseshoe crab. In: Ezekowitz RAB, Hoffmann JA eds. Innate Immunity, Totowa: Humana Press, 2003:109-125.

43. Miura Y, Kawabata S, Iwanaga S. A limulus intracellular coagulation inhibitor with characteristics of the serpin superfamily, purification, characterization and cDNA cloning. J Biol Chem 1994; 269:542-547.
44. Miura Y, Kawabata S, Wakamiya Y et al. A limulus intracellular coagulation inhibitor type 2. Purification, characterization, cDNA cloning and tissue localization. J Biol Chem 1995; 270:558-565.
45. Lal Agarwala K, Kawabata S, Miura Y et al. Limulus intracellular coagulation inhibitor type 3. Purification, characterization, cDNA cloning and tissue localization. J Biol Chem 1996; 271:23768-23774.
46. Kanost MR, Jiang H, Yu W-Q. Innate immune response of a lepidopteran insect, Manduca sexta. Immunol Rev 2004; 198:97-105.
47. Cerenius L, Söderhäll K. The prophenoloxidase-activating system in invertebrates. Immunol Rev 2004; 198:116-126.
48. Brown GD. Dectin-1: a signaling nonTLR pattern-recognition receptor. Nat Rev Immunol 2006; 6:33-43.
49. Seki N, Muta T, Oda T et al. Horseshoe crab (1,3)-β-D-glucan-sensitive coagulation factor G. J Biol Chem 1994; 269:1370-1374.
50. Takaki Y, Seki N, Kawabata S et al. Duplicated binding sites for (l→3)-β-D-glucan in the horseshoe crab coagulation factor G. J Biol Chem 2002; 277:14281-14287.
51. Fontes CMGA, Clarke JH, Hazlewood GP et al. Identification of tandemly repeated type VI cellulose-binding modules in an endoglucanase from the aerobic soil bacterium Cellvibrio mixtus. Appl Microbiol Biotechnol 1998; 49:552-559.
52. Ueda Y, Ohwada S, Abe Y et al. Factor G utilizes a carbohydrate-binding cleft that is conserved between horseshoe crab and bacteria for the recognition of β-1,3-D-glucans. J Immunol 2009; 183:3810-3818.
53. Brown J, O'Callaghan CA, Marshall AS et al. Structure of the fungal β-glucan-binding immune receptor dectin-1: Implications for function. Protein Sci 2007; 16:1042-1052.
54. Saito T, Kawabata S, Hirata M et al. A novel type of limulus lectin-L6. J Biol Chem 1995; 270:14493-14499.
55. Okino N, Kawabata S, Saito T et al. Purification, characterization and cDNA cloning of a 27-kDa lectin (L10) from horseshoe crab hemocytes. J Biol Chem 1995; 270:31008-31015.
56. Saito T, Hatada S, Iwanaga S et al. A newly identified horseshoe crab lectin with specificity for blood group A antigen recognizes specific O-antigens of bacterial lipopolysaccharides. J Biol Chem 1997; 272:30703-30708.
57. Inamori K, Saito T, Iwaki D et al. A newly identified horseshoe crab lectin with specificity for blood group A antigen recognizes specific O-antigens of bacterial lipopolysaccharides. J Biol Chem 1999; 274:3272-3278.
58. Chiou ST, Chen YW et al. Isolation and characterization of proteins that bind to galactose, lipopolysaccharide of Escherichia coli and protein A of Staphylococcus aureus from the hemolymph of Tachypleus tridentatus. J Biol Chem 2000; 275:1630-1634.
59. Chen SC, Yen CH, Yeh MS et al. Biochemical properties and cDNA cloning of two new lectins from the plasma of Tachypleus tridentatus-Tachypleus plasma lectin 1 and 2. J Biol Chem 2001; 276:9631-9639.
60. Iwaki D, Osaki T, Yoshimitsu M et al. Functional and structural diversities of C-reactive proteins present in horseshoe crab hemolymph plasma. Eur J Biochem 1999; 264:314-326.
61. Ng PML, Jin ZX, Tan SSH et al. C-reactive protein: a predominant LPS-binding acute phase protein responsive to Pseudomonas infection. J Endotoxin Res 2004; 10:163-174.
62. Harrington JM, Chou H-T, Gutsmann T et al. Membrane activity of a C-reactive protein. FEBS Lett 2009; 583:1001-1005.
63. Nagai T, Kawabata S, Shishikura F et al. Purification, characterization and amino acid sequence of an embryonic lectin in perivitelline fluid of the horseshoe crab. J Biol Chem 1999; 274:37673-37678.
64. Beisel H-G, Kawabata S, Iwanaga S et al. Tachylectin-2: crystal structure of a specific GlcNAc/GalNAc-binding lectin involved in the innate immunity host defense of the Japanese horseshoe crab Tachypleus tridentatus. EMBO J 1999; 18:2312-2322.
65. Kairies N, Beisel H-G, Fuentes-Prior P et al. The 2.0-Å crystal structure of tachylectin 5 provides evidence for the common origin of the innate immunity and the blood coagulation systems. Proc Natl Acad Sci USA 2001; 98:13519-13524.
66. Saito T, Kawabata S, Shigenaga T et al. A novel big defensin identified in horseshoe crab hemocytes: isolation, amino acid sequence and antibacterial activity. J Biochem 1995; 117:1131-1137.
67. Kawabata S, Nagayama R, Hirata M et al. Tachycitin, a small granular component in horseshoe crab hemocytes, is an antimicrobial protein with chitin-binding activity. J Biochem 1996; 120:1253-1260.
68. Kawabata S, Saito T, Saeki K et al cDNA cloning, tissue distribution and subcellular localization of horseshoe crab big defensin. Biol Chem 1997; 378:289-292.
69. Osaki T, Omotezako M, Nagayama R et al. Horseshoe crab hemocyte-derived antimicrobial peptides, tachystatins, with sequence similarity to spider neurotoxins. J Biol Chem 1999; 274:26172-26178.

70. Suetake T, Tsuda S, Kawabata S et al. Chitin-binding proteins in invertebrates and plants comprise a common chitin-binding structural motif. J Biol Chem 2000; 275:17929-17932.
71. Fujitani N, Kawabata S, Osaki T et al. Structure of the antimicrobial peptide tachystatin A. J Biol Chem 2002; 277:23651-23657.
72. Fujitani N, Kouno T, Nakahara T et al. The solution structure of horseshoe crab antimicrobial peptide tachystatin B with an inhibitory cystine-knot motif. J Pep Sci 2007; 13:269-279.
73. Kouno T, Fujitani N, Mizuguchi M et al. A novel β-defensin structure: a potential strategy of big defensin for overcoming resistance by Gram-positive bacteria. Biochemistry 2008; 47:10611-10619.
74. Solomon EI, Sundaram UM, Machonkin TE. Multicopper oxidases and oxygenases. Chemical Rev 1996; 96:2563-2605.
75. Ashida M, Yamazaki H. Biochemistry of the phenoloxidase system in insects: with special reference to its activation. In: Ohnishi E, Ishizaki H, eds. Molting and Metamorphosis. Berlin: Springer-Verlag, 1990; 239-265.
76. Nellaiappan K, Sugumaran M. On the presence of prophenoloxidase in the hemolymph of the horseshoe crab, Limulus. Comp Biochem Physiol Part B 1996; 113:163-168.
77. Nagai T, Kawabata S. A link between blood coagulation and prophenoloxidase activation in arthropod host defense. J Biol Chem 2000; 275:29264-29267.
78. Burmester T, Scheller K. Common origin of arthropod tyrosinase, arthropod hemocyanin, insect hexamerin and dipteran arylphorin receptor. J Mol Evol 1996; 42:713-728.
79. Aspan A, Huang TS, Cerenius L et al cDNA cloning of prophenoloxidase from the freshwater crayfish Pacifastacus leniusculus and its activation. Proc Natl Acad Sci USA 1995; 92:939-943.
80. Decker H, Rimke T. Tarantula hemocyanin shows phenoloxidase activity. J Biol Chem 1998; 273:25889-25892.
81. Lee SY, Lee BL, Söderhäll K. Processing of an antibacterial peptide from hemocyanin of the freshwater crayfish Pacifastacus leniusculus. J Biol Chem 2003; 278:7927-7933.
82. Jiang N, Tan NS, Ho B et al. Respiratory protein-generated reactive oxygen species as an antimicrobial strategy. Nature Immunol 2007; 8:1114-1122.
83. Doolittle RF, Riley M. The amino-terminal sequence of lobster fibrinogen reveals common ancestry with vitellogenins. Biochem Biophys Res Commun 1990; 167:16-19.
84. Hall M, Wang R, van Antwerpen R et al. The crayfish plasma clotting protein: a vitellogenin-related protein responsible for clot formation in crustacean blood. Proc Natl Acad Sci USA 199; 96:1965-1970.
85. Lindgren M, Riazi R, Lesch C et al. Fondue and transglutaminase in the Drosophila larval clot. J Insect Physiol 2008; 54:586-598.
86. Kawasaki H, Nose T, Muta et al. Head-to-tail polymerization of coagulogen, a clottable protein of the horseshoe crab. J Biol Chem 2000; 275:35297-35301.
87. Tokunaga T, Yamada M, Miyata T et al. Limulus hemocyte transglutaminase: its purification and characterization and identification of the intracellular substrates. J Biol Chem 1993; 268:252-261.
88. Tokunaga T, Muta T, Iwanaga S et al. Limulus hemocyte transglutaminase: cDNA cloning, amino acid sequence and tissue localization. J Biol Chem 1993; 268:262-268.
89. Matsuda Y, Osaki T, Hashii T et al. A cysteine-rich protein from an arthropod stabilizes clotting mesh and immobilizes bacteria at injured sites. J Biol Chem 2007; 282:33545-33552.
90. Osaki T, Okino N, Tokunaga F et al. Proline-rich cell surface antigens of horseshoe crab hemocytes are substrates for protein cross-linking with a clotting protein coagulin. J Biol Chem 2002; 277:40084-40090.
91. Iijima M, Hashimoto T, Matusda Y et al. Comprehensive sequence analysis of horseshoe crab cuticular proteins and their involvement in transglutaminase-dependent cross-linking. FEBS J 2005; 272:4774-4786.
92. Matusda Y, Koshiba T, Osaki T et al. An arthropod cuticular chitin-binding protein endows injured sites with transglutaminase-dependent mesh. J Biol Chem 2007; 282:37316-37324.
93. Armstrong PB. Adhesion and motility of the blood cells of Limulus. In: Cohen WD, ed. Blood Cells of Marine Invertebrates: Experimental Systems in Cell Biology and Comparative Physiology. New York: Alan R. Liss, Inc., 1985:77-124.
94. Bursey CR. Histological response to injury in the horseshoe crab, Limulus polyphemus. Can J Zool 1977; 55:1158-1165.
95. Iwaki D, Kawabata S, Miura et al. Molecular cloning of Limulus α_2-macroglobulin. Eur J Biochem 1996; 242:822-831.
96. Zhu Y, Thangamani S, Ho B et al. The ancient origin of the complement system. EMBO J 2005; 24:382-394.
97. Ariki S, Takahara S, Shibata T et al. Factor C acts as a lipopolysaccharide-responsive C3 convertase in horseshoe crab complement activation. J Immunol 2008; 181:7794-8001.
98. Srimal S, Miyata T, Kawabata S et al. The complete amino acid sequence of coagulogen isolated from Southeast Asian horseshoe crab, Carcinoscorpius rotundicauda. J Biochem 1985; 98:305-318.
99. Saux AL, Ng PML, Koh JJY et al. The macromolecular assembly of pathogen-recognition receptors is impelled by serine proteases, via their complement control protein modules. J Mol Biol 2008; 377:902-913.

CHAPTER 8

TICK INNATE IMMUNITY

Petr Kopáček,*[,1] Ondřej Hajdušek,[1] Veronika Burešová[1] and Sirlei Daffre[2]

[1]*Institute of Parasitology, Biology Centre of the Academy of Sciences of the Czech Republic and Faculty of Science, University of South Bohemia, Branišovská, České Budějovice, Czech Republic;* [2]*Departamento de Parasitologia, Instituto de Ciências Biomédicas, Universidade de São Paulo. São Paulo, SP, Brazil*

**Corresponding Author: Petr Kopáček—Email: kopajz@paru.cas.cz*

Abstract: Ticks are blood feeding parasites transmitting a wide variety of pathogens to their vertebrate hosts. The vector competence of ticks is tightly linked with their immune system. Despite its importance, our knowledge of tick innate immunity is still inadequate and the limited number of sufficiently characterized immune molecules and cellular reactions are dispersed across numerous tick species. The phagocytosis of microbes by tick hemocytes seems to be coupled with a primitive complement-like system, which possibly involves self/nonself recognition by fibrinogen-related lectins and the action of thioester-containing proteins. Ticks do not seem to possess a pro-phenoloxidase system leading to melanization and also coagulation of tick hemolymph has not been experimentally proven. They are capable of defending themselves against microbial infection with a variety of antimicrobial peptides comprising lysozymes, defensins and molecules not found in other invertebrates. Virtually nothing is known about the signaling cascades involved in the regulation of tick antimicrobial immune responses. Midgut immunity is apparently the decisive factor of tick vector competence. The gut content is a hostile environment for ingested microbes, which is mainly due to the antimicrobial activity of hemoglobin fragments generated by the digestion of the host blood as well as other antimicrobial peptides. Reactive oxygen species possibly also play an important role in the tick-pathogen interaction. The recent release of the *Ixodes scapularis* genome and the feasibility of RNA interference in ticks promise imminent and substantial progress in tick innate immunity research.

Invertebrate Immunity, edited by Kenneth Söderhäll.

INTRODUCTION

Ticks (Acari: Ixodida) are parasitic mites that suck blood from their vertebrate hosts. Systematically, they belong to the Phylum Arthropoda, Class Arachnida, Subclass Acari, Order Parasitoformes and Suborder Ixodida.[1] To date, more than 900 tick species have been described and divided into two major families—Ixodidae (generally referred to as hard ticks) and Argasidae (also known as soft ticks).[2] As this common nomenclature implies, the hard ticks possess a sclerotized dorsal body surface plate (scutum), whereas the soft ticks have only a leathery cuticle.[3] This most conspicuous morphological feature of Ixodidae and Argasidae is just one hallmark for the other numerous biological, physiological and ecological differences between these two families,[3] out of which we mention just the most relevant to the topics of this chapter. Soft ticks, represented e.g., by the genera *Argas* or *Ornithodoros*, are multi-host parasites, having several nymphal stages (2-8) that all feed rapidly on their hosts (within minutes to hours). The adult ticks can feed repeatedly and the females, which mate away from the host, deposit just a limited number of eggs (few hundreds) after each feeding. The life span of Argasidae is quite long (up to several years) and they are capable to live without food for months between individual blood meals as nymphs or adults. In contrast, hard ticks possess only three developmental stages—larvae, nymphs and adults. Most Ixodidae (e.g., genera *Ixodes, Amblyomma, Dermacentor*, *Haemaphysalis*) are three-host ticks, where each stage can feed on a different host. The hard ticks feed slowly for several days, depending on the stage and species. The adult females mate usually on the host and after full engorgement they drop off. The blood meal is digested within several weeks and the female oviposits thousands of eggs and dies. Few, veterinary important species, such as the cattle tick *Rhipicephalus* (formerly *Boophilus*) *microplus*, are one-host ticks where larvae, nymphs and adults remain and feed on the same bovine host. The obvious differences in physiology and feeding strategy strongly support the theory that adaptation to blood-feeding in hard and soft ticks has evolved independently.[4]

Ticks do harm to their hosts as obligatory ectoparasites, sometimes causing severe blood loss. More importantly, the main danger ticks present to their hosts is the enormous variety of pathogens they can transmit, including viruses, bacteria and protozoa.[5] A comprehensive survey of tick-borne pathogens causing serious diseases to humans or domestic animals was reviewed recently.[6] It demonstrates that the most recognized human illnesses (e.g., Lyme disease, tick-borne encephalitis, granular anaplasmosis) or animal diseases (e.g., anaplasmosis, babesiosis, theileriosis, African swine fever) represent just the tip of the iceberg of the entire problem of ticks as disease vectors.[6]

The success of pathogen transmission by ticks is due to the long period of co-evolution between pathogen and vector and also because of specific aspects of tick biology. Pathogens take advantage of the modulation of host haemostatic, inflammatory and immune responses mediated by the inexhaustible pharmacology of molecules present in tick saliva.[7] Moreover, ingested microbes are not exposed to immediate proteolytic attack by digestive enzymes within the gut lumen, since ticks digest blood intracellularly in digestive lysosomal vesicles.[3] On the other hand, it is obvious that ticks have to possess efficient defense mechanisms to eliminate microbial infections, or at least to mitigate it to a level tolerable for the vector. Despite its importance, we still know very little about how transmitted pathogens can evade or withstand the innate immune responses of ticks since, the vector-pathogen interface is the least understood site of the pathogen-tick-host interaction triangle.

Invertebrate animals lack an adaptive immune system and their defense against potential pathogens has to rely on a network of cellular immune reactions (e.g., phagocytosis, encapsulation and nodulation) and humoral factors involved in pathogen recognition and elimination (pattern recognition receptors, lectins, complement-like system, pro-phenoloxidase activation, hemolymph coagulation, antimicrobial peptides, reactive oxygen species, etc.). During the past two decades, our knowledge of innate immunity in invertebrates has developed rapidly thanks to the considerable body of work carried out on the model insect *Drosophila melanogaster*, other arthropods, such as the horseshoe crab, freshwater crayfish and prochordates such as ascidia (for review, see ref. 8 and this book). The detailed model of innate immune system based on that from *Drosophila*[9] provided a solid basis for further investigation of immune responses and parasite transmission in insect disease vectors, such as mosquitoes[10] or tse-tse flies.[11]

In contrast to the model invertebrates and blood-sucking insects, our understanding of innate immunity in ticks is still inadequate. Only a limited number of described cellular immune reactions and sufficiently characterized immune molecules (see the overview in Table 1 and refs. 12, 13 for the recent reviews) allow extrapolation to the immune system of other model arthropods, especially the horseshoe crab, which is phylogenetically the closest relative of ticks within the subphylum Chelicerata. A few years ago, it seemed that research on tick immune system would probably remain forever at a purely descriptive level. However, several recent advances reversed this pessimistic outlook. The availability of large scale EST data sets on different tick species, including the first version of the American deer tick *Ixodes scapularis* genome, released in December 2008,[14] the feasibility of RNA interference (RNAi) in ticks[15] and the establishment of several tick in vitro cell culture lines as valuable models for analyzing the immunological interactions with tick-borne pathogens[16] opened the gate to the postgenomic era in tick-host-pathogen interaction research.

In this chapter, we review the current knowledge on tick cellular events and molecules involved in defense against ingested and transmitted microbes. We also highlight further perspectives in the research of tick innate immunity relying mainly on the recently implemented reverse genetics and functional genomics approaches.

DEFENSE MECHANISMS IN THE TICK HAEMOCOEL

Cellular Immune Responses

Similar to other arthropods, ticks possess an open circulatory system surrounding their internal organs.[3] It was demonstrated in the tick *Dermacentor andersoni* that during feeding the hemolymph volume increases linearly from about to 2-3 μl in unfed to almost 150 μl in fully engorged females. However, the ratio of hemolymph content to the total body weight remains constant at about 23%.[17]

Based on morphological and physiological studies of hemolymph cells in different tick species, several hemocyte types have been described.[3,18-20] Most reports are consistent in the description of three basic classes of tick hemocytes, namely **plasmatocytes**, **granulocytes I** and **granulocytes II**. Phagocytic activity is mainly attributed to plasmatocytes and granulocytes I, whereas granulocytes II are nonphagocytic cells. In some tick species, another hemocyte type resembling insect **spherulocytes** has been described.[3,20] The

Table 1. Overview of characterized tick immune molecules

Defense System	Molecule	Tick Species[a]	Method[b]/ Starting Material[c,d]	Molecular Characteristics	Protein Localization/ Gene Expression[d]	Function and/ or Other Characteristics	References GenBank
Lectin-complement system							
Lectins-agglutinins	Lectin (DorinM)	*O. moubata*	Classical/Hl	37 kDa, 3 × glycosylated noncovalent oligomer	Hl/H, SG	Lectin, sialic acid-specific Fibrinogen related protein	36,37 AY333989
	OMFREP	*O. moubata*	Reverse genetics/H	28.9 kDa	/H, SG	Fibrinogen related protein	40 AF527411
	Ixoderin A	*I. ricinus*	Reverse genetics/H	30.8 kDa	/H, SG, MG	Fibrinogen related protein	40 AY341424
	Ixoderin B	*I. ricinus*	Reverse genetics/SG	32.6 kDa	/SG	Fibrinogen related protein	40 AY643518
	Galectin (OmGa-lectin)	*O. moubata*	Reverse genetics/multiple tissues	37.4 kDa	/Multiple tissues	Intracellular protein	41 AB255165
Complement-like molecules	α2-macroglobulin (TAM)	*O. moubata*	Classical/Hl	2 × 160 kDa, glycosylated noncovalent dimer	Hl/H, SG	Universal protease inhibitor	42,43 AF538967
	α2-macroglobulin (IrAM)	*I. ricinus*	Reverse genetics/SG	2 × 160 kDa, glycosylated noncovalent dimer	Hl/H, SG	Function in phagocytosis of *C. indologenes*	23 EU835901

continued on next page

Table 1. Continued

Defense System	Molecule	Tick Species[a]	Method[b]/ Starting Material[c,d]	Molecular Characteristics	Protein Localization/ Gene Expression[d]	Function and/ or Other Characteristics	References GenBank
	Factor D-like	*D. variabilis*	Reverse genetics/H	39 kDa, clip domain, protease homolog	/H	*E. coli* inducible	46 AY161279
Antimicrobial peptides							
Lysozymes	c-type lysozyme	*O. moubata*	Classical/Gut contents	14 kDa, pI 9.7	Gut contents/ MG	Active against *M. luteus*, bloodmeal inducible	75,76 AAL17868
	c-type lysozyme	*D. variabilis*	Reverse genetics/H	14 kDa, pI 9.97	/H, MG, FB	*E. coli* and *R. montanensi* inducible	55,77 AY183671
	c-type lysozyme	*D. andersoni*	Reverse genetics/Cell line DEA100	14 kDa, pI 9.91			55 AY207371
Defensins	Defensin A,B,C,D	*O. moubata*	Reverse genetics/4th instar nymphs	4 kDa mature peptide	Gut contents (DefA)/ MG	Active against Gram[+]	78-80 (DefA) AB041816
	Defensin 1 (varisin)	*D. variabilis*	Classical/Hl	4.2 kDa mature peptide	H	Active against Rickettsia	59-62 AY181027
	Defensin	*R. microplus*	Classical/H	4,291 Da mature peptide	H	Active against *M. luteus*	65 AY233213
	Defensin (amercin)	*A. americanum*	Reverse genetics/H	4,050 Da mature peptide	/MG, H, FB, SG		58 DQ864986

continued on next page

Table 1. Continued

Defense System	Molecule	Tick Species[a]	Method[b]/ Starting Material[c,d]	Molecular Characteristics	Protein Localization/ Gene Expression[d]	Function and/ or Other Characteristics	References GenBank
	Defensin (longicin)	*H. longicornis*	Reverse genetics/MG	5,820 Da mature peptide	MG	Active against Gram$^+$, Gram$^-$, fungi and *Babesia sp.*	81 AB105544
	Defensin (Hl-gut)	*H. longicornis*	Reverse genetics/MG	5.8 kDa pro-defensin	/MG	LPS inducible	82 EF432731
	Defensin	*I. persulcatus*	Reverse genetics/ Whole nymhs	4,199 Da mature peptide	/MG	Active against Gram$^+$	56 AB469201
	Defensin (scapularisin)	*I. scapularis*	Reverse genetics/ Mixed tissuses	4,180 Da mature peptide	/MG, (H,FB)		84 AY660970
	Defensin	*I. ricinus*	Reverse genetics/ Whole females	4,496 Da mature peptide	/MG	Upregulated by Borrelia-infected blood meal	83 AY335442
Defensin–like	Noncationic defensins (ADP1, ADP2)	*A. hebraeum*	Reverse genetics/ Syngaglium	4,612 mature ADP2	Hl (ADP2)	Active against *S. aureus* > *E. coli* (ADP2)	63 AY437137 AY437138
(No RVRR cleavage motif)	Defensin -2	*D. variabilis*	Reverse genetics	4.7 kDa pro-defensin	/FB, MG, Ov		77 AY159879
	Defensin (Hl-sal)	*H. longicornis*	Reverse genetics/SG	6.5 kDa pro-defensin	/SG	LPS inducible	82 EF432732

continued on next page

Table 1. Continued

Defense System	Molecule	Tick Species[a]	Method[b]/ Starting Material[c,d]	Molecular Characteristics	Protein Localization/ Gene Expression[d]	Function and/ or Other Characteristics	References GenBank
Other AMPs	Hebraein	*A. hebraeum*	Reverse genetics/ Syngaglium	11 kDa histidine-rich	Hl	Active against *S. aureus, E. coli,* yeast *C. glabrato*	64 AY437139
	Microplusin	*R. microplus*	Classical/Hl, eggs	10,204 Da, histidine-rich	Hl, Ov	Active against *M. luteus*; divalent metal chelator	65,66,109 AY233212
	Ixodidin	*R. microplus*	Classical/H	7,103 Da	H	Active against *M. luteus*; chy-motrypsin-elastase inhibitor	67 P83516
	Ixosin	*I. sinensis*	Classical/SG	2,870.5 Da, mature peptide	SG	Activity against *E. coli, S. aureus, C. albicans*	105 DQ100462
	Ixosin B	*I. sinensis*	Classical/SG	3,814.5 Da, mature peptide	SG	Activity against *E. coli, S. aureus, C. albicans*	106 EU047746
	IsAMP	*I. scapularis*	Classical/ Saliva	5,300 Da, mature peptide	Saliva/H, SG, FB	Activity against Gram⁺, Gram-, up-regulated by Borrelia infection	108 AF483734
	Hemoglobin fragment	*R. microplus*	Classical/Gut contents	3,206 Da		Activity against Gram⁺, fungi	68,71

continued on next page

Table 1. Continued

Defense System	Molecule	Tick Species[a]	Method[b]/ Starting Material[c,d]	Molecular Characteristics	Protein Localization/ Gene Expression[d]	Function and/ or Other Characteristics	References GenBank
		O. moubata	Classical/Gut contents			Activity against *S. aureus*	69
		D. variabilis	Classical/Gut contents			Activity against *M. luteus*	70
Protease inhibitors							
	Kunitz-type serine protease inhibitor (KPI)	*D. variabilis*	Reverse genetics/MG	35.5 kDa, 5 Kunitz BPTI domains	/MG	Trypsin inhibitor, anticoagulant; active against Rickettsia	89 ABY26517
	Hl-cystatin 2	*H. longicornis*	Reverse genetics/MG	12.9 kDa	/MG	Cysteine peptidase inhibitor babesiacidal	88 DQ364159
	BmSI-7, BmSI-6—subtilisin inhibitors	*R. microplus*	Classical				
			/eggs	7,408 Da (BmSI-7) 7,271 Da (BmSI-6)	Eggs/Ov, MG, SG	Active against PR1 protease from the fungus *Metharizium anisoplae*	110 (BmSI-7) GQ452779

continued on next page

Table 1. Continued

Defense System	Molecule	Tick Species[a]	Method[b]/ Starting Material[c,d]	Molecular Characteristics	Protein Localization/ Gene Expression[d]	Function and/ or Other Characteristics	References GenBank
Antioxidant system							
Antioxidant							
	Catalase	*R. microplus*	Classical/MG	inhibited by aminotriazol			95
	Glutathion peroxi-dase (HlPrx)	*H. longicornis*	Reverse genetics/SG	26.5 kDa	SG/SG	Antioxidant activity	99 AB038382
	Phospholipid-hy-droperoxideglu-tath. peroxidase	*R. microplus*	Reverse genetics/cell line C34, coumaphos resistant	18.9 kDa contains selenocystein		Increased expression in acaricide-resistant strains	102 DQ172828
	Glutathion S-transferase	*R. microplus*	Classical/ Whole larvae	25.6 kDa, pI 8.53			98 AAD15991
	Glutathion S-transferase	*R. microplus*	Reverse genetics/SG	25.6 kDa	SG (MG)/ SG, MG		100 AF366931
	Glutathion S-transferase DvGST1, DvGST2	*D. variabilis*	Reverse genetics/H	24 kDa, pI 9.16 (1) 24 kDa, pI 5.27 (2)	/MG (1); MG, Ov (2)	Blood meal inducible	97 (1) DQ224235 (2) AY241958

continued on next page

Table 1. Continued

Defense System	Molecule	Tick Species[a]	Method[b]/ Starting Material[c,d]	Molecular Characteristics	Protein Localization/ Gene Expression[d]	Function and/ or Other Characteristics	References GenBank
Signaling cascade							
Transcription factor	Akirin (Subolesin, protective antigen 4D8)	*I. scapularis*	Reverse genetics	20.6 kDa		Intracellular transcription factor	115 AY652654

[a] *A.–Amblyomma*; *D.–Dermacentor*; *H.–Haemaphysalis*; *I.–Ixodes*; *O.–Ornithodoros*; *R.–Rhipicephalus.*
[b] Classical—from protein to gene, Reverse genetics—from gene to protein.
[c] Tissue used for protein purification or RNA isolation, respectively.
[d] H–hemocytes; Hl–hemolymph; SG–salivary glands; MG–midgut; Ov–ovaries; FB–fat body.

hemocytes are believed to differentiate from stem cells called **prohemocytes**, which are only rarely detected in the hemolymph.[20]

Phagocytosis of foreign material and microbes by tick hemocytes has been studied both in vivo and in vitro in several tick species. Fluorescent polystyrene beads inoculated into the haemocoel of *Ornithodoros moubata* were phagocytosed by plasmatocytes twice as much as by granulocytes.[19] Hemocytes of *O. moubata* were also reported to phagocytose the yeast *Candida haemulonii*[21] or the yellow pigmented Gram$^-$ bacteria *Chryseobacterium indologenes*.[22] Despite the active cellular response against these microbes, infection with these pathogens caused high mortality rates of these soft ticks.[21,22] Recent RNAi experiments revealed that phagocytosis of *C. indologenes* by hemocytes of the hard tick *Ixodes ricinus* is mediated by a plasma α_2-macroglobulin[23] and possibly also by other thioester-containing proteins[24] (see also the later section on complement-related molecules). Pereira et al[25] investigated the phagocytic activity of the *R. microplus* hemocytes upon injection of the yeast *Saccharomyces cerevisiae*. They demonstrated that plasmatocytes is the major cell type involved in phagocytosis of the yeast. Moreover, they were able to show in an in vitro assay that an oxidative burst in the phagocytic hemocytes occurred through the production of the reactive oxygen species superoxide and H_2O_2.

The process of hemocytic **encapsulation** was examined using small pieces of Epon-Araldite implanted under the cuticle of the American dog tick, *Dermacentor variabilis*.[26] This study showed the successive involvement of Type I and Type II granulocytes as well as plasmatocytes in the gradual forming of a multiple cell layer capsule around the implant. In addition, the authors observed Epon-Araldite particles surrounded by a fibrous matrix, which they interpreted as hemolymph coagulation. On the other hand, they did not observe any sign of melanin formation on the surface of the implant. Another report on **nodulation** in ticks was also demonstrated in *D. variabilis*.[27] The challenge of ticks by direct inoculation of *Escherichia coli* into the haemocoel led to the immediate clearance of bacteria from the hemolymph. A direct fluorescence assay revealed that a rapid formation of bacterial clumps surrounded by aggregated and disintegrated hemocytes is responsible for this phenomenon. The aggregated bacteria were not viable, possibly because of the involvement of other immune mechanisms, e.g., bacteriolysis or phagocytosis.[27]

Thus far, no cell line originating from tick hemocytes has been available. However, some tick cell lines derived from embryonic cells[16] display remarkable phagocytic activity and have been used as useful models for deeper insight into tick cell-pathogen interaction.[28,29] An elegant example of this approach was provided in a recent work showing how cell lines derived from *I. scapularis* (IDE12) and *Amblyomma americanum* (AAE2) respond to infection by the entomopathogenic fungi *Metarhizium anisoplae*.[30]

Certainly, much attention has been given to the response of tick cells against the human pathogen *Borrelia burgdorferi*, the causative agent of Lyme disease. The hemocytes from *I. ricinus* were shown to use the mechanism of "coiling" phagocytosis to engulf *B. burgdorferi* spirochetes, as performed by vertebrate phagocytic cells.[31] Coleman et al demonstrated that *B. burgdorferi* overcome the midgut barrier of the tick *I. scapularis* via a surface bound plasmin, originating from the host blood, which facilitates spirochete dissemination into the tick hemolymph and migration towards the salivary glands.[32] At least some of the spirochetes were phagocytosed and the process of their internalization by tick hemocytes was examined by confocal microscopy.[32] The differences in the immune responses to the cultivated *B. burgdorferi* upon direct injection into the haemocoel were compared using the natural vector,

I. scapularis and *D. variabilis,* which is unable to transmit the pathogen.[33] The authors found that the spirochetes were rapidly cleared from the hemolymph of *D. variabilis* in contrast to *I. scapularis*, which appeared to provide a benign environment for the pathogen. The immunocompetent *D. variabils* displayed a faster phagocytic response and a higher increase in hemocytic cells following challenge with the *B. burgdorferi*, as well as a much higher borreliacidal activity in the plasma, as compared to the immunotolerant *I. scapularis.*[33,34] These results are also in accordance with the study of *B. burgdorferi* phagocytosis by cell lines from *I. scapularis* (IDE12) and *D. andersoni* (DAE15).[28]

Lectins—The Potential Recognition Molecules

Research on tick lectins/hemagglutinins has been the subject of constant attention since the specific protein-carbohydrate interaction is believed to be a key to understand self/nonself recognition and how the transmitted pathogens evade tick immune responses. The hemagglutination activity, conferring the ability to agglutinate mammalian red blood cells, has been reported for a variety of tick species mainly in the hemolymph, gut and salivary glands. The binding specificity of tick lectins were mainly observed towards N-acetyl-D-hexosamines, sialic acid and glycoconjugates like fetuin (see ref. 35 for a recent review). The only sufficiently characterized tick lectin referred to as Dorin M was purified from the plasma of the soft tick *O. moubata* (Table 1).[36] Dorin M is mainly expressed in tick hemocytes and salivary glands and its C-terminal domain is highly homologous to the γ-chain of vertebrate fibrinogens and fibrinogen-related proteins (FREPs), such as mammalian ficolins.[37] Unlike ficolins, Dorin M lacks the N-terminal collagen-like domain. This feature is shared by the closely related tachylectins 5A and 5B from the horseshoe crab *Tachypleus tridentatus*,[38] which function as pattern recognition molecules.[39] Another FREP has been cloned from *O. moubata* (OMFREP) (Table 1) and found to have a similar sequence and expression profile as Dorin M.[40] By contrast, two FREPs identified in the hard tick *I. ricinus*, namely the Ixoderin A and B, are phylogenetically more distinct and differ also in their tissue expression profile. Ixoderin A gene expression was found in hemocytes, salivary glands and midgut, whereas Ixoderin B was specifically expressed in the salivary glands (Table 1).[40] Our search through the *I. scapularis* genome revealed that at least three isoforms of Ixoderin A could be identified, multiple (≥14) isoforms of Ixoderin B and a still uncharacterized type named Ixoderin C which was predicted to be intracellular protein (Hajdušek, unpublished results).

The gene coding for a protein related to galectins has been isolated from the whole body homogenate of *O. moubata* 4th instar nymphs and was named OmGalec (Table 1).[41] Its gene was found to be expressed in all developmental stages, including eggs and in variety of tissues. The protein is likely to be present intracellularly and it lacks a transmembrane domain. OmGalec sequence contains tandem-repeated carbohydrate recognition domains. The sugar binding specificity of recombinant OmGalec was tested for a broad spectrum of glycans using an automated frontal affinity chromatography system. Its affinity profile corresponds to the general features of the galectin family, e.g., specificity for lactosamine-like disaccharides. The authors speculate that OmGalec may have multiple functions, including in innate immunity.[41]

Complement-Related Molecules

Two α_2-macroglobulins (α_2M) described in the soft tick *O. moubata* and the hard tick *I. ricinus* designated TAM and IrAM, respectively (Table 1),[23,42,43] are the first tick representatives of the evolutionary oldest and best conserved family of thioester-containing proteins (α_2M superfamily), which have an important role in both invertebrate and vertebrate innate immunity. This superfamily comprises in invertebrates three main, phylogenetically distinct groups (i) α_2-macroglobulins—the universal macromolecular protease inhibitors; (ii) the components C3/C4/C5 of the complement system; (iii) insect thioester proteins (TEPs).[44] TAM and IrAM share high sequence homology, in addition to disulfide bridge and glycosylation patterns. However, several structural features differentiate them from the majority of vertebrate and invertebrate α_2Ms: the absence of disulfide bridges covalently binding their subunits, posttranslational processing of their precursors resembling cleavage events typical for C3 and C4 complement components and variability within their bait regions (site of protease attack) generated by alternative splicing.[23,43] A recent functional study based on RNAi-silencing linked with an in vitro phagocytic assay revealed that IrAM is involved in phagocytosis of the tick pathogen *C. indologenes* by *I. ricinus* hemocytes. The phagocytosis activity was dependent on an active metalloprotease secreted by the bacteria, indicating that interaction of tick α_2M with a protease from an invading pathogen is linked with a cellular immune response.[23] Interestingly, phagocytosis of the Lyme disease agent *B. burgorferi* was not affected either by IrAM silencing or methylamine treatment indicating that engulfment of the spirochetes is mediated by a different mechanism independent on thioester-containing proteins.[23] We searched for other thioester-containing proteins in the *I. scapularis* genome and found nine different molecules belonging to the α_2M-superfamily (tentatively named as IsAM1-9).[24] The phylogenetic analysis (Fig. 1) revealed that three IsAMs are related to α_2Ms, three are of the C3-complement type, one is related to insect TEPs and two IsAMs belong to a fourth branch recently renamed macroglobulin complement related (MCR) proteins.[45] Hence, the tick *I. scapularis* is so far the only organism within the animal kingdom known to us to possess representatives of all major groups of invertebrate thioester-containing proteins.[24]

Factor D-like clip-domain serine proteinase homologues have been identified from *D. variabilis* (Dv FacD-L) and also from the *D. andersoni* DEA100 cell line (Da FacD-L) (Table 1).[46] The serine residue of the catalytic triad is substituted with a glycine as in the closely related Limulus factor D, having a microbial activity,[47] or other arthropod clip-domain serine protease homologues playing various roles in immune response.[46] Dv FacD-L is expressed mainly in tick hemocytes and its mRNA level increased significantly upon intrahaemocoelic inoculation of *E. coli.* A clear ortholog of Dv(a) FacD-L is also present in the *I. scapularis* genome. A better understanding of the role of the immune–responsive FacD-L in tick immunity, as well as other molecules potentially related to a tick complement-like system, awaits appropriate functional genomic studies.

Do Ticks Have a Coagulation or Pro-Phenoloxidase Cascade?

Although some authors admit that tick hemolymph is capable to clotting or melanization,[3,12,13] there are actually only a few reports dealing with this aspect of tick innate immunity. As mentioned previously, a process resembling hemolymph coagulation has

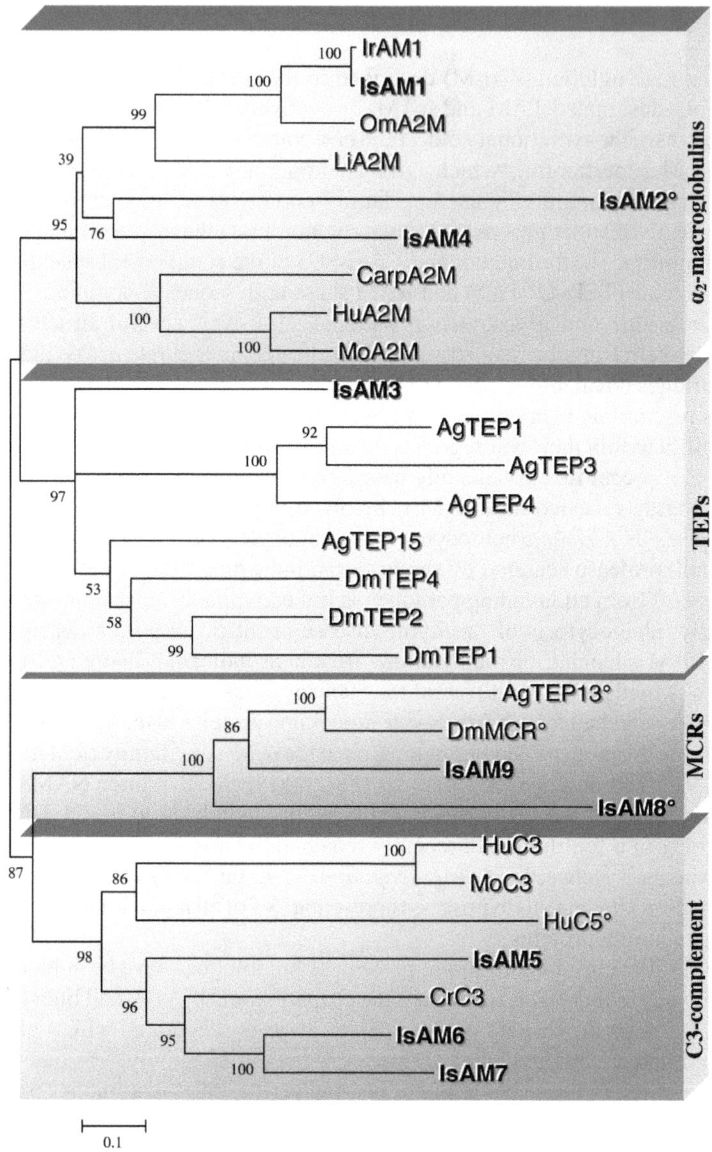

Figure 1. Phylogenetic tree of tick and selected invertebrate and vertebrate thioester-containing proteins (α_2-macroglobulin superfamily). The tree was constructed by the neighbour-joining method using amino acid sequence of ~700 residues spanning across the conserved α_2-M superfamily and thioester domains. Numbers at the branches display bootsrap support. **IsAM**(1-9): thioester proteins found in the *Ixodes scapularis* genome; **Ir**: *Ixodes ricinus* (hard tick); **Om**: *Ornithodoros moubata* (soft tick); **Li**: *Limulus polyphemus* (horseshoe crab); **Cr**: *Carcinoscorpius rotundicauda* (horseshoe crab); **Dm**: *Drosophila melanogaster*; **Ag**: *Anopheles gambiae;* **Carp**: *Cyprinus carpio*; **Hu**: human, **Mo**: mouse; The circlets indicate the absent thioester bond. **MCR** stands for macroglobulin-complement-related. Adapted from Burešová.[24]

been observed in the form of a mass around Epon-Araldite particles implanted under the *D. variabilis* cuticle, although without identification of any of the molecules involved.[26] In addition, we were not able to find any gene related either to the horseshoe crab coagulogen or crustacean clotting protein.[8] On the other hand, the *I. scapularis* genome contains putative transglutaminases. Therefore, a possibility of hemolymph coagulation based on crosslinking of other hemolymph proteins (e.g., the abundant heme lipoproteins[48,49]) can not completely be ruled out. Considering melanization, there are controversial data about the PPO activity in the hemolymph of ticks. It has been shown no PPO activity in the hemolymph of the hard ticks *A. americanum, D. variabilis* and *I. scapularis*,[50] while such activity was detected in the hemolymph of the soft tick *O. moubata*.[51] Since neither PPO nor hemocyanin, which displays PPO activity in the horseshoe crab,[52] seem to be present in the tick genome, it would be essential to characterize the molecule responsible for this activity.

Antimicrobial Peptides in the Tick Hemolymph

The existence of a potent antimicrobial activity against various Gram$^+$, Gram$^-$ bacteria and rickettsiae has been described by Russian researchers in the early 1990s for a relative large set of ticks representing both the Argasidae and Ixodidae families.[53] The antibacterial activities were mainly attributed to the c-type **lysozymes** that were purified and partially characterized.[53,54] Later, two c-type lysozyme orthologues were cloned and sequenced from *D. variabilis* hemocytes and the *D. andersoni* DAE100 cell line (Table 1) and their gene expression was shown to be upregulated upon *E. coli* challenge.[55]

The greatest part of literature on tick antimicrobial peptides (AMPs) deals with **defensins.** This family of small cationic peptides is synthesized as prepropeptides of about 8 kDa, from which approximately 4 kDa mature peptides are cleaved C-terminally via a conserved furin cleavage motif (RVRR). Tick defensins (often having several isoforms) were so far identified in nearly twenty hard and soft tick species.[12,56-58] In addition to the hemolymph, tick defensins were also reported to be expressed in the midgut and other tissues, including salivary glands and fat body (see below and the Table 1). Typical tick defensins with the furin cleavage site and conserved pattern of six paired cysteines within the mature active peptide can be exemplified by **varisin** from *D. variabilis*. Varisin was purified by reverse-phase high performance chromatography (RP-HPLC) from the hemolymph of *D. variabilis* females challenged with *B. burgdorferi* and was shown to be active against the Gram$^+$ *Bacillus subtilis*.[59,60] This defensin or chicken lysozyme alone have low activity against cultured borrelia spirochetes. However, their combination significantly increased borreliacidal activity, indicating that a synergistic effect of these molecules together with efficient phagocytosis (mentioned above) is involved in the immediate clearance of spirochetes from the *D. variabilis* hemolymph.[33,59] The role of varisin in *D. variabilis* innate imunity was also studied by RNAi. Although varisin was shown to be depleted from the hemolymph of the varisin-KD ticks, the activity against the Gram$^+$ *Micrococcus luteus* was only reduced by 50%. This result indicates that varisin contributes significantly but not exclusively to the overall antimicrobial activity in the *D. variabilis* hemolymph.[61] A study of the effect of varisin silencing on the transmission of a relevant pathogen, the rickettsia *Anaplasma marginale*, presented a surprising finding. In opposite to authors' expectations, the *A. marginale* infection was significantly reduced in the varisin-KD ticks for yet unknown reasons.[62]

Two anionic **defensin-like peptides** (ADP1 and ADP2), sharing the conserved cysteine residues of typical defensins but lacking the typical furin cleavage motif, were identified as transcripts in a substractive cDNA library originating from the syngaglion (central nervous system) dissected from fed vs unfed female *Amblyoma hebraeum* ticks (Table 1).[63] The authors succeeded in the purification of ADP2 from tick hemolymph using RP-HPLC and demonstrated its antimicrobial activity against both Gram$^+$ and Gram$^-$ bacteria but not against fungi. The same fed vs unfed *A. hebraeum* syngaglion subtraction cDNA library was used in the identification of a novel type of antimicrobial peptide, named **hebraein** (Table 1).[64] It consists of 102 amino acids, including six cysteins. However, its pattern of disulfide bridges, anionic character and secondary structure with prevailing α-helices clearly differentiate this AMP from arthropod defensins. The most striking feature of hebraein is its C-terminal histidine-rich domain. Isolated as well as recombinant hebraein were active against Gram$^+$ as well as Gram$^-$ bacteria (*Staphylococcus aureus* and *E. coli*) and against the yeast *Candida glabrato*, but not against *C. albicans*. The antibacterial activity of the histidine-deficient mutant was significantly reduced especially against *C. glabrato*. The authors also demonstrated that C-terminal histidine-rich fragment alone does not display any intrinsic antimicrobial activity.[64] At about the same time, a similarly histidine-rich molecule called **microplusin** (Table 1) was purified by RP-HPLC from the hemolymph of *R. microplus*.[65] Recently, the molecular structure of microplusin has been solved using NMR, leading to a model of its mode of action.[66] In addition to *M. luteus* and other Gram$^+$ bacteria, microplusin was also found to be very efficient against some filamentous fungi like *Aspergillus niger* and very specific against the yeast *Candida neoformans* but not against *C. albicans*.[66] In contrast to defensins, microplusin does not affect the permeability of the microbial membrane. Instead, microplusin was found to chelate copper ions by binding them most probably to the histidine residues located at both N-terminal (H2) and C-terminal (H74) ends of the molecule (Fig. 2). The bacteriostatic effect of microplusin against *M. luteus* seems to be due to its capacity to sequester copper ions needed for bacterial respiration, a cooper-dependent process. The growth of *M. luteus* could be restored by supplementing the medium with copper.[66]

Besides defensin and microplusin,[65] a third cystein-rich antimicrobial peptide, designated **ixodidin** (Table 1), has been isolated by RP-HPLC from an acidic extract of *R. microplus* hemocytes.[67] Mature ixodidin possesses ten cysteins forming five internal disulfide bridges.[67] The ixodidin activity against *M. luteus* was in the submicromolar range and five times less efficient against *E. coli*. The sequence of ixodidin displayed a significant similarity with several serine protease inhibitors from different invertebrates and accordingly the peptide exerted inhibitory activity against elastase and chymotrypsin. Whether the antibacterial activity of ixodidin is due to the proteinase inhibition or to a direct effect on the bacterial membrane remains an unresolved issue.[67]

IMMUNITY IN THE TICK GUT

The tick gut is the primary interface of the tick-pathogen interaction. Thus, the ability of a pathogen to survive in the gut lumen, penetrate the gut epithelia and eventually multiply before reaching the haemocoel are intuitively considered as the determinants of the tick vector competence. Ticks differ from other hematophagous arthropods in that they digest blood intracellularly and the protein-rich nutrients are only gradually absorbed by the gut epithelium.[3] Since tick feeding is hardly an aseptic process, there has to be

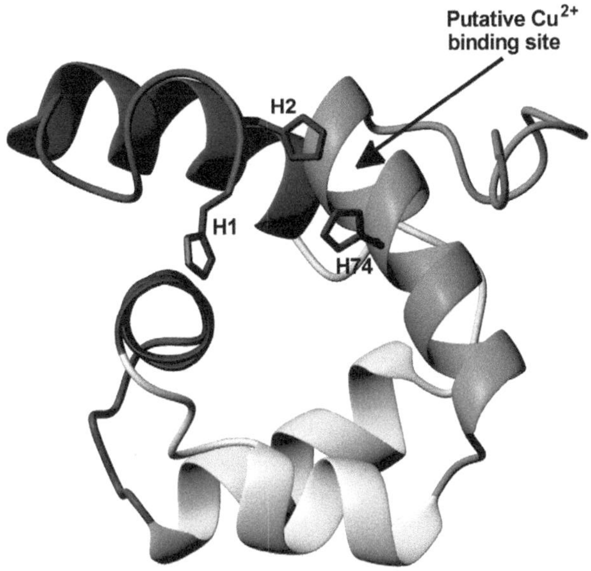

Figure 2. Ribbon representation of the microplusin structure obtained by solution NMR (PDB ID 2knj).[66] Microplusin consists of five α-helices, having histidine residues located at both N-terminal (H2) and C-terminal (H74) ends of the molecule as putative binding site for Cu^{2+} ions.

an active and efficient defense mechanism preventing bacterial growth in the tick gut lumen. The failure of such a mechanism may lead to high mortality, such as reported for the laboratory colony of the soft tick *O. moubata* accidentally infected via membrane feeding with *C. indologenes*.[22] By contrast, the hard tick *I. ricinus* was quite resistant to the per os infection with this bacterium, suggesting that the families Argasidae and Ixodidae may differ also in midgut defense strategies.[22]

The pivotal and constitutive role in microbial control in the tick gut of both families is apparently played by the antimicrobial activity of the host **hemoglobin fragments**. Fogaca et al[68] were the first who demonstrated that a 3,205 Da fragment of the bovine α-hemoglobin purified from the gut contents of *R. microplus* is responsible for activity against several Gram+ bacteria and fungi. A similar purification approach, assaying for activity against *S. aureus* or *M. luteus*, led to the identification of antibacterial hemoglobin fragments in the gut lumen of the soft tick *O. moubata*[69] and the hard tick *D. variabilis*,[70] respectively. The structural analysis of the bovine α-hemoglobin fragment (Hb 33-61) made it possible to explain how the N- and C-terminus of the molecule contribute to the disruption of the *M. luteus*[71] and *C. albicans*[72] membrane. The large hemoglobin fragments are generated in the initial phase of hemoglobin digestion, mainly by the action of an aspartic peptidase of the cathepsin D type.[73] However, the mechanism of their secretion from the digestive vesicles to the gut contents remains obscure. Interestingly, the hemoglobin-derived antimicrobial fragments seem to have a conserved and universal role in innate immunity since they have been also reported as defense molecules in the human placenta.[74]

Apart from the antimicrobial peptides of host origin, several types of endogenous molecules demonstrably or presumably contributing to immunity within the tick gut have been described. A c-type **lysozyme** was purified from the gut contents of the soft tick *O. moubata* showing activity against *M. luteus* (Table 1).[75] The sequence of the *O. moubata* gut lysozyme shared some motives with lysozymes found in the digestive tract of insects or stomachs of ruminants and its expression was strongly up-regulated by blood meals.[76] The lysozyme of the hard tick *D. variabilis*, originally identified in hemocytes,[55] was later found to be also markedly expressed in the midgut of this species.[77] In contrast to the *O. moubata* gene, the expression *D. variabilis* lysozyme mRNA did not increase upon feeding but seemed to be up-regulated in ticks infected *per os* with the causative agent of spotted fever, *Rickettsia montanensis*, by capillary feeding.[77]

Four isoforms of typical cationic **defensin**-A, B, C, D (Table 1) that became up-regulated after blood feeding were described in *O. moubata* ticks by Nakajima et al[78,79] Isoforms A, B and C were preferentially expressed in the midgut and isoform D in the fat body. The authors succeeded in purification defensin A from the gut contents by RP-HPLC and demonstrated its activity against *S. aureus.*[79] The antibacterial assays performed with synthetic *O. moubata* defensin A confirmed its bactericidal activity against a panel of $Gram^{+}$ but not $Gram^{-}$ bacteria. The mode of action of defensin A peptide was demonstrated by the permeabilization of the bacterial membrane.[80]

Of special interest is a report on a defensin-related molecule termed **longicin** (Table 1), expressed mainly in the midgut of the babesial vector *Haemaphysalis longicornis.*[81] Although longicin possesses the typical furin cleavage motif, the authors have shown that the mature protein also contains the N-terminal part of typical tick pro-defensins. Recombinant longicin was active against a variety of fungi, $Gram^{+}$ and $Gram^{-}$ bacteria, including multidrug-resistant strains. Most strikingly, the recombinant protein exerted babesiacidal activity in vitro. In an in vivo experiment, the inoculation of longicin significantly reduced parasitemia in mice infected with *Babesia microti.* RNAi experiments performed in this work further supported the role of longicin in *Babesia* transmission and regulation of *H. longicornis* vectorial capacity. Since this work points to an exceptional activity of a defensin-related molecule against a protozoan apicomplexan parasite, which is related to the causative agent of malaria, any follow-up studies would be quite significant.

Another typical tick-defensin named Hl-gut defensin (Table 1) was found to be specifically expressed in the *H. longicornis* midgut and strongly up-regulated upon intrahaemocoelic injection of LPS.[82] Defensins predominantly expressed in the midgut tissue were also reported for the genera *Ixodes* (Table 1), namely *I. ricinus*,[83] *I. scapularis*[84] and *I. persulcatus.*[56] The most recent comparison of the antibacterial activity of synthetic peptides corresponding to the *I. ricinus* defensin, *O. moubata* defensin D and the above mentioned longicin from *H. longicornis*, revealed that all defensins were active against $Gram^{+}$ *S. aureus.* On the other hand, the Lyme disease spirochete *Borrelia garinii* and the symbiotic bacteria *Stenotrophomonas maltophila* isolated from *I. persulcatus* were resistant to the tick defensins, suggesting an adaptation of these microbes to the tick antimicrobial response.[85]

In contrast to the soft ticks, neither defensins nor lysozymes have been unambiguously proven to be secreted from the midgut epithelium into the gut lumen of hard ticks. An attempt to detect these immune molecules, next to the hemoglobin antimicrobial fragments in the *D. varibilis* gut contents, was not successful.[70] Defensin and lysozyme transcripts were surprisingly absent in the *D. variabilis* midgut transcriptome.[86] These antibacterial peptides were also not found among almost 150 proteins identified by a recent extensive proteomic analysis of the midgut from partially fed *R. microplus* females.[87] Given these data, the ultimate evidence that defensins and/or lysozymes really contribute to the control of microbial growth in the midgut of hard ticks awaits further experimental verification.

Protease inhibitors may play a role in tick midgut immunity via the inhibition of microbe-specific proteases. Zhou et al reported that the expression of a cysteine protease inhibitor of the cystatin type (Hlcyst-2) (Table 1) in the midgut of *H. longicornis* was increased in response to blood feeding, LPS injection and experimental infection with *Babesia gibsoni.* They have further shown that recombinant cystatin partially limited the growth of *Babesia bovis* in culture.[88] Another example was provided for a Kunitz-type serine protease inhibitor from *D. variabilis*, referred to as KPI.[89] In contrast to the majority of related tick Kunitz-type serine protease inhibitors expressed in the salivary glands to control homeostasis in the host, KPI was highly expressed in the midgut in response to feeding and its mRNA level was also induced after *per os* infection with rickettsia. Beside a trypsin inhibitory activity and possible role as an anticoagulant of the imbibed blood, the authors demonstrated that *D. variabilis* KPI limits rickettsial colonization of mouse fibroblasts.[89]

The paradigm that *Drosophila* midgut epithelial immunity efficiently and specifically controls intestinal microbes by the fine-tuned generation and elimination of reactive oxygen species (see refs. 90, 91) raises the question as to whether **redox homeostasis** contributes to tick midgut immunity as well. Ticks defend themselves against the oxidative stress caused by an enormous amount of heme liberated from the digested hemoglobin by complex heme detoxification mechanisms (see ref. 92 for a review). Ticks get rid of the heme mainly by its aggregation into hemosomes–special organelles of digestive cells that are subsequently released to the gut lumen.[93] A small portion of heme binds to a heme-binding protein named HeLp,[53] which serves both as heme transporter and antioxidant enzyme in the hemolymph (Table 1).[48,94] Citteli et al demonstrated that catalase, an enzyme responsible for H_2O_2 detoxification, plays an important role in regulation of oxidative stress in the *R. microplus* midgut.[95] Analysis of the *D. variabilis* midgut transcriptome revealed other putative antioxidant enzymes and radical scavengers, such as superoxide dismutase, glutathione peroxidase, thioredoxin, glutathion S-transferase (GST), selenoprotein M and others.[86] Interestingly, several of these antioxidant enzymes were reported to be up-regulated in different tick species in response to infection with relevant pathogens. In the hard tick *I. ricinus,* GST and thioredoxin peroxidase were found by the method of subtractive hybridization among the genes induced after *B. burgdorferi*-infected blood meal.[83] GST was also identified to be significantly up-regulated in the gut of *D. variabilis* males fed on calf infected with *A. marginale.* RNAi experiments suggested that GST is involved in the initial phase of midgut infection and possibly also in trafficking of the pathogen and its multiplication in the tick salivary glands.[96] Another study on two midgut specific *D. variabilis* GSTs demonstrated that expression of these enzymes in the tick gut did not respond to *E. coli* haemocoelic injection but was primarily up-regulated by blood feeding.[97] Several putative antioxidant enzymes have been cloned and partially characterized in different tick species, but their function in tick redox balance or immune response has not yet been demonstrated (Table 1).[98-102] The acquisition of *B. burgdorferi* by *I. scapularis* was reported to be facilitated in part by Salp25D, a protein homologous to glutathione peroxidases (peroxiredoxins) that are thiol-specific antioxidants.[103] The authors were able to perform a tissue-specific silencing of Salp25D, which allowed them to demonstrate that Salp25D from salivary glands but not the midgut is important for successful entry of the pathogen into the midgut. Furthermore, they showed that spirochetes exploit the salivary Salp25D to protect themselves against reactive oxygen intermediates generated by the mammalian neutrophils at the vector-host interface.[103] This study elegantly illustrates the complexity of the molecular interplay between the host, vector and pathogen.

IMMUNE MOLECULES IN OTHER TICK TISSUES

Beside defensin and microplusin transcripts,[104] only three antibacterial peptides have been thus far identified among the numerous molecules secreted by tick salivary glands (Table 1). Two short peptides of different sequence, designated as Ixosin and Ixosin B were isolated from the salivary glands of the hard tick *Ixodes sinensis*.[105,106] Both mature Ixosins lack cysteine residues and display a similar antimicrobial activity mainly against *S. aureus* and the yeast *C. albicans* and to a lesser extent against *E. coli*. Recently, a novel salivary antimicrobial peptide named IsAMP has been isolated from the saliva of *I. scapularis*.[107] Interestingly, it was found to belong to a family of secreted 5.3 kDa proteins previously identified to be specifically up-regulated in *B. burgdorferi*-infected nymphs.[108] The mature IsAMP is composed of 47 amino acid residues including six cysteine residues. IsAMP is active against $Gram^+$ and $Gram^-$ bacteria. In addition to salivary glands, IsAMP is expressed in hemocytes and the fat body but not in the midgut.[107]

Esteves et al. detected several antimicrobial activities in the egg homogenate and surface material from the tick *R. microplus*.[109] One of the antimicrobial peptides identified in the egg homogenates was attributed to the previously described microplusin.[65] Moreover, the regulation of microplusin mRNA transcription seems to be stage-dependent, during ovary and egg development. The authors suggest that microplusin acts as a protective factor of both the female reproductive tract and embryo.[109] In contrast, the compound(s) responsible for antimicrobial activity on the egg surface remain to be identified. A defense role is possibly played also by two subtilisin inhibitors named as BmSI-7 and BmSI-6 (Table 1) purified from the *R. microplus* eggs since they strongly inhibit Pr1 protease from the entomopathogenic fungus *Metarhizium anisopliae*.[110]

Several transcripts of genes putatively involved in immune response were identified in phagocytic cell line BME26, derived from *R. microplus* embryos.[111] They comprise e.g., microplusin, hebraein, α_2-macroglobulin, factor D-like protein, GST, peroxidase and NADPH oxidase. This result suggests that this cell line has the potential to be a valuable tool for further studies on the interaction of tick cells with relevant pathogens, as recently demonstrated for the *A. marginale*.[112]

Regulation of Immune Response in Ticks

Although several of the above mentioned immune molecules are induced by microbial challenge, there is virtually nothing known about the regulation of the immune response in ticks. The *I. scapularis* genome contains a number of genes automatically annotated as putative components of the immune sensing and signaling pathway(s),[9] comprising e.g., putative Toll-like receptors with leucine-rich repeats domains, a nuclear factor κB (NF-κB) related to Dorsal and also an ankyrin-repeat containing protein resembling Cactus, a NF-κB inhibitor. This finding suggests that the Toll signaling pathway is present in ticks. Several components putatively related to the *Drosophila* Imd pathway, such as Caspar and Relish, are also found in the *I. scapularis* genome, but an ortholog of Imd itself seems to be missing.

The only sufficiently characterized tick molecule that plays a role in regulation of immune gene transcription is **subolesin** (Table 1).[113] This molecule (originally named 4D8) was discovered in *I. scapularis* as a tick protective antigen, since immunization with recombinant subolesin effectively protected hosts against infestation and had a deleterious effect on tick survival and further development.[114] Recently, it was demonstrated that tick subolesin is an ortholog of insect and vertebrate akirin,[115] a highly conserved nuclear protein involved in the transcription of immune genes associated with NF-κB.[116]

CONCLUSION

Research on tick innate immunity and the molecular interplay at the tick-pathogen interface is just entering its postgenomic era. Reverse genetics and functional genomics based on RNAi will continue to substantially extend our knowledge that can potentially be exploited to combat tick-borne pathogens. In addition, the introduction of these techniques may make ticks to an additional model for the study of innate immunity, at least for the class Arachnida. Certainly, the main effort will be always targeted to understand how the transmitted pathogens evade the tick immune response. Data from other invertebrate models can be used to fish out putative functional homologues in tick genomes or EST databases and ultimately verify their immune functions. We believe that ticks will prove to be a suitable model system for discovering molecules involved in pathogen recognition, the action of the primordial complement system and regulation of antimicrobial responses, among other aspects of innate immunity. Further research on tick midgut immunity and its relation to blood digestion and/or redox homeostasis may be a key to efficient control of ticks and tick-borne pathogens.

ACKNOWLEDGEMENTS

The Czech authors were supported by a grant P506/10/2136 to P.K. from the Grant Agency of the Czech Republic, by Research Center No. LC06009 and research projects Nos. Z60220518 and MSMT 6007665801 from the Ministry of Education, Youth and Sports of the Czech Republic. Sirlei Daffre has financial support from FAPESP (Fundação de Amparo a Pesquisa do Estado de São Paulo, Brazil) and CNPq (Conselho Nacional de Desenvolvimento Científico e Tecnológico, Brazil).

REFERENCES

1. Nava S, Guglielmone AA, Mangold AJ. An overview of systematics and evolution of ticks. Front Biosci 2009; 14:2857-2877.
2. Barker SC, Murell A. Systematics and evolution of ticks with a list of valid genus and species names. In: Bowman AS, Nuttall PA, eds. Ticks: Biology, Disease and Control. New York: Cambridge University Press; 2008:1-39.
3. Sonenshine DE. Biology of ticks. Vol 1. New York: Oxford University Press; 1991.
4. Mans BJ, Neitz AW. Adaptation of ticks to a blood-feeding environment: evolution from a functional perspective. Insect Biochem Mol Biol 2004; 34:1-17.
5. Jongejan F, Uilenberg G. The global importance of ticks. Parasitology 2004; 129 Suppl:S3-14.
6. de la Fuente J, Estrada-Pena A, Venzal JM et al. Overview: Ticks as vectors of pathogens that cause disease in humans and animals. Front Biosci 2008; 13:6938-6946.

7. Francischetti IM, Sa-Nunes A, Mans BJ et al. The role of saliva in tick feeding. Front Biosci 2009; 14:2051-2088.
8. Iwanaga S, Lee BL. Recent advances in the innate immunity of invertebrate animals. J Biochem Mol Biol 2005; 38:128-150.
9. Ferrandon D, Imler JL, Hetru C et al. The Drosophila systemic immune response: sensing and signalling during bacterial and fungal infections. Nat Rev Immunol 2007; 7:862-874.
10. Osta MA, Christophides GK, Vlachou D et al. Innate immunity in the malaria vector Anopheles gambiae: comparative and functional genomics. J Exp Biol 2004; 207:2551-2563.
11. Lehane MJ, Aksoy S, Levashina E. Immune responses and parasite transmission in blood-feeding insects. Trends Parasitol 2004; 20:433-439.
12. Sonenshine DE, Hynes WL. Molecular characterization and related aspects of the innate immune response in ticks. Front Biosci 2008; 13:7046-7063.
13. Taylor D. Innate immunity in ticks. J Acarol Soc Jpn 2006; 15:109-127.
14. Nene V. Tick genomics—coming of age. Front Biosci 2009; 14:2666-2673.
15. de la Fuente J, Kocan KM, Almazan C et al. RNA interference for the study and genetic manipulation of ticks. Trends Parasitol 2007; 23:427-433.
16. Bell-Sakyi L, Zweygarth E, Blouin EF et al. Tick cell lines: tools for tick and tick-borne disease research. Trends Parasitol 2007; 23:450-457.
17. Kaufman WR, Phillips JE. Ion and Water-Balance in Ixodid Tick Dermacentor-Andersoni. 1. Routes of Ion and Water Excretion. Journal of Experimental Biology 1973; 58:523-536.
18. Kuhn KH, Haug T. Ultrastructural, Cytochemical and Immunocytochemical Characterization of Hemocytes of the Hard Tick Ixodes-Ricinus (Acari Chelicerata). Cell and Tissue Research 1994; 277:493-504.
19. Inoue N, Hanada K, Tsuji N et al. Characterization of phagocytic hemocytes in Ornithodoros moubata (Acari: Ixodidae). J Med Entomol 2001; 38:514-519.
20. Borovickova B, Hypsa V. Ontogeny of tick hemocytes: a comparative analysis of Ixodes ricinus and Ornithodoros moubata. Exp Appl Acarol 2005; 35:317-333.
21. Loosova G, Jindrak L, Kopacek P. Mortality caused by experimental infection with the yeast Candida haemulonii in the adults of Ornithodoros moubata (Acarina: Argasidae). Folia Parasitol (Praha) 2001; 48:149-153.
22. Buresova V, Franta Z, Kopacek P. A comparison of Chryseobacterium indologenes pathogenicity to the soft tick Ornithodoros moubata and hard tick Ixodes ricinus. J Invertebr Pathol 2006; 93:96-104.
23. Buresova V, Hajdusek O, Franta Z et al. IrAM-An alpha2-macroglobulin from the hard tick Ixodes ricinus: characterization and function in phagocytosis of a potential pathogen Chryseobacterium indologenes. Dev Comp Immunol 2009; 33:489-498.
24. Burešová V. Function of the α2-macroglobulin protein family in the immune response of the tick Ixodes ricinus. [PhD.]. České Budějovice: Faculty of Science, University of South Bohemia; 2009.
25. Pereira LS, Oliveira PL, Barja-Fidalgo C et al. Production of reactive oxygen species by hemocytes from the cattle tick Boophilus microplus. Exp Parasitol 2001; 99:66-72.
26. Eggenberger LR, Lamoreaux WJ, Coons LB. Hemocytic encapsulation of implants in the tick Dermacentor variabilis. Exp Appl Acarol 1990; 9:279-287.
27. Ceraul SM, Sonenshine DE, Hynes WL. Resistance of the tick dermacentor variabilis (Acari: Ixodidae) following challenge with the bacterium Escherichia coli (Enterobacteriales: Enterobacteriaceae). J Med Entomol 2002; 39:376-383.
28. Mattila JT, Munderloh UG, Kurtti TJ. Phagocytosis of the Lyme disease spirochete, Borrelia burgdorferi, by cells from the ticks, Ixodes scapularis and Dermacentor andersoni, infected with an endosymbiont, Rickettsia peacockii. J Insect Sci 2007; 7:58.
29. Blouin EF, de la Fuente J, Garcia-Garcia JC et al. Applications of a cell culture system for studying the interaction of Anaplasma marginale with tick cells. Anim Health Res Rev 2002; 3:57-68.
30. Kurtti TJ, Keyhani NO. Intracellular infection of tick cell lines by the entomopathogenic fungus Metarhizium anisopliae. Microbiology 2008; 154:1700-1709.
31. Rittig MG, Kuhn KH, Dechant CA et al. Phagocytes from both vertebrate and invertebrate species use "coiling" phagocytosis. Developmental and Comparative Immunology 1996; 20:393-406.
32. Coleman JL, Gebbia JA, Piesman J et al. Plasminogen is required for efficient dissemination of B. burgdorferi in ticks and for enhancement of spirochetemia in mice. Cell 1997; 89:1111-1119.
33. Johns R, Ohnishi J, Broadwater A et al. Contrasts in tick innate immune responses to Borrelia burgdorferi challenge: immunotolerance in Ixodes scapularis versus immunocompetence in Dermacentor variabilis (Acari: Ixodidae). J Med Entomol 2001; 38:99-107.
34. Johns R, Sonenshine DE, Hynes WL. Response of the tick Dermacentor variabilis (Acari: Ixodidae) to hemocoelic inoculation of Borrelia burgdorferi (Spirochetales). J Med Entomol 2000; 37:265-270.

35. Grubhoffer L, Rego ROM, Hajdušek O et al. Tick lectins and fibrinogen-related proteins. In: Bowman AS, Nuttall PA, eds. Ticks: Biology, Disease and Control. New York: Cambridge University Press; 2008:127-142.
36. Kovar V, Kopacek P, Grubhoffer L. Isolation and characterization of Dorin M, a lectin from plasma of the soft tick Ornithodoros moubata. Insect Biochem Mol Biol 2000; 30:195-205.
37. Rego RO, Kovar V, Kopacek P et al. The tick plasma lectin, Dorin M, is a fibrinogen-related molecule. Insect Biochem Mol Biol 2006; 36:291-299.
38. Gokudan S, Muta T, Tsuda R et al. Horseshoe crab acetyl group-recognizing lectins involved in innate immunity are structurally related to fibrinogen. Proc Natl Acad Sci USA 1999; 96:10086-10091.
39. Kawabata S, Tsuda R. Molecular basis of nonself recognition by the horseshoe crab tachylectins. Biochim Biophys Acta 2002; 1572:414-421.
40. Rego RO, Hajdusek O, Kovar V et al. Molecular cloning and comparative analysis of fibrinogen-related proteins from the soft tick Ornithodoros moubata and the hard tick Ixodes ricinus. Insect Biochem Mol Biol 2005; 35:991-1004.
41. Huang X, Tsuji N, Miyoshi T et al. Molecular characterization and oligosaccharide-binding properties of a galectin from the argasid tick Ornithodoros moubata. Glycobiology 2007; 17:313-323.
42. Kopacek P, Weise C, Saravanan T et al. Characterization of an alpha-macroglobulin-like glycoprotein isolated from the plasma of the soft tick Ornithodoros moubata. Eur J Biochem 2000; 267:465-475.
43. Saravanan T, Weise C, Sojka D et al. Molecular cloning, structure and bait region splice variants of alpha2-macroglobulin from the soft tick Ornithodoros moubata. Insect Biochem Mol Biol 2003; 33:841-851.
44. Blandin S, Levashina EA. Thioester-containing proteins and insect immunity. Mol Immunol 2004; 40:903-908.
45. Stroschein-Stevenson SL, Foley E, O'Farrell PH et al. Identification of Drosophila gene products required for phagocytosis of Candida albicans. PLoS Biol 2006; 4:e4.
46. Simser JA, Mulenga A, Macaluso KR et al. An immune responsive factor D-like serine proteinase homologue identified from the American dog tick, Dermacentor variabilis. Insect Mol Biol 2004; 13:25-35.
47. Kawabata S, Tokunaga F, Kugi Y et al. Limulus factor D, a 43-kDa protein isolated from horseshoe crab hemocytes, is a serine protease homologue with antimicrobial activity. FEBS Lett 1996; 398:146-150.
48. Maya-Monteiro CM, Daffre S, Logullo C et al. HeLp, a heme lipoprotein from the hemolymph of the cattle tick, Boophilus microplus. J Biol Chem 2000; 275:36584-36589.
49. Gudderra NP, Sonenshine DE, Apperson CS et al. Hemolymph proteins in ticks. J Insect Physiol 2002; 48:269-278.
50. Zhioua E, Yeh MT, LeBrun RA. Assay for phenoloxidase activity in Amblyomma americanum, Dermacentor variabilis and Ixodes scapularis. J Parasitol 1997; 83:553-554.
51. Kadota K, Satoh E, Ochiai M et al. Existence of phenol oxidase in the argasid tick Ornithodoros moubata. Parasitol Res 2002; 88:781-784.
52. Nagai T, Kawabata S. A link between blood coagulation and prophenol oxidase activation in arthropod host defense. J Biol Chem 2000; 275:29264-29267.
53. Podboronov VM, Berdyev A. Defense Mechanisms of Ixodid Ticks and Their Hosts (Parasite-Host Interactions) (in Russian) Aschabad: Ylym; 1991.
54. Podboronov VM. Antibacterial protective mechanisms of Ixodoid ticks. In: Dusbabek F, Bukva V, eds. Modern Acarology. Vol 2. Prague: Academia; 1991:375-380.
55. Simser JA, Macaluso KR, Mulenga A et al. Immune-responsive lysozymes from hemocytes of the American dog tick, Dermacentor variabilis and an embryonic cell line of the Rocky Mountain wood tick, D. andersoni. Insect Biochem Mol Biol 2004; 34:1235-1246.
56. Saito Y, Konnai S, Yamada S et al. Identification and characterization of antimicrobial peptide, defensin, in the taiga tick, Ixodes persulcatus. Insect Mol Biol 2009; 18:531-539.
57. Chrudimska T, Chrudimsky T, Golovchenko M et al. New defensins from hard and soft ticks: Similarities, differences and phylogenetic analyses. Vet Parasitol 2009.
58. Todd SM, Sonenshine DE, Hynes WL. Tissue and life-stage distribution of a defensin gene in the Lone Star tick, Amblyomma americanum. Med Vet Entomol 2007; 21:141-147.
59. Johns R, Sonenshine DE, Hynes WL. Identification of a defensin from the hemolymph of the American dog tick, Dermacentor variabilis. Insect Biochem Mol Biol 2001; 31:857-865.
60. Ceraul SM, Sonenshine DE, Ratzlaff RE et al. An arthropod defensin expressed by the hemocytes of the American dog tick, Dermacentor variabilis (Acari: Ixodidae). Insect Biochem Mol Biol 2003; 33:1099-1103.
61. Hynes WL, Stokes MM, Hensley SM et al. Using RNA interference to determine the role of varisin in the innate immune system of the hard tick Dermacentor variabilis (Acari: Ixodidae). Exp Appl Acarol 2008; 46:7-15.

62. Kocan KM, de la Fuente J, Manzano-Roman R et al. Silencing expression of the defensin, varisin, in male Dermacentor variabilis by RNA interference results in reduced Anaplasma marginale infections. Exp Appl Acarol 2008; 46:17-28.
63. Lai R, Lomas LO, Jonczy J et al. Two novel noncationic defensin-like antimicrobial peptides from haemolymph of the female tick, Amblyomma hebraeum. Biochem J 2004; 379:681-685.
64. Lai R, Takeuchi H, Lomas LO et al. A new type of antimicrobial protein with multiple histidines from the hard tick, Amblyomma hebraeum. FASEB J 2004; 18:1447-1449.
65. Fogaca AC, Lorenzini DM, Kaku LM et al. Cysteine-rich antimicrobial peptides of the cattle tick Boophilus microplus: isolation, structural characterization and tissue expression profile. Dev Comp Immunol 2004; 28:191-200.
66. Silva FD, Rezende CA, Rossi DC et al. Structure and mode of action of microplusin, a copper II-chelating antimicrobial peptide from the cattle tick Rhipicephalus (Boophilus) microplus. J Biol Chem 2009; 284:34735-34746.
67. Fogaca AC, Almeida IC, Eberlin MN et al. Ixodidin, a novel antimicrobial peptide from the hemocytes of the cattle tick Boophilus microplus with inhibitory activity against serine proteinases. Peptides 2006; 27:667-674.
68. Fogaca AC, da Silva PI, Jr., Miranda MT et al. Antimicrobial activity of a bovine hemoglobin fragment in the tick Boophilus microplus. J Biol Chem 1999; 274:25330-25334.
69. Nakajima Y, Ogihara K, Taylor D et al. Antibacterial hemoglobin fragments from the midgut of the soft tick, Ornithodoros moubata (Acari: Argasidae). J Med Entomol 2003; 40:78-81.
70. Sonenshine DE, Hynes WL, Ceraul SM et al. Host blood proteins and peptides in the midgut of the tick Dermacentor variabilis contribute to bacterial control. Exp Appl Acarol 2005; 36:207-223.
71. Sforca ML, Machado A, Figueredo RC et al. The micelle-bound structure of an antimicrobial peptide derived from the alpha-chain of bovine hemoglobin isolated from the tick Boophilus microplus. Biochemistry 2005; 44:6440-6451.
72. Machado A, Sforca ML, Miranda A et al. Truncation of amidated fragment 33-61 of bovine alpha-hemoglobin: effects on the structure and anticandidal activity. Biopolymers 2007; 88:413-426.
73. Horn M, Nussbaumerova M, Sanda M et al. Hemoglobin digestion in blood-feeding ticks: mapping a multipeptidase pathway by functional proteomics. Chem Biol 2009; 16:1053-1063.
74. Liepke C, Baxmann S, Heine C et al. Human hemoglobin-derived peptides exhibit antimicrobial activity: a class of host defense peptides. J Chromatogr B Analyt Technol Biomed Life Sci 2003; 791:345-356.
75. Kopacek P, Vogt R, Jindrak L et al. Purification and characterization of the lysozyme from the gut of the soft tick Ornithodoros moubata. Insect Biochem Mol Biol 1999; 29:989-997.
76. Grunclova L, Fouquier H, Hypsa V et al. Lysozyme from the gut of the soft tick Ornithodoros moubata: the sequence, phylogeny and postfeeding regulation. Dev Comp Immunol 2003; 27:651-660.
77. Ceraul SM, Dreher-Lesnick SM, Gillespie JJ et al. New tick defensin isoform and antimicrobial gene expression in response to Rickettsia montanensis challenge. Infect Immun 2007; 75:1973-1983.
78. Nakajima Y, van der Goes van Naters-Yasui A, Taylor D et al. Two isoforms of a member of the arthropod defensin family from the soft tick, Ornithodoros moubata (Acari: Argasidae). Insect Biochem Mol Biol 2001; 31:747-751.
79. Nakajima Y, van der Goes van Naters-Yasui A, Taylor D et al. Antibacterial peptide defensin is involved in midgut immunity of the soft tick, Ornithodoros moubata. Insect Mol Biol 2002; 11:611-618.
80. Nakajima Y, Ishibashi J, Yukuhiro F et al. Antibacterial activity and mechanism of action of tick defensin against Gram-positive bacteria. Biochim Biophys Acta 2003; 1624:125-130.
81. Tsuji N, Battsetseg B, Boldbaatar D et al. Babesial vector tick defensin against Babesia sp. parasites. Infect Immun 2007; 75:3633-3640.
82. Zhou J, Liao M, Ueda M et al. Sequence characterization and expression patterns of two defensin-like antimicrobial peptides from the tick Haemaphysalis longicornis. Peptides 2007; 28:1304-1310.
83. Rudenko N, Golovchenko M, Edwards MJ et al. Differential expression of Ixodes ricinus tick genes induced by blood feeding or Borrelia burgdorferi infection. J Med Entomol 2005; 42:36-41.
84. Hynes WL, Ceraul SM, Todd SM et al. A defensin-like gene expressed in the black-legged tick, Ixodes scapularis. Med Vet Entomol 2005; 19:339-344.
85. Isogai E, Isogai H, Takahashi K et al. Antimicrobial activity of three tick defensins and four mammalian cathelicidin-derived synthetic peptides against Lyme disease spirochetes and bacteria isolated from the midgut. Exp Appl Acarol 2009; 49:221-228.
86. Anderson JM, Sonenshine DE, Valenzuela JG. Exploring the mialome of ticks: an annotated catalogue of midgut transcripts from the hard tick, Dermacentor variabilis (Acari: Ixodidae). BMC Genomics 2008; 9:552.
87. Kongsuwan K, Josh P, Zhu Y et al. Exploring the midgut proteome of partially fed female cattle tick (Rhipicephalus (Boophilus) microplus). J Insect Physiol 2010; 56:212-226.

88. Zhou J, Ueda M, Umemiya R et al. A secreted cystatin from the tick Haemaphysalis longicornis and its distinct expression patterns in relation to innate immunity. Insect Biochem Mol Biol 2006; 36:527-535.
89. Ceraul SM, Dreher-Lesnick SM, Mulenga A et al. Functional characterization and novel rickettsiostatic effects of a Kunitz-type serine protease inhibitor from the tick Dermacentor variabilis. Infect Immun 2008; 76:5429-5435.
90. Ha EM, Oh CT, Bae YS et al. A direct role for dual oxidase in Drosophila gut immunity. Science 2005; 310:847-850.
91. Ha EM, Lee KA, Seo YY et al. Coordination of multiple dual oxidase-regulatory pathways in responses to commensal and infectious microbes in drosophila gut. Nat Immunol 2009; 10:949-957.
92. Graca-Souza AV, Maya-Monteiro C, Paiva-Silva GO et al. Adaptations against heme toxicity in blood-feeding arthropods. Insect Biochem Mol Biol 2006; 36:322-335.
93. Lara FA, Lins U, Paiva-Silva G et al. A new intracellular pathway of haem detoxification in the midgut of the cattle tick Boophilus microplus: aggregation inside a specialized organelle, the hemosome. J Exp Biol 2003; 206:1707-1715.
94. Maya-Monteiro CM, Alves LR, Pinhal N et al. HeLp, a heme-transporting lipoprotein with an antioxidant role. Insect Biochem Mol Biol 2004; 34:81-88.
95. Citelli M, Lara FA, da Silva Vaz I Jr et al. Oxidative stress impairs heme detoxification in the midgut of the cattle tick, Rhipicephalus (Boophilus) microplus. Mol Biochem Parasitol 2007; 151:81-88.
96. Kocan KM, Zivkovic Z, Blouin EF et al. Silencing of genes involved in Anaplasma marginale-tick interactions affects the pathogen developmental cycle in Dermacentor variabilis. BMC Dev Biol 2009; 9:42.
97. Dreher-Lesnick SM, Mulenga A, Simser JA et al. Differential expression of two glutathione S-transferases identified from the American dog tick, Dermacentor variabilis. Insect Mol Biol 2006; 15:445-453.
98. He H, Chen AC, Davey RB et al. Characterization and molecular cloning of a glutathione S-transferase gene from the tick, Boophilus microplus (Acari: Ixodidae). Insect Biochem Mol Biol 1999; 29:737-743.
99. Tsuji N, Kamio T, Isobe T et al. Molecular characterization of a peroxiredoxin from the hard tick Haemaphysalis longicornis. Insect Mol Biol 2001; 10:121-129.
100. Rosa de Lima MF, Sanchez Ferreira CA, Joaquim de Freitas DR et al. Cloning and partial characterization of a Boophilus microplus (Acari: Ixodidae) glutathione S-transferase. Insect Biochem Mol Biol 2002; 32:747-754.
101. da Silva Vaz Jnr I, Imamura S, Ohashi K et al. Cloning, expression and partial characterization of a Haemaphysalis longicornis and a Rhipicephalus appendiculatus glutathione S-transferase. Insect Mol Biol 2004; 13:329-335.
102. Cossio-Bayugar R, Miranda E, Holman PJ. Molecular cloning of a phospholipid-hydroperoxide glutathione peroxidase gene from the tick, Boophilus microplus (Acari: Ixodidae). Insect Biochem Mol Biol 2005; 35:1378-1387.
103. Narasimhan S, Sukumaran B, Bozdogan U et al. A tick antioxidant facilitates the Lyme disease agent's successful migration from the mammalian host to the arthropod vector. Cell Host Microbe 2007; 2:7-18.
104. Mans BJ, Andersen JF, Francischetti IM et al. Comparative sialomics between hard and soft ticks: implications for the evolution of blood-feeding behavior. Insect Biochem Mol Biol 2008; 38:42-58.
105. Yu D, Sheng Z, Xu X et al. A novel antimicrobial peptide from salivary glands of the hard tick, Ixodes sinensis. Peptides 2006; 27:31-35.
106. Liu Z, Liu H, Liu X et al. Purification and cloning of a novel antimicrobial peptide from salivary glands of the hard tick, Ixodes sinensis. Comp Biochem Physiol B Biochem Mol Biol 2008; 149:557-561.
107. Pichu S, Ribeiro JM, Mather TN. Purification and characterization of a novel salivary antimicrobial peptide from the tick, Ixodes scapularis. Biochem Biophys Res Commun 2009; 390:511-515.
108. Ribeiro JM, Alarcon-Chaidez F, Francischetti IM et al. An annotated catalog of salivary gland transcripts from Ixodes scapularis ticks. Insect Biochem Mol Biol 2006; 36:111-129.
109. Esteves E, Fogaca AC, Maldonado R et al. Antimicrobial activity in the tick Rhipicephalus (Boophilus) microplus eggs: Cellular localization and temporal expression of microplusin during oogenesis and embryogenesis. Dev Comp Immunol 2009; 33:913-919.
110. Sasaki SD, de Lima CA, Lovato DV et al. BmSI-7, a novel subtilisin inhibitor from Boophilus microplus, with activity toward Pr1 proteases from the fungus Metarhizium anisopliae. Exp Parasitol 2008; 118:214-220.
111. Esteves E, Lara FA, Lorenzini DM et al. Cellular and molecular characterization of an embryonic cell line (BME26) from the tick Rhipicephalus (Boophilus) microplus. Insect Biochem Mol Biol 2008; 38:568-580.
112. Esteves E, Bastos CV, Zivkovic Z et al. Propagation of a Brazilian isolate of Anaplasma marginale with appendage in a tick cell line (BME26) derived from Rhipicephalus (Boophilus) microplus. Vet Parasitol 2009; 161:150-153.

113. de la Fuente J, Maritz-Olivier C, Naranjo V et al. Evidence of the role of tick subolesin in gene expression. Bmc Genomics 2008; 9:372.
114. de la Fuente J, Almazan C, Blas-Machado U et al. The tick protective antigen, 4D8, is a conserved protein involved in modulation of tick blood ingestion and reproduction. Vaccine 2006; 24:4082-4095.
115. Galindo RC, Doncel-Perez E, Zivkovic Z et al. Tick subolesin is an ortholog of the akirins described in insects and vertebrates. Dev Comp Immunol 2009; 33:612-617.
116. Goto A, Matsushita K, Gesellchen V et al. Akirins are highly conserved nuclear proteins required for NF-kappaB-dependent gene expression in drosophila and mice. Nat Immunol 2008; 9:97-104.

CHAPTER 9

BEETLE IMMUNITY

Ji-Won Park,[1] Chan-Hee Kim,[1] Jiang Rui,[1] Keun-Hwa Park,[1] Kyung-Hwa Ryu,[1] Jun-Ho Chai,[1] Hyun-Ok Hwang,[1] Kenji Kurokawa,[1] Nam-Chul Ha,[1] Irene Söderhäll,[2] Kenneth Söderhäll[2] and Bok Luel Lee*,[1]

[1]*National Research Laboratory of Defense Proteins, College of Pharmacy, Pusan National University, Busan, Korea;* [2]*Department of Comparative Physiology, Uppsala University, Uppsala, Sweden*
**Corresponding Author: Bok Luel Lee—Email: brlee@pusan.ac.kr*

Abstract: Genetic studies have elegantly characterized the innate immune response in *Drosophila melanogaster*. However, these studies have a limited ability to reveal the biochemical mechanisms underlying the innate immune response. To investigate the biochemical basis of how insects recognize invading microbes and how these recognition signals activate the innate immune response, it is necessary to use insects, from which larger amounts of hemolymph can be extracted. Using the larvae from two species of beetle, *Tenebrio molitor* and *Holotrichia diomphalia*, we elucidated the mechanisms underlying pathogenic microbe recognition. In addition, we studied the mechanism of host defense molecule amplification. In particular, we identified several pattern recognition proteins, serine proteases, serpins and antimicrobial peptides and examined how these molecules affect innate immunity.

INTRODUCTION

Beetles (Coleoptera) such as fireflies and bombardier beetles, are one of the most evolutionarily conserved metazoan families of insects. The red flour beetle, *Tribolium castaneum* (Coleoptera, Tenebrionidae), is a common model organism used for developmental, evolution, comparative genomics and pest science. In addition, *Tribolium* has been the subject of studies investigating host interactions with a variety of pathogenic bacteria, sporozoa, cestoda, nematoda, mites and hymenopterous parasites.[1] The *Tribolium* genome sequence has recently been determined by the Human Genome Sequencing Center, Baylor College of Medicine.[2] This information has made *Tribolium* an excellent model for

Invertebrate Immunity, edited by Kenneth Söderhäll.

Figure 1. The pictures of beetles. (A) *Tenebrio molitor*; (B) *Holotrichia diomphalia*; (C) *Tribolium castaneum*. To image the real sizes of beetles, measures are inserted. Reproduced with permission from Iwanaga S, Lee BL. J Biochem Mol Biol 2005; 38:128-50, ©2005.[36]

studying insect immunity. *Tenebrio molitor* (mealworm) and *Holotrichia diomphalia* are from the same family as *Tribolium* and are larger than *Tribolium* (Fig. 1). These insects reproduce rapidly and up to 500 μl of hemolymph (insect blood) can be extracted from a single larva. Because the *Tenebrio* and *Holotrichia* larvae elicit a high antimicrobial activity when challenged with pathogenic microbes, we used *Tenebrio* and *Holotrichia* larvae as a model system to elucidate the biochemical mechanisms underlying the host defense system.

BIOCHEMICAL CHARACTERIZATION OF PATTERN RECOGNITION PROTEINS IN BEETLES

The innate immune response is a crucial host defense system to defend against microbial infection.[3] The ability of a host to distinguish between self and nonself remains a central hallmark of innate immunity.[4] The pathogenic microbes possess distinct pathogen-associated molecular patterns (PAMPs), such as lipopolysaccharides (LPSs) of Gram-negative bacteria, peptidoglycans (PGs) of Gram-positive bacteria and β-1,3-glucans of fungi.[5] The recognition of PAMPs is achieved by a group of germ line-encoded receptors and soluble proteins.[6] The *Drosophila* Toll receptor signaling pathway is responsible for the defense against Gram-positive bacteria and fungi, while the immune deficiency (Imd) pathway is activated primarily in response to Gram-negative bacteria. Both of these pathways lead to the expression of antimicrobial peptides (AMPs) via the activation of NF-κB-like transcription factors.[7-10] In insects, bacterial lysine (Lys)-type PGs are recognized by the PG recognition protein-SA (PGRP-SA) and Gram-negative-binding protein 1 (GNBP1).[11-13] These proteins are believed to mediate the activation of a serine protease (SP) cascade and ultimately, the cleavage of Spätzle pro-protein. Upon cleavage, Spätzle serves as a ligand for the Toll receptor and increases the production of AMPs.[10,14] GNBP3 is required for the detection of fungal cell wall components.[15] These recognition signals are amplified in hemolymph by a proteolytic SP cascade; this protein cascade is conserved in the vertebrate system.[10,16] The amplification of these recognition signals results in an efficient host defense strategy in insects, which are devoid of an acquired immune system.

The prophenoloxidase (proPO) activation pathway, similar to the vertebrate complement system, is a proteolytic molecular cascade comprised of pattern recognition proteins and several SPs and their inhibitors. The cascade terminates with the zymogen, proPO.[17,18] Microbial carbohydrates, such as LPS, PG and β-1,3-glucan, are initially recognized by pattern recognition proteins and induce SP-dependent activation of the proPO system. The proPO-activating enzyme (PPAE) and proPO-activating factor (PPAF), which are similar to the *Drosophila* Easter-type SP, cleave proPO to generate the active enzyme, phenoloxidase (PO).[19-22] This enzyme produces compounds that are toxic to microorganisms by the oxidation of phenols and the subsequent formation of melanin. In addition, this enzyme participates in the sclerotization of cuticles, a process that is vital for the survival of insects.[23] Numerous reports have been published about invertebrate proPO and its activation mechanism.[18] A key question regarding the proPO system is: how do pattern recognition molecules activate this system in response to microbial infection?

Although information about the function of distinct PAMP recognition proteins, such as PGRP-SA and GNBP3, from genetic studies in *Drosophila* gives insight into the Toll cascade, the activation mechanism of the extracellular SP cascade in the Toll pathway has not been clearly elucidated in *Drosophila*. Because *Tenebrio* and *Holotrichia* larvae have a high antimicrobial activity against bacteria and fungi,[24-28] we sought to biochemically determine the molecular activation mechanism of the Toll cascade in beetles. Initially, we purified and cloned GNBP3[29] from *Tenebrio* hemolymph using a 1,3-β-glucan affinity column. The purified protein bound to β-1,3-glucan but not to bacterial PG. Subsequent molecular cloning revealed that GNBP3 contains a region with a sequence similar to bacterial glucanases.[29] Interestingly, two catalytically important residues in the glucanases had been replaced with nonhomologous amino acids in *Tenebrio* GNBP3, suggesting that GNBP3 had evolved from an ancestral glucanase gene but retained only the ability to recognize β-1,3-glucan. Next, we purified and cloned the PGRP proteins,[30,31] termed *Holotrichia* PGRP-SA[30] and *Tenebrio* PGRP-SA.[31] The purified 20-kDa and 17-kDa *Holotrichia* proteins (referred to as PGRP-1 and PGRP-2, respectively) had a high sequence identity to *Drosophila* PGRP-SA. We purified *Tenebrio* PGRP-SA using a novel synthetic Lys-type PG peptide [(GlcNAc-Mur-NAc-L-Ala-D-isoGln-L-Lys-D-Ala)$_2$]. This synthetic Lys-type PG fragment is a competitive inhibitor of PG-dependent proPO cascade and thus, did not activate the proPO system. The purified *Tenebrio* PGRP-SA recognized both Lys-type PG and diaminopimel:c(DAP)-type PG. In vitro reconstitution experiments demonstrated that *Tenebrio* PGRP-SA is a common recognition molecule of Lys-PG- and DAP-PG-dependent proPO cascades.[31]

We purified a novel 40-kDa LPS recognizing protein, termed LPS recognition protein (LRP), from the cell-free plasma of *H. diomphalia* larvae.[32] LRP exhibited agglutinating activity in *Escherichia coli,* but not in *Staphylococcus aureus* and *Candida albicans*. The *E. coli* agglutinating activity was preferentially inhibited by the rough-type LPS, which contains an oligosaccharide core. The LRP consisted of 317 amino acid residues and six repeats of an epidermal growth factor (EGF)-like domain. Recombinant LRP also had *E. coli* agglutination activity in vitro and was able to neutralize LPS by the inhibition of LPS-induced interleukin 6 (IL-6) production in mouse bone marrow mast cells. Furthermore, the *E. coli* coated with the purified LRP were cleared more rapidly in the *Holotrichia larvae* than in the noncoated *E. coli*, indicating that this protein participates in the clearance of *E. coli*. Three N-terminal EGF-like domains of LRP, but not the three C-terminal EGF-like domains, were involved in the LPS-binding activity. In conclusion, the data suggest that LRP is a pattern recognition protein for LPS and is a protein central to the innate immune response.

SERINE PROTEASE ZYMOGENS REGULATE THE proPO CASCADE IN BEETLES

Initially, we identified and cloned two proPO-activating factors, termed PPAF-I and PPAF-II. These proteins directly activate proPO in the *Holotrichia* hemolymph.[19,33,34] The overall structure of PPAF-I is very similar to that of the *Drosophila* Easter SP (42.9% identity),[35] an essential SP zymogen in pattern formation during embryonic development. The locations of disulfide linkages in the prosegment of PPAF-I are similar to those of the proclotting enzyme from *Tachypleus*.[36] Furthermore, [^{3}H]diisopropylphosphate (iPr2P)-labeled PPAF-I was produced in the crude preparation of the PPAF-I zymogen when incubated with LPS or β–1,3-glucan. However, [^{3}H]iPr2P-labeled PPAF-I was not produced under the same conditions in the absence of these microbial polysaccharides. These results indicate that microbial polysaccharide-dependent activation of the proPO system in *Holotrichia* larvae activates the PPAF-I zymogen. The structure of the PPAF-II is similar to that of the masquerade protein,[37] a SP homologue expressed during embryogenesis and *Drosophila* larval and pupal development. The 45-kDa PPAF-II protein contained a trypsin-like SP domain at the C-terminus. PPAF-II differs from trypsin in that there is a substitution of a serine residue in the active site triad to a glycine and this protein has a disulfide-knotted clip domain at the N-terminus. Using in vitro reconstitution experiments, we demonstrated that PPAF-II, active PPAF-I and proPO were necessary for showing PO activity in the *Holotrichia* proPO system, suggesting that PPAF-II and PPAF-I are essential factors in the *Holotrichia* proPO cascade.

The crystal structure of PPAF-II (Fig. 2) revealed that the clip domain adopts a novel fold containing a central cleft, which is distinct from the structures of defensins that have a similar arrangement of cysteine residues.[38] The PPAF-II forms a homo-oligomer upon cleavage by the upstream SP and the central cleft in the clip domain of PPAF-II functions as a module for binding PO. Similarly, the clip domain in the catalytically active Easter-type SP plays an essential role in the rapid activation of the protease domain.[38] In addition, we

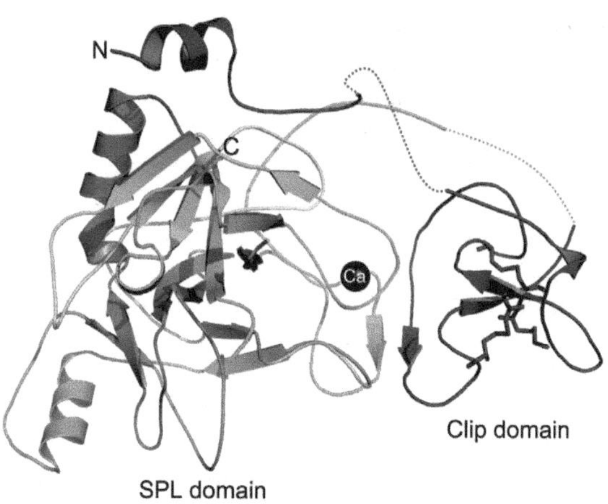

Figure 2. Overall structure of *Holotrichia* PPAF-II. A ribbon representation of PPAF-II structure. The clip domain and the PPAF-II family-specific N-terminal segment are shown in blue. Bound calcium is in black. The disordered regions are shown in dashed lines. A color version of this image is available at www.landesbioscience.com/curie.

previously reported that the crystal structure of the SP domain of the Easter-type SP is structurally similar to PPAF-I. This analysis revealed several important structural features including a Ca^{2+} binding region, an additional loop with unique disulfide linkages, a canyon-like deep active site and an exposed activation loop.[39] We subsequently elucidated the function of the bound Ca^{2+} and discovered that the activation loop is susceptible to proteolytic degradation, which did not involve the clip domain.

We purified and cloned the 27-kDa PPAF-III and two proPOs from the plasma of *Holotrichia* larvae.[40] The *PPAF-III* gene encoded an Easter-type SP zymogen with a mass of 40 kDa. The active form of the 27-kDa PPAF-III specifically cleaved a 55-kDa proPPAF-II in a Ca^{2+}-independent manner to generate a 45-kDa PPAF-II. Furthermore, *Holotrichia* proPO-I and proPO-II have been characterized and the activation-dependent structural changes were examined using in vitro reconstitution experiments. When the proPOs were incubated with PPAF-I, the 79-kDa proPOs were converted to 76-kDa proPOs, which did not exhibit any PO activity. However, when the proPOs were incubated simultaneously with PPAF-I, proPPAF-II and PPAF-III in the presence of Ca^{2+}, a 60-kDa protein (PO-1) with PO activity was detected in addition to the 76-kDa proPO-II protein. These results indicate that the conversion of *Holotrichia* proPOs to enzymatically active PO was accomplished by PPAF-I, PPAF-II and PPAF-III through a two-step Ca^{2+}-dependent proteolytic process. Finally, we purified and cloned the *Tenebrio* SP homologue-1 (SPH1) and proPO[41] from the *Tenebrio* hemolymph.[42] Using an in vitro reconstitution experiment, we demonstrated that the cleavage of the 55-kDa SPH1 zymogen by an unidentified SP was necessary for PO activity and that SPH1 functioned as a PPAF.[42]

A MODULAR SERINE PROTEASE IS THE IMMEDIATE DOWNSTREAM PROTEASE OF THE PGRP-SA/GNBP1 COMPLEX

Drosophila genetic studies suggest that PGRP-SA and GNBP1 form a complex that recruits SPs and activates the Toll signaling cascade.[43] However, the identification of the SPs immediately downstream of the PGRP-SA/GNBP1 complex is not easily performed in *Drosophila* due to protein redundancy. Since we observed a high antimicrobial activity against a challenge with Gram-positive bacteria or fungi in *Tenebrio* larvae, we hypothesized that *Tenebrio* has the essential components necessary for the activation of the Toll signaling cascade. Therefore, we purified the immediate downstream SP of the PGRP-SA/GNBP1 complex in *Tenebrio* using a biochemical approach.[13]

We hypothesize that GNBP1 and an SP would be recruited to the PG/PGRP-SA complex when the soluble Lys-type PG/*Tenebrio* PGRP-SA complex was incubated with PGRP-SA-depleted *Tenebrio* hemolymph. Indeed, the *Tenebrio* PGRP-SA/Lys-type PG complex enriched a 50-kDa protein and a 35-kDa protein. The sequencing of the N-terminal residues identified the 50-kDa protein as *Tenebrio* GNBP1 and the 35-kDa protein as *Tenebrio* modular serine protease (MSP). However, these proteins do not contain a clip domain, which is commonly found in proteases upstream of the Toll and proPO cascades.[38] GNBP1 physically interacts with PGRP-SA and activates the Toll pathway in *Drosophila*.[43] However, an interaction between GNBP1 and PGRP-SA has not been observed in vitro. Our observation suggests that the binding of PGRP-SA to PG enhanced the interaction between PGRP-SA and GNBP1 and the active form of MSP was recruited to the Lys-type PG/PGRP-SA/GNBP1 complex. Subsequent cDNA cloning demonstrated that *Tenebrio* MSP contains four low-density lipoprotein receptor A repeat (LDL) domains, one complement control protein (CCP) domain and a trypsin-like SP domain.[44] We demonstrated that the Lys-type PG/PGRP-SA complex recruited GNBP1 and MSP and proposed that this Lys-type PG/PGRP-SA/GNBP1/MSP complex initiates the SP cascade in the Toll and proPO pathways in beetles.

Manduca sexta hemolymph protease 14 (Ms-HP-14), which contains a domain arrangement similar to that of the *Tenebrio* MSP, initiates the activation of the proPO system.[45,46] In addition, Lemaitre et al[47] reported that the *Drosophila* MSP plays an essential role in the activation of the Toll pathway by Gram-positive bacteria and fungi in *Drosophila.* Their analysis showed that the MSP integrates signals that originate from the circulating recognition molecules, GNBP3 and PGRP-SA and relays them to the Grass-Spätzle processing enzyme (SPE) extracellular pathway, which is upstream SPs of the Toll receptor.

THREE SERINE PROTEASE ZYMOGENS ARE INVOLVED IN LYS-TYPE PG-DEPENDENT ACTIVATION OF THE TOLL SIGNALING CASCADE

Although we found that MSP is recruited into the Lys-type PG recognition complex, the lack of information regarding the identity of the immediate downstream factor(s) of MSP limits our knowledge of the molecular details of PG recognition and the involvement of SPs in the Toll signaling pathway. Therefore, the identification and determination of the biological functions of SPs involved in the *Tenebrio* Toll signaling pathway are essential to the elucidation of the host defense system. SPE, a terminal SP that converts Spätzle pro-protein into a processed form capable of binding to the Toll receptor, was identified in *Drosophila.*[48] We purified the immediately downstream SPs of *Tenebrio* MSP and *Tenebrio* SPE in order to obtain biochemical information regarding the Lys-type PG recognition signal-dependent activation of the Toll cascade.

We purified three SPs, *Tenebrio* MSP and a 41-kDa and 44-kDa protein, to homogeneity by column chromatography and cloned their cDNAs (Fig. 3).[44] We sought to identify the SPE molecule in *Tenebrio*. Because the 44-kDa zymogen protein is similar to *Drosophila* SPE and Easter, we initially studied the active form of the 44-kDa protease. Since the cDNAs of *Tenebrio* Spätzle and the Toll proteins have not been determined, we expressed the *Tribolium* Spätzle pro-protein and the Toll ectodomain in a baculovirus system (Fig. 3). To address whether the purified active 44-kDa protease cleaves *Tribolium* Spätzle pro-protein in vitro, we incubated the Spätzle pro-protein with trypsin, the active forms of *Tenebrio* MSP and the 41-kDa and 44-kDa proteases. Only the active 44-kDa protease cleaved the Spätzle pro-protein. Under the same conditions, trypsin cleaved the Spätzle pro-protein nonspecifically and the active forms of *Tenebrio* MSP and the 41-kDa protease did not cleave the Spätzle pro-protein. In addition, we determined the N-terminal amino acid sequence of the cleaved 24-kDa Spätzle pro-protein to be: Phe-Asn-Thr-Asp-Glu-Met-Ser-Leu. This sequence is identical to residues 125-132 of the *Tribolium* Spätzle protein. Therefore, we named the 44-kDa protease as *Tenebrio* SPE.

We next tried to identify the upstream activator of *Tenebrio* SPE. Because *Tenebrio* SPE has a trypsin cleavage site, we hypothesized that the upstream SP is a trypsin-like SP. As mentioned above, the MSP and the 41-kDa zymogens were identified as chymotrypsin-like and trypsin-like SPs, respectively, suggesting that the active form of the 41-kDa protease may cleave the SPE zymogen. To test this hypothesis, we incubated the active form of

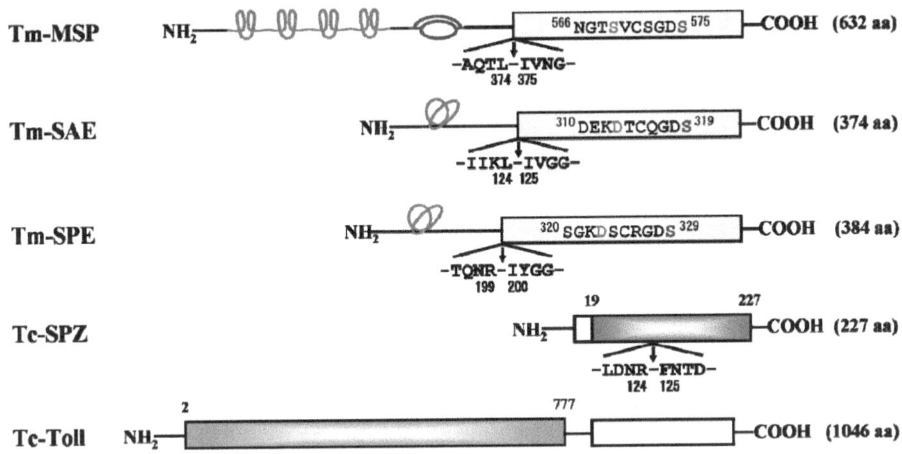

Figure 3. The domain organizations based on deduced amino acid sequences of cDNAs of *T. molitor*-MSP (Tm-MSP), Tm-41 kDa SP (Tm-SAE), Tm-44 kDa SP (Tm-SPE) and *T. castaneum*-Spätzle (TC-SPZ) and Tc-ecotodomain Toll. *Rabbit ears*, *half-double circle* and *rectangular symbols* indicate the domains of LDL, CCP and SP domains of MSP, SAE and SPE, respectively. *Moon* symbols indicate the clip domain of SPs. *Arrows* represent the cleavage sites of SP zymogens during activation. The *red* and *blue* residues in the boxes indicate the specificity-related residue and catalytic triad Ser residue, respectively. Reproduced from Kim CH et al. J Biol Chem 2008; 283:7599-07,[44] ©2008, with permission from the American Society for Bioschemistry and Molecular Biology. A color version of this image is available at www.landesbioscience.com/curie.

41-kDa protein with the purified SPE zymogen. As expected, the SPE zymogen was hydrolyzed into a 35-kDa SP domain and a 15-kDa clip domain. The 35-kDa band was identified as the SP domain of SPE. Therefore, we designated the 41-kDa protease as *Tenebrio* SPE-activating enzyme (SAE).

Because the sequence of the cleavage site in the SAE zymogen is Leu-124-Ile-125, the upstream SP of SAE is probably similar to chymotrypsin. This result suggests that the SAE zymogen is cleaved by the MSP. Therefore, activated MSP was incubated with the recombinant SAE zymogen. The SAE zymogen was hydrolyzed into a 35-kDa SP domain and an 11-kDa clip domain. The N-terminal amino acid sequence of the 35-kDa band was Ile-Val-Gly-Gly-Thr-Asn. This sequence is identical to the amino acid sequence of the SAE zymogen from Ile-125 to Asn-130, demonstrating that the MSP induced a limited proteolytic cleavage between the clip domain and the catalytic SP domain of the SAE zymogen. Thus, the SAE protease is an immediate downstream target of MSP.

Finally, to confirm whether the Spätzle pro-protein is cleaved, Lys-type-PG, PGRP-SA, GNBP1, zymogens of MSP, SAE, SPE and pro-Spätzle were incubated together in the presence of Ca^{2+} and the cleavage of pro-Spätzle was detected by western blot analysis. As predicted, a 14-kDa band corresponding to cleaved Spätzle was observed (Fig. 4, column 9). However, if any one of the components was omitted from the incubation mixture, no cleavage of the pro-Spätzle occurred (columns 2-8). In conclusion, these experiments demonstrated that the PGRP-SA/GNBP1-mediated Lys-type PG recognition signal is transferred by three different SPs; the initiating enzyme is the 82-kDa chymotrypsin-like MSP and the other two enzymes are the 41-kDa SAE and 44-kDa SPE clip domain-containing trypsin-like SPs.

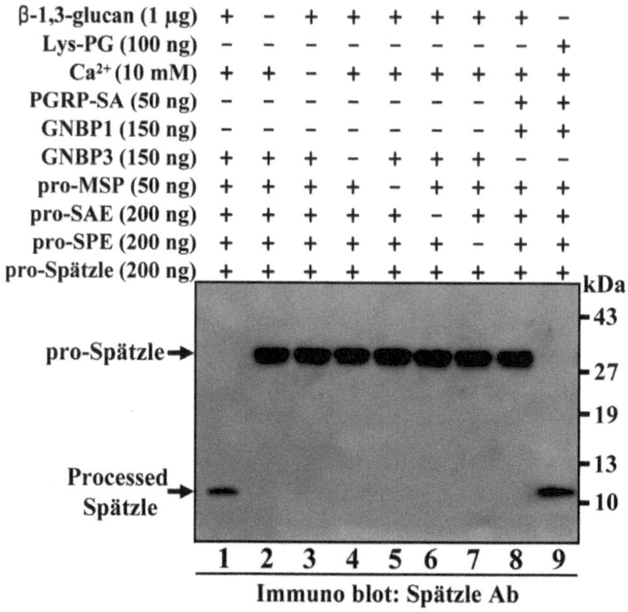

Figure 4. In vitro reconstitution experiments for the activation of pro-Spätzle by β-1,3-glucan and Lys-type PG. The mixture of *Tenebrio* pro-MSP, GNBP3, pro-MSP, pro-SAE, pro-SPE and *Tribolium* pro-Spätzle in the presence of β-1,3-glucan and Ca^{2+} was incubated for 60 min and analyzed by Western blotting with an affinity purified anti-Spätzle antibody (*lane 1*). The 30-kDa pro-Spätzle and the 12-kDa-processed Spätzle are indicated with *arrows*. As a control, when eight components, such as Lys-type PG/PGRP-SA/GNBP1/MSP/SAE/SPE/Spätzle were incubated together, the cleaved 12-kDa Spätzle was generated (*lane 9*). In the absence of any one of these components, pro-Spätzle was not converted to the processed Spätzle (*lanes 2-8*). Reproduced from Roh KB et al. J Biol Chem 2009; 284:19474-81,[49] ©2009, with permission from the American Society for Biochemistry and Molecular Biology.

IDENTIFICATION OF THE ESSENTIAL COMPONENTS REGULATING THE β-1,3-GLUCAN-DEPENDENT TOLL SIGNALING CASCADE

We identified three SPs that are involved in the Lys-type PG-dependent Toll signaling pathway. Because the activation mechanism of Toll pathway by β-1,3-glucan is not clearly characterized in insects, we aimed to identify the downstream SPs of GNBP3, a receptor for β-1,3-glucan from fungi, and their order of activation.[49] To identify the immediate downstream molecule(s) that is recruited by the β-1,3-glucan/GNBP3 complex, the insoluble β-1,3-glucan and the purified *Tenebrio* GNBP3 complex were incubated with the hemolymph fraction. A 35-kDa band was enriched from the GNBP3/insoluble β-1,3-glucan complex but not from the insoluble β-1,3-glucan lacking GNBP3. The N-terminal amino acid sequence of the 35-kDa protein was identical to the catalytic SP domain of *Tenebrio* MSP, suggesting that the MSP was recruited to the β-1,3-glucan/GNBP3 complex. *Tenebrio* MSP is an apical SP that is activated downstream of PGRP-SA/GNBP1 in response to Lys-type PG treatment.[13]

To confirm that the 35-kDa band was generated by *Tenebrio* pro-MSP in the presence of β-1,3-glucan, we performed western blot analysis using an MSP antibody on hemolymph that was incubated with β-1,3-glucan and GNBP3. *Tenebrio* pro-MSP in the hemolymph was not activated by GNBP3 but was processed to the 35-kDa active form of MSP when both GNBP3 and β-1,3-glucan were added to the hemolymph. These results suggest that the pro-MSP is processed into its active 35-kDa form in the presence of GNBP3 and that β-1,3-glucan and MSP binds to the β-1,3-glucan/GNBP3 complex. These data indicate that MSP is an apical SP that is immediately downstream of the β-1,3-glucan/GNBP3 complex.

Because the *Tenebrio* MSP zymogen is activated in the presence of the β-1,3-glucan/GNBP3 complex and the Lys-type PGN/PGRP-SA/GNBP1 complex, we hypothesized that the SPs downstream of the β-1,3-glucan/GNBP3 complex are identical to *Tenebrio* SAE and SPE, which are activated in response to Lys-type PG treatment.[44] To address this hypothesis, we performed an in vitro reconstitution experiment using five purified proteins: GNBP3, pro-MSP, pro-SAE, pro-SPE and pro-Spätzle. A western blot analysis revealed that the processed Spätzle was generated upon incubation of the five proteins with β-1,3-glucan and Ca^{2+} (Fig. 4, column 1). The elimination of any of the components resulted in the loss of pro-Spätzle cleavage (columns 2-8). Processing of pro-Spätzle to the processed Spätzle was also observed when PGRP-SA, GNBP1, MSP, SAE, SPE and Spätzle were incubated with Lys-type PG and Ca^{2+}. These results clearly demonstrate that GNBP3, in the presence of β-1,3-glucan, activates a three-step proteolytic cascade involving MSP, SAE and SPE. The cascade activation results in the formation of processed Spätzle, which functions as a ligand for the Toll receptor. In addition, our data demonstrate that the β-1,3-glucan and Lys-type PG recognition signals share a three-step proteolytic cascade that relays their signals to the Toll receptor (Fig. 5).

Although we have provided biochemical evidence elucidating the mechanism by which the β-1,3-glucan recognition signal is transferred to the Toll receptor, we have not demonstrated whether this SP cascade is present in vivo. We hypothesized that if this cascade is present in vivo, the same AMP(s) will be produced in the insect hemolymph when the pathway molecules are injected into the *Tenebrio* larvae. Furthermore, we anticipate that AMP(s) expression will increase upon injection of Lys-type PG. To address this hypothesis, we injected four molecules, β-1,3-glucan, Lys-type PG, activated SAE and processed Spätzle, into the *Tenebrio* larvae. The hemolymph samples had high antimicrobial activities against *S. aureus*, *E. coli* and *S. cerevisiae*. We purified two AMPs, tenecin 1 and tenecin 2, by column chromatography from four hemolymph samples. Tenecin 1 had a bactericidal activity against Gram-positive bacteria and was previously identified by our group.[25] The amino acid sequence of tenecin 1 and its disulfide bond arrangement are similar to the defensin protein from *Drosophila*.[50] Tenecin 2 is highly homologous to the antimicrobial peptide, coleoptericin,[51] which was purified from the coleopteran insect, *Zophobas atratus*. Tenecin 2 has a high sequence identity with coleoptericin and holotricin 2 (65% and 36%, respectively).[24] Holotricin 2, previously identified by our group, is an inducible antibacterial peptide purified from the *Holotrichia* larvae. Tenecin 2 had a bactericidal activity against *E. coli* and *S. cerevisiae*. These results demonstrated that tenecin 1 and tenecin 2 are induced by treatment with β-1,3-glucan, Lys-type PG, SAE and Spätzle. These results support the notion that β-1,3-glucan and Lys-type PG activate Toll receptors by the same three-step proteolytic cascade, which results in the production of tenecin 1 and 2 (Fig. 5).

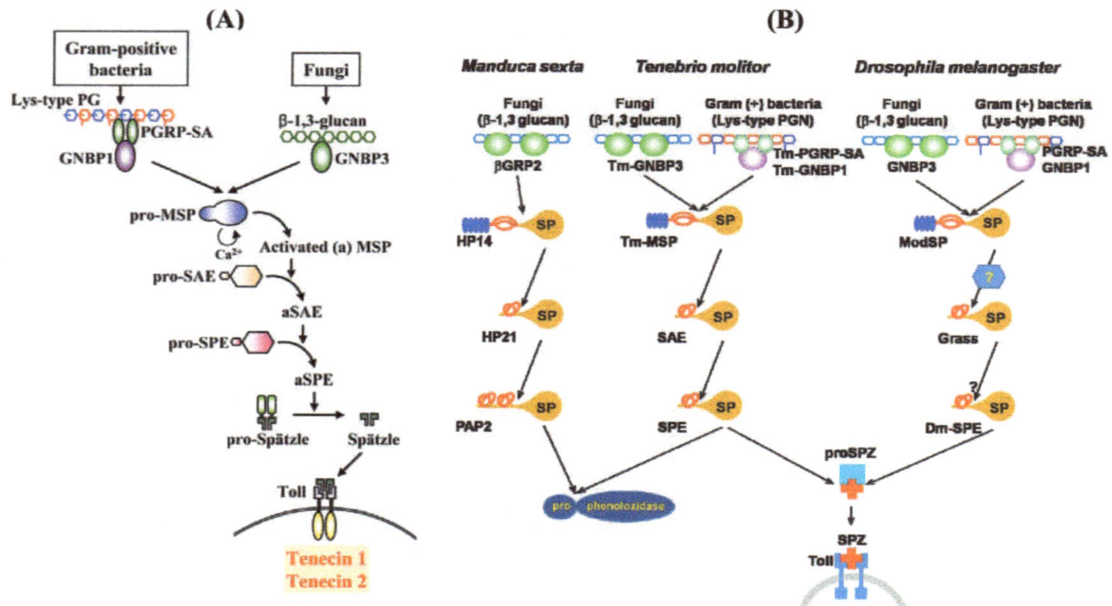

Figure 5. A model of the serine proteases cascades regulating the Toll pathway in *Tenebrio* (A) and in other insects (B). A) When the Lys-type-PG of Gram-positive bacteria or β-1,3-glucan of fungi is exposed to the insect, *Tenebrio* PGRP-SA/GNBP1 complex or GNBP3 binds to Lys-type PG or β-1,3-glucan, respectively and then recruits pro-MSP. In the presence of Ca^{2+}, the Lys-type PG/PGRP-SA/GNBP1/or β-1,3-glucan/GNBP3 complex induces activation of pro-MSP to activated MSP (*aMSP*). The aMSP converts pro-SAE to activated SAE (*aSAE*), which subsequently converts the pro-SPE to activated SPE (*aSPE*). The aSPE cleaves pro-Spätzle to processed Spätzle leading to the production of AMPs such as tenecin 1 and tenecin 2 via the Toll receptor. Reproduced from Roh KB et al. J Biol Chem 2009; 284:19474-81,[49] ©2009, with permission from the American Society for Biochemistry and Molecular Biology. B) Biochemical studies performed in *T. molitor* and *M. sexta* and genetic analysis in *D. melanogaster*, reveal striking similarities in the mechanisms underlying SP activation by pattern recognition proteins. In *M. sexta*, HP-14 is converted to a 2-chain active form in the presence of β-1,3-glucan and the β-1,3-glucan recognition protein 2 (β-GRP2, a protein related to GNBP3).[46,69] Binding of β-1,3-glucan and β-GRP2 triggers the autoactivation of proHP14 into HP14 that then processes HP21. In turn, HP21 cleaves proPAP-2 into PAP2 that finally activates proPO. Hence, the mechanism of activation of HP14 in response to glucan is highly reminiscent of the mode of activation of Tenebrio-MSP, with the exception that HP14 can be activated upon direct binding to PG during bacterial infection.[69] This suggests that the PGRP-SA/GNBP1 complex involved in sensing Gram-positive bacteria has been added on to a more ancient cascade core during evolution.

THREE PROTEASE-SERPIN COMPLEXES COOPERATIVELY REGULATE THE INNATE IMMUNE RESPONSES IN BEETLES

The serpins belong to a superfamily of SP inhibitors and act as suicide substrates by binding covalently to their target proteases.[52] Serpins are known to regulate various physiological processes and molecular defense systems in mammals.[53] To date, the four *Drosophila* serpins involved in innate immunity, SPN43Ac, SPN27A, SPN77Ba and SPN28D, have been extensively analyzed using a genetic approach. *SPN43Ac* mutant flies accumulate cleaved Spätzle, resulting in the constitutive activation of the Toll pathway and expression of AMPs.[14] *SPN27A* and *SPN28D* are known to regulate the Toll pathway during early development[54-56] and are involved in melanin biosynthesis.[57,58] Another serpin, *SPN77Ba*, was identified as a negative regulator of melanization in the respiratory system (the trachea) in *Drosophila*.[59] Six serpins have been identified in *M. sexta* and characterized as negative regulators of the proPO cascade.[60] These serpins are thought to regulate the Toll and proPO cascades. However, the molecular identity of the serpin target SPs and the biochemical regulatory mechanisms of these serpins in relation to the Toll and proPO cascades are not clearly understood. Because we have already identified three SPs that directly activate the *Tenebrio* Toll cascade, we used the *Tenebrio* model system to identify and characterize the novel serpin targets. Recently, we identified three novel serpin-protease pairs that negatively regulate the Toll and proPO cascades.[61]

A hallmark of serpin-dependent SP inhibition is the formation of a covalently bonded SP-serpin complex that is visualized as a high-molecular-weight band by SDS-PAGE.[62] The identity of this band is easily confirmed by western blot analysis using SP antibodies.[63] Recently, we obtained the active forms of three *Tenebrio* Toll cascade-activating SPs and raised polyclonal antibodies to these proteins.[44] These tools were used to monitor the generation of SP-serpin complexes and to purify the serpins. Three novel serpins with a molecular mass of 40 kDa, 55 kDa and 48 kDa (termed *Tenebrio* SPN40, SPN55 and SPN48, respectively) were purified and formed complexes with MSP, SAE and SPE, respectively. To examine the specificity of each serpin for its protease target in vitro, we incubated each purified serpin with the SPs and performed a western blot analysis (Fig. 6). As expected, SPN40 formed a complex with the MSP (lane 2 in Fig. 6B); it did not form a complex with SAE or SPE (lanes 3 and 4). In addition, SPN55 and SPN48 formed complexes with SAE and SPE, respectively (lanes 7 and 12). As a control, purified recombinant *Tenebrio* SPN1, which has a high sequence identity (37%) with *Drosophila* SPN1,[64] was incubated with the proteases under the same conditions. SPN1 did not bind to the SPs (lanes 14-16). These results suggest that the three purified serpins bind specifically to the Toll cascade-activating SPs. We then cloned the serpin genes and expressed the recombinant serpins with *Tenebrio* SPN1 in *E. coli*. *Tenebrio* SPN40, SPN55, SPN48 and SPN1 consisted of 392, 461, 389 and 394 amino acid residues, respectively. A sequence comparison of the four serpins determined that these proteins share a reactive center loop (RCL) domain that acts as bait for trapping the target SP. This RCL domain contains a putative cleavage site (between the P1 and P1' residues) that is targeted by the proteases during complex formation (Fig. 7).

SAE and SPE cleave after Arg residues in the SPE zymogen (Arg199-Ile200) and pro-Spätzle (Arg124-Phe125) during the activation of the *Tenebrio* Toll cascade, respectively.[44] Therefore, we hypothesized that SPN55 and SPN48 might have an Arg residue in their P1 position. However, SPN55 was cleaved between the putative P2 Tyr and P1 Met residues by SAE; the cleavage site of SPN48, which is targeted by SPE,

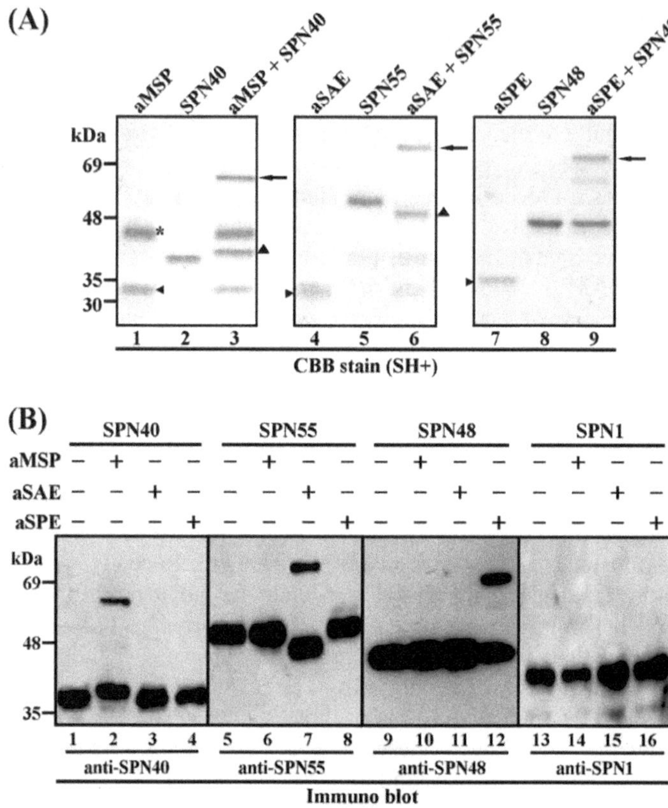

Figure 6. Three *Tenebrio* serpins formed specific complexes with their target SPs. (A) The specific SP-serpin complexes were observed using SDS-PAGE. The purified activated Toll cascade-activating SPs (500 ng, *lanes 1*, *4* and *7*) and purified serpins (1 μg, *lanes 2*, *5* and *8*) were incubated for 1 h at 30°C and then analyzed by SDS-PAGE under reducing conditions. The gels were stained with a solution containing 0.1% Coomassie Brilliant Blue (CBB) R-250. *Arrows* in *lanes 3*, *6* and *9* indicate the SP-serpin complex. *Star* and *arrowheads* indicate the N-terminal domain of MSP and the C-terminal catalytic SP domains, respectively. *Triangles* indicate cleaved SPN40 and SPN55. (B) Each purified serpin (20 pmol), including SPN40 (*lanes 1-4*), SPN55 (*lanes 5-8*), SPN48 (*lanes 9-12*) and SPN1 as a control (*lanes 13-16*), was incubated with a Toll cascade-activating SP (5 pmol), MSP, SAE, or SPE, under the same conditions as in *A*. The reaction mixtures were analyzed by immunoblotting to visualize the SP-serpin complexes. Reproduced from Jiang R et al. J Biol Chem 2009; 284:35652-58,[61] ©2009, with permission from the American Society for Biochemistry and Molecular Biology.

was between the putative P1 Glu and P2 Met residues. This unusual P1 specificity of the Arg-targeting SPs is similar to several mammalian serpins.[52,65,66] For example, kallistatin, a serpin that inhibits tissue kallikrein, has a P1 Phe residue, despite the fact that kallikrein targets Arg-specific substrates.[66] In addition, the protein Z-dependent protease inhibitor, which has a P1 Tyr residue, is a specific inhibitor of the membrane bound factor Xa, an Arg-specific SP.[52] These unexpected findings provide more evidence supporting the specificity and tight regulation of the Toll cascade by serpins in insects. These proteins colocalize with various digestive SPs in the insect circulatory system. The combined activity of the three serpins described in our study results in the enhancement of the Toll

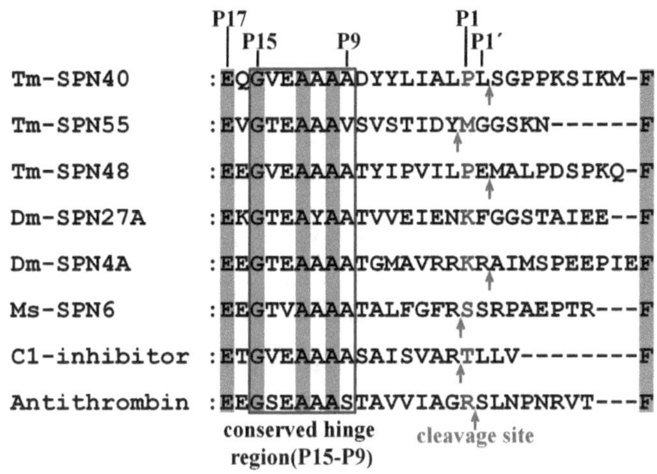

Figure 7. Sequence alignment of the putative and experimentally determined cleavage sites (P1-P1′) in the serpins RCL domains. *Blue boxes* indicate the hinge region (P15-P9) of the RCL domains. *Blue highlighted residues* are the putative P1 sites cleaved by target SPs. *Red arrows* indicate the experimentally determined cleavage sites of Dm-SPN4,[70] Ms-SPN6,[71] human antithrombin,[72] and C1 inhibitor[73] by their target SPs. *Tm-*, *Dm-* and *Ms-* are abbreviations for *T. molitor*, *D. melanogaster* and *M. sexta*, respectively. Reproduced from Jiang R et al. J Biol Chem 2009; 284:35652-58,[61] ©2009, with permission from the American Society for Biochemistry and Molecular Biology. A color version of this image is available at www.landesbioscience.com/curie.

signaling cascade in a specific and a tightly-regulated manner. Our study also provides novel insights into the regulation of pathogen recognition mechanisms by the innate immune response.

When invading pathogens are recognized by the soluble pattern recognition proteins in the hemolymph, SP zymogens, which are positioned upstream of the Toll cascade, are sequentially activated and amplify the pathogen recognition signals.[6,44,49] We hypothesized that serpins, which are present in low abundance in un-stimulated conditions, may act as a negative feedback inhibitor coincident with Toll cascade activation. To address this hypothesis, we injected β-1,3-glucan, pro-Spätzle or processed Spätzle, a ligand of the Toll receptor, into the *Tenebrio* larvae and determined serpin expression by western blot analysis (Fig. 8). As expected, the levels of SPN40 and SPN55, but not SPN48 or SPN1, were dramatically increased 24 h after injection of β-1,3-glucan and processed Spätzle. These results support the hypothesis; the induction of SPN40 and SPN55 expression may decrease MSP and SAE activation in order to regulate the Toll cascade.

A balance between activation and inhibition of the SP-mediated innate immune response must be maintained to avoid damage to the host.[67] In our study, we identified three serpins that target three Toll cascade-activating SPs, which act as negative regulators of Toll signaling and melanin synthesis. This is the first functional study of the SP-serpin pairs, which directly regulate the pattern recognition protein-dependent Toll signaling cascade. Surprisingly, each protease in the proteolytic cascade is selectively regulated by a serpin inhibitor. It had previously been hypothesized that these cascades were regulated by a single "bottleneck" protease. There is increasing evidence, in both mammals and *Drosophila*, that the control of proteolytic signaling cascades is

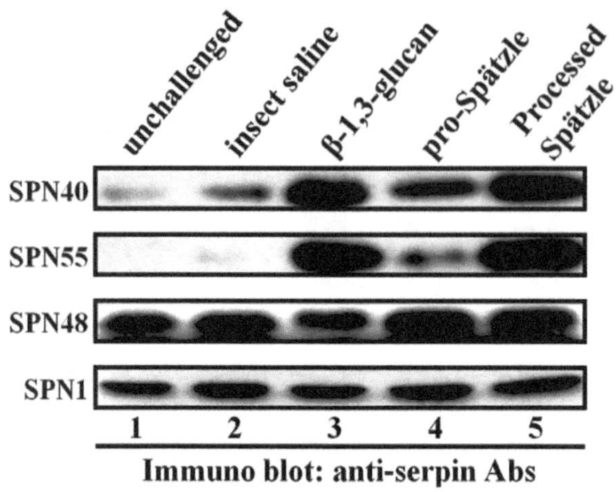

Figure 8. The amounts of SPN40 and SPN55 in the hemolymph increased after injection of β-1,3-glucan or processed Spätzle into *Tenebrio* larvae. Four microliters of insect saline (*lane 2*), 50 ng of β-1,3-glucan (*lane 3*), 60 ng of pro-Spätzle (*lane 4*) and 60 ng of processed Spätzle (*lane 5*) were injected into 10 *Tenebrio* larvae and hemolymph was collected after 24 h. A portion of each sample (40 μg) was analyzed by immunoblotting using affinity-purified antiserpin antibodies. Reproduced from Jiang R et al. J Biol Chem 2009; 284:35652-58,[61] ©2009, with permission from the American Society for Biochemistry and Molecular Biology.

precisely regulated. Our data indicate that each protease in a cascade may be regulated by a specific serpin. Furthermore, we have shown that SPN40 and SPN55 function as inducible negative feedback regulators in vivo. This study, in combination with our other reports,[44,49,68] supports a model in which the Lys-type PG- and β-1,3-glucan-dependent Toll signaling and proPO cascades are negatively regulated by three serpins (Fig. 9). Furthermore, our study highlights the elaborate regulatory mechanism of invertebrate molecular defense systems.

CONCLUSION

During last 15 years, we have used two beetles for biochemical studies of invertebrate's innate immunity. The main reason to use beetles as a model system is because they can be used for biochemical studies since fairly large amounts of hemolymph can be obtained. A weak point is that the genomic sequence information of beetles except for *T. castaneum* is not available until now. When we purified new proteins that are involved in innate immune responses, their putative biological functions have been predicted based on amino acid sequence comparison between partial amino acid sequence of the purified beetles' proteins and amino acid sequences obtained from *Tribolium* and *Drosophila* genomic sequence. Furthermore, in order to determine the exact biological functions of these proteins, we have spent long time to obtain their cDNAs and to express the recombinant proteins. Therefore, the determination of the whole genomic sequences of *Tenebrio* larvae and adults will be essential and efficient for further understanding of invertebrate's innate immune reactions. Finally, a greater understanding of molecular

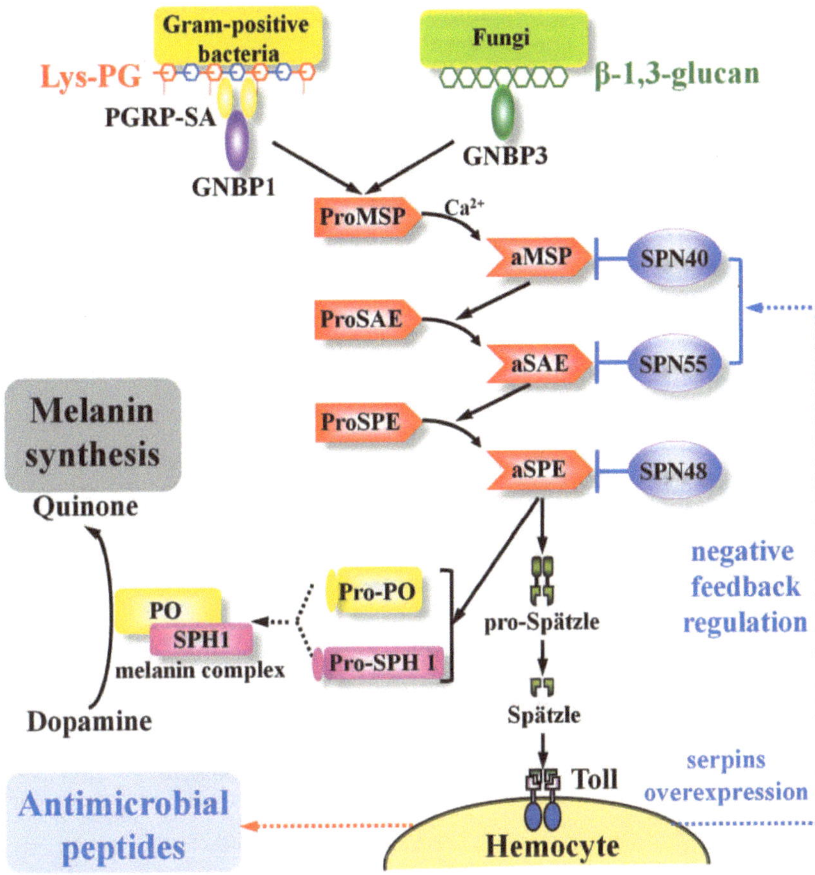

Figure 9. Model summarizing the molecular events in the regulation of the Toll and melanin synthesis cascades. The molecular activation mechanism of the *Tenebrio* Toll cascade leading to production of AMPs[44,49] and of the pro-PO cascade leading to melanin synthesis[68] were reported. Processed Spätzle induces the production of endogenous SPN40 and SPN55 as a negative feedback regulator of the Toll cascade. Also, the three serpins blocked activation of proPO, leading to inhibition of melanin synthesis. Reproduced from Jiang R et al. J Biol Chem 2009; 284:35652-58,[61] ©2009, with permission from the American Society for Biochemistry and Molecular Biology.

pathogen recognition mechanisms and their signaling pathways of proteolytic cascade reactions will also facilitate the development of novel kits to rapidly and sensitively detect bacterial PG or β-1,3-glucan in blood and food products.

ACKNOWLEDGEMENTS

We thank all members of present and former members of the National Research Laboratory of Defense Proteins, College of Pharmacy, Pusan National University for stimulating discussions and contribution of ideas and information to this chapter. This work was supported by the programs (M10400000028-04J0000-02, 2009-032-2000, 2008-8057-6001 and BK21) of National Research Foundation of Korea.

REFERENCES

1. Sokoloff A. The Biology of Tribolium with Special Emphasis on Genetic Aspects. Oxford:Oxford Universty Press, 1974.
2. Richards S, Gibbs RA, Weinstock GM et al. The genome of the model beetle and pest Tribolium castaneum. Nature 2008; 452:949-55.
3. Akira S, Uematsu S, Takeuchi O. Pathogen recognition and innate immunity. Cell 2006; 124:783-01.
4. Medzhitov R, Janeway CA, Jr. Innate immunity: the virtues of a nonclonal system of recognition. Cell 1997; 91:295-98.
5. Medzhitov R, Janeway CA, Jr. Decoding the patterns of self and nonself by the innate immune system. Science 2002; 296:298-00.
6. Hoffmann JA, Kafatos FC, Janeway CA et al. Phylogenetic perspectives in innate immunity. Science 1999; 284:1313-18.
7. Anderson KV. Toll signaling pathways in the innate immune response. Curr Opin Immunol 2000; 12:13-19.
8. Royet J, Reichhart JM, Hoffmann JA. Sensing and signaling during infection in Drosophila. Curr Opin Immunol 2005; 17:11-17.
9. Cherry S, Silverman N. Host-pathogen interactions in drosophila: new tricks from an old friend. Nat Immunol 2006; 7:911-17.
10. Lemaitre B, Hoffmann J. The host defense of Drosophila melanogaster. Annu Rev Immunol 2007; 25:697-43.
11. Michel T, Reichhart JM, Hoffmann J et al. Drosophila Toll is activated by Gram-positive bacteria through a circulating peptidoglycan recognition protein. Nature 2001; 414:756-59.
12. Leulier F, Parquet C, Pili-Floury S et al. The Drosophila immune system detects bacteria through specific peptidoglycan recognition. Nat Immunol 2003; 4:478-84.
13. Park JW, Kim CH, Kim JH et al. Clustering of peptidoglycan recognition protein-SA is required for sensing lysine-type peptidoglycan in insects. Proc Natl Acad Sci USA 2007; 104:6602-07.
14. Levashina EA, Langley E, Green C et al. Constitutive activation of toll-mediated antifungal defense in serpin-deficient Drosophila. Science 1999; 285:1917-19.
15. Gottar M, Gobert V, Matskevich AA et al. Dual detection of fungal infections in Drosophila via recognition of glucans and sensing of virulence factors. Cell 2006; 127:1425-37.
16. Holmskov U, Thiel S, Jensenius JC. Collections and ficolins: humoral lectins of the innate immune defense. Annu Rev Immunol 2003; 21:547-78.
17. Gillespie JP, Kanost MR, Trenczek T. Biological mediators of insect immunity. Annu Rev Entomol 1997; 42:611-43.
18. Cerenius L, Lee BL, Söderhäll K. The proPO-system: pros and cons for its role in invertebrate immunity. Trends Immunol 2008; 29:263-71.
19. Lee SY, Cho MY, Hyun JH et al. Molecular cloning of cDNA for pro-phenol-oxidase-activating factor I, a serine protease is induced by lipopolysaccharide or 1,3-beta-glucan in coleopteran insect, Holotrichia diomphalia larvae. Eur J Biochem 1998; 257:615-21.
20. Jiang H, Wang Y, Kanost MR. Pro-phenol oxidase activating proteinase from an insect, Manducasexta: a bacteria-inducible protein similar to Drosophila easter. Proc Natl Acad Sci USA 1998; 95:12220-25.
21. Satoh D, Horii A, Ochiai M et al. Prophenoloxidase-activating enzyme of the silkworm, Bombyx mori. Purification, characterization and cDNA cloning. J Biol Chem 1999; 274:7441-53.
22. Wang R, Lee SY, Cerenius L et al. Properties of the prophenoloxidase activating enzyme of the freshwater crayfish, Pacifastacus leniusculus. Eur J Biochem 2001; 268:895-02.
23. Sugumaran M. Molecular mechanisms for mammalian melanogenesis. Comparison with insect cuticular sclerotization. FEBS Lett 1991; 295:233-39.
24. Lee SY, Moon HJ, Kurata S et al. Purification and molecular cloning of cDNA for an inducible antibacterial protein of larvae of a coleopteran insect, Holotrichia diomphalia. J Biochem 1994; 115:82-86.
25. Moon HJ, Lee SY, Kurata S et al. Purification and molecular cloning of cDNA for an inducible antibacterial protein from larvae of the coleopteran, Tenebrio molitor. J Biochem 1994; 116:53-58.
26. Lee SY, Moon HJ, Kawabata S et al. A sapecin homologue of Holotrichia diomphalia: purification, sequencing and determination of disulfide pairs. Biol Pharm Bull 1995; 18:457-59.
27. Lee SY, Moon HJ, Kurata S et al. Purification and cDNA cloning of an antifungal protein from the hemolymph of Holotrichia diomphalia larvae. Biol Pharm Bull 1995; 18:1049-52.

28. Lee YJ, Chung TJ, Park CW et al. Structure and expression of the tenecin 3 gene in Tenebrio molitor. Biochem Biophys Res Commun 1996; 218:6-11.
29. Zhang R, Cho HY, Kim HS et al. Characterization and properties of a 1,3-beta-D-glucan pattern recognition protein of Tenebrio molitor larvae that is specifically degraded by serine protease during prophenoloxidase activation. J Biol Chem 2003; 278:42072-79.
30. Lee MH, Osaki T, Lee JY et al. Peptidoglycan recognition proteins involved in 1,3-beta-D-glucan-dependent prophenoloxidase activation system of insect. J Biol Chem 2004; 279:3218-27.
31. Park JW, Je BR, Piao S et al. A synthetic peptidoglycan fragment as a competitive inhibitor of the melanization cascade. J Biol Chem 2006; 281:7747-55.
32. Ju JS, Cho MH, Brade L et al. A novel 40-kDa protein containing six repeats of an epidermal growth factor-like domain functions as a pattern recognition protein for lipopolysaccharide. J Immunol 2006; 177:1838-45.
33. Lee SY, Kwon TH, Hyun JH et al. In vitro activation of pro-phenol-oxidase by two kinds of pro-phenol-oxidase-activating factors isolated from hemolymph of coleopteran, Holotrichia diomphalia larvae. Eur J Biochem 1998; 254:50-57.
34. Kwon TH, Kim MS, Choi HW et al. A masquerade-like serine proteinase homologue is necessary for phenoloxidase activity in the coleopteran insect, Holotrichia diomphalia larvae. Eur J Biochem 2000; 267:6188-96.
35. Chasan R, Anderson KV. The role of easter, an apparent serine protease, in organizing the dorsal-ventral pattern of the Drosophila embryo. Cell 1989; 56:391-00.
36. Iwanaga S, Lee BL. Recent advances in the innate immunity of invertebrate animals. J Biochem Mol Biol 2005; 38:128-50.
37. Murugasu-Oei B, Rodrigues V, Yang X et al. Masquerade: a novel secreted serine protease-like molecule is required for somatic muscle attachment in the Drosophila embryo. Genes Dev 1995; 9:139-154.
38. Piao S, Song YL, Kim JH et al. Crystal structure of a clip-domain serine protease and functional roles of the clip domains. EMBO J 2005; 24:4404-14.
39. Piao S, Kim S, Kim JH et al. Crystal structure of the serine protease domain of prophenoloxidase activating factor-I. J Biol Chem 2007; 282:10783-91.
40. Kim MS, Baek MJ, Lee MH et al. A new easter-type serine protease cleaves a masquerade-like protein during prophenoloxidase activation in Holotrichia diomphalia larvae. J Biol Chem 2002; 277:39999-04.
41. Lee HS, Cho MY, Lee KM et al. The pro-phenoloxidase of coleopteran insect, Tenebrio molitor, larvae was activated during cell clump/cell adhesion of insect cellular defense reactions. FEBS Lett 1999; 444:255-59.
42. Lee KY, Zhang R, Kim MS et al. A zymogen form of masquerade-like serine proteinase homologue is cleaved during pro-phenoloxidase activation by Ca^{2+} in coleopteran and Tenebrio molitor larvae. Eur J Biochem 2002; 269:4375-83.
43. Gobert V, Gottar M, Matskevich AA et al. Dual activation of the Drosophila toll pathway by two pattern recognition receptors. Science 2003; 302:2126-30.
44. Kim CH, Kim SJ, Kan H et al. A Three-step Proteolytic Cascade Mediates the Activation of the Peptidoglycan-induced Toll Pathway in an Insect. J Biol Chem 2008; 283:7599-07.
45. Ji C, Wang Y, Guo X et al. A pattern recognition serine proteinase triggers the prophenoloxidase activation cascade in the tobacco hornworm, Manduca sexta. J Biol Chem 2004; 279:34101-06.
46. Wang Y, Jiang H. Interaction of beta-1,3-glucan with its recognition protein activates hemolymph proteinase 14, an initiation enzyme of the prophenoloxidase activation system in Manduca sexta. J Biol Chem 2006; 281:9271-78.
47. Buchon N, Poidevin M, Kwon HM et al. A single modular serine protease integrates signals from pattern-recognition receptors upstream of the Drosophila Toll pathway. Proc Natl Acad Sci USA 2009; 106:12442-47.
48. Jang IH, Chosa N, Kim SH et al. A Spätzle-processing enzyme required for toll signaling activation in Drosophila innate immunity. Dev Cell 2006; 10:45-55.
49. Roh KB, Kim CH, Lee H et al. Proteolytic cascade for the activation of the insect toll pathway induced by the fungal cell wall component. J Biol Chem 2009; 284:19474-81.
50. Bulet P, Hetru C, Dimarcq JL et al. Antimicrobial peptides in insects; structure and function. Dev Comp Immunol 1999; 23:329-44.
51. Bulet P, Cociancich S, Dimarcq JL et al. Insect immunity. Isolation from a coleopteran insect of a novel inducible antibacterial peptide and of new members of the insect defensin family. J Biol Chem 1991; 266:24520-25.
52. Gettins PG. Serpin structure, mechanism and function. Chem Rev 2002; 102:4751-04.
53. Gooptu B, Lomas DA. Conformational pathology of the serpins: themes, variations and therapeutic strategies. Annu Rev Biochem 2009; 78:147-176.

54. Ligoxygakis P, Roth S, Reichhart JM. A serpin regulates dorsal-ventral axis formation in the Drosophila embryo. Curr Biol 2003; 13:2097-02.
55. Hashimoto C, Kim DR, Weiss LA et al. Spatial regulation of developmental signaling by a serpin. Dev Cell 2003; 5:945-50.
56. Scherfer C, Tang H, Kambris Z et al. Drosophila Serpin-28D regulates hemolymph phenoloxidase activity and adult pigmentation. Dev Biol 2008; 323:189-96.
57. De Gregorio E, Han SJ, Lee WJ et al. An immune-responsive Serpin regulates the melanization cascade in Drosophila. Dev Cell 2002; 3:581-92.
58. Ligoxygakis P, Pelte N, Ji C et al. A serpin mutant links Toll activation to melanization in the host defence of Drosophila. EMBO J 2002; 21:6330-37.
59. Tang H, Kambris Z, Lemaitre B et al. A serpin that regulates immune melanization in the respiratory system of Drosophila. Dev Cell 2008; 15:617-26.
60. Kanost MR, Jiang H, Yu XQ. Innate immune responses of a lepidopteran insect, Manduca sexta. Immunol Rev 2004; 198:97-05.
61. Jiang R, Kim EH, Gong JH et al. Three pairs of protease-serpin complexes cooperatively regulate the insect innate immune responses. J Biol Chem 2009; 284:35652-58.
62. Nash P, Whitty A, Handwerker J et al. Inhibitory specificity of the anti-inflammatory myxoma virus serpin, SERP-1. J Biol Chem 1998; 273:20982-91.
63. Jesty J. The kinetics of formation and dissociation of the bovine thrombin antithrombin III complex. J Biol Chem 1979; 254:10044-50.
64. Han J, Zhang H, Min G et al. A novel Drosophila serpin that inhibits serine proteases. FEBS Lett 2000; 468:194-98.
65. Gettins PG, Olson ST. Exosite determinants of serpin specificity. J Biol Chem 2009; 284:20441-45.
66. Chen VC, Chao L, Chao J. Roles of the P1, P2 and P3 residues in determining inhibitory specificity of kallistatin toward human tissue kallikrein. J Biol Chem 2000; 275:38457-66.
67. Ferrandon D, Imler JL, Hetru C et al. The Drosophila systemic immune response: sensing and signalling during bacterial and fungal infections. Nat Rev Immunol 2007; 7:862-74.
68. Kan H, Kim CH, Kwon HM et al. Molecular control of phenoloxidase-induced melanin synthesis in an insect. J Biol Chem 2008; 283:25316-23.
69. Wang Y, Jiang H. Reconstitution of a branch of the Manduca sexta prophenoloxidase activation cascade in vitro: Snake-like hemolymph proteinase 21 (HP21) cleaved by HP14 activates prophenoloxidase-activating proteinase-2 precursor. Insect Biochem Mol Biol 2007; 37:1015-25.
70. Richer MJ, Keays CA, Waterhouse J et al. The Spn4 gene of Drosophila encodes a potent furin-directed secretory pathway serpin. Proc Natl Acad Sci USA 2004; 101:10560-65.
71. Zou Z, Jiang H. Manduca sexta serpin-6 regulates immune serine proteinases PAP-3 and HP8. cDNA cloning, protein expression, inhibition kinetics and function elucidation. J Biol Chem 2005; 280:14341-48.
72. Bjork I, Jackson CM, Jornvall H et al. The active site of antithrombin. Release of the same proteolytically cleaved form of the inhibitor from complexes with factor IXa, factor Xa and thrombin. J Biol Chem 1982; 257:2406-11.
73. Salvesen GS, Catanese JJ, Kress LF et al. Primary structure of the reactive site of human C1-inhibitor. J Biol Chem 1985; 260:2432-36.

CHAPTER 10

IMMUNITY IN LEPIDOPTERAN INSECTS

Haobo Jiang,[1] Andreas Vilcinskas[2] and Michael R. Kanost*,[3]

[1]*Department of Entomology and Plant Pathology, Oklahoma State University, Stillwater, Oklahoma, USA;*
[2]*Institut für Phytopathologie und Angewandte Zoologie, Justus-Liebig-Universität Gießen, Gießen, Germany;*
[3]*Department of Biochemistry, Kansas State University, Manhattan, Kansas, USA*
**Corresponding Author: Michael R. Kanost—Email: kanost@ksu.edu*

Abstract: Lepidopteran insects provide important model systems for innate immunity of insects, particularly for cell biology of hemocytes and biochemical analyses of plasma proteins. Caterpillars are also among the most serious agricultural pests, and understanding of their immune systems has potential practical significance. An early response to infection in lepidopteran larvae is the activation of hemocyte adhesion, leading to phagocytosis, nodule formation, or encapsulation. Plasmatocytes and granular cells are the hemocyte types involved in these responses. Infectious microorganisms are recognized by binding of hemolymph plasma proteins to microbial surface components. This "pattern recognition" triggers phagocytosis and nodule formation, activation of prophenoloxidase and melanization and the synthesis of antimicrobial proteins that are secreted into the hemolymph. Many hemolymph proteins that function in such innate immune responses of insects were first discovered in lepidopterans. Microbial proteinases and nucleic acids released from lysed host cells may also activate lepidopteran immune responses. Hemolymph antimicrobial peptides and proteins can reach high concentrations and may have activity against a broad spectrum of microorganisms, contributing significantly to clearing of infections. Serine proteinase cascade pathways triggered by microbial components interacting with pattern recognition proteins stimulate activation of the cytokine Spätzle, which initiates the Toll pathway for expression of antimicrobial peptides. A proteinase cascade also results in proteolytic activation of phenoloxidase and production of melanin coatings that trap and kill parasites and pathogens. The proteinases in hemolymph are regulated by specific inhibitors, including members of the serpin superfamily. New developments in lepidopteran functional genomics should lead to much more complete understanding of the immune systems of this insect group.

Invertebrate Immunity, edited by Kenneth Söderhäll.

INTRODUCTION

Moth larvae have proven to be extremely useful for experiments providing insights on the innate immune systems of insects. Many hemolymph proteins with immune functions were first studied in caterpillars by biochemical methods.[1-11] Lepidopteran larvae have also been important for experiments aimed at characterizing immune functions of insect hemocytes.[12] Much of the research on lepidopteran immunity has made use of large moth species, including the tobacco hornworm, *Manduca sexta* and wild silkmoths such as *Hyalophora cecropia*. The domestic silkmoth, *Bombyx mori* has also provided significant discoveries about immunity in moths and with the advantage of a sequenced genome,[13] it continues to serve as an important model organism. The wax moth, *Galleria mellonella*, was one of the earliest species used for research in insect immunity[14,15] and is the subject of much current research. In addition to these laboratory model species, research on immunity is also underway to investigate the immune systems of moths whose larvae are among the most destructive agricultural pests worldwide, particularly including species in the family Noctuidae. In comparison, there has been relatively little research on immunity in butterflies, perhaps because they can be more difficult to rear or due to lower agricultural impact.

In this chapter, we describe selected developments in research on innate immunity in lepidopteran insects, with an emphasis on aspects discovered through biochemical and cell biological approaches and potential new insights that may be gained from functional genomics methods. Several recent reviews with a focus on lepidopteran immunity are available.[16-23]

HEMOCYTES

With a large hemolymph volume in many lepidopteran larvae and pupae, it is possible to collect 10^4-10^6 hemocytes from a single individual, making feasible the use of cell biological techniques such as flow cytometry and cell sorting that have advanced mammalian immunology. The most abundant hemocyte types typically described in Lepidopteran larvae are granular cells and plasmatocytes, which are capable of adhesion and phagocytosis.[12] Nonadherant hemocyte types include oenocytoids, which synthesize prophenoloxidase (proPO), and spherule cells, whose functions are poorly understood. Monoclonal antibodies are very useful reagents for distinguishing lepidopteran hemocyte populations based on antigenicity rather than morphology,[24-26] which can vary considerably, especially for plasmatocytes. Two subpopulations of plasmatocytes have been distinguished based on monoclonal antibody markers in *Pseudoplusia includens*[26] and in *M. sexta*,[27] which may indicate different stages in differentiation or functional specialization. The composition of hemocyte populations varies through larval development.[28,29] Hematopoesis in lepidopteran larvae[28,31] and the embryonic origin granular cells and plasmatocytes[30] have been described.

Granular cells and plasmatocytes function in immunity through responses that involve adherence of cells to foreign surfaces or to other hemocytes.[22,32] Both granular cells and plasmatocytes can be phagocytic and in different lepidopteran species either cell type may be the predominant contributor to phagocytosis as a defense.[12] A hyperphagocytic cell type, very large cells capable of phagocytosing 500 bacteria, has been described in *M. sexta*.[33] These cells are morphologically similar to a neuroglian-positive subpopulation of plasmatocytes, which can act as a focus for attachment of hemocytes to foreign surfaces[34] and perhaps are the same hemocyte type.

Adhesion of granular cells and plasmatocytes leads to two related responses to infection: nodule formation, in which hemocytes cluster to entrap aggregated microorganisms, and encapsulation, in which hemocytes form a multi-layered cellular capsule around a larger eukaryotic parasite.[12] Adhesion of hemocytes to injured body wall also probably helps to seal wounds to prevent bleeding.[32] Cytokines have been identified in lepidopterans, which promote these hemocyte functions. A hemocyte chemotactic peptide from *Pseudaletia separata* stimulates directed movement and aggregation of hemocytes.[35] This peptide is structurally similar to lepidopteran cytokines called ENF peptides, which have multiple biological activities, including the stimulation of plasmatocyte adhesion and spreading, reduced bleeding and the stimulation of oenocytoid lysis to release proPO.[36-40] RNAi results indicate that the *M. sexta* plasmatocyte spreading peptide from this family promotes hemocyte nodule formation as a protective response to bacterial infection.[41] The active ENF peptides are produced by proteolytic processing of a larger protein present in hemolymph.[36,37] In *M. sexta* hemolymph, a serine proteinase with trypsin-like specificity is responsible for this activation, but has not yet been identified due to its instability (Kanost et al, unpublished results). Eicosanoids such as prostaglandins can also simulate hemocytes to aggregate to form nodules.[42,43] Prostaglandins can also elicit the lysis of oenocytoids, releasing proPO into the plasma.[44] ProPO is activated by plasma proteinases and participates in formation of melanin, which coats nodules and encapsulated objects. This response is discussed in more detail below.

Hemocyte attachment during encapsulation and nodulation is mediated by cell surface adhesion proteins. Lepidopteran hemocytes have cell surface integrins, which function as adhesion molecules.[45-50] Plasmatocytes of *M. sexta* express a specific integrin that is required for efficient encapsulation.[47,50] The adhesive properties of this hemocyte-specific integrin derive at least in part from binding to neuroglian and a tetraspanin on neighboring hemocytes.[49,51]

RECOGNITION OF MICROORGANISMS

Plasma proteins that bind to components on the surface of microorganisms are a key component of the innate immune system of insects. Such proteins stimulate responses including phagocytosis and activation of proteinase signaling cascades. Some of these "pattern recognition" proteins were first discovered in silkworms and then found to occur in immune systems of other insect groups, whereas others appear to be unique to lepidopterans.

Hemolin is a 48 kDa plasma protein first identified in *Hyalophora cecropia*[52,53] and *Manduca sexta*.[54,55] Hemolin is composed of four I-set immunoglobulin (Ig) domains commonly found in cell adhesion proteins of vertebrates and invertebrates. Hemolin also exists in other lepidopteran species including *B. mori, Hyphantria cunea, Lymantria dispar, Antheraea pernyi, Antheraea mylitta, Plutella xylostella, Samia cynthia*.[13,56-61] Hemolin has not been identified in insects from any other order, suggesting that it may have evolved after the split of the Lepidoptera from other insect groups. Hemolin shares structural features with neuroglian,[62] a transmembrane Ig-domain protein located on the surface of *Manduca* glial and neuronal cells[63] and the developing embryonic prothoracic gland,[64] as well as a subpopulation of *M. sexta* plasmatocytes.[34,51] Hemolin expression is induced by bacteria or their surface components injected into the hemocoel. In *H. cecropia*, this transcriptional activation is controlled via an intronic enhancer that contains a κB motif.[65] Developmental and hormonal signals (e.g., 20-hydroxyecdysone) also affect hemolin production.[58,66,67] It has been speculated that hemolin may have an antiviral function.[68] Baculovirus exposure up-regulates hemolin transcription in *A. pernyi*,[69] but not in *B. mori*, *Helicoverpa zea*, or *Heliothis virescens*.[13,70] A possible role for hemolin in antiviral responses remains to be established. On the other hand, several lines of evidence support the idea that hemolin functions in immune responses to bacterial infection. Hemolin binds to bacterial lipopolysaccharide (LPS) and lipoteichoic acid[71,72] and associates with hemocytes.[55,73,74] The horseshoe-shape structure of hemolin suggests that one or more of its Ig domains may interact through domain swapping with Ig domains of cell adhesion molecules such as neuroglian on hemocyte surfaces.[75] Interaction of hemolin with molecules on the surface of bacteria and with hemocyte membranes suggests that it may bring microorganisms to hemocyte surfaces, promoting phagocytosis or nodule formation. RNAi knockdown of hemolin expression in *M. sexta* larvae significantly reduced phagocytosis and nodulation of *E. coli*.[76]

Peptidoglycan recognition proteins (PGRPs) associate with bacterial peptidoglycans through a conserved domain homologous to T4 bacteriophage lysozyme.[77] PGRPs were first discovered in lepidopterans. *B. mori* PGRP-S1 was isolated from plasma of silkworm larvae as a 19 kDa protein that binds to *Micrococcus luteus* peptidoglycan and triggers the proPO activation system.[8] Molecular cloning of PGRPs from *Trichoplusia ni*,[10] *B. mori*,[13,78] *M. sexta*,[79-81] *G. mellonella*,[82] *S. cynthia*,[83,84] *P. xylostella*,[60] and *Ostrinia nubilalis*[85] suggests that multiple PGRPs are present in every lepidopteran species. In addition to binding to peptidoglycan, these proteins may hydrolyze peptidoglycan if the residues for Zn^{2+}-binding and amidase activity are present.[86] The *B. mori* genome has six short (S) and six long (L) PGRP genes: S3 through S6 are putative amidases; L1 and L4 possess a transmembrane region; L6 is likely cytosolic.[13] Constitutive expression of PGRP-S1 is up-regulated in larvae after injection of *Enterobacter cloacae*.[78] *E. coli* or *Bacillus subtilis* treatment increases the mRNA levels for PGRP-S1, -S2 and -S5, whereas *Staphylococcus aureus* injection enhances transcription of *B. mori* PGRP-S1, -S5, -L1, but not -S2.[13] Microbe-induced transcription of PGRPs also occurs in other lepidopteran species.[10,79,82,84] The *T. ni*, *S. cynthia* and *M. sexta* PGRPs bind to *Bacillus meso*-diaminopimelic acid-type peptidoglycans and *M. luteus* Lys-type peptidoglycans.[10,83,87] This *M. sexta* PGRP can stimulate proPO activation.[88] Knockdown of *M. sexta* PGRP1 synthesis increased larval susceptibility to infection by *Photorhabdus luminescens*.[89,90] So far, the level of knowledge on structures and functions of moth PGRPs is significantly less than for *Drosophila* PGRPs.[91,92] Furthermore, molecular details of how peptidoglycan binding by PGRPs promotes the activation of serine proteinases involved in proPO and spätzle activation are unknown.

Insect β-1,3-glucan recognition proteins (βGRPs) and Gram-negative bacteria binding proteins (GNBPs) are a family of ~55 kDa plasma proteins with an amino-terminal glucan-binding domain and a carboxyl-terminal region similar to β-1,3-glucanases. *B. mori* βGRP1, *M. sexta* βGRP1 and βGRP2 and *Plodia interpunctella* βGRP bind to β-1,3-glucans and to bacteria and stimulate the proPO activation cascade.[9,93-97] The amino-terminal domain of *B. mori* βGRP1 adopts an Ig-like β-sandwich fold and residues have been identified that may form a glucan-binding site.[98] *M. sexta* βGRP1 and βGRP2 gene expression in fat body is differentially regulated: βGRP1 is constitutively expressed, whereas βGRP2 transcripts become highly abundant in the early wandering stage prior to pupation or after an immune challenge.[94,95] *M. sexta* βGRP2 is also present in cuticle of wandering stage larvae.[95] Binding of βGRP1 or βGRP2 to curdlan and *M. sexta* hemolymph proteinase-14 precursor (proHP14) stimulates autoactivation of HP14 to initiate an immune proteinase cascade leading to proPO activation.[99,100] *B. mori* GNBP binds to *E. cloacae* but not *Bacillus licheniformis* or curdlan.[101] An orthologous *M. sexta* GNBP recognizes LPS and laminarin (a β-1,3-glucan with β-1,6 branches) and initiates melanization (Y. Wang and H. Jiang, unpublished results). An active glucanase related in sequence to the GRPs was isolated from midgut extract of *Helicoverpa armigera* larvae. It hydrolyzes β-1,3-glucan but not β-1,4-glucan or glucans with mixed β-1,3 and β-1,4 linkages[102] and probably functions as a digestive enzyme.

C-type lectins (CTLs) from animals are a large group of carbohydrate-recognition molecules that bind ligands in a calcium-dependent manner.[103] *B. mori* LPS-binding protein (LBP or CTL20),[104,105] *Hyphantria cunea* Hdd15,[106] and *M. sexta* immulectin-1[107] were among the first C-type lectins identified in plasma of lepidopteran larvae. They all contain two carbohydrate-recognition domains. CTLs with this dual-domain structure also include: *M. sexta* immulectins-2, -3 and -4,[108-110] *B. mori* MBP (CTL10),[111] immulectin (CTL11),[112] LEL-1 and -2,[113] CTL19, CTL21,[13] and *H. armigera* Ha-lectin.[114] GenBank currently contains more than 30 additional similar CTL cDNA sequences from eleven other lepidopteran species in five families. The only other dual-domain CTL identified in any other insect species is TcCTL3[115] from a beetle, *Tribolium castaneum.* Hence, such tandem domain CTL genes appear to be fairly unique to Lepidoptera and their emergence and expansion may have occurred early in evolution of this insect group. Expression of at least some of these CTL genes is induced microbial infection.[106,108-110,114]

Many of the lepidopteran CTLs bind to bacterial LPS and some to lipoteichoic acid.[105,106,108-110] They can cause agglutination of bacteria and yeast,[107,109-111] presumably due to each of the molecule's two carbohydrate-binding domains binding to carbohydrates on the surface of adjacent microbial cells. This activity most often requires the presence of calcium. Such clustering of microorganisms may result in more efficient clearance of pathogens by hemocytes through nodule formation.[105] Experiments have demonstrated that *M. sexta* immulectin-2 enhances clearance of *Serratia marcesens*,[116] and suppression of immulectin-2 expression by RNA interference reduced larval survival of a *Photorhabdus* infection.[90] In addition, the immulectins can promote proPO activation and melanin deposition at the surface of objects encapsulated by hemocytes,[107,108,110,117,118] which may also promote killing of pathogens and parasites.

Lipophorins are insect hemolymph proteins that transport lipids between tissues.[119] The lipophorin particle contains two protein subunits, apolipophorin-I and apophorin-II. When the neutral lipid load is high, an exchangeable plasma protein, apolipophorin-III (apoLpIII) also associates with low density lipophorin to cover hydrophobic surfaces. In lepidopterans, apoLpIII concentration in hemolymph is generally much lower in larvae than in adults, where it functions in transporting lipids from fat body to flight muscles during prolonged flight. Lipophorins and apoLpIII have been implicated in several defense mechanisms.[32] Lipophorin appears to be involved in hemolymph clotting in at least some species,[20] and may in this way participate in physical trapping of invading microorganisms. There may also be an association of proteins from the melanization cascade with lipophorin or other clotting components, which could contribute to defense.[120-122] The affinity of lipophorin and apoLpIII for hydrophobic ligands is consistent with their reported binding of bacterial LPS and lipoteichoic acids,[123-128] and the partitioning of these microbial membrane lipids into complexes with lipophorin or apoLpIII may contribute to reducing their toxicity to insect hosts. Lipophorin and apoLpIII have been reported to stimulate other humoral or cellular immune responses in *G. mellonella*,[129-133] and research toward understanding molecular mechanisms underlying such observations is needed.

Pathogenic bacteria and fungi produce proteinases to utilize lepidopteran host proteins as a source of nutrients and to degrade immunity-related defense molecules such as antimicrobial peptides. Thermolysin-like metalloproteinases associated with entomopathogenic bacteria and fungi are essential virulence factors.[134-137] However, the presence of microbial proteinases may also be recognized as a signal of infection and stimulate immune defenses in the host. Evidence for sensing of microbial proteinases and their regulation during innate immune responses has been reported from *G. mellonella*.[138] Thermolysin is a potent activator of the serine proteinase cascade that controls proPO activation leading to melanization. Thus, this virulence factor also directly triggers an immune response. In addition, the activity of microbial metalloproteinases within the body of *G. mellonella* generates peptide fragments that strongly elicit the synthesis of antimicrobial peptides.[139] The immune-stimulatory peptides were identified as collagen IV fragments containing the RGD/RGE motif, which bind to integrins of immune-compentent hemocytes.[140] The stimulation of two innate immune responses, proPO activation and antimicrobial peptide synthesis, in response to microbial metalloproteinases provides evidence that lepidopteran innate immune responses include reaction to danger-associated molecules produced by microbial virulence factors.[141] Furthermore, *G. mellonella* hemolymph contains an inducible metalloproteinase inhibitor (IMPI), the only specific peptidic inhibitor of thermolysin-like metalloproteinases reported to date from any animal.[142] The IMPI gene encodes two distinct metalloproteinase inhibitors that putatively contribute to the regulation of metalloproteinases associated with invading pathogens.[143,144]

Nucleic acids released from damaged or necrotic cells form another danger signal to enhance insect immune responses. Injection of synthetic oligonucleotides induced attacin expression in *B. mori* larvae.[145] Co-injection of purified host nucleic acids with heat-inactivated *P. luminescens* into *G. mellonella* larvae synergistically elevated the level of antimicrobial activity, reduced the total number of hemocytes (a consequence of the attachment of immune-competent cells to tissues during cellular responses) and prolonged the survival of insects infected by *P. luminescens*.[146] DNA and RNA released from damaged cells may interact with lipophorin to trigger clot formation and entrap invading pathogens in hemolymph. In *Pseudaletia separata*, nucleic acids as well as cytoplasmic proteins (*e.g.*, proPO) are released from oenocytoids through cell lysis induced by the growth-blocking peptide.[40]

ANTIMICROBIAL PEPTIDES AND PROTEINS

Many insect antimicrobial plasma proteins were first identified and isolated from lepidopteran hemolymph. Expression of antimicrobial peptides and proteins is often induced by microbial infection, with strongest expression usually occurring in fat body, although hemocytes also contribute to the pool of antimicrobial peptides secreted into hemolymph plasma.[147] Antimicrobial peptides are also expressed in the midgut of lepidopteran prepupae and secreted into the lumen, perhaps as a prophylaxis against infection during metamorphosis.[148] Antimicrobial peptides can also be expressed in extraembryonic tissues of lepidopteran eggs, providing protection against infection for the developing embryo.[149]

Lysozyme, the first antimicrobial protein reported from insects, was identified more than forty years ago in *G. mellonella*[150] and, like other insect lysozymes, shares structural similarity with C-type (chicken) lysozyme.[151] Its activity against Gram-positive bacteria has been attributed to its ability to degrade cell wall peptidoglycan by hydrolysis of the β-1,4 linkages between N-acetylglucosamine and N-acetylmuramic acid residues.[2] Besides moderate activity against Gram-negative bacteria,[152-154] lepidopteran lysozyme also exhibits antifungal activity.[155] Lyozyme also appears to negatively regulate activation of proPO in *M. sexta*.[156]

Insects produce a variety of amphipathic peptides with antimicrobial activity attributed to their ability to damage cell membranes of pathogens. The first antimicrobial peptide from insects, isolated from the hemolymph of the silkmoth *H. cecropia*, was named cecropin.[4] Families of cecropin genes have now been found in many lepidopteran (and dipteran) species. Thirteen cecropin genes were identified in the *B. mori* genome.[13] Cecropins are typically ~4 kDa basic peptides, which have an amphipathic α-helical structure. The moricins constitute another group of amphipathic α-helical antimicrobial peptides,[157,158] first discovered in *B. mori*.[159] Nine moricin genes are present in the *B. mori* genome.[13] Eight moricin homologs identified in *G. mellonella* have activity against Gram-negative and Gram-positive bacteria, as well as against yeast and filamentous fungi.[160]

Lepidoptera possess glycine-rich AMPs (attacins and gloverins) and proline-rich AMPs (lebocins). *H. cecropia* has two 20 kDa attacin isoforms, an acidic and a basic attacin, with 80% sequence identity.[5,161] The *B. mori* genome also contains two attacin genes,[13] and attacin cDNAs have now been cloned from many lepidopteran species. Treatment of *E. coli* with *H. cecropia* attacins leads to an increase in outer-membrane permeability, preceding any increase in inner-membrane permeability by at least one generation time. Inhibition of outer-membrane protein synthesis is achieved on the transcriptional level and triggered by binding of attacin to the cell surface without entering the inner membrane or the cytoplasm. Primary binding occurs on LPS, explaining why basic attacin is more active against *E. coli* than the acidic form. Another family of proline-rich AMPs, the gloverins, appears to exert a similar mechanism of inhibition of outer-membrane protein synthesis.[162] Expression and evolution of four *B. mori* gloverin genes have been investigated,[163,164] and gloverins have been studied in several other lepidopteran species,[16,165-168] but no homologs have been identified to date in insects from other orders. A family of proline-rich AMPs called lebocins has been characterized in lepidopterans.[16,147,169-173] A somewhat puzzling aspect of this family is that the 3.5 kDa active lebocin peptide is processed from a larger precursor and in some members of the family, the amino-terminal pro-region of the protein is conserved in sequence, but the carboxyl-terminal sequence corresponding to the original antimicrobial peptide is not, suggesting that the pro-region may have a function not yet discovered.

AMPs stabilized by intramolecular disulfide bonds are widely distributed in insects. Those with three or four disulfide bonds are commonly referred to as insect defensins because of overall structural similarities to mammalian α and β defensins.[174] Insect defensins can be grouped into peptides with an α-helix/β-sheet mixed structure and peptides forming triple-stranded antiparallel β-sheets. Defensin-like AMPs with antibacterial and antifungal activities from several lepidopteran species have been investigated.[175-180] Two cysteine-rich defensin-like peptides from *G. mellonella* specifically inhibit growth of filamentous fungi.[179,180] A group of atypical defensin-like peptide named x-tox identified in *G. mellonella* and two *Spodoptera* species is characterized by imperfectly conserved tandem repeats of cysteine-stabilized αβ motifs, the structural scaffold characteristic of invertebrate defensins and scorpion toxins.[181-183] They are induced upon activation of the immune system but lack detectable antimicrobial activity, suggesting that they may have an immune function yet to be discovered.

EXTRACELLUAR AND INTRACELLULAR SIGNAL TRANSDUCTION STIMULATING ANTIMICROBIAL PEPTIDE SYNTHESIS

Genetic investigations in *Drosophila* have revealed three major immune signaling pathways (Toll, Imd-JNK and JAK-STAT) that are conserved in mammals.[184] Genomic analyses suggest similar pathways exist in other holometabolous insects, including *B. mori*.[13] Additional bioinformatic and experimental evidence described below supports the existence of functional Toll and Imd pathways in lepidopterans.

There are fourteen Toll-like receptor genes in the silkworm genome: six in group A with *Drosophila* Toll and eight in group B with *Drosophila* 18-wheeler.[13] *BmToll* and *BmToll2* are group B genes highly expressed in ovary and their transcripts become more abundant in fat body after injection of microorganisms,[185] suggesting possible involvement in embryonic development and immune response. *BmToll3*, *BmToll4*, *BmToll9* and *BmToll10* mRNA levels in fat body also increase after injection of some microbes. In *M. sexta*, a Toll-like receptor is present in hemocytes, fat body, epidermis, midgut and Malpighian tubules.[186]. Its mRNA level increased in hemocytes, but not in fat body, after injection of microorganisms.

Spätzle, the ligand which activates Toll, has been identified and functionally characterized in *B. mori* and *M. sexta*.[187,188] Spätzle is synthesized as an inactive precursor, proSpätzle, which is secreted as a disulfide-linked homodimer into the hemolymph and requires proteolytic processing to form the active Toll ligand. Expression of proSpätzle is significantly greater in *M. sexta* hemocytes than in fat body.[188] In *B. mori*, expression was detected primarily in fat body and midgut, but hemocytes were not tested.[187] *B. mori* and *M. sexta* proSpätzle proteins are only ~20% identical to *Drosophila* proSpätzle, but have slightly greater similarity (~26% identity) in the carboxyl-terminal 108 residues corresponding to the active form of *Drosophila* Spätzle known to bind to Toll. Recombinant *B. mori* and *M. sexta* Spätzle were active when injected into larvae, inducing fat body expression of attacins, cecropins, gloverin, moricin and lebocin in *B. mori*[187] and attacin, cecropin, moricin and hemolin in *M. sexta*, with corresponding strong induction of plasma antimicrobial activity.[188] Howevever, injection of proSpätzle had little effect, indicating the need for proteolytic activation of this cytokine in response to infection.

Hemolymph proteinase-8 (HP8) is a clip-domain proteinase demonstrated to activate proSpätzle in *M. sexta* by specific cleavage to produce the carboxyl-terminal 108 residue fragment, as a disulfide-linked homodimer.[188] The *Drosophila* proteinases most similar to HP8 are Easter and Spätzle processing enzyme, both of which function to activate proSpätzle.[189] Injection of active HP8 into larvae stimulates expression of attacin, cecropin, gloverin and moricin and elevates plasma antibacterial activity, consistent with a role for HP8 as an activator of proSpätzle.[189] HP8 is present in plasma as a zymogen, proHP8 and is activated by another clip-domain proteinase, HP6, an apparent ortholog of *Drosophila* Persephone. Injection of recombinant HP6 also promoted expression of antimicrobial peptides in larvae.[189] ProHP6 is activated in plasma exposed to bacteria or the β-1,3-glucan curdlan, but a hemolymph proteinase responsible for activation of HP6 has not been identified yet. It is apparent that recognition of microbial pattern

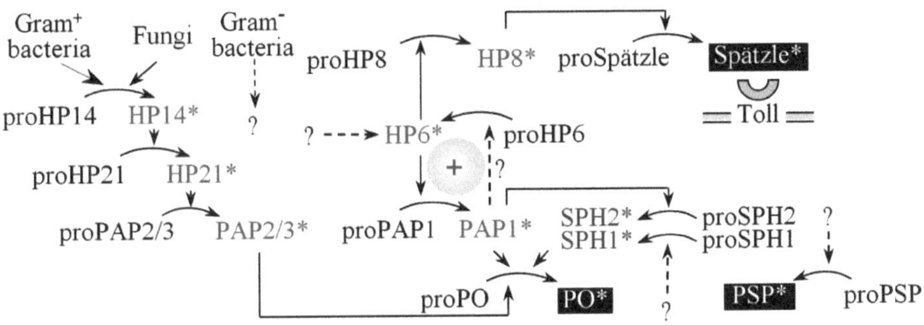

Figure 1. A current model of the hemolymph proteinase system in *M. sexta* larvae. An initiation proteinase precursor, proHP14, is autoactivated in response to Gram-positive bacterial or fungal infection. HP14 activates proHP21; HP21 activates proPAP2 or proPAP3; PAP2 or PAP3 then cleaves proPO to form active PO in the presence of SPH1 and SPH2. Activation of proPO can also be catalyzed by PAP1 when the high M_r SPH complex is present simultaneously. PAP1 also activates proSPH2 directly and can indirectly lead to proHP6 activation. HP6, whose direct activator is unknown, cleaves proPAP1 and proHP8. PAP1 and HP6 form a positive feedback loop, in which PAP1 indirectly stimulates activation of HP6. HP8 activates Spätzle to induce antimicrobial peptide synthesis via Toll receptor. Active proteins, including HPs, PAPs, SPHs, PO, Spätzle and plasmatocyte-spreading peptide (PSP), are labeled "*" and unknown HPs are marked with "?".

molecules triggers activation of a proteinase cascade to generate the cytokine Spätzle, leading to expression of a suite of antimicrobial peptides as an innate immune response in this moth (Fig. 1).

Two rel-family transcription factors participate in immunity-related gene expression in *B. mori*.[190-192] *BmRel* encodes, via alternative splicing, RelA and RelB, which are orthologous to *Drosophila* Dorsal. *BmRelish* also encodes two splicing isoforms, Relish1 and Relish2. RelA activates the lebocin-4 gene strongly and an attacin gene weakly.[190] RelB, lacking the first 52 residues of RelA, activates the attacin gene strongly and other genes to a lesser extent. The Rel homology domain in RelA and RelB binds specifically to κB sites in attacin and lebocin-4 genes. In transgenic silkworms whose *BmRel* expression is knocked down, expression of antibacterial peptide genes fails to be induced by *M. luteus*. Knockdown of *BmRelish* expression abolishes antimicrobial peptide production elicited by *E. coli*.[191] Intact Relish1 and Relish2 do not activate promoters of *B. mori* attacin, cecropin B1, lebocin-3 and lebocin-4 genes. Removal of the ankyrin repeats in Relish1 is necessary for its transcriptional activation of antibacterial peptide genes. However, Relish2, which lacks the repeats and the transactivation domain, serves as a dominant negative factor to suppress the function of active Relish1. Relish1 binds to the κB sites in attacin and cecropin B1 genes, while the sites for activating lebocin-4 promotor differ between Relish1 active form and RelA. *Meso*-diaminopimelic acid- and Lys-type peptidoglycans can stimulate differential expression of antimicrobial peptide genes in the silkworm, which is affected by the level, binding affinity and transactivation activity of Relishes and Rels.[192] Relish1 is inhibited by its own ankyrin repeats and RelA and RelB are negatively regulated by *B. mori* Cactus.[193] Cactus interacts with the DNA-binding domain in the Rels but not with Relish1 or Relish2. Taken together, these data strongly suggest that the Toll and Imd pathways are functional in the silkworm to regulate immunity-related gene expression.

PROPHENOLOXIDASE ACTIVATION SYSTEM

PO-catalyzed quinone and melanin formation is a universal response in arthropods for killing and entrapping pathogens or parasites.[194,195] PO in insect hemolymph has tyrosinase-like activities, including *o*-hydroxylation of monophenols and oxidation of *o*-diphenols to quinones.[18] Tyrosine, DOPA and dopamine are substrates in insect hemolymph that contribute to PO-catalyzed quinone formation and subsequent melanin synthesis. Tyrosine hydroxylase and dopa decarboxylase are upregulated in fat body after injection of bacteria and probably contribute to provision of hemolymph dopamine for the innate immune response.[80,196-198] Oxidation of dopamine by PO leads to production of 5,6-dihydroxyindole, which has antimicrobial activity toward bacteria and fungi.[199] PO is produced as a zymogen that requires a specific proteolytic cleavage to gain activity. This regulatory mechanism protects the host insect from potentially harmful effects of the reactive chemicals produced by PO, as the enzyme is activated only when elicited by wounding or infection. Understanding of insect PO and its activation was pioneered through detailed biochemical investigations with *B. mori,*[11] and the *M. sexta* model system is providing new insights into lepidopteran PO function and regulation.[17,19]

ProPO from *B. mori* and *M. sexta* exists in plasma as a heterodimer of two related subunits, each ~80 kDa.[200,201] Insect proPO sequences are related to arthropod hemocyanins, and copper-binding motifs in the two groups of proteins are conserved.[202] ProPO is synthesized constitutively by oenocytoids.[197,201,203] ProPOs lack secretion signal peptides and are released from oenocytoids by lysis of the cells.[40,44] Activation of the proPO zymogen requires cleavage of a conserved Arg-Phe bond about 50 residues from the amino-terminus.[11,18] The crystal structure of the *M. sexta* proPO heterodimer suggests that the proteolysis between Arg51 and Phe52 induces a conformational change to dislodge a specific Phe residue in each subunit and open up the active site for substrate binding.[204] The active site contains a canonical Type-3 di-nuclear copper center, with each copper ion coordinated by three conserved His residues. Glu395 of the subunit-2 may act as a general base to deprotonate monophenols, a key step in the *o*-hydroxylation of tyrosine by PO.

Extracellular serine proteinase pathways have evolved in animals to stimulate rapid responses to tissue damage, pathogen invasion, or physiological cues.[205] A few eliciting molecules, via specific recognition and cascade amplification, lead to sequential proteolytic activation of a large number of pathway components within minutes. This type of proteinase pathway results in activation of proPO in response to infection. Many of the proteinases that function in such cascade pathways in arthropods contain a carboxyl-terminal serine proteinase domain similar trypsin or chymotrypsin and one or two amino-terminal clip domains, which have likely regulatory functions.[206] The proteinases from lepidopterans known to activate proPO and most of the proteinases upstream in the activation pathway are clip domain proteinases. Fifteen clip domain proteinase genes were identified in the *B. mori* genome,[13] and fourteen such enzymes are expressed in fat body or hemocytes of *M. sexta*.[207]

In lepidopterans, proPO activating proteinases (PAPs) have been well characterized in *M. sexta*[208-211] and *B. mori*.[212] They are present in hemolymph as zymogens at low concentration in naïve larvae, and their expression in fat body is upregulated in response to injection of bacteria. *M. sexta* PAP1 contains a single clip domain, whereas PAP2 and PAP3 and *B. mori* proPO activating enzyme each contain two clip domains. The solution structure of the region of PAP2 containing the dual clip domains suggests a potential proPO-binding site, a bacteria interacting region and a surface for activator/adaptor docking in each domain.[213] Purified *M. sexta* PAP1, PAP2 and PAP3 do not efficiently generate active PO activity, even after a significant amount of proPO is cleaved at Arg51, without the presence of protein cofactors from hemolymph, identified as serine proteinase homologs (SPHs). SPHs also contain clip domains but lack proteinase activity due to substitution of the active site Ser residue with Gly.[214,215] *M. sexta* SPH1 and SPH2 also require proteolytic processing to gain function, which leads to their assembly into the active, high M_r cofactor required in the reaction with proPO and PAP to generate high levels of PO activity.[215,216] This interesting interaction, which does not seem to be required for the silkworm proPO activating enzyme,[212] is not well understood and requires further investigation.

The *M. sexta* proPO activation system includes at least four other serine proteinases (Fig. 1). An initiatiing hemolymph proteinase (HP14) contains five low-density lipoprotein receptor class A repeats, one Sushi domain, one Wonton domain and one proteinase catalytic domain.[217] Adding recombinant proHP14 to larval plasma greatly enhances proPO activation in response to *M. luteus*. The HP14 proenzyme, with its first domain truncated, was isolated from plasma of larvae injected with bacteria.[99] After incubation with β-1,3-glucan and βGRP1 or βGRP2, the proHP14 was converted to a two-chain active form, which significantly enhanced plasma proPO activation. The activation of proHP14 results from an autoactivation cleavage after Leu387, occurring when proHP14 interacts with β-1,3-glucan and βGRP. Characterization of individual domains and truncation mutants of HP14 showed that the amino-terminal regulatory region of HP14 participates in the specific binding of microbial polysaccharides and βGRP1.[100] Proteins orthologous to *M. sexta* HP14 also function at the top of proteinase cascades in immune responses of *Drosophila* and a beetle, *Tenebrio molitor*.[218,219]

HP14 activates a clip domain proteinase HP21, which can then activate proPAP2 and proPAP3,[220,221] resulting in activation of proPO. ProPAP1, which differs from proPAP2 and proPAP3 in having only one clip domain, is activated by HP6.[189] HP6 also functions in the proSpätzle activation pathway,[188,189] providing cross-talk between these two immune cascades in *M. sexta*. Addition of active PAP1 to hemolymph stimulates the proteolytic activation of HP6, HP8, SPH1 and SPH2.[222] PAP1 directly activates proSPH2, but processing of the other precursors is probably indirect, depending on other plasma factors. Consequently, a minute amount of PAP1 added to plasma from naïve larvae stimulates a remarkably high level of PO activity in a short period of time, as a result of a positive feedback loop (Fig. 1). Some gaps in this pathway still need to be filled. To date, it is not clear which HP generates active HP6, leading to both PAP1-mediated melanization and HP8-mediated Toll pathway activation. The proteinase which activates SPH1 has not been identified and the possible involvement of HPs in Gram-negative bacteria-induced defense responses is not yet well understood.

INHIBITORY REGULATION OF HEMOLYMPH PROTEINASES BY SERPINS

Immune responses can produce molecules that are harmful to the host. Serine proteinases and the molecules they activate have potentially toxic effects. Proteinase inhibitors of different families can exist constitutively at relatively high levels in plasma of naïve insects and may also be produced in response to physiological or pathological stimuli.[223] Serpins are ~50 kDa proteins, many of which are irreversible inhibitors and key modulators of immune proteinase pathways.[224] Serpins occur as plasma proteins in vertebrates and invertebrates. They have been purified and studied by molecular cloning from lepidopteran insects including *B. mori*,[225,226] *M. sexta*,[7,227-232] *H. cunea*,[56] *Mythimna unipuncta*,[233] and *Mamestra configurata*.[234] The *B. mori* genome contains 34 serpin genes, which have been analyzed with regard to molecular evolution of this gene family.[13,235] Serpin biochemical and physiological functions in lepidopterans have been characterized most extensively in *M. sexta*.[17,236]

The *M. sexta* serpin-1gene encodes twelve protein isoforms, each having the same amino-terminal 336 residues and a variable region consisting of the carboxyl-terminal ~40 residues, including the reactive center loop that interacts with a serine proteinase during an inhibition reaction. The variable region is produced by mutually exclusive alternative splicing of twelve different versions of the ninth exon of the gene,[237,238] resulting in a group of serpin proteins with diverse inhibitory selectivity.[227] Serpin genes that employ alternative splicing at the same position to generate multiple serpin isoforms have been studied in other moth species. These include *B. mori* serpin genes 1 (3 isoforms) and 28 (4 isoforms),[235] *M. configurata* serpin-1 (9 isoforms),[239] and *Choristoneura fumiferana* serpin-1 (at least 4 isoforms).[240] This mechanism for expanding serpin functional diversity, first discovered in Lepidoptera, been observed in other insect orders and in nematodes.[241]

Physiological functions have been identified for a few of the *M. sexta* serpin-1 isoforms. Serpin-1A, -1E and -1J can inhibit HP8, and serpin-1J appears to be a physiologically relevant regulator of HP8 activity during immune responses[87] (An, Ragan, Kanost, unpublished results). Serpin-1J also inhibits all three PAPs to regulate proPO activation[210,216] (Jiang, unpublished data). Serpin-1I can inhibit HP14.[99] Serpin-1K was identified in hemolymph in a compex with a midgut chymotrypsin,[87] suggesting a potential role for serpin-1 proteins in protection from digestive proteinases that escape into the hemocoel.

A putative orthologous group of serpins including *M. sexta* serpin-3,[229] *B. mori* serpin-3,[235] and an *H. cunea* serpin[56] are synthesized in response to infection and form a clade with *Drosophila* serpin-27A and *Anopheles gambiae* and *Aedes aegypti* serpin-2, which have immune regulatory functions.[235] *M. sexta* serpin-3 contains a reactive site sequence (Asn-Lys-Phe-Gly) highly similar to the proteolytic activation site (Asn-Arg-Phe-Gly) in both proPO subunits. By mimicking these natural substrates, serpin-3 acts as an efficient inhibitor of all three PAPs.[229] *M. sexta* serpin-4 and serpin-5[230] are closely related to each other and form a clade with *B. mori* serpins 4, 5, 7, 8, 14, 31 and 32.[235] Serpin-4 suppresses proPO activation by inhibiting HPs upstream of the PAPs, such as HP1, HP6 and HP21, while serpin-5 forms complexes with HP1 and HP6.[231] *M. sexta* and *B. mori* serpin-6 are apparent orthologs. *M. sexta* serpin-6 can inhibit PAP3 to block proPO activation and it also inhibits HP8 to potentially regulate the Toll pathway.[242]

LEPIDOPTERAN IMMUNE RESPONSES TO DIFFERENT TYPES OF INFECTION

Herbivorous lepidopteran caterpillars consume enormous amounts of plant diet and are capable of increasing their body weight up to 20% per day. Plant leaves harbor microbial communities, which enter the alimentary canal with the ingested food. The midgut of caterpillars can sense bacterial contamination of the diet and trigger immune responses, which are accompanied with life history tradeoffs.[213,244] Bacteria can also naturally enter and infect lepidopterans through wounds. Phagocytosis by granular hemocytes or plasmatocytes, depending on the species is probably the earliest response,[245] and when numbers of bacteria are relatively low, can efficiently clear infection, particularly with bacteria of low virulence. Larger numbers of bacteria lead to hemocyte aggregation and formation of hemocyte nodules, probably aided by plasma pattern recognition proteins that agglutinate bacteria. Activation of lepidopteran hemocytes to become adhesive involves cytokines from the ENF family[245] and eicosanoid signaling.[42] Hemocyte nodules often become melanized, as products of PO polymerize to form a melanin coat around the aggregated hemocytes and bacteria. This response also generates quinones and reactive oxygen species that may help to kill the entrapped bacteria. The humoral response, synthesis of antibacterial peptides, occurs more slowly than the initial hemocyte response, requiring several hours before significant concentrations of antimicrobial molecules accumulate. This broad-spectrum antibacterial activity, comprised of a mixture of different antibacterial peptides, is an effective protective response that can last up to a few days.

Entomopathogenic fungi can invade insect hosts directly via their sclerotized chitinous integument. Penetration and lateral growth within the inner part of the integument is achieved by joint action of physical pressure and secreted enzymes among which proteinases play a predominant role.[246] Most, if not all, entomopathogenic fungi develop in the hemocoel as cells known as protoplasts or hyphal bodies, which lack a fully developed cell wall. The absence of typical fungal cell wall components such as β-1,3-glucan may allow these fungi to evade the host immune surveillance. However, hyphal bodies of the entomopathogenic fungus *Metarhizium anisopliae* are ingested by plasmatocytes in *G. mellonella* during an early phase of infection,[247] even though they lack β-1,3-glucan on their surface. Ingested hyphal bodies are not killed, but propagate and grow within phagocytic vacuoles of the plasmatocytes, which are likely occupied as a vehicle for dispersal within the hemocoel. Survival of hyphal bodies within the hemocytes as well as overcoming of multicellular encapsulation have been attributed to fungal secondary metabolites (toxins), such as destruxins and cyclosporins which suppress cellular and humoral responses within the infected hosts.[248,249] Similar to bacteria, fungal cells are recognized, phagocytosed or, if too large or too numerous, encapasulated by hemocytes in the hemcoel of Lepidoptera.[250] Eicosanoids have been implicated as mediators in cellular antifungal defense.[251] The lepidopteran antifungal response also encompasses proPO activation, production of reactive oxygen species,[252] and synthesis of potent antifungal peptides including cecropins[253] and gallerimycin.[254] However, the humoral responses upon infection with parasitic fungi such as *Beauveria bassiana* is different from that observed after challenge with bacteria.[255]

Larger eukaryotic parasites such as nematodes and parasitic wasps provoke hemocytic encapsulation and melanization,[245] but little is understood about molecular mechanisms for recognizing such parasites as foreign.[256] Successful parasites are able to disrupt or suppress the host insect's immune response. Entomopathogenic nematodes have a mutualistic relationship with virulent bacterial pathogens of insects, in which the bacteria produce virulence factors that disable cellular and humoral immune responses of their insect host.[257] Parasitoid wasps that use lepidopteran larvae as hosts inject venom and accessory fluid components that disrupt the host immune system. In braconid and ichneumonid parasitoid wasps, this adaptation includes the injection of polydnaviruses, which infect host hemocytes and express immunity-disrupting proteins, but do not replicate.[258] Among these are gene products which cause apoptosis of hemocytes, disrupt signal transduction pathways in the humoral immune response and block melanization.[259-261]

Some aspects of immunity to viral infection in lepidopterans are now becoming understood.[262] Baculoviruses are the most commonly studied viral pathogens of these insects. These viruses enter the larva by first infecting the gut epithelial cells. A response that can protect caterpillars from these infections is apoptosis of infected midgut cells. Infected cells die before viral replication can be completed, thus preventing spread of the virus to other cells or tissues.[263] Baculoviruses encode gene products that inhibit caspases responsible for initiating apoptosis, allowing the infection cycle to proceed.[263] Hemocytic encapsulation of infected tracheal cells is another immune response to baculoviruses that has been observed in lepidopterans.[262] In addition, hemolymph PO is correlated with virucidal activity toward baculoviruses,[264] and thus, the proPO activation cascade may help protect against baculoviral infection.

CONCLUSION

Lepidopteran insects have some important advantages as model systems for immunological research, including a depth of knowledge developed so far and the availability of large hemolymph samples from individual insects for studies of hemocytes and plasma proteins. With the recent exception of *B. mori*, studies on moths and butterflies have been hampered by a lack of genomic information, which would facilitate proteomics investigations and also lead to more ready identification of candidate genes for experimental study. This situation is likely to change dramatically in the next few years, as it is anticipated that genome sequences for several additional lepidopteran species will soon become available. Transgenic technology for silkworms is now well developed and may yield new fundamental information on immunity and perhaps strains with improved disease resistance. Furthermore, more complete understanding of lepidopteran immune responses could lead to future developments of enhanced strategies for regulating insect pest populations through use of specific pathogens and parasites.

ACKNOWLEDGEMENTS

Research in the authors' laboratories was supported by NIH grants GM58634 (HJ) and GM41247 (MRK) and by grants from the Deutsche Forschungsgemeinschaft (AV) VI 219/3-1.

REFERENCES

1. Faye I, Pye A, Rasmuson T et al. Insect immunity II. Simultaneous induction of antibacterial activity and selective synthesis of some hemolymph proteins in diapausing pupae of Hyalophora cecropia and Samia cynthia. Infect Immun 1975; 12:1426-38.
2. Powning RF, Davidson WJ. Studies on the insect bacteriolytic enzymes-II. Some physical and enzymatic properties of lysozyme from haemolymph of Galleria mellonella. Comp Biochem Physiol 1976; 55:221-28.
3. Hultmark D, Steiner H, Rasmuson T et al. Insect immunity. Purification and properties of three inducible bactericidal proteins from hemolymph of immunized pupae of Hyalophora cecropia. Eur J Biochem 1980; 106:7-16.
4. Steiner H, Hultmark D, Engström A et al. Sequence and specificity of two antibacterial proteins involved in insect immunity. Nature 1981; 292:246-48.

5. Hultmark D, Engström A, Andersson K et al. Insect immunity. Attacins, a family of antibacterial proteins from Hyalophora cecropia. EMBO J 1983; 2:571-76.
6. Yoshida H, Ochai M, Ashida M. β-1,3-Glucan receptor and peptidoglycan receptor are present as separate entities within insect prophenoloxidase activating system. Biochem Biophys Res Commun 1986; 114:1177-83.
7. Kanost MR, Prasad SV, Wells MA. Primary structure of a member of the serpin superfamily of proteinase inhibitors from an insect, Manduca sexta. J Biol Chem 1989; 264:965-72.
8. Yoshida H, Kinoshita K, Ashida M. Purification of a peptidoglycan recognition protein from hemolymph of the silkworm, Bombyx mori. J Biol Chem 1996; 271:13854-60.
9. Ochiai M, Ashida M. Purification of a β-1,3-glucan recognition protein in the prophenoloxidase system from hemolymph of the silkworm, Bombyx mori. J Biol Chem 1998; 263:12056-62.
10. Kang D, Liu G, Lundström A et al. A peptidoglycan recognition protein in innate immunity conserved from insects to humans. Proc Natl Acad Sci USA 1998; 95:10078-82.
11. Ashida M, Brey PT. Recent advances on the research of the insect prophenoloxidase cascade. In: Brey PT Hultmark D, eds. Molecular Mechanisms of Immune Responses in Insects. London:Chapman and Hall, 1998:135-72.
12. Lavine MD, Strand MR. Insect hemocytes and their role in immunity. Insect Biochem Mol Biol 2002; 32:1295-309.
13. Tanaka H, Ishibashi J, Fujita K et al. A genome-wide analysis of genes and gene families involved in innate immunity of Bombyx mori. Insect Biochem Mol Biol 2008; 38:1087-110.
14. Stevens JM. Bactericidal activity of the blood of actively immunized wax moth larvae. Canadian J Microbiol 1962; 8:491-9.
15. Powning RF, Davidson WJ. Studies on the insect bacteriolytic enzymes-II. Some physical and enzymatic properties of lysozyme from haemolymph of Galleria mellonella. Comp Biochem Physiol 1976; 55:221–28.
16. Kanost MR, Jiang H, Yu XQ. Innate immune responses of a lepidopteran insect, Manduca sexta. Immunol Rev 2004; 198:97-105.
17. Jiang H. The biochemical basis of antimicrobial responses in Manduca sexta. Insect Science 2008; 15:53-66.
18. Kanost MR, Gorman MJ. Phenoloxidases in insect immunity. In: Beckage NE ed. Insect Immunology. San Diego: Elsevier, 2008:69-96.
19. Ragan EJ, An C, Jiang H et al. Roles of hemolymph proteins in antimicrobial defences of Manduca sexta. In: Insect Infection and Immunity. Reynolds S, Rolff J eds. Oxford University Press, 2009:34-48.
20. Dushay MS. Insect hemolymph clotting. Cell Mol Life Sci 2009; 66:2643-50.
21. Vilcinskas A. Lepidopterans as model mini-hosts for human pathogens and as a source for peptide antibiotics. In: Goldsmith M and Marec F, eds. Molecular Biology and Genetics of the Lepidoptera Boca Raton:CRC Press, 2010:293-305.
22. Kanost MR, Nardi JB. Innate immune responses of Manduca sexta. In: Goldsmith M and Marec F, eds. Molecular Biology and Genetics of the Lepidoptera Boca Raton:CRC Press, 2010:271-291.
23. Mukherjee K, Altincicek B, Hain T et al. Galleria mellonella as a model system for studying Listeria pathogenesis. Appl Environ Microbiol. 2010; 76:310-7.
24. Mullett H, Ratcliffe NA, Rowley AF. The generation and characterization of anti-insect blood cell monoclonal antibodies. J. Cell Sci 1992; 105:93-100.
25. Willott E, Trenczek T, Thrower LW et al. Immunochemical identification of insect hemocyte populations: monoclonal antibodies distinguish four major hemocyte types in Manduca sexta. Eur J Cell Biol 1994; 65:417-23.
26. Gardiner EM, Strand MR. Monoclonal antibodies bind distinct classes of hemocytes in the moth Pseudoplusia includens. J Insect Physiol 1999; 45:113-126.
27. Nardi JB, Pilas B, Bee CM et al. Neuroglian-positive plasmatocytes of Manduca sexta and the initiation of hemocyte attachment to foreign surfaces. Dev Comp Immunol 2006; 30:447-62.
28. Gardiner EM, Strand MR. Hematopoiesis in larval Pseudoplusia includens and Spodoptera frugiperda. Arch Insect Biochem Physiol 2000; 43:147-64.
29. Beetz S, Holthusen TK, Koolman J et al. Correlation of hemocyte counts with different developmental parameters during the last larval instar of the tobacco hornworm, Manduca sexta. Arch Insect Biochem Physiol 2008; 67:63-75.
30. Nardi JB. Embryonic origins of the two main classes of hemocytes—granular cells and plasmatocytes—in Manduca sexta. Dev Genes Evol 2004; 214:19-28.
31. Nardi JB, Pilas B, Ujhelyi E et al. Hematopoietic organs of Manduca sexta and hemocyte lineages. Dev Genes Evol 2003; 213(10):477-91.
32. Schmidt O, Söderhäll K, Theopold U et al. Role of adhesion in arthropod immune recognition. Annu Rev Entomol 2010; 55:485-504.

33. Dean P, Potter U, Richards EH et al. Hyperphagocytic haemocytes in Manduca sexta. J Insect Physiol 2004; 50:1027-36.
34. Nardi JB, Pilas B, Bee CM et al. Neuroglian-positive plasmatocytes of Manduca sexta and the initiation of hemocyte attachment to foreign surfaces. Dev Comp Immunol 2006; 30:447-62.
35. Nakatogawa S, Oda Y, Kamiya M et al. A novel peptide mediates aggregation and migration of hemocytes from an insect. Curr Biol 2009; 19:779-85.
36. Clark KD, Pech LL, Strand MR. Isolation and identification of a plasmatocyte-spreading peptide from the hemolymph of the lepidopteran insect Pseudoplusia includens. J Biol Chem 1997; 272:23440-7.
37. Wang Y, Jiang H, Kanost MR. Biological activity of Manduca sexta paralytic and plasmatocyte spreading peptide and primary structure of its hemolymph precursor. Insect Biochem Mol Biol 1999; 29:1075-86.
38. Strand MR, Hayakawa Y, Clark KD. Plasmatocyte spreading peptide (PSP1) and growth blocking peptide (GBP) are multifunctional homologs. J Insect Physiol 2000; 46:817-24.
39. Volkman BF, Anderson ME, Clark KD et al. Structure of the insect cytokine peptide plasmatocyte-spreading peptide 1 from Pseudoplusia includens. J Biol Chem 1999; 274:4493-6.
40. Matsumoto Y, Oda Y, Uryu M et al. Insect cytokine growth-blocking peptide triggers a termination system of cellular immunity by inducing its binding protein. J Biol Chem 2003; 278:38579-85.
41. Eleftherianos I, Xu M, Yadi H et al. Plasmatocyte-spreading peptide (PSP) plays a central role in insect cellular immune defenses against bacterial infection. J Exp Biol 2009; 212(Pt 12):1840-8.
42. Stanley D. Prostaglandins and other eicosanoids in insects: biological significance. Annu Rev Entomol 2006; 51:25-44.
43. Shrestha S, Kim Y. Various eicosanoids modulate the cellular and humoral immune responses of the beet armyworm, Spodoptera exigua. Biosci Biotechnol Biochem 2009; 73:2077-84.
44. Shrestha S, Kim Y. Eicosanoids mediate prophenoloxidase release from oenocytoids in the beet armyworm Spodoptera exigua. Insect Biochem Mol Biol 2008; 38:99-112.
45. Lavine MD, Strand MR. Haemocytes from Pseudoplusia includens express multiple alpha and beta integrin subunits. Insect Mol Biol 2003; 12:441-52.
46. Wiegand C, Levin D, Gillespie J et al. Monoclonal antibody MS13 identifies a plasmatocyte membrane protein and inhibits encapsulation and spreading reactions of Manduca sexta hemocytes. Arch Insect Biochem Physiol 2000; 45:95-108.
47. Levin DM, Breuer LN, Zhuang S et al. A hemocyte-specific integrin required for hemocytic encapsulation in the tobacco hornworm, Manduca sexta. Insect Biochem Mol Biol 2005; 35:369-80.
48. Nardi JB, Zhuang S, Pilas B et al. Clustering of adhesion receptors following exposure of insect blood cells to foreign surfaces. J Insect Physiol 2005; 51:555-64.
49. Zhuang S, Kelo L, Nardi JB. An integrin-tetraspanin interaction required for cellular innate immune responses of an insect, Manduca sexta. J Biol Chem 2007; 282:22563-72.
50. Zhuang S, Kelo L, Nardi JB et al. Multiple alpha subunits of integrin are involved in cell-mediated responses of the Manduca immune system. Dev Comp Immunol 2008; 32:365-79.
51. Zhuang S, Kelo L, Nardi JB et al. Neuroglian on hemocyte surfaces is involved in homophilic and heterophilic interactions of the innate immune system of Manduca sexta. Dev Comp Immunol 2007; 31:1159-67.
52. Andersson K, Steiner H. Structure and properties of protein P4, the major bacteria-inducible protein in pupae of Hyalophora cecropia. Insect Biochem 1987; 17:133-40.
53. Sun SC, Lindstöm I, Boman HG et al. Hemolin: an insect immune protein belonging to the immunoglobulin superfamily. Science 1990; 250:1729-32.
54. Ladendorff NE, Kanost MR. Isolation and characterization of bacteria induced protein P4 from hemolymph of Manduca sexta. Arch Insect Biochem Physiol 1990; 15:33-41.
55. Ladendorff NE, Kanost MR. Bacteria-induced protein P4 (hemolin) from Manduca sexta: a member of the immunoglobulin superfamily which can inhibit hemocyte aggregation. Arch Insect Biochem Physiol 1991; 18:285-300.
56. Shin SW, Park SS, Park DS et al. Isolation and characterization of immune-related genes from the fall webworm, Hyphantria cunea, using PCR-based differential display and subtractive cloning. Insect Biochem Mol Biol 1998; 28:827-37.
57. Lee KY, Horodyski FM, Valaitis AP et al. Molecular characterization of the insect immune protein hemolin and its high induction during embryonic diapause in the gypsy moth, Lymantria dispar. Insect Biochem Mol Biol 2002; 32:1457-67.
58. Li W, Terenius O, Hirai M et al. Cloning, expression and phylogenetic analysis of hemolin, from the Chinese oak silk moth, Antheraea pernyi. Dev Comp Immunol 2005; 29:853-64.
59. Gandhe AS, Arunkumar KP, John SH et al. Analysis of bacteria-challenged wild silkmoth, Antheraea mylitta (Lepidoptera) transcriptome reveals potential immune genes. BMC Genomics 2006; 7:184.
60. Eum JH, Seo YR, Yoe SM et al. Analysis of the immune-inducible genes of Plutella xylostella using expressed sequence tags and cDNA microarray. Dev Comp Immunol 2007; 31:1107-20

61. Bao Y, Yamano Y, Morishima I. Induction of hemolin gene expression by bacterial cell wall components in eri-silkworm, Samia cynthia ricini. Comp Biochem Physiol Part B 2007; 146:147-51.
62. Faye I, Kanost MR. Function and regulation of hemolin. In: Brey PT, Hultmark D eds. Molecular Mechanisms of Immune Responses in Insects. London: Chapman and Hall, 1998:173-88.
63. Gibson NJ, Tolbert LP. Activation of epidermal growth factor receptor mediates receptor axon sorting and extension in the developing olfactory system of the moth Manduca sexta. J Comp Neurol 2006; 495:554-72.
64. Chen CL, Lampe DJ, Robertson HM et al. Neuroglian is expressed on cells destined to form the prothoracic glands of Manduca embryos as they segregate from surrounding cells and rearrange during morphogenesis. Dev Biol 1997; 181:1-13.
65. Roxström-Lindquist K, Lindström-Dinnetz I, Olesen J et al. An intron enhancer activates the immunoglobulin-related hemolin gene in Hyalophora cecropia. Insect Mol Biol 2002; 11:505-15.
66. Yu XQ, Kanost MR. Developmental expression of Manduca sexta hemolin. Arch Insect Biochem Physiol 1999; 42:198-212.
67. Roxström-Lindquist K, Assefaw-Redda Y, Rosinska K et al. 20-hydroxyecdysone indirectly regulates hemolin gene expression in Hyalophora cecropia. Insect Mol Biol 2005; 14:645-52.
68. Terenius O. Hemolin-A lepidopteran anti-viral defense factor? Dev Comp Immunol 2008; 32:311-6.
69. Hirai M, Terenius O, Li W et al. Baculovirus and dsRNA induce hemolin, but no antibacterial activity in Antheraea pernyi. Insect Mol Biol 2004; 13:399-405.
70. Terenius O, Popham HJ, Shelby KS. Bacterial, but not baculoviral infections stimulate hemolin expression in noctuid moths. Dev Comp Immunol 2009; 33:1176-85.
71. Daffre S, Faye I. Lipopolysaccharide interaction with hemolin, an insect member of the Ig-superfamily. FEBS Lett 1997; 408:127-30.
72. Yu XQ, Kanost MR. Binding of hemolin to bacterial lipopolysaccharide and lipoteichoic acid. An immunoglobulin superfamily member from insects as a pattern-recognition receptor. Eur J Biochem 2002; 269:1827-34.
73. Zhao L, Kanost MR. In search of a function for hemolin, a hemolymph protein from the immunoglobulin superfamily. J Insect Physiol 1996; 42:73-9.
74. Bettencourt R, Lanz-Mendoza H, Lindquist KR et al. Cell adhesion properties of hemolin, an insect immune protein in the Ig superfamily. Eur J Biochem 1997; 250:630-7.
75. Su XD, Gastinel LN, Vaughn DE et al. Crystal structure of hemolin: a horseshoe shape with implications for homophilic adhesion. Science 1998; 281:991-5.
76. Eleftherianos I, Gokcen F, Felfoldi G et al. The immunoglobulin family protein hemolin mediates cellular immune responses to bacteria in the insect Manduca sexta. Cellular Microbiol 2007; 9:1137-47.
77. Dziarski R, Gupta D. The peptidoglycan recognition proteins (PGRPs). Genome Biol 2006; 7:232.
78. Ochiai M, Ashida M. A pattern recognition protein for peptidoglycan. Cloning the cDNA and the gene of the silkworm, Bombyx mori. J Biol Chem 1999; 274:11854-8.
79. Yu XQ, Zhu YF, Ma C et al. Pattern recognition proteins in Manduca sexta plasma. Insect Biochem Mol Biol 2002; 32:1287-93.
80. Zhu Y, Johnson T, Kanost MR. Identification of differentially expressed genes in the immune response of the tobacco hornworm, Manduca sexta. Insect Biochem Mol Biol 2003; 33:541-59.
81. Zou Z, Najar F, Wang Y et al. Pyrosequence analysis of expressed sequence tags for Manduca sexta hemolymph proteins involved in immune responses. Insect Biochem Mol Biol 2008; 38:677-82.
82. Seitz V, Clermont A, Wedde M et al. Identification of immunorelevant genes from greater wax moth (Galleria mellonella) by a subtractive hybridization approach. Dev Comp Immunol 2003; 27:207-15.
83. Onoe H, Matsumoto A, Hashimoto K et al. Peptidoglycan recognition protein (PGRP) from eri-silkworm, Samia cynthia ricini; protein purification and induction of the gene expression. Comp Biochem Physiol B 2007; 147:512-19.
84. Hashimoto K, Mega K, Matsumoto Y et al. Three peptidoglycan recognition protein (PGRP) genes encoding potential amidase from eri-silkworm, Samia cynthia ricini. Comp Biochem Physiol B Biochem Mol Biol 2007; 148:322-8.
85. Coates BS, Sumerford DV, Hellmich RL et al. Mining an Ostrinia nubilalis midgut expressed sequence tag (EST) library for candidate genes and single nucleotide polymorphisms (SNPs). Insect Mol Biol 2008; 17:607-20.
86. Mellroth P, Karlsson J, Steiner H. A scavenger function for a Drosophila peptidoglycan recognition protein. J Biol Chem 2003; 278:7059-64.
87. Ragan EJ. Immune-related protein complexes and serpin-1 isoforms in Manduca sexta plasma. PhD Dissertation, Kansas State University. 2008.
88. Sumathipala N. Expression, purification and characterization of peptidoglycan recognition protein 1 from Manduca sexta (L.). MS Thesis, Oklahoma State University. 2009.

89. Eleftherianos I, Marokhazi J, Millichap PJ et al. Prior infection of Manduca sexta with nonpathogenic Escherichia coli elicits immunity to pathogenic Photorhabdus luminescens: roles of immune-related proteins shown by RNA interference. Insect Biochem Mol Biol 2006; 36:517-25.
90. Eleftherianos I, Millichap PJ, ffrench-Constant RH et al. RNAi suppression of recognition protein mediated immune responses in the tobacco hornworm Manduca sexta causes increased susceptibility to the insect pathogen Photorhabdus. Dev Comp Immunol 2006; 30:1099-107.
91. Royet J, Dziarski R. Peptidoglycan recognition proteins: pleiotropic sensors and effectors of antimicrobial defences. Nat Rev Microbiol 2007; 5:264-77.
92. Charroux B, Rival T, Narbonne-Reveau K et al. Bacterial detection by Drosophila peptidoglycan recognition proteins. Microbes Infect 2009; 11:631-636.
93. Ochiai M, Ashida M. A pattern-recognition protein for β-1,3-glucan: the binding domain and the cDNA cloning of β-1,3-glucan recognition protein from the silkworm, Bombyx mori. J Biol Chem 2000; 275:4995-5002.
94. Ma C, Kanost MR. A β-1,3-glucan recognition protein from an insect, Manduca sexta, agglutinates microorganisms and activates the phenoloxidase cascade. J Biol Chem 2000; 275:7505-14.
95. Jiang H, Ma C, Lu ZQ et al β-1,3-glucan recognition protein-2 (βGRP-2) from Manduca sexta: an acute-phase protein that binds β-1,3-glucan and lipoteichoic acid to aggregate fungi and bacteria and stimulate prophenoloxidase activation. Insect Biochem Mol Biol 2004; 34:89-100.
96. Fabrick JA, Baker JE, Kanost MR. cDNA cloning, purification, properties and function of a beta-1,3-glucan recognition protein from a pyralid moth, Plodia interpunctella. Insect Biochem Mol Biol 2003; 33:579-94.
97. Fabrick JA, Baker JE, Kanost MR. Innate immunity in a pyralid moth: functional evaluation of domains from a β-1,3-glucan recognition protein. J Biol Chem 2004; 279:26605-11.
98. Takahasi K, Ochiai M, Horiuchi M et al. Solution structure of the silkworm βGRP/GNBP3 N-terminal domain reveals the mechanism for β-1,3-glucan-specific recognition. Proc Natl Acad Sci USA 2009; 106:11679-84.
99. Wang Y, Jiang H. Interaction of β-1,3-glucan with its recognition protein activates hemolymph proteinase 14, an initiation enzyme of the prophenoloxidase activation system in Manduca sexta. J Biol Chem 2006; 281:9271-8.
100. Wang Y, Jiang H. Binding properties of the regulatory domains in Manduca sexta hemolymph proteinase-14, an initiation enzyme of the prophenoloxidase activation system. Dev Comp Immunol 2010; 34:316-22.
101. Lee WJ, Lee JD, Kravchenko VV et al. Purification and molecular cloning of an inducible gram-negative bacteria-binding protein from the silkworm, Bombyx mori. Proc Natl Acad Sci USA 1996; 93:7888-93.
102. Pauchet Y, Freitak D, Heidel-Fischer HM et al. Immunity or digestion: glucanase activity in a glucan-binding protein family from Lepidoptera. J Biol Chem 2009; 284:2214-24.
103. Vasta GR, Quesenberry M, Ahmed H et al. C-type lectins and galectins mediate innate and adaptive immune functions: their roles in the complement activation pathway. Dev Comp Immunol 1999; 23,401-20.
104. Koizumi N, Morozumi A, Imamura M et al. Lipopolysaccharide-binding proteins and their involvement in the bacterial clearance from the hemolymph of the silkworm Bombyx mori. Eur J Biochem 1997; 248:217-24.
105. Koizumi N, Imai Y, Morozumi A et al. Lipopolysaccharide-binding protein of Bombyx mori participates in a hemocyte-mediated defense reaction against Gram-negative bacteria. J Insect Physiol 1999; 45:853-9.
106. Shin SW, Park DS, Kim SC et al. Two carbohydrate recognition domains of Hyphantria cunea lectin bind to bacterial lipopolysaccharides through O-specific chain. FEBS Lett 2000; 467:70-4.
107. Yu XQ, Gan H, Kanost MR. Immulectin, an inducible C-type lectin from an insect, Manduca sexta, stimulates activation of plasma prophenol oxidase. Insect Biochem Mol Biol 1999; 29:585-97.
108. Yu XQ, Kanost MR. Immulectin-2, a lipopolysaccharide-specific lectin from an insect, Manduca sexta, is induced in response to Gram-negative bacteria. J Biol Chem 2000; 275:37373-81.
109. Yu XQ, Tracy ME, Ling E et al. A novel C-type immulectin-3 from Manduca sexta is translocated from hemolymph into the cytoplasm of hemocytes. Insect Biochem Mol Biol 2005; 35:285-95.
110. Yu XQ, Ling E, Tracy ME et al. Immulectin-4 from the tobacco hornworm Manduca sexta binds to lipopolysaccharide and lipoteichoic acid. Insect Mol Biol 2006; 15:119-28.
111. Watanabe A, Miyazawa S, Kitami M et al. Characterization of a novel C-type lectin, Bombyx mori multibinding protein, from the B. mori hemolymph. J Immunol 2006; 177:4594-604.
112. Kim SR, Lee KS, Kim I et al. cDNA sequence of a novel immulectin homologue from the silkworm, Bombyx mori. Int J Indust Entomol 2003; 6:99-102.
113. Takase H, Watanabe A, Yoshizawa Y et al. Identification and comparative analysis of three novel C-type lectins from the silkworm with functional implications in pathogen recognition. Dev Comp Immunol 2009; 33:789-800.

114. Chai LQ, Tian YY, Yang DT et al. Molecular cloning and characterization of a C-type lectin from the cotton bollworm, Helicoverpa armigera. Dev Comp Immunol 2008; 32:71-83.
115. Zou Z, Evans J, Lu Z et al. Comparative genome analysis of the Tribolium immune system. Genome Biol 2007; 8:R177.
116. Yu XQ, Kanost MR. Manduca sexta lipopolysaccharide-specific immulectin-2 protects larvae from bacterial infection. Dev Comp Immunol 2003; 27:189-96.
117. Yu XQ, Kanost MR. Immulectin-2, a pattern recognition receptor that stimulates hemocyte encapsulation and melanization in the tobacco hornworm, Manduca sexta. Dev Comp Immunol 2004; 28:891-900.
118. Ling E, Yu XQ. Cellular encapsulation and melanization are enhanced by immulectins, pattern recognition receptors from the tobacco hornworm Manduca sexta. Dev Comp Immunol 2006; 30:289-99.
119. Van der Horst DJ, Roosendaal SD, Rodenburg KW. Circulatory lipid transport: lipoprotein assembly and function from an evolutionary perspective. Mol Cell Biochem. 2009; 326:105-19.
120. Li D, Scherfer C, Korayem AM et al. Insect hemolymph clotting: evidence for interaction between the coagulation system and the prophenoloxidase activating cascade. Insect Biochem Mol Biol 2002; 32:919-28.
121. Ma G, Hay D, Li D et al. Recognition and inactivation of LPS by lipophorin particles. Dev Comp Immunol 2006; 30:619-26.
122. Rahman MM, Ma G, Roberts HL et al. Cell-free immune reactions in insects. J Insect Physiol 2006; 52:754-62.
123. Kato Y, Motoi Y, Taniai K et al. Lipopolysaccharide-lipophorin complex formation in insect hemolymph: a common pathway of lipopolysaccharide detoxification both in insects and in mammals. Insect Biochem Mol Biol 1994; 24:547-55.
124. Dunphy G, Halwani A. Haemolymph proteins of larvae of Galleria mellonella detoxify endotoxins of the insect pathogenic bacteria Xenhorabdus nematophilus (Enterobacteriaceae). J Insect Physiol 1997; 43:1023-9.
125. Halwani AE, Niven DF, Dunphy GB. Apolipophorin-III and the interactions of lipoteichoic acids with the immediate immune responses of Galleria mellonella. J Invertebr Pathol 2000; 76:233-41.
126. Pratt CC, Weers PM. Lipopolysaccharide binding of an exchangeable apolipoprotein, apolipophorin III, from Galleria mellonella. Biol Chem 2004; 385:1113-9.
127. Kim HJ, Je HJ, Park SY et al. Immune activation of apolipophorin-III and its distribution in hemocyte from Hyphantria cunea. Insect Biochem Mol Biol 2004; 34:1011-23.
128. Leon LJ, Idangodage H, Wan CP et al. Apolipophorin III: lipopolysaccharide binding requires helix bundle opening. Biochem Biophys Res Commun 2006; 348:1328-33.
129. Weisner A, Losen S, Kopacek P et al. Isolated apolipophorin III from Galleria mellonella stimulates the immune reaction of this insect. J Insect Physiol 1997; 43:383-91.
130. Halwani AE, Dunphy GB. Apolipophorin-III in Galleria mellonella potentiates hemolymph lytic activity. Dev Comp Immunol 1999; 23:563-70.
131. Iimura Y, Ishikawa H, Yamamoto K et al. Hemagglutinating properties of apolipophorin III from the hemolymph of Galleria mellonella larvae. Arch Insect Biochem Physiol 1998; 38:119-25.
132. Halwani AE, Niven DF, Dunphy GB. Apolipophorin-III in the greater wax moth, Galleria mellonella (Lepidoptera: Pyralidae). Arch Insect Biochem Physiol 2001; 48:135-43.
133. Whitten MM, Tew IF, Lee BL et al. A novel role for an insect apolipoprotein (apolipophorin III) in β-1,3-glucan pattern recognition and cellular encapsulation reactions. J Immunol 2004; 172:2177-85.
134. St. Leger RJ, Bidochka MJ, Roberts DW. Isoforms of the cuticle-degrading Pr1 proteinase and production of a metalloproteinase by Metarhizium anisopliae. Arch Biochem Biophys 1994; 313:1-7.
135. Fedhila S, Nel P, Lereclus D. The InhA2 metalloprotease of Bacillus thuringiensis strain 407 is required for pathogenicity in insects infected via the oral route. J Bacteriol. 2002; 184:3296-304.
136. Held KG, LaRock CN, D'Argenio DA et al. A metalloprotease secreted by the insect pathogen Photorhabdus luminescens induces melanization. Appl Environ Microbiol 2007; 73(23):7622-8.
137. Qazi SS, Khachatourians GG. Hydrated conidia of Metarhizium anisopliae release a family of metalloproteases. J Invertebr Pathol 2007; 95:48-59.
138. Griesch J, Wedde M, Vilcinskas A. Recognition and regulation of metalloproteinase activity in the haemolymph of Galleria mellonella: a new pathway mediating induction of humoral immune responses. Insect Biochem Mol Biol 2000; 30:461-72.
139. Altincicek B, Linder M, Linder D et al. Microbial metalloproteinases mediate sensing of invading pathogens and activate innate immune responses in the lepidopteran model host Galleria mellonella. Infect Immun 2007; 75:175-83.
140. Altincicek B, Berisha A, Mukherjee K et al. Identification of collagen IV derived danger/alarm signals in insect immunity by nanoLC-FTICR MS. Biological Chemistry 2009; 390:1303-11.
141. Matzinger P. The danger model: a renewed sense of self. Science 2002; 296:301-05.

142. Vilcinskas A, Wedde M. Inhibition of Beauveria bassiana proteases and fungal development by inducible protease inhibitors in the haemolymph of Galleria mellonella larvae. Biocontrol Sci Technol 1997; 7:591-601.
143. Clermont A, Wedde M, Seitz V et al. Cloning and expression of an inhibitor against microbial metalloproteinases from insects (IMPI) contributing to innate immunity. Biochem J 2004; 382:315-22.
144. Wedde M, Weise C, Nuck C et al. The insect metallo-proteinase inhibitor gene of the lepidopteran Galleria mellonella encodes two distinct inhibitors. Biol Chem 2007; 388:119-27.
145. Kim I, Kim SH, Lee YS et al. Immune stimulation in the silkworm, Bombyx mori L., by CpG oligodeoxy-nucleotides. Arch Insect Biochem Physiol 2004; 55:43-8.
146. Altincicek B, Stötzel S, Wygrecka M et al. Host-derived extracellular nucleic acids enhance innate immune responses, induce coagulation and prolong survival upon infection in insects. J Immunol 2008; 181:2705-12.
147. Lavine MD, Chen G, Strand MR. Immune challenge differentially affects transcript abundance of three antimicrobial peptides in hemocytes from the moth Pseudoplusia includens. Insect Biochem Mol Biol 2005; 35:1335-46.
148. Dunn PE, Bohnert TJ, Russell V. Regulation of antibacterial protein synthesis following infection and during metamorphosis of Manduca sexta. Ann N Y Acad Sci 1994; 712:117-30.
149. Gorman MJ, Kankanala P, Kanost MR. Bacterial challenge stimulates innate immune responses in extra-embryonic tissues of tobacco hornworm eggs. Insect Mol Biol 2004; 13:19-24.
150. Mohrig W, Messner B. Immunreaktionen bei Insekten. I. Lysozyme als grundlegender antibakterieller Faktor im humoralen Abwehrgeschehen. Biol Zentralbl 1968; 87:439-47.
151. Jolles J, Schoentgen F, Croizier G et al. Insect lysozymes from three species of Lepidoptera: Their structural relatedness to the C (chicken) type lysozyme. J Mol Evol 1979; 14:267-71.
152. Yu KH, Kim KN, Lee JH et al. Comparative study on characteristics of lysozymes from the hemolymph of three lepidopteran larvae, Galleria mellonella, Bombyx mori, Agrius convolvuli. Dev Comp Immunol 2002; 26:707-13.
153. Chapelle M, Girard PA, Cousserans F et al. Lysozymes and lysozyme-like proteins from the fall armyworm, Spodoptera frugiperda. Mol Immunol 2009; 47:261-9.
154. Wang WX, Wang YP, Deng XJ et al. Molecular and functional characterization of a c-type lysozyme from the Asian corn borer, Ostrinia furnacalis. J Insect Sci 2009; 9:17.
155. Vilcinskas A, Matha V. Effect of the entomopathogenic fungus Beauveria bassiana on humoral immune response of Galleria mellonella larvae (Lepidoptera: Pyralidae). Eur J Entomol 1997; 94:461-72.
156. Rao XJ, Ling E, Yu XQ. The role of lysozyme in the prophenoloxidase activation system of Manduca sexta: an in vitro approach. Dev Comp Immunol 2010; 34:264-71.
157. Hemmi H, Ishibashi J, Hara S et al. Solution structure of moricin, an antibacterial peptide, isolated from the silkworm Bombyx mori. FEBS Lett 2002; 518:33-8.
158. Dai H, Rayaprolu S, Gong Y et al. Solution structure, antibacterial activity and expression profile of Manduca sexta moricin. J Pept Sci 2008; 14:855-63.
159. Hara S, Yamakawa M. Moricin, a novel antibacterial peptide family isolated from the silkworm, Bombyx mori. Biochem J 1995; 310:651-56.
160. Brown S, Howard A, Kasprzak AB et al. The discovery and analysis of a diverged family of novel antifungal moricin-like peptides in the wax moth Galleria mellonella. Insect Biochem Mol Biol 2008; 38:201-12.
161. Kockum K, Faye I, Hofsten PV et al. Insect immunity. Isolation and sequence of two cDNA clones corresponding to acidic and basic attacins from Hyalophora cecropia. EMBO J 1984; 3:2071-75.
162. Axén A, Carlsson A, Engström A et al. Gloverin, an antibacterial protein from the immune hemolymph of Hyalophora pupae. Eur J Biochem 1997; 247:614-619.
163. Mrinal N, Nagaraju J. Intron loss is associated with gain of function in the evolution of the gloverin family of antibacterial genes in Bombyx mori. J Biol Chem 2008; 283:23376-87.
164. Kawaoka S, Katsuma S, Daimon T et al. Functional analysis of four Gloverin-like genes in the silkworm, Bombyx mori. Arch Insect Biochem Physiol 2008; 67:87-96.
165. Mackintosh JA, Gooley AA, Karuso PH et al. A gloverin-like antibacterial protein is synthesized in Helicoverpa armigera following bacterial challenge. Dev Comp Immunol 1998; 22:387-99.
166. Lundström A, Liu G, Kang D et al. Trichoplusia ni gloverin, an inducible immune gene encoding an antibacterial insect protein. Insect Biochem Mol Biol 2002; 32:795-801.
167. Seitz V, Clermont A, Wedde M et al. Identification of immunorelevant genes from greater wax moth (Galleria mellonella) by a subtractive hybridization approach. Dev Comp Immunol 2003; 27:207-15.
168. Wang Q, Liu Y, He HJ et al. Immune responses of Helicoverpa armigera to different kinds of pathogens. BMC Immunol 2010; 11:9.
169. Hara S, Yamakawa M. A novel antibacterial peptide family isolated from the silkworm, Bombyx mori. Biochem J 1995; 310:651-6.
170. Furukawa S, Taniai K, Ishibashi J et al. A novel member of lebocin gene family from the silkworm, Bombyx mori. Biochem Biophys Res Commun 1997; 238:769-74.

171. Liu G, Kang D, Steiner H. Trichoplusia ni lebocin, an inducible immune gene with a downstream insertion element. Biochem Biophys Res Commun 2000; 269:803-7.
172. Bao Y, Yamano Y, Morishima I. A novel lebocin-like gene from eri-silkworm, Samia cynthia ricini, that does not encode the antibacterial peptide lebocin. Comp Biochem Physiol B Biochem Mol Biol 2005; 140:127-31
173. Rayaprolu S, Wang Y, Kanost MR et al. Functional analysis of four processing products from multiple precursors encoded by a lebocin-related gene from Manduca sexta. Dev Comp Immunol 2010; 34:638-47.
174. Bulet P, Stöcklin R, Menin L. Anti-microbial peptides: from invertebrates to vertebrates. Immunol Rev 2004; 198:169-84.
175. Lamberty M, Ades S, Uttenweiler-Joseph S et al. Insect immunity. Isolation from the lepidopteran Heliothis virescens of a novel insect defensin with potent antifungal activity. J Biol Chem 1999; 274:9320-6.
176. Lamberty M, Caille A, Landon C et al. Solution structures of the antifungal heliomicin and a selected variant with both antibacterial and antifungal activities. Biochemistry 2001; 40:11995-2003.
177. Mandrioli M, Bugli S, Saltini S et al. Molecular characterization of a defensin in the IZD-MB-0503 cell line derived from immunocytes of the insect Mamestra brassicae (Lepidoptera). Biol Cell 2003; 95:53-7.
178. Volkoff AN, Rocher J, d'Alençon E et al. Characterization and transcriptional profiles of three Spodoptera frugiperda genes encoding cysteine-rich peptides. A new class of defensin-like genes from lepidopteran insects? Gene 2003; 319:43-53.
179. Schuhmann B, Seitz, V, Vilcinskas A et al. Cloning and expression of gallerimycin, an antifungal peptide expressed in immune response by the greater wax moth, Galleria mellonella. Arch Insect Biochem Physiol 2003; 53:125-33.
180. Lee YS, Yun EK, Jang WS et al. Purification, cDNA cloning and expression of an insect defensin from the great wax moth, Galleria mellonella. Insect Mol Biol 2004; 13:65-72.
181. Seitz V, Clermont A, Wedde M et al. Identification of immunorelevant genes from greater wax moth (Galleria mellonella) by a subtractive hybridization approach. Dev Comp Immunol 2003; 27:207-15.
182. Girard P, Boublik Y, Wheat CW et al. X-tox: an atypical defensin derived family of immune-related proteins specific to Lepidoptera. Dev Comp Immunol 2008; 32:575-84.
183. Destoumieux-Garzón D, Brehelin M, Bulet P et al. Spodoptera frugiperda X-tox protein, an immune related defensin rosary, has lost the function of ancestral defensins. PLoS One 2009; 4(8):e6795.
184. Lemaitre B, Hoffmann J. The host defense of Drosophila melanogaster. Ann Rev Immunol 2007; 25:697-743.
185. Cheng TC, Zhang YL, Liu C et al. Identification and analysis of Toll-related genes in the domesticated silkworm, Bombyx mori. Dev Comp Immunol 2008; 32:464-75.
186. Ao JQ, Ling E, Yu XQ. A Toll receptor from Manduca sexta is in response to Escherichia coli infection. Mol Immunol 2008; 45:543-52.
187. Wang Y, Cheng T, Rayaprolu S et al. Proteolytic activation of pro-spätzle is required for the induced transcription of antimicrobial peptide genes in lepidopteran insects. Dev Comp Immunol 2007; 31:1002-12.
188. An C, Jiang H, Kanost MR. Proteolytic activation and function of the cytokine Spätzle in innate immune response of a lepidopteran insect, Manduca sexta. FEBS J 2010; 277:148-62.
189. An C, Ishibashi J, Ragan EJ et al. Functions of Manduca sexta hemolymph proteinases HP6 and HP8 in two innate immune pathways. J Biol Chem 2009; 284:19716-26.
190. Tanaka H, Yamamoto M, Moriyama Y et al. A novel Rel protein and shortened isoform that differentially regulate antibacterial peptide genes in the silkworm Bombyx mori. Biochim Biophys Acta 2005; 1730:10-21.
191. Tanaka H, Matsuki H, Furukawa S et al. Identification and functional analysis of Relish homologs in the silkworm, Bombyx mori. Biochim Biophys Acta 2007; 1769:559-68.
192. Tanaka H, Sagisaka A, Nakajima Y et al. Correlation of differential expression of silkworm antimicrobial peptide genes with different amounts of rel family proteins and their gene transcriptional activity. Biosci Biotechnol Biochem 2009; 73:599-606.
193. Furukawa S, Tanaka H, Ishibashi J et al. Functional characterization of a cactus homolog from the silkworm Bombyx mori. Biosci Biotechnol Biochem 2009; 73:2665-70.
194. Nappi AJ, Christensen BM. Melanogenesis and associated cytotoxic reactions: applications to insect innate immunity. Insect Biochem Mol Biol 2005; 35:443-59.
195. Cerenius L, Lee BL, Söderhäll K. The proPO-system: pros and cons for its role in invertebrate immunity. Trends Immunol 2008; 29:263-71.
196. Noguchi H, Tsuzuki S, Tanaka K et al. Isolation and characterization of a dopa decarboxylase cDNA and the induction of its expression by an insect cytokine, growth-blocking peptide in Pseudaletia separata. Insect Biochem Mol Biol 2003; 33:209-17.

197. Gorman MJ, An C, Kanost MR. Characterization of tyrosine hydroxylase from Manduca sexta. Insect Biochem Mol Biol 2007; 37:1327-37.
198. Hashimoto K, Yamano Y, Morishima I. Induction of tyrosine hydroxylase gene expression by bacteria in the fat body of eri-silkworm, Samia cynthia ricini. Comp Biochem Physiol B 2008; 149:501-6.
199. Zhao P, Li J, Wang Y et al. Broad-spectrum antimicrobial activity of the reactive compounds generated in vitro by Manduca sexta phenoloxidase. Insect Biochem Mol Biol 2007; 37:952-9.
200. Yasuhara Y, Koizumi Y, Katagiri C et al. Reexamination of properties of prophenoloxidase isolated from larval hemolymph of the silkworm Bombyx mori. Arch Biochem Biophys 1995; 320:14-23.
201. Jiang H, Wang Y, Ma C et al. Subunit composition of pro-phenol oxidase from Manduca sexta: molecular cloning of subunit proPO-p1. Insect Biochem Mol Biol 1997; 27:835-50.
202. Burmester T. Origin and evolution of arthropod hemocyanins and related proteins. J Comp Physiol B 2002; 172:95-107.
203. Iwama R, Ashida M. Biosynthesis of prophenoloxidase in hemocytes of larval hemolymph of the silkworm, Bombyx mori. Insect Biochem 16:547-55.
204. Li Y, Wang Y, Jiang H et al. Crystal structure of Manduca sexta prophenoloxidase provides insights into the mechanism of type-3 copper enzymes. Proc Natl Acad Sci USA 2009; 106:17001-5.
205. Krem MM, Di Cera E. Evolution of enzyme cascades from embryonic development to blood coagulation. Trends Biochem Sci 2002; 27:67-74.
206. Jiang H, Kanost MR. The clip-domain family of serine proteinases in arthropods. Insect Biochem Mol Biol 2000; 30:95-105.
207. Jiang H, Wang Y, Gu Y et al. Molecular identification of a bevy of serine proteinases in Manduca sexta hemolymph. Insect Biochem Mol Biol 2005; 35:931-43.
208. Jiang H, Wang Y, Kanost MR. Pro-phenol oxidase activating proteinase from an insect, Manduca sexta: a bacteria-inducible protein similar to Drosophila Easter. Proc Natl Acad Sci USA 1998; 95:12220-5.
209. Jiang H, Wang Y, Yu XQ et al. Prophenoloxidase-activating proteinase-2 from hemolymph of Manduca sexta. A bacteria-inducible serine proteinase containing two clip domains. J Biol Chem 2003; 278:3552-61.
210. Jiang H, Wang Y, Yu XQ et al. Prophenoloxidase-activating proteinase-3 (PAP-3) from Manduca sexta hemolymph: a clip-domain serine proteinase regulated by serpin-1J and serine proteinase homologs. Insect Biochem Mol Biol 2003; 33:1049-60.
211. Gupta S, Wang Y, Jiang H. Purification and characterization of Manduca sexta prophenoloxidase -activating proteinase-1 (PAP-1), an enzyme involved in insect immune responses. Protein Exp Purif 2005a; 39:261-8.
212. Satoh D, Horii A, Ochiai M et al. Prophenoloxidase-activating enzyme of the silkworm, Bombyx mori: purification, characterization and cDNA cloning. J Biol Chem 1999; 274:7441-53.
213. Huang R, Lu Z, Dai H et al. The solution structure of clip domains from Manduca sexta prophenoloxidase activating proteinase-2. Biochemistry 2007; 46:11431-9.
214. Yu XQ, Jiang H, Wang Y et al. Nonproteolytic serine proteinase homologs involved in phenoloxidase activation in the tobacco hornworm, Manduca sexta. Insect Biochem Mol Biol 2003; 33:197-208.
215. Wang Y, Jiang H. Prophenoloxidase (proPO) activation in Manduca sexta: an analysis of molecular interactions among proPO, proPO-activating proteinase-3 and a cofactor. Insect Biochem Mol Biol 2004; 34:731-42.
216. Gupta S, Wang Y, Jiang H. Manduca sexta prophenoloxidase (proPO) activation requires proPO-activating proteinase (PAP) and serine proteinase homologs (SPHs) simultaneously. Insect Biochem Mol Biol 2005; 35:241-8.
217. Ji C, Wang Y, Guo X et al. A pattern recognition serine proteinase triggers the prophenoloxidase activation cascade in the tobacco hornworm, Manduca sexta. J Biol Chem 2004; 279:34101-6.
218. Buchon N, Poidevin M, Kwon HM et al. A single modular serine protease integrates signals from pattern-recognition receptors upstream of the Drosophila Toll pathway. Proc Natl Acad Sci USA 2009; 106:12442-7.
219. Roh KB, Kim CH, Lee H et al. Proteolytic cascade for the activation of the insect toll pathway induced by the fungal cell wall component. J Biol Chem 2009; 284:19474-81.
220. Wang Y, Jiang H. Reconstitution of a branch of the Manduca sexta prophenoloxidase activation cascade in vitro: snake-like hemolymph proteinase 21 (HP21) cleaved by HP14 activates prophenoloxidase-activating proteinase-2 precursor. Insect Biochem Mol Biol 2007; 37:1015-25.
221. Gorman MJ, Wang Y, Jiang H et al. Manduca sexta hemolymph proteinase 21 activates prophenoloxidase-activating proteinase 3 in an insect innate immune response proteinase cascade. J Biol Chem 2007b; 282:11742-9.
222. Wang Y, Jiang H. A positive feedback mechanism in the Manduca sexta prophenoloxidase activation system. Insect Biochem Mol Biol 2008; 38:763-9.

223. Kanost MR. Serine proteinase inhibitors in arthropod immunity. Dev Comp Immunol 1999; 23:291-301.
224. Gettins PG. Serpin structure, mechanism and function. Chem Rev 2002; 102:4751-804.
225. Sasaki T, Kobayashi K. Isolation of two novel proteinase inhibitors from hemolymph of silkworm larva, Bombyx mori: comparison with human serum proteinase inhibitors. J Biochem 1984; 95:1009-17.
226. Sasaki T. Patchwork-structure serpins from silkworm (Bombyx mori) larval hemolymph. Eur J Biochem 1991; 202:255-61.
227. Jiang H, Kanost MR. Characterization and functional analysis of 12 naturally occurring reactive site variants of serpin-1 from Manduca sexta. J Biol Chem 1997; 272:1082-7.
228. Gan H, Wang Y, Jiang H et al. A bacteria-induced, intracellular serpin in granular hemocytes of Manduca sexta. Insect Biochem Mol Biol 2001; 31:887-98.
229. Zhu Y, Wang Y, Gorman MJ et al. Manduca sexta serpin-3 regulates prophenoloxidase activation in response to infection by inhibiting prophenoloxidase-activating proteinases. J Biol Chem 2003b; 278:46556-64.
230. Tong Y, Kanost MR. Manduca sexta serpin-4 and serpin-5 inhibit the prophenol oxidase activation pathway: cDNA cloning, protein expression and characterization. J Biol Chem 2005; 280:14923-31.
231. Tong Y, Jiang H, Kanost MR. Identification of plasma proteases inhibited by Manduca sexta serpin-4 and serpin-5 and their association with components of the prophenol oxidase activation pathway. J Biol Chem 2005; 280:14932-42.
232. Wang Y, Jiang H. Purification and characterization of Manduca sexta serpin-6: a serine proteinase inhibitor that selectively inhibits prophenoloxidase-activating proteinase-3. Insect Biochem Mol Biol 2004; 34, 387-95.
233. Cherqui A, Cruz N, Simões N. Purification and characterization of two serine protease inhibitors from the hemolymph of Mythimna unipuncta. Insect Biochem Mol Biol 2001; 31:761-9.
234. Chamankhah M, Braun L, Visal-Shah S et al. Mamestra configurata serpin-1 homologs: cloning, localization and developmental regulation. Insect Biochem Mol Biol 2003; 33:355-69.
235. Zou Z, Zhao P, Weng H et al. A comparative analysis of serpin genes in the silkworm genome. Genomics 2009; 93:367-75.
236. Kanost MR. Serpins in a Lepidopteran insect, Manduca sexta. In: Silverman GA, Lomas DA eds. The Serpinopathies: Molecular and Cellular Aspects of Serpins and their Disorders. Hackensack, NJ: World Scientific Publishing. 2007:229-42.
237. Jiang H, Wang Y, Kanost MR. Mutually exclusive exon use and reactive center diversity in insect serpins. J Biol Chem 1994; 269:55-8.
238. Jiang H, Wang Y, Huang Y et al. Organization of serpin gene-1 from Manduca sexta: evolution of a family of alternate exons encoding the reactive site loop. J Biol Chem 1996; 271:28017-23.
239. Hegedus DD, Erlandson M, Baldwin D et al. Differential expansion and evolution of the exon family encoding the Serpin-1 reactive centre loop has resulted in divergent serpin repertoires among the Lepidoptera. Gene 2008; 418:15-21.
240. Zheng YP, He WY, Béliveau C et al. Cloning, expression and characterization of four serpin-1 cDNA variants from the spruce budworm, Choristoneura fumiferana. Comp Biochem Physiol B Biochem Mol Biol 2009; 154:165-73.
241. Krüger O, Ladewig J, Köster K et al. Widespread occurrence of serpin genes with multiple reactive centre-containing exon cassettes in insects and nematodes. Gene 2002; 29:97-105.
242. Zou Z, Jiang H. Manduca sexta serpin-6 regulates immune serine proteinases PAP-3 and HP8. cDNA cloning, protein expression, inhibition kinetics and function elucidation. J Biol Chem 2005; 280:14341-8.
243. Freitak D, Wheat CW, Heckel DG et al. Immune system responses and fitness costs associated with consumption of bacteria in larvae of Trichoplusia ni. BMC Biol 2007; 5:56.
244. Freitak D, Heckel DG, Vogel H. Bacterial feeding induces changes in immune-related gene expression and has trans-generational impacts in the cabbage looper (Trichoplusia ni). Front Zool 2009a; 6:7.
245. Strand MR. Insect hemocytes and their role in immunity. In: Beckage, NE, ed. Insect Immunology. San Diego, CA: Academic Press. 2008:25-47.
246. St. Leger RJ, Charnley AK, Cooper RM. Cuticle-degrading enzymes of entomopathogenic fungi: synthesis in culture on cuticle. J Invertebr Pathol 1986; 48:85-95.
247. Vilcinskas A, Matha V, Götz P. Effects of the entomopathogenic fungus Metarhizium anisopliae and its secondary metabolites on morphology and cytoskeleton of plasmatocytes isolated from Galleria mellonella. J Insect Physiol 1997; 43:1149-59.
248. Vilcinskas A, Götz P. Parasitic fungi and their interactions with the insect immune system. Adv Parasitol 1999; 43:267-13.
249. Fiolka MJ. Immunosuppressive effect of cyclosporin A on insect humoral immune response. J Invertebr Pathol 2008; 98:287-92.
250. Gillespie J, Bailey A, Cobb B et al. Fungal elicitors of insect immune responses. Arch Insect Biochem Physiol 2000; 44:49-68.

251. Lord JC, Anderson S, Stanley DW. Eicosanoids mediate Manduca sexta cellular response to the fungal pathogen Beauveria bassiana: a role for the lipoxygenase pathway. Arch Insect Biochem Physiol 2002; 51:46-54.
252. Bergin D, Reeves E, Renwick J et al. Superoxide production in Galleria mellonella hemocytes: Identification of proteins homologous to the NADPH oxidase complex of human neutrophils. Infect Immun 2005; 73:4161-4170.
253. Ekengren S, Hultmark D. Drosophila cecropin as an antifungal agent. Insect Biochem Mol Biol 1999; 29(11):965-72.
254. Langen G, Imani J, Altincicek B et al. Transgenic expression of gallerimycin, a novel antifungal insect defensin from the greater wax moth Galleria mellonella, confers resistance against pathogenic fungi in tobacco. Biol Chem 2006; 387:549-57.
255. Wojda I, Kowalski P, Jakubowicz T. Humoral immune response of Galleria mellonella larvae after infection by Beauveria bassiana under optimal and heat-shock conditions. J Insect Physiol 2009; 55:525-31.
256. Schmidt O. Insect immune recognition and suppression. In: Beckage, NE, ed. Insect Immunology. San Diego, CA: Academic Press. 2008; 271-294.
257. Clarke DJ. Photorhabdus: a model for the analysis of pathogenicity and mutualism. Cell Microbiol 2008; 10:2159-67.
258. Beckage NE. Parasitoid polydnaviruses and insect immunity. In: Beckage, NE, ed. Insect Immunology. San Diego, CA: Academic Press. 2008; 243-270.
259. Suderman RJ, Pruijssers AJ, Strand MR. Protein tyrosine phosphatase-H2 from a polydnavirus induces apoptosis of insect cells. J Gen Virol 2008; 89:1411-20.
260. Thoetkiattikul H, Beck MH, Strand MR. Inhibitor kappaB-like proteins from a polydnavirus inhibit NF-kappaB activation and suppress the insect immune response. Proc Natl Acad Sci USA 2005; 102:11426-31.
261. Lu Z, Beck MH, Wang Y et al. The viral protein Egf1.0 is a dual activity inhibitor of prophenoloxidase-activating proteinases 1 and 3 from Manduca sexta. J Biol Chem 2008; 283:21325-33.
262. Sparks WO, Bartholomay LC, Bonning BC. Insect immunity to viruses. Beckage, NE Parasitoid polydnaviruses and insect immunity. In: Beckage, NE, ed. Insect Immunology. San Diego, CA: Academic Press. 2008; 209242-270.
263. Clem RJ. Baculoviruses and apoptosis: a diversity of genes and responses. Curr Drug Targets 2007; 8:1069-74.
264. Shelby KS, Popham HJ. Plasma phenoloxidase of the larval tobacco budworm, Heliothis virescens, is virucidal. J Insect Sci 2006; 6:1-12.

CHAPTER 11

FLY IMMUNITY:
Recognition of Pathogens and Induction of Immune Responses

Shoichiro Kurata
Graduate School of Pharmaceutical Sciences, Tohoku University, Sendai, Japan
Email: kurata@mail.pharm.tohoku.ac.jp

Abstract: Despite the lack of adaptive immunity based on gene rearrangement such as that in higher vertebrates, flies are able to defend themselves from a wide array of pathogens using multiple innate immune responses whose molecular mechanisms are strikingly similar to those of the innate immune responses of other multicellular organisms, including humans. Invading pathogens passing through the epithelial barriers, the first line of self-defense, are detected by pattern recognition receptors that identify pathogen-associated molecular patterns in the hemolymph or on the immune cell surface and are eliminated by humoral and cellular responses. Some pathogens escape recognition and elimination in the hemolymph by invading the host cell cytoplasm. Some of these intracellular pathogens, however, such as *Listeria monocytogenes*, are identified by pattern recognition receptors in the cytoplasm and are eliminated by intracellular responses, including autophagy, an intracellular degradation system. Although some of these pattern recognition receptors are encoded in the germ-line as protein families, another type of receptor in the immunoglobulin-superfamily is extensively diversified by alternative splicing in somatic immune cells in *Drosophila*.

INTRODUCTION

Diseases of the silkworm, *Bombyx mori*, were already described in the classical Chinese literature “Guanzi” by Master Guan in the 7th century BCE and honeybee diseases were described in Greek mythology as early as the 6th century BCE. The analysis of insect immune responses, however, did not begin until the 19th century. In 1870, Louis Pasteur pioneered the study of insect immune responses in *Bombyx*.[1] His work was followed in

Invertebrate Immunity, edited by Kenneth Söderhäll.

1892 by Elie Metchnikoff's investigations on cellular immunity in insects.[2] Some of their observations on insect immune responses have been driven further and in some cases the molecular mechanisms explaining their observations have been elucidated.[3] Studies of a model organism, the fruit fly *Drosophila melanogaster*, have greatly contributed to a deeper understanding of the detailed molecular mechanisms of innate immune responses due to the powerful genetic tools available for this organism and these investigations have revealed fundamental conservation between insect immunity and innate immunity of other multicellular organisms, including humans.

The frontline of fly host defense are the epithelial tissues, such as the epidermis, gut and trachea, which act not only as mechanical barriers but also produce defense molecules such as lysozymes, antimicrobial peptides and reactive oxygen species.[4] Because flies have open vascular systems, pathogens passing through the epithelial barrier encounter cellular and humoral defense responses in the hemolymph.[5] As a cellular response, immune cells called hemocytes phagocytose and encapsulate pathogens in the hemolymph. Humoral reactions are induced by the activation of cascades of constitutive proteins present in the hemolymph, such as the prophenoloxidase (proPO) cascade and the coagulation cascade and the activation of intracellular signaling pathways that produce defense proteins such as antimicrobial peptides in immune-responsive tissues and cells, such as hemocytes and the fat body, the functional equivalent of the vertebrate liver. This chapter describes the molecular mechanisms of fly immune responses, mainly the invading pathogen-recognition mechanisms in the hemolymph and in the cytoplasm of *Drosophila* immune cells. The immune responses are activated via specific recognition of pathogen-associated molecular patterns by germ-line encoding pattern recognition receptor families[6] such as the peptidoglycan recognition protein (PGRP) family.

PRIMARY HUMORAL RESPONSES: ACTIVATION OF CONSTITUTIVE PROTEIN CASCADES IN THE HEMOLYMPH

In his study of *Bombyx* disease, Pasteur observed melanization of the insect cuticle caused by the protozoan pathogen, *Nosema bombysis*.[1] Melanization of the hemolymph of the wax moth larvae *Galleria mellonella* has been reported by several authors, including Serge Metalnikov in 1927,[7] who speculated that the dark pigment was formed by the action of a special enzyme on a chromogenic substrate in the insect hemolymph, which is now supported by molecular level studies. The enzyme is phenoloxidase and the chromogenic substrates in the insect hemolymph are tyrosine and dopa. Phenoloxidase is a copper-containing enzyme that catalyzes two types of reactions: the oxidation of monophenols to orthodiphenols and the oxidation of orthodiphenols to orthoquinones, which are key reactions required for melanin formation.[8] Melanization has an important role in arthropod immune reactions such as wound healing, surrounding invading pathogens and production of intermediates toxic to invading microorganisms.[9] Detailed biochemical mechanisms of the proPO cascade in moth, beetle and crustacean are discussed in chapters by Kanost, Lee, and Cerenius.

Phenoloxidase was purified from several arthropods, including *Bombyx mori* and is synthesized as inactive zymogens called proPO.[10,11] *Drosophila* has three genes that encode proPO, *Black cells*/*Dox-A1*, *Dox-A3* and *CG8193*.[12,13] Activation of proPO in the hemolymph is mediated by the proPO cascade, which consists of some serine proteases and some pattern-recognition receptors that recognize invading microorganisms.[8] As pattern recognition receptors, PGRP and β-1,3-glucan binding protein (βGRP) were first purified from *Bombyx* hemolymph by Masaaki Ashida and his colleagues.[14,15] PGRP and βGRP specifically bind to peptidoglycan and β-1,3-glucans, cell-wall components of bacteria and fungi, respectively and activate the proPO cascade in vitro. cDNA cloning of these proteins revealed that these proteins are evolutionarily conserved; e.g., there are 13 PGRP family members in *Drosophila*, 7 in *Anopheles* and 4 in both mouse and humans.[16-21] The PGRP family has a PGRP-domain in the C-terminal region that has some amino acid sequence similarity to peptidoglycan-degrading enzymes with *N*-acetylmuramyl-alanine amidase activity, such as bacteriophage lysozymes. βGRP has sequence similarity with bacterial β-1,3-glucanase in the C-terminal region, which is also in Gram-negative binding protein (GNBP) identified from *Bombyx* hemolymph.[22,23] In *Drosophila*, one PGRP family member, PGRP-LE, which recognizes diaminopimelic acid (DAP)-containing peptidoglycans, is involved in activation of the proPO cascade upstream of a serine protease called prophenoloxidase activating enzyme (PPAE), which is regulated by a serine protease inhibitor, Serpin 27A.[24-26] Two Clip domain-containing serine proteases,[27] melanization protease (MP)1 and serine protease (Sp) 7/MP2/PAE1, activate the proPO cascade upstream of PPAE and are also regulated by Serpin 27A.[28-30] MP1 is required to activate the proPO cascade in response to both bacterial and fungal infections, whereas Sp7 is mainly involved in response to fungal infections.[29] Coagulation to initiate wound healing and limit hemolymph loss is also induced at the site of injury in a proPO cascade-independent manner.[31,32] In *Drosophila*, a hemocyte expressing protein, Hemolectin and an abundant hemolymph protein, Fondue, are required for efficient coagulation.[33-35] Transglutaminase acts in coagulation.[36] Hemolectin has similarity with the human von Willebrand factor, which is defective in some coagulation diseases in humans, suggesting a conserved coaglution cascade in fly and humans.[37]

SECONDARY HUMORAL RESPONSES: INDUCTION OF SELF-DEFENSE MOLECULES

In 1906, Metalnikov observed bacteriolysis of *Mycobacterium tuberculosis* in the hemolymph of *Galleria*,[38] which was followed by studies on insect bacteriolysins by Andre Paillot.[39] We now know that bacteriolysis is explained by the activity of lysozymes and antimicrobial peptides. A lysozyme is an enzyme that catalyzes the cleavage of the β-1,4-glycoside bonds in the peptidoglycans of the bacterial cell wall, which is the main antibacterial factor in insects as well as in vertebrates. Since 1981 when antimicrobial peptides were first identified in the cecropia moth, *Hyalophora cecropia,* by Hans Boman and coworkers,[40] several hundreds of antimicrobial peptides have been isolated from various invertebrates, plants and mammals, including humans.[41] *Drosophila* mutants that fail to produce antimicrobial peptides are susceptible to bacterial and fungal infections, indicating that the induction of antimicrobial peptides is crucial for host defense in fly.[42]

Shunji Natori and his coworkers isolated several antimicrobial peptides from the flesh fly *Sarcophaga peregrina*, which is ideally suited for biochemical analysis. In response to infection or injury, four different types of antibacterial peptides are synthesized in the fat body and secreted into the hemolymph.[43] These are the Sarcotoxin I, II, III families and the Sapecin family. The Sarcotoxin I family has potent bactericidal activity against mainly Gram-negative bacteria.[44-46] The Sarcotoxin I family comprises five subtypes, each consisting of 39 or 40 amino acids, which have slightly different target specificities.[47]

Table 1. Summary of antimicrobial peptides in *Sarcophaga*

Effector Molecules	Molecular Mass	Subtypes	Primary Target Specificity	*Drosophila* Homologue
Sarcotoxin I	4000	5	Gram-negative	Cecropin
Sarcotoxin II	24000	3	Gram-negative	Attacin
Sarcotoxin III	8000	>2	Gram-negative	Diptericin
Sapecin	4000	3	Gram-positive	Defensin
Antifungal protein	7000	1	Fungi	Non

The characteristics of *Sarcophaga* antimicrobial peptides and their homologues in *Drosophila* are summarized in reference 108.

Many insect antimicrobial peptide families have several subtypes, suggesting that insects utilize antimicrobial peptide family subtypes with different target specificities to eliminate various microbes (Table 1). Sarcotoxin IA consists of two amphiphilic alpha-helical regions with a hinge region.[48] The amphiphilic helical segments are important for the expression of the antibacterial activity, suggesting that Sarcotoxin IA acts primarily on the bacterial membrane.[49] Cholesterol-containing liposomes are less sensitive to Sarcotoxin IA, explaining the selective toxicity of Sarcotoxin IA to bacteria because the membranes of the host cells contain cholesterol whereas those of bacteria do not. Although the Sarcotoxin II and III families also act on Gram-negative bacteria, Sarcotoxin families have different target specificities. The Sarcotoxin I family is similar to Cecropin, initially discovered in the hemolymph of bacteria-challenged diapausing pupae of *Hyalophora*.[40] The Sarcotoxin II and III families are similar to *Drosophila* Attacin and Diptericin, respectively (Table 1). In contrast to the Sarcotoxin families, Sapecin acts mainly on Gram-positive bacteria, including methicillin-resistant *Staphylococcus aureus*.[50] Sapecin has similarity to *Drosophila* Defensin. Sapecin comprises 40 amino acids with three internal disulfide bridges and has high affinity for cardiolipin, which is a major phospholipid of *S. aureus*. A mutant of *Escherichia coli* defective in cardiolipin synthesis is more resistant to Sapecin than wild-type *E. coli*, suggesting that cardiolipin is a target of Sapecin.[51] Sapecin was originally purified from the culture medium of a *Sarcophaga* embryonic cell line and is expressed during the embryonic stage of development, suggesting a role of Sapecin in development.[52] Consistent with this idea, Sapecin is involved in the proliferation of embryonic cells of *Sarcophaga*.[53] In addition to the inducible antibacterial peptides, the constitutive protein Antifungal protein, which has no similarity to other known proteins, is also present in the hemolymph.[54] The amino acid composition of Antifungal protein is unique: glycine and histidine account for more than half of the total amino acids. Antifungal protein has no bactericidal activity, but it has lethal activity against fungi such as *Schizosaccharomyces pombe*. Antifungal protein has less activity against *Candida albicans*, but the activity is greatly enhanced by the presence of Sarcotoxin IA, an inducible antibacterial peptide, suggesting that *Sarcophaga* constructs an efficient system with constitutive Antifungal protein and inducible antibacterial peptides to eliminate various microbes.

In *Drosophila*, seven classes of antimicrobial peptides have been identified. Diptericin, Attacin and Drosocin are effective against Gram-negative bacteria, whereas Defensin is active against Gram-positive bacteria. Cecropin has wide specificity, acting on both

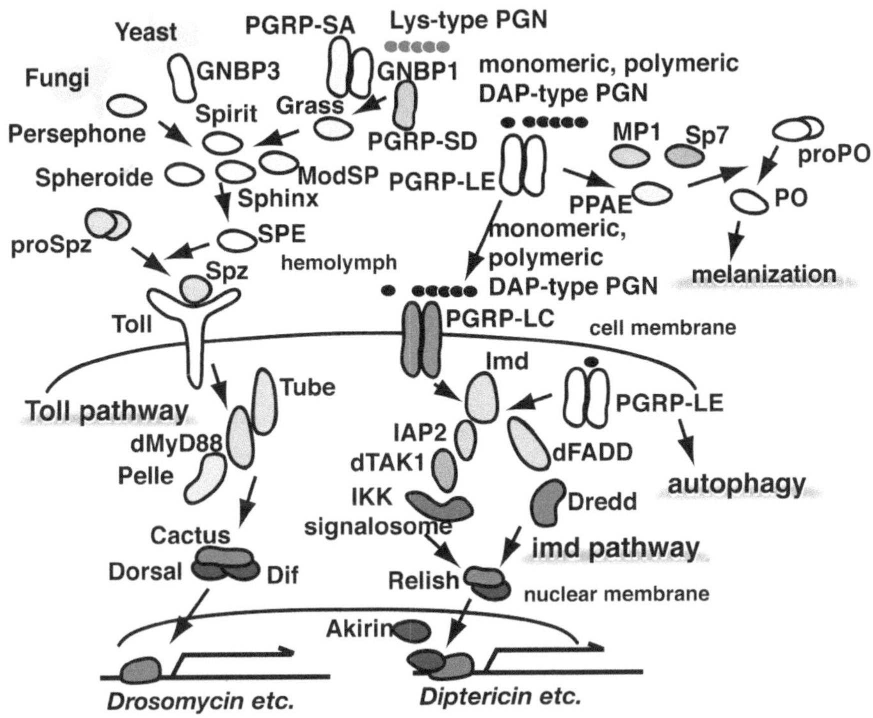

Figure 1. Recognition of various pathogens and induction of immune responses in *Drosophila*. Activation of the proPO cascade leading to melanization is induced by the cleavage of proPO to PO by a serine protease, PPAE, which is regulated by two serine proteases, MP1 and Sp7. A constitutive hemolymph protein, PGRP-LE, binds to monomeric and polymeric DAP-type peptidoglycans (PGNs), which are components of many Gram-negative and some Gram-positive bacteria and activates the proPO cascade upstream of PPAE. PGRP-LE is also involved in activating the imd pathway-dependent induction of antibacterial peptides in the hemolymph and inside the immune cells. Membrane PGRP-LC is required for monomeric and polymeric DAP-type PGN-mediated activation of the imd pathway. The imd pathway comprises Imd, IAP2, dFADD, Dredd, dTAK1, IKK signalosome, Akirin and Relish. PGRP-LE induces autophagy through intracellular recognition of DAP-type peptidoglycans in an imd independent manner. PGRP-SA in the hemolymph binds to Lys-type PGN of Gram-positive bacteria and is required for activation of the Toll pathway in cooperation with GNBP1. PGRP-SD has some redundant functions with PGRP-SA and GNBP1. GNBP3 is involved in yeast-mediated activation of the Toll pathway. The activation of the Toll pathway is mediated by its active ligand, Spz, cleaved from proSpz by the serine protease cascade, including the SPE, Spheroide, Spirit, Sphinx, Persephone, Grass and ModSP. The Toll pathway consists of Toll, dMyD88, Tube, Pelle, Dorsal and Dif. The transcriptional factors, Dorsal and Dif, are negatively regulated by Cactus. This figure was modified from Kurata S., Research Signpost, Copyright 2008.[108]

bacteria and fungi. Drosomycin and Metchnikowin, identified so far only in Drosophilidae, act on fungi.[55] Induction of antimicrobial peptides is mediated by two distinct pathways, the imd pathway and the Toll pathway (Fig. 1),[4,5] which are mechanistically similar to the mammalian tumor necrosis factor-α receptor signaling pathway and Toll-like receptor (TLR)/interleukin-1 receptor signaling pathway, respectively.[56] The imd pathway regulates the synthesis of several antibacterial peptides such as Diptericin, which is mediated by cell surface PGRP-LC, Imd adaptor protein, IAP2, dTAK1 kinase, IκB kinase (IKK) complex, FADD, Dredd caspase, Akirin and a Rel transcriptional factor, Relish.[57-65] The tumor necrosis factor-α signaling pathway comprises receptor-interacting protein, TAK1 kinase, IKK complex, caspase-8 and nuclear factor κB (NF-κB), similar to the Imd, dTAK1, dIKK complex, Dredd and Relish, respectively, in the *Drosophila* imd pathway.[56] The Toll pathway predominantly regulates induction of the antifungal peptide Drosomycin, which is mediated by the Toll receptor, dMyD88 adaptor protein and Dorsal related immune factor, similar to TLR, MyD88 and NF-κB in the mammalian TLR signaling pathway.[66-69] The important roles of these signaling pathways in host defense systems are demonstrated by the fact that mutations in the genes that regulate these pathways dramatically decrease resistance to microbial infections, e.g., *Toll* mutants are susceptible to fungal infections and *Relish* mutants lose resistance to Gram-negative bacterial infections.[63,66]

RECOGNITION OF VARIOUS PATHOGENS IN THE HEMOLYMPH

The Toll pathway is activated predominantly in response to fungal and some Gram-positive bacterial infections, whereas the imd pathway is activated predominantly in response to Gram-negative and other Gram-positive bacterial infections.[70] Therefore, *Drosophila* distinguishes different pathogens by specific mechanisms. In contrast to mammalian TLRs, which recognize pathogen components such as lipopolysaccharides,[71] the ligand of the *Drosophila* Toll receptor is Spätzle (Spz), which is produced by proteolysis of an endogenous protein (proSpz) in response to infection.[72] Therefore, *Drosophila* Toll does not act as a pattern recognition receptor, but rather mediates the downstream signaling of such receptors. In *Drosophila*, some PGRP and GNBP family members act as pattern recognition receptors upstream of both the Toll and imd pathways.[4,73]

PGRP-SA, PGRP-SD, GNBP1 and GNBP3 act as pattern recognition receptors upstream of the Toll pathway in *Drosophila*. PGRP-SA, a hemolymph protein that was identified from a loss-of-function mutant screen, is required for activation of the Toll pathway in response to Gram-positive bacterial infections.[74] PGRP-SA cooperates with GNBP1 to activate the Toll pathway in response to Gram-positive bacterial infections.[75] PGRP-SD has some redundancy and recognizes Gram-positive bacteria with PGRP-SA and GNBP1.[76] GNBP3 is required for activation of the Toll pathway in response to fungal infections.[77] Downstream of these recognition receptors, pro-Spz is cleaved by a serine protease, Spätzle processing enzyme (SPE).[78] The SPE-mediated processing is regulated by several serine proteases such as Persephone, Spirit, Spheroide, Sphinx, Grass and ModSP.[79-82] Persephone and Grass are specifically required for the signaling from fungi and Gram-positive bacteria, respectively and other proteases are required for the signaling from both microorganisms.[80]

Upstream of the imd pathway, PGRP-LC and PGRP-LE act as pattern recognition receptors. PGRP-LC, a membrane-associated PGRP-family member, is required for activation of the imd pathway in response to Gram-negative bacterial infections.[83-85] Susceptibility of the *PGRP-LC* mutant to Gram-negative bacterial infections is less dramatic than that of an IKK-complex mutant, in which signaling is impaired downstream of imd, suggesting that there is an activator of the imd pathway in addition to PGRP-LC.[83] Consistent with this finding, PGRP-LE, a constitutive hemolymph protein, selectively activates the imd pathway.[24] The *PGRP-LC*/*PGRP-LE* double-mutant is much more susceptible to *E. coli* infection than either single mutant alone, suggesting that PGRP-LC and PGRP-LE are major recognition receptors for Gram-negative bacteria.[25] As described above, PGRP-LE activates the proPO cascade, which branches from the PGRP-LE-mediated activation of the imd pathway upstream of PPAE.[25] In addition to the extracellular functions of PGRP-LE in the hemolymph, PGRP-LE acts as a coreceptor of the membrane receptor PGRP-LC on the surface of immune cells and as an intracellular receptor in immune reactive cells to induce antibacterial peptides by activating the imd pathway.[86]

Peptidoglycans are an essential cell wall component of almost all bacteria except mycoplasma, which lack a cell wall. Peptidoglycan is a polymer of β-1,4-linked *N*-acetylglucosamine (GlcNAc) and *N*-acetylmuramic acid (MurNAc) crosslinked by short stem peptides. The glycan chain is relatively conserved in all bacteria, whereas the amino acid composition of the crosslinking stem peptides and the linkages between stem peptides are diversified, depending on the bacterial species.[87] Many Gram-negative bacteria and some Gram-positive bacteria, such as *Bacillus* species, directly crosslink DAP-containing peptidoglycans and preferentially activate the imd pathway in *Drosophila*. Many Gram-positive bacteria have peptidoglycans containing lysine instead of DAP with crosslinking peptides between stem peptides and these bacteria preferentially activate the Toll pathway. Consistent with the finding that PGRP-LE selectively activates the imd pathway in vivo, PGRP-LE selectively binds DAP-type peptidoglycans, but not Lys-containing peptidoglycans in vitro.[24] *PGRP-LC* is required for DAP-type peptidoglycan-mediated activation of the imd pathway in vivo and in vitro.[88,89] The minimum structure of the DAP-type peptidoglycan required for PGRP-LC-mediated activation of the imd pathway is GlcNAc-MurNAc with an internal 1,6-anhydro bond attached to a tripeptide containing DAP, known as tracheal cytotoxin.[90] These results are consistent with the findings that PGRP-LC and PGRP-LE act synergistically to resist bacteria with DAP-type peptidoglycans such as *E. coli* and *Bacillus megaterium*.[25] The minimum structure of the Lys-type peptidoglycan required for PGRP-SA-mediated activation of the Toll pathway comprises two units consisting of GlcNAc-MurNAc attached to Lys-containing tetrapeptides that are covalently dimerized by an interpeptide.[91] Therefore, *Drosophila* PGRP family members distinguish the structural diversity of peptidoglycans and activate the appropriate immune responses. In addition to PGRPs acting as pattern recognition receptors, there are PGRP family members such as PGRP-SC and PGRP-LB that have enzyme activity that hydrolyzes the lactylamide bond between the glycan strand and the stem peptides of peptidoglycans.[92-94] The *N*-acetylmuramyl-alanine amidase activity is consistent with the structural similarity of the PGRP-domain to *N*-acetylmuramyl-alanine amidase. The degraded peptidoglycans lose their elicitor activity, suggesting a scavenger function of the enzyme PGRP family members and negative feedback regulation of immune responses by enzyme PGRP family members.[95]

RECOGNITION AND ELIMINATION OF INTRACELLULAR BACTERIA IN THE CYTOPLASM

Intracellular pathogens, a diverse group of organisms that cause serious diseases such as tuberculosis and malaria, invade the cells and escape serum and cell surface innate immune receptors. The intracellular function of PGRP-LE is suggested by the findings that the delivery of tracheal cytotoxin into S2 cells, the *Drosophila* macrophage-like cell line, induces antimicrobial peptides in a PGRP-LE dependent manner and PGRP-LE induces antimicrobial peptides in a cell autonomous manner in the Malpighian tubules, the functional equivalent of the mammalian kidney. Consistent with these results, the *PGRP-LE* mutant is susceptible to infection by *Listeria monocytogenes*, intracellular bacteria with DAP-containing peptidoglycans recognized by PGRP-LE.[96] In contrast to Gram-negative bacterial infections, the survival rate of the double mutant *PGRP-LE/PGRP-LC* infected with *L. monocytogenes* is similar to that of the *PGRP-LE* mutant, suggesting that PGRP-LE and PGRP-LC do not have redundant functions in producing resistance to *L. monocytogenes*.[96] In vitro infection experiments using hemocytes and PGRP-LE expressing S2 cells revealed that PGRP-LE colocalizes with *L. monocytogenes* in the cytoplasm and is essential for resistance against intracellular bacterial growth, but, interestingly, inhibition of intracellular bacterial growth is not dependent on the known innate immune signaling pathways, the Toll and imd pathways.

Autophagy was initially reported as a bulk self-degradation mechanism for the turnover of proteins and organelles and is conserved from yeast to humans.[97] Recent studies, however, revealed that autophagy has a role in many biological events, such as nutrient supply upon starvation, cell death and cancer. Autophagy also functions as an innate immune response against intracellular bacteria, viruses and parasites in cultured cells.[98-104] In *Drosophila*, autophagy is induced in hemocytes in response to *L. monocytogenes* infection in a PGRP-LE dependent manner, which is crucial for host survival against *Listeria* infection as well as for the inhibition of bacterial growth in hemocytes. PGRP-LE is also required for DAP-type peptidoglycan-mediated autophagy, but not for Lys-type peptidoglycan-mediated autophagy in hemocytes, suggesting the existence of another intracellular sensor for bacteria with Lys-type peptidoglycans. Therefore, PGRP-LE has an essential role in detecting intracellular bacteria through DAP-type peptidoglycans to induce autophagy as an innate immune response, which indicates a direct link between pathogen recognition and the induction of autophagy in the cytoplasm (Fig. 1).

EXTENSIVE SOMATIC DIVERSIFICATION OF A RECOGNITION MOLECULE

Pattern recognition receptors acting on innate immunity, such as PGRP family members and GNBP family members, are encoded in the germ line as protein families and distinguish between different types of microorganisms. On the other hand, in *Drosophila*, the diversity of an immunoglobulin-superfamily receptor Down syndrome cell adhesion molecule (Dscam) is generated by alternative splicing in hemocytes.[105] The *Dscam* gene contains three arrays of alternative exons, exons 4, 6 and 9 and two alternative transmembrane domains, in addition to 20 constant exons, which have the potential to express Dscam with 19008 different extracellular domains.[106] Many of these different isomers are expressed in the fat body, hemocytes and brain and different Dscam isomers bind to bacteria with different affinities, suggesting a role for Dscam isomers in the recognition of different types of pathogens.[105] Moreover, *Dscam* is required for bacterial phagocytosis by hemocytes.[105] The *Dscam* gene in the mosquito *Anopheles gambiae* also has a complex organization with 101 exons and the potential to express AgDscam with 15960 different extracellular domains.[107] *AgDscam* is involved in host defense against bacterial infections by producing pathogen-specific splice form repertoires.[107] Therefore, the diversity in insect Dscam and in vertebrate antigen receptors is generated by alternative splicing and gene rearrangements to achieve similar goals, which is the recognition of a multitude of pathogens with adaptive characteristics.

CONCLUSION

In fly immunity, invading pathogens are recognized in the hemolymph, on the immune cell surface, and in the host cell cytoplasm by pattern recognition receptors, such as PGRP family members that distinguish the structural diversity of bacterial peptidoglycans, and eliminated by humoral and cellular responses, such as induction of antimicrobial peptides and autophagy, respectively.

ACKNOWLEDGEMENT

This work was supported by Grants-in-Aid for Scientific Research from the Ministry of Education, Culture, Sports, Science and Technology of Japan; the Japan Society for the Promotion of Science; the Program for the Promotion of Basic Research Activities for Innovative Biosciences (PROBRAIN); an NIH grant (AI074958); the Naito Foundation; and the Uehara Memorial Foundation.

REFERENCES

1. Pasteur L. Etudes sur la maladie des vers à soie, Vol. 1, Gauthier-Villars, Paris, 1870.
2. Metchnikoff E. Leçons sur la pathologie comparée de l'inflamation, Masson et Cie., Paris, 1892.
3. Brey PT. The contributions of the Pasteur school of insect immunity. In: Brey PT, Hultmark D, eds. Molecular Mechanisms of Immune Responses in Insects. London:Chapman and Hall, 1998
4. Lemaitre B, Hoffmann J. The host defense of Drosophila melanogaster. Annu Rev Immunol 2007; 25:697-743.
5. Hultmark D. Drosophila immunity: paths and patterns. Curr Opin Immunol 2003; 15:12-19.
6. Janeway CA Jr. Approaching the asymptote? Evolution and revolution in immunology. Cold Spring Harb Symp Quant Biol 1989; 54:1.
7. Metalnikov S. Infection microbienne et l'immunité chezla mite des abeilles Galleria mellonella, Masson et Cie., Paris, 1927.
8. Ashida M, Brey PT. Recent advances in research on the insect prophenoloxidase cascade. In: Brey PT, Hultmark D, eds. Molecular Mechanisms of Immune Responses in Insects. London:Chapman and Hall, 1998.
9. Söderhäll K, Cerenius L. Role of the prophenoloxidase-activating system in invertebrate immunity. Curr Opin Immunol 1998; 10:23-28.
10. Ashida M, Yamazaki HI. Biochemistry of the phenoloxidase system in insect: with special reference to its activation. In: Ohnishi E, Ishizaki H, eds. Molting and Metamorphosis. Berlin:Springer-Verlag, 1990.
11. Johansson MW, Söderhäll K. The prophenoloxidase activating system and associated proteins in invertebrates. Prog Mol Subcell Biol. 1996;15:46-66. Review.
12. Fujimoto K, Okino N, Kawabata S-I et al. Nucleotide sequence of the cDNA encoding the proenzyme of phenol oxidase A1 of Drosophila melanogaster. Proc Natl Acad Sci USA 1995; 92:7769-7773.
13. Irving P, Ubeda JM, Doucet D et al. New insights into Drosophila larval haemocyte functions through genome-wide analysis. Cell Microbiol 2005; 7:335-350.
14. Yoshida H, Kinoshita K, Ashida M. Purification of a peptidoglycan recognition protein from hemolymph of the silkworm, Bombyx mori. J Biol Chem 1996; 271:13854-13860.
15. Ochiai M, Ashida M. Purification of a beta-1,3-glucan recognition protein in the prophenoloxidase activating system from hemolymph of the silkworm, Bombyx mori. J Biol Chem 1988; 263:12056-12062.
16. Ochiai M, Ashida M. A pattern recognition protein for peptidoglycan. Cloning the cDNA and the gene of the silkworm, Bombyx mori. J Biol Chem 1999; 274:11854-11858.
17. Kang D, Liu G, Lundstrom A et al. A peptidoglycan recognition protein in innate immunity conserved from insects to humans. Proc Natl Acad Sci USA. 1998; 95:10078-10082.
18. Werner T, Liu G, Kang D et al. A family of peptidoglycan recognition proteins in the fruit fly Drosophila melanogaster. Proc Natl Acad Sci USA 2000; 97:13772-13777.

19. Christophides GK et al. Immunity-related genes and gene families in Anopheles gambiae. Science 2002; 298:159-165.
20. Ma C, Kanost MR. A beta1,3-glucan recognition protein from an insect, Manduca sexta, agglutinates microorganisms and activates the phenoloxidase cascade. J Biol Chem 2000; 275:7505-7514.
21. Zhang R, Cho HY, Kim HS et al. Characterization and properties of a 1,3-beta-D-glucan pattern recognition protein of Tenebrio molitor larvae that is specifically degraded by serine protease during prophenoloxidase activation. J Biol Chem 2003; 278:42072-42079.
22. Ochiai M, Ashida M. A pattern-recognition protein for beta-1,3-glucan. The binding domain and the cDNA cloning of beta-1,3-glucan recognition protein from the silkworm, Bombyx mori. J Biol Chem 2000; 275:4995-5002.
23. Lee WJ, Lee JD, Kravchenko VV et al. Purification and molecular cloning of an inducible gram-negative bacteria-binding protein from the silkworm, Bombyx mori. Proc Natl Acad Sci USA 1996; 93:7888-7893.
24. Takehana A, Katsuyama T, Yano T et al. Overexpression of a pattern-recognition receptor, peptidoglycan-recognition protein-LE, activates imd/relish-mediated antibacterial defense and the prophenoloxidase cascade in Drosophila larvae. Proc Natl Acad Sci USA 2002; 99:13705-13710.
25. Takehana A, Yano T, Mita S et al. Peptidoglycan recognition protein (PGRP)-LE and PGRP-LC act synergistically in Drosophila immunity. EMBO J 2004; 23:4690-4700.
26. De Gregorio E et al. An immune-responsive Serpin regulates the melanization cascade in Drosophila. Dev Cell 2002; 3:581-592.
27. Muta T, Hashimoto R, Miyata T et al. Proclotting enzyme from horseshoe crab hemocytes. cDNA cloning, disulfide locations and subcellular localization. J Biol Chem 1990; 265:22426-22433.
28. Castillejo-Lopez C, Hacker U. The serine protease Sp7 is expressed in blood cells and regulates the melanization reaction in Drosophila. Biochem Biophys Res Commun 2005; 338:1075-1082.
29. Tang H, Kambris Z, Lemaitre B et al. Two proteases defining a melanization cascade in the immune system of Drosophila. J Biol Chem 2006; 281:28097-28104.
30. Leclerc V, Pelte NEl, Chamy L et al. Prophenoloxidase activation is not required for survival to microbial infections in Drosophila. EMBO Rep 2006; 7:231-235.
31. Scherfer C, Scherfer C, Karlsson C et al. Isolation and characterization of hemolymph clotting factors in Drosophila melanogaster by a pullout method. Curr Biol 2004; 14:625-629.
32. Karlsson C, Karlsson C, Korayem AM et al. Proteomic analysis of the Drosophila larval hemolymph clot. J Biol Chem 2004; 279:52033-52041.
33. Goto A, Kadowaki T, Kitagawa Y. Drosophila hemolectin gene is expressed in embryonic and larval hemocytes and its knock down causes bleeding defects. Dev Biol 2003; 264:582-591.
34. Lesch C, Goto A, Lindgren M et al. A role for Hemolectin in coagulation and immunity in Drosophila melanogaster. Dev Comp Immunol 2007; 31:1255-1263.
35. Scherfer C, Qazi MR, Takahashi K et al. The Toll immune-regulated Drosophila protein Fondue is involved in hemolymph clotting and puparium formation. Dev Biol 2006; 295:156-163.
36. Lindgren M, Riazi R, Lesch C et al. Fondue and transglutaminase in the Drosophila larval clot. J Insect Physiol 2007; 54:586-592.
37. Goto A, Aoki M, Ichihara S et al. A Drosophila haemocyte-specific protein, hemolectin, similar to human von Willebrand factor. Biochem J 2001; 359:99-108.
38. Metalnikov SCR. Soc Biol 1906; 60:518.
39. Paillot ACR. Acad Sci 1919; 169:202.
40. Steiner H, Hultmark D, Engstrom A et al. Sequence and specificity of two antibacterial proteins involved in insect immunity. Nature 1981; 292:246-248.
41. Bulet P, Stocklin R, Menin L. Anti-microbial peptides: from invertebrates to vertebrates. Immunol Rev 2004; 198:169-184.
42. Tzou P, Reichhart JM, Lemaitre B. Constitutive expression of a single antimicrobial peptide can restore wild-type resistance to infection in immunodeficient Drosophila mutants. Proc Natl Acad Sci USA. 2002; 99:2152-2157.
43. Natori S. Bactericidal substance induced in the hemolymph of Sarcophaga peregrina larvae. J Insect Physiol 1977; 23:1169-1173.
44. Okada M, Natori S. Purification and characterization of an antibacterial protein from haemolymph of Sarcophaga peregrina (flesh-fly) larvae. Biochem J 1983; 211:727-734.
45. Ando K, Okada M, Natori S. Purification of sarcotoxin II, antibacterial proteins of Sarcophaga peregrina (flesh fly) larvae. Biochemistry 1987; 26:226-230.
46. Baba K, Okada M, Kawano T et al. Purification of sarcotoxin III, a new antibacterial protein of Sarcophaga peregrina. J. Biochem (Tokyo) 1987; 102:69-74.
47. Kanai A, Natori S. Cloning of gene cluster for sarcotoxin I, antibacterial proteins of Sarcophaga peregrina. FEBS Lett 1989; 258:199-202.

48. Iwai H, Nakajima Y, Natori S et al. Solution conformation of an antibacterial peptide, sarcotoxin IA, as determined by 1H-NMR. Eur J Biochem 1993; 217:639-644.
49. Nakajima Y, Qu XM, Natori S. Interaction between liposomes and sarcotoxin IA, a potent antibacterial protein of Sarcophaga peregrina (flesh fly). J Biol Chem 1987; 262:1665-1669.
50. Matsuyama K, Natori S. Purification of three antibacterial proteins from the culture medium of NIH-Sape-4, an embryonic cell line of Sarcophaga peregrina. J Biol Chem 1988; 263:17112-17116.
51. Matsuyama K, Natori S. Mode of action of sapecin, a novel antibacterial protein of Sarcophaga peregrina (flesh fly). J. Biochem (Tokyo), 1990; 108:128-132.
52. Matsuyama K, Natori S. Molecular cloning of cDNA for sapecin and unique expression of the sapecin gene during the development of Sarcophaga peregrina. J Biol Chem 1988; 263:17117-17121.
53. Komano H, Homma K, Natori S. Involvement of sapecin in embryonic cell proliferation of Sarcophaga peregrina (flesh fly). FEBS Lett 1991; 289:167-170.
54. Iijima R, Kurata S, Natori S. Purification, characterization and cDNA cloning of an antifungal protein from the hemolymph of Sarcophaga peregrina (flesh fly) larvae. J Biol Chem 1993; 268:12055-12061.
55. Bulet P, Charlet m, Hetru C. Innate immunity, Ezekowitz, RAB, Hoffmann, JA (Eds.), Humana Press, New Jersey. 2003.
56. Silverman N, Maniatis T. NF-kappaB signaling pathways in mammalian and insect innate immunity. Genes Dev 2001; 15:2321-2342.
57. Georgel P et al. Drosophila immune deficiency (IMD) is a death domain protein that activates antibacterial defense and can promote apoptosis. Dev Cell 2001; 1:503-514.
58. Vidal S, Khush RS, Leulier F et al. Mutations in the Drosophila dTAK1 gene reveal a conserved function for MAPKKKs in the control of rel/NF-kappaB-dependent innate immune responses. Genes Dev 2001; 15:1900-1912.
59. Silverman N, Zhou R, Stoven S et al. A Drosophila IkappaB kinase complex required for Relish cleavage and antibacterial immunity. Genes Dev 2000; 14, 2461-2471.
60. Rutschmann S, Jung AC, Zhou R et al. Role of Drosophila IKK gamma in a toll-independent antibacterial immune response. Nat Immunol 2000; 1:342-347.
61. Leulier F, Rodriguez A, Khush RS et al. The Drosophila caspase Dredd is required to resist gram-negative bacterial infection. EMBO Rep 2000; 1:353-358.
62. Stoven S, Ando I, Kadalayil L et al. Activation of the Drosophila NF-kappaB factor Relish by rapid endoproteolytic cleavage. EMBO Rep 2000; 1:347-352.
63. Hedengren M, Asling B, Dushay MS et al. Relish, a central factor in the control of humoral but not cellular immunity in Drosophila. Mol Cell 1999; 4:827-837.
64. Leulier F, Lhocine N, Lemaitre B et al. The Drosophila inhibitor of apoptosis protein DIAP2 functions in innate immunity and is essential to resist gram-negative bacterial infection. Mol Cell Biol 2006; 26:7821-7831.
65. Goto A et al. Akirins are highly conserved nuclear proteins required for NF-kappaB-dependent gene expression in drosophila and mice. Nat Immunol 2008; 9:97-104.
66. Lemaitre B, Nicolas E, Michaut L et al. The dorsoventral regulatory gene cassette spätzle/Toll/cactus controls the potent antifungal response in Drosophila adults. Cell 1996; 86:973-983.
67. Tauszig-Delamasure S, Bilak H, Capovilla M et al. Drosophila MyD88 is required for the response to fungal and Gram-positive bacterial infections. Nat Immunol 2002; 3:91-97.
68. Ip YT, Reach M, Engstrom Y et al. Dif, a dorsal-related gene that mediates an immune response in Drosophila. Cell 1993; 75:753-763.
69. Wu LP, Anderson KV. Regulated nuclear import of Rel proteins in the Drosophila immune response. Nature 1998; 392:93-97.
70. Lemaitre B, Reichhart JM, Hoffmann JA. Drosophila host defense: differential induction of antimicrobial peptide genes after infection by various classes of microorganisms. Proc Natl Acad Sci USA 1997; 94:14614-14619.
71. Akira S, Takeda K, Kaisho T. Toll-like receptors: critical proteins linking innate and acquired immunity. Nat Immunol 2001; 2:675-680.
72. Weber AN, Tauszig-Delamasure S, Hoffmann JA et al. Binding of the Drosophila cytokine Spätzle to Toll is direct and establishes signaling. Nat Immunol 2003; 4:794-800.
73. Kurata S. Recognition of infectious nonself and activation of immune responses by peptidoglycan recognition protein (PGRP)-family members in Drosophila. Dev Comp Immunol 2004; 28:89-95.
74. Michel T, Reichhart JM, Hoffmann JA et al. Drosophila Toll is activated by Gram-positive bacteria through a circulating peptidoglycan recognition protein. Nature 2001; 414:756-759.
75. Gobert V, Gottar M, Matskevich AA et al. Dual activation of the Drosophila toll pathway by two pattern recognition receptors. Science 2003; 302:2126-2130.
76. Bischoff V, Vignal C, Boneca IG et al. Function of the drosophila pattern-recognition receptor PGRP-SD in the detection of Gram-positive bacteria. Nat Immunol 2004; 5:1175-1180.

77. Gottar M, Gobert V, Matskevich AA et al. Dual detection of fungal infections in Drosophila via recognition of glucans and sensing of virulence factors. Cell 2006; 127:1425-1437.
78. Jang IH et al. A Spätzle-processing enzyme required for toll signaling activation in Drosophila innate immunity. Dev Cell 2006; 10:45-55.
79. Ligoxygakis P, Pelte N, Hoffmann JA et al. Activation of Drosophila Toll during fungal infection by a blood serine protease. Science 2002; 297:114-116.
80. Kambris Z, Brun S, Jang IH et al. Drosophila immunity: a large-scale in vivo RNAi screen identifies five serine proteases required for Toll activation. Curr Biol 2006; 16:808-813.
81. El Chamy L, Leclerc V, Caldelari I et al. Sensing of 'danger signals' and pathogen-associated molecular patterns defines binary signaling pathways 'upstream' of Toll. Nat Immunol 2008; 9:1165-1170.
82. Buchon N, Poidevin M, Kwon HM et al. A single modular serine protease integrates signals from pattern-recognition receptors upstream of the Drosophila Toll pathway. Proc Natl Acad Sci USA 2009; 106:12442-12447.
83. Gottar M, Gobert V, Michel T et al. The Drosophila immune response against Gram-negative bacteria is mediated by a peptidoglycan recognition protein. Nature 2002; 416:640-644.
84. Choe KM, Werner T, Stoven S et al. Requirement for a peptidoglycan recognition protein (PGRP) in Relish activation and antibacterial immune responses in Drosophila. Science 2002; 296:359-362.
85. Rämet M, Manfruelli P, Pearson A et al. Functional genomic analysis of phagocytosis and identification of a Drosophila receptor for E. coli. Nature 2002; 416:644-648.
86. Kaneko T et al. PGRP-LC and PGRP-LE have essential yet distinct functions in the drosophila immune response to monomeric DAP-type peptidoglycan. Nat Immunol 2006; 7:715-723.
87. Schleifer KH, Kandler O. Peptidoglycan types of bacterial cell walls and their taxonomic implications. Bacteriol Rev 1972; 36:407-477.
88. Leulier F, Parquet C, Pili-Floury S et al. The Drosophila immune system detects bacteria through specific peptidoglycan recognition. Nat Immunol 2003; 4:478-484.
89. Kaneko T et al. Monomeric and polymeric gram-negative peptidoglycan but not purified LPS stimulate the Drosophila IMD pathway. Immunity 2004; 20:637-649.
90. Stenbak CR et al. Peptidoglycan molecular requirements allowing detection by the Drosophila immune deficiency pathway. J Immunol 2004; 173:7339-7348.
91. Filipe SR, Tomasz A, Ligoxygakis P. Requirements of peptidoglycan structure that allow detection by the Drosophila Toll pathway. EMBO Rep 2005; 6:327-333.
92. Mellroth P, Karlsson J, Steiner H. A scavenger function for a Drosophila peptidoglycan recognition protein. J Biol Chem 2003; 278:7059-7064.
93. Zaidman-Remy A et al. The Drosophila amidase PGRP-LB modulates the immune response to bacterial infection. Immunity 2006; 24:463-473.
94. Bischoff V, Vignal C, Duvic B et al. Downregulation of the Drosophila immune response by peptidoglycan-recognition proteins SC1 and SC2. PloS Pathog 2006; 2:e14.
95. Steiner H. Peptidoglycan recognition proteins: on and off switches for innate immunity. Immunol Rev 2004; 198:83-96.
96. Yano T et al. Autophagic control of Listeria through intracellular innate immune recognition in Drosophila. Nat Immunol 2008; 9:908-916.
97. Mizushima N, Levine B, Cuervo AM et al. Autophagy fights disease though cellular self-digestion. Nature 2008; 451:1069-1075.
98. Levine B, Deretic V. Unveiling the roles of autophagy in innate and adaptive immunity. Nat Rev Immunol 2007; 7:767-777.
99. Nakagawa I, Amano A, Mizushima N et al. Autophagy defends cells against invading Group A Streptococcus. Science 2004; 306:1037-1040.
100. Gutierrez MG., Master SS, Singh SB et al. Autophagy is a defense mechanism inhibiting BCG and Mycobacterium tuberculosis survival in infected macrophages. Cell 2004; 119:753-766.
101. Ogawa M, Yoshimori T, Suzuki T et al. Escape of intracellular Shigella from autophagy. Science 2005; 307:727-731.
102. Ling YM, Shaw MH, Ayala C et al. Vacuolar and plasma membrane stripping and autophagic elimination of Toxoplasma gondii in primed effector macrophages. J Exp Med 2006; 203:2063-2071.
103. Andrade RM, Weaaendarp M, Gubbels MJ et al. CD40 induces macrophage anti-Toxoplasma gondii activity by triggering autophagy-dependent fusion of pathogen-containing vacuoles and lysosomes. J Clin Invest 2006; 116:2366-2377.
104. Singh SB, Davis AS, Taylor GA et al. Human IRGM induces autophagy to eliminate intracellular mycobacteria. Science 2006; 313:1438-1441.
105. Watson FL, Puttmann-Holgado R, Thomas F et al. Extensive diversity of Ig-superfamily proteins in the immune system of insects. Science 2005; 309:1874-1878.

106. Schmucker D, Clemens JC, Shu H et al. Drosophila Dscam is an axon guidance receptor exhibiting extraordinary molecular diversity. Cell 2000; 101:671-684.
107. Dong Y, Taylor HE, Dimopoulos G. AgDscam, a hypervariable immunoglobulin domain-containing receptor of the Anopheles gambiae innate immune system. PloS Biol 2006; 4:e229.
108. Kurata, S. Molecular mechanisms of immune responses in insects. In: Liu N, ed. Recent Advances in Insect Physiology, Toxicology and Molecular Biology. Kerala, India: Research Signpost, 2008.

CHAPTER 12

MOSQUITO IMMUNITY

Julián F. Hillyer

Department of Biological Sciences and Institute for Global Health, Vanderbilt University, Nashville, Tennessee, USA
Email: julian.hillyer@vanderbilt.edu

Abstract: Throughout their lifetime, mosquitoes are exposed to pathogens during feeding, through breaks in their cuticle and following pathogen-driven cuticular degradation. To resist infection, mosquitoes mount innate cellular and humoral immune responses that are elicited within minutes of exposure and can lead to pathogen death via three broadly defined mechanisms: lysis, melanization and hemocyte-mediated phagocytosis. This chapter reviews our current understanding of the mosquito immune system, with an emphasis on the physical barriers that prevent pathogens from entering the body, the organs and tissues that regulate immune responses and the mechanistic and molecular bases of immunity.

INTRODUCTION

Mosquitoes (Diptera: Culicidae), like all organisms, are under constant threat of infection. For the continuation of their life cycles, females of all anautogenous species are required to take a blood meal for the production of eggs. This act of blood feeding often exposes mosquitoes to blood-borne pathogens that aim to undergo complex developmental, reproductive and/or migrational processes inside a mosquito host before they can be transmitted during a subsequent blood meal. In addition to risking infection through blood feeding, mosquitoes often acquire pathogens through sugar feeding, through breaks in their cuticle that are created after physical injury and following pathogen-driven cuticular degradation. Whereas pathogen acquisition through blood feeding occurs exclusively during the adult life stage, infection through the cuticle is likely most prevalent during the aquatic developmental stages, when mosquitoes live in environments rife with bacteria.

Although culicine and anopheline mosquitoes are effective vectors of human and animal pathogens, susceptibility or resistance to infection is often the result of intricate co-evolutionary

Invertebrate Immunity, edited by Kenneth Söderhäll.

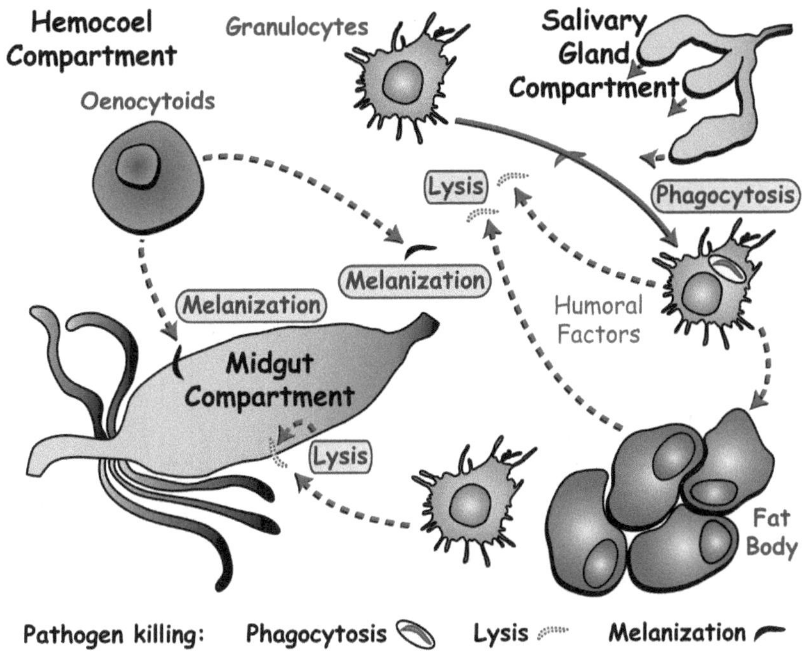

Figure 1. Mosquito immune responses in the three major immune compartments. In the hemocoel, granulocyte and oenocytoid hemocytes, as well as fat body, kill pathogens via phagocytosis, lysis and melanization. In the midgut, immune factors produced by epithelial cells, hemocytes and possibly fat body, kill pathogens via lytic and melanization pathways. Little is know about the role salivary glands play in immune responses, but they produce immune factors in response to infection.

processes in which mosquitoes and pathogens engage in counter-adaptations for survival and infection. As a consequence of this evolutionary arms race, only mosquitoes of the genus *Aedes* are capable of transmitting dengue fever virus and only *Culex* mosquitoes transmit Japanese encephalitis virus.[1] Similarly, of the greater than 3,000 known species of mosquitoes, only a subset of species from the genus *Anopheles* is capable of transmitting human malaria.[2] Even within the susceptible *Anopheles gambiae* species, some individuals are resistant to infection and others, while unable to eliminate the infection, are capable of drastically reducing pathogen numbers.[3,4] The specificity of mosquito-pathogen associations also varies among species of parasites. For example, the mosquito *Armigeres subalbatus* effectively transmits the filarial nematode *Brugia pahangi* but is resistant to a close relative, *Brugia malayi*.[5] Several factors account for the ability of pathogens to survive inside mosquitoes, including behavior (e.g., will the mosquito encounter the pathogen?) and physiological compatibility (e.g., are the correct conditions present in the host that allow the pathogen to complete its life cycle?). Another major factor that determines whether a pathogen can survive inside mosquitoes rests on the strong innate immune responses mounted by the host and on whether the pathogens have evolved mechanisms to evade these defenses.

In broad terms, pathogen killing by mosquitoes is accomplished by three primary mechanisms: cell-mediated phagocytosis, melanization and lysis (Fig. 1). Each is initiated by pattern recognition receptors and the factors leading to killing can be subdivided into cellular and humoral components (Fig. 2). The cellular response includes phagocytosis

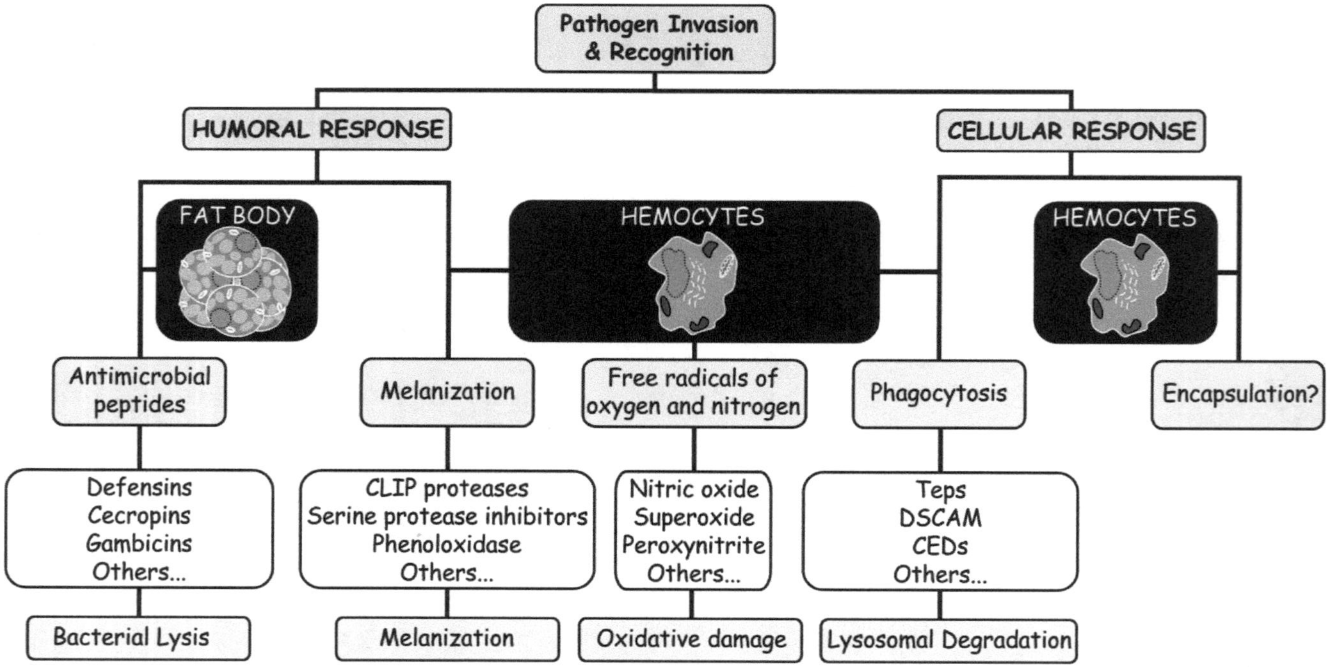

Figure 2. Immune responses in the mosquito hemocoel. Antimicrobial responses in the hemocoel are mediated by hemocytes and fat body and include antimicrobial peptides, melanization, reactive oxygen species, reactive nitrogen species and phagocytosis.

and encapsulation by hemocytes and pericardial cells.[6-9] The humoral response includes pattern recognition receptors, inducible antimicrobial peptides, the phenoloxidase cascade system of melanization and wound healing, and reactive oxygen and reactive nitrogen intermediates.[10-15] Regardless of this conceptual organization, the line between cellular and humoral immunity is blurred because many humoral components are produced by hemocytes and participate in cellular immune responses.[16,17] This chapter reviews the interactions between mosquitoes and pathogens, with emphasis placed on the physical barriers that prevent pathogens from entering the body, the organs and tissues that regulate immune responses, the mechanistic manifestations of immunity and our current understanding of the molecular basis of immunity.

BIOLOGY OF PATHOGENS INSIDE MOSQUITOES

Mosquitoes are subject to infection by viral, bacterial, fungal, protozoan and metazoan pathogens and initial entry into the host generally occurs either through breaks in the cuticle or by ingestion. The biology of pathogens inside the mosquito is dependent on their mode of transmission. By and large, bacteria and fungi enter mosquitoes through wounds in their outer cuticle or through the midgut epithelium after feeding. They quickly replicate in the host's gut or hemocoel (body cavity) and can be transmitted to a subsequent host while the initial host is alive or after its death. The time between colonization and transmission can be very short; some of these infectious agents are highly pathogenic and are lethal within hours of infection.[18] For that reason, bacterial and fungal pathogens are currently being used in the development of novel pest control strategies.[19,20]

Pathogens acquired and transmitted through blood feeding, on the other hand, must undergo obligatory processes that require their interaction with multiple tissue types and require that the host survives for days or weeks before transmission can take place.[21] *Plasmodium* parasites and arboviruses, for example, must cross the midgut epithelium, replicate, migrate through the hemocoel and invade the salivary glands before the mosquito can infect a subsequent host during her next blood feeding. Similarly, filarial nematodes must leave the midgut, develop in the thoracic musculature or Malpighian tubules and migrate to the mouthparts for the mosquito to become infectious. Unlike fungal and bacterial pathogens, the transmission of blood-borne pathogens requires mosquito viability throughout the entirety of the pathogen's life cycle inside the insect host. Nevertheless, these pathogens decrease fitness, reduce fecundity and, if acquired in large enough numbers, can be lethal to the mosquito.[22-25]

MOSQUITO COMPARTMENTS AND BARRIERS TO INFECTION

Pathogens inhabit three primary compartments in the mosquito: the midgut, the hemocoel and the salivary glands (Fig. 1). All three of these compartments include physical and physiological barriers that limit or reduce pathogen development. In addition, cells in all of these compartments produce immune factors with antimicrobial activity. These responses, though powerful, are innate and lack the properties of somatic hypermutation that are hallmarks of vertebrate adaptive immunity.[26]

The Gut Compartment

When pathogens enter mosquitoes via ingestion, the initial barrier faced is physical destruction by the cibarial armature.[27] This barrier, composed of sclerotized teeth and spines that protrude into the lumen of the foregut, slices large pathogens during the initial stages of ingestion and before they reach the midgut. The cibarial armature is effective in limiting infection by large metazoan parasites such as filarial nematodes but does little to destroy protozoan, bacterial and viral pathogens. Once in the midgut lumen, pathogens must survive digestive enzymes and invade the midgut epithelium by either digesting a thick acellular chitinaceous peritrophic matrix formed in response to blood feeding or by initiating epithelium invasion prior to the formation of this matrix.[28-30] The mechanisms leading to midgut penetration are not well understood, but depending on the pathogen may involve receptor-ligand interactions, physical burrowing through the epithelium and/or digestion of host cells.[31-33]

Mosquitoes drastically limit pathogen development in the midgut. For *Plasmodium* parasites, the causative agents of malaria, ookinete development in the gut results in a 500 to 100,000-fold reduction in parasites numbers and the ookinete to oocyst transformation that occurs on the basal side of the midgut experiences parasite losses of 5 to 100-fold.[34] The bases for these parasite reductions are complex, but include lytic and melanization events that are controlled by pattern recognition receptors, serine proteases and their inhibitors, and enzymatic cascades.[35-40] Indeed, molecular and biochemical studies have shown that the midgut rapidly produces a vast milieu of antimicrobial proteins in response to pathogen exposure.[41-45] In addition to these midgut-produced immune factors, infection of this organ triggers the production of immune proteins in other tissues, some of which are transported into the midgut where they exert their antimicrobial activities.[36,38,46,47]

The Hemocoel Compartment

The hemocoel compartment is an open body cavity that contains all visceral organs and is delineated by the outer cuticle and the basal lamina surrounding internal tissues. For many reasons, including preventing infection, insects have developed a tightly-sealed hydrophobic outer cuticle that shields internal organs from the outside environment. Breaks in this cuticle form temporary openings through which pathogens may enter. While these wounds commonly occur in nature, coagulation and melanization responses involving wound contraction, hemocyte degranulation and scar formation rapidly close these lesions.[48,49] As described above, pathogens also enter the hemocoel through ingestion followed by midgut penetration. Regardless of the mode of entry, pathogens disseminate throughout the hemocoel by either pathogen-driven active motility or the natural flow of hemolymph.[50-53]

Once in the hemocoel, pathogens are immersed in a nutrient-rich medium that contains immune cells and humoral immune factors produced by hemocytes, pericardial cells and fat body. Hemocytes are immunosurveillance cells that initiate innate immune responses and are found circulating with the hemolymph or attached to visceral tissues (Fig. 3). They are involved in the killing and sequestration of pathogens via phagocytosis, nodulation and the secretion of humoral immune factors.[6-9,12,54] Transcriptomic analyses in several mosquito genera have repeatedly shown the broad range of immune factors produced by hemocytes, which include pattern recognition receptors, proteins involved in phagocytosis, melanization modulators and enzymes, signal transduction proteins, stress response proteins

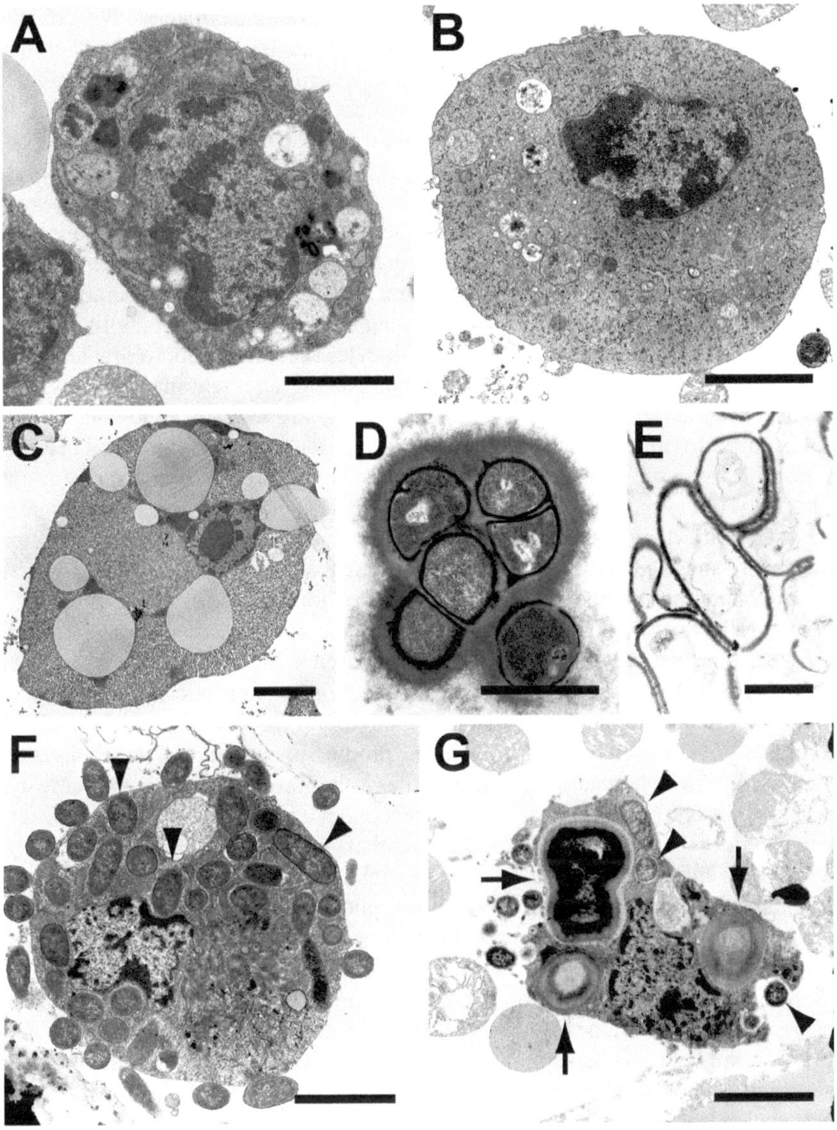

Figure 3. Transmission electron micrographs of immune cells and immune responses in the mosquito hemocoel. A) Circulating granulocyte. B) Circulating oenocytoid. C) Fat body. D) Melanization of *Staphylococcus aureus* in the hemocoel. E) Lysis of *S. aureus* in the hemocoel. F) Phagocytosis of *Escherichia coli* (e.g., arrowheads) by a circulating granulocyte. G) Phagocytosis of unmelanized *E. coli* (e.g., arrowheads) and melanized *Micrococcus luteus* (e.g., arrows) by a circulating granulocyte. Scale bars: A-B, 2 μm; C, 5 μm; D-E, 1 μm; F-G, 3 μm.

and antimicrobial peptides.[16,17,47,55] RNA interference-based transcriptional knockdown of several hemocyte-produced immunity genes results in increased susceptibility to infection with bacteria and malaria parasites, indicating that hemocytes and the effector molecules they produce are essential for efficient immune responses.[36,38,46,47,56]

On the basis of morphology, lectin binding properties and enzymatic activity, mosquito hemocytes were initially subdivided into granulocytes, oenocytoids, adipohemocytes and thrombocytoids.[57] The latter two subtypes were later determined not to be true circulating hemocytes but instead fat body and pericardial cells also collected during the extraction process.[9] A subsequent study confirmed the initial hemocyte classification system and postulated that a third cell type, progenitor prohemocytes, also circulates with the hemolymph.[6]

The number of circulating hemocytes in an adult mosquito is limited and mitotic activity has not been observed in these cells, suggesting that after eclosion no new hemocytes are produced. Adult mosquitoes contain slightly more than 1,000 circulating hemocytes at the time of emergence and the number of circulating cells drops with age, falling to 800 or less by their sixth day after eclosion.[6,58] Granulocytes account for approximately 95% of the circulating hemocyte population. They contain membrane-delimited vesicles, exhibit acid phosphatase activity and strongly adhere to artificial substrates following extraction from the mosquito (Fig. 3A).[8,57] Granulocytes also produce proteins involved in humoral immune pathways, such as nitric oxide synthase, serine proteases and serine protease inhibitors.[6,12] However, the most striking characteristic of granulocytes is their phagocytic capacity: granulocytes engage in the phagocytosis of bacterial pathogens within 5 minutes after exposure and as infections progress individual hemocytes dramatically grow in size to accommodate the internalization of hundreds of foreign entities (Fig. 3F-G).[9,12,54,58] Granulocytes have also been observed to phagocytose *Plasmodium* sporozoites, but the rate of sporozoite phagocytosis is low when compared to bacteria and the importance of this immune process in limiting *Plasmodium* infection remains unclear.[8,59]

The other major hemocyte class is the oenocytoid (Fig. 3B). These cells account for approximately 5% of circulating hemocytes and produce phenoloxidase and phenylalanine hydroxylase,[9,57,60] which are rate-limiting enzymes in the humoral melanization pathway.[61,62] These enzymes are present in the cytosol of oenocytoids and their sequences do not contain classical signal peptides,[60,63] indicating that enzyme release into the hemolymph must occur by either nonclassical secretion mechanisms or cell rupture. Although this process has not been resolved in mosquitoes, phenoloxidase release in lepidopterans and brachyceran dipterans has been shown to occur by cell rupture in an eicosanoid dependent manner.[64,65]

While mosquitoes mount strong and rapid immune responses following pathogen exposure, the type of immune response can vary depending on the pathogen. For example, the primary immune mechanism against *Escherichia coli* is phagocytosis, but against *Micrococcus luteus* is melanization.[8,9] Studies on the immune response mounted against a large panel of bacteria determined that the strength of phagocytosis versus melanization responses is not dependent on Gram-type, but the variable response illustrates that mosquitoes discriminate between pathogens.[54]

Additional evidence supporting the importance of hemocytes in immunity comes from a study showing that there is age-associated mortality in *Aedes aegypti* following *E. coli* immune challenge.[58] This mortality correlates with a decrease in the number of circulating hemocytes and a decrease in the ability to kill *E. coli*, but age has no effect on the transcription of the antimicrobial peptides cecropin, defensin, or gambicin.[58] These findings suggest that the increase in susceptibility is not due to antimicrobial peptide production but instead to a decrease in the number of circulating hemocytes available to quell the infection and are in agreement with a recent report in *D. melanogaster* showing that targeted ablation of hemocytes renders flies incapable of surviving bacterial infections.[66]

In addition to hemocytes, pericardial cells have been relentlessly implicated in immune surveillance, but no direct evidence has been published. Pericardial cells are large binucleate cells that flank the mosquito heart. In *Anopheles*, these cells have been reported to vary in number from 56 to greater than 300, depending on the observer, and also to uptake ammonia carmine dye.[67,68] More recent studies have suggested that pericardial cells contain several immune-related molecules,[69,70] but whether these cells actually produce these proteins or sequester them from the hemolymph is not known. In addition, intense phagocytic activity has been reported near the surface of the mosquito heart, an area densely populated by pericardial cells. However, it remains unclear whether this phagocytic activity is carried out by pericardial cells, sessile hemocytes, or yet unidentified immune cells.[59] Nevertheless, the location of effector cells in the vicinity of the heart is advantageous for immune surveillance and pathogen destruction, as their position in areas of high hemolymph flow increases their probability of encountering invading pathogens.[51]

While hemocytes and possibly pericardial cells mediate both cellular and humoral responses, the fat body's role in immunity is exclusively humoral. The fat body is a multifunctional organ consisting of loosely assembled cells that are rich in glycogen and lipids and line the mosquito integument (Fig. 3C).[57,71] Among its many functions, fat body synthesizes vitellogenin precursors required for the production of eggs, serves in energy storage and produces numerous hemolymph components.[72] Specifically, immune activity in the fat body includes infection-induced production of antimicrobial peptides, reactive oxygen species and reactive nitrogen species.[12,15,73-75]

The Salivary Gland Compartment

Transmission of many viral and protozoan parasites to a vertebrate host requires their injection with the mosquito saliva during blood feeding. Hence, invasion of the salivary gland epithelium and migration into the salivary duct is a requirement for the continuation of the life cycle of these pathogens. The salivary gland epithelium forms a physical barrier that pathogens must cross and *Plasmodium* parasites as well as other pathogens have evolved proteins that drive invasion by first binding to specific mosquito salivary gland surface factors.[76-78] To date, little is known about the role of the salivary glands in antimicrobial responses, but several gene expression studies have described the production of immune proteins in the salivary glands of naïve mosquitoes, *Plasmodium*-infected mosquitoes and mosquitoes that have ingested a noninfectious blood meal.[79-86] Empirical evidence showing that the salivary glands serve as an active immune organ is largely lacking, with the exception of a single publication showing that a serine protease inhibitor (SRPN6) produced in the salivary epithelium limits gland invasion by *Plasmodium* sporozoites.[87]

MOLECULAR BASIS OF MOSQUITO IMMUNITY

The publication of the *An. gambiae* and *Ae. aegypti* genomes has led to an explosion in the number of studies focusing on mosquito immunity.[88-91] Many of these studies have employed homology searches to identify putative immune genes and infer their function. These bioinformatic observations, together with gene expression data and transcriptional manipulations using RNA interference or transgenesis, have allowed researchers to conclusively identify genes that are required for pathogen suppression.

Conversely, similar approaches have been used to identify mosquito genes that facilitate pathogen survival.

Pattern Recognition Receptors

Invading pathogens are recognized by the molecular interaction between host-derived pattern recognition receptors (PRRs) and pathogen associated molecular patterns (PAMPs). Bioinformatic analysis of the *An. gambiae* genome has identified approximately 150 putative PRRs.[92] Most are secreted proteins that contain adhesive domains capable of interacting with PAMPs and cluster as members of large gene families. While experimental evidence has shown that many are involved in immune responses, their actual role as PRRs has not been cemented, as their recognized PAMPs have not been identified.

Thioester containing proteins (TEPs) are hemolymph proteins involved in the killing of bacteria and *Plasmodium* ookinetes.[41,46,93,94] Members of the TEP gene family share structural similarities with α2-macroglobulins and vertebrate complement components C3, C4 and C5. By and large, most studies on mosquito TEPs have focused on the hemocyte-produced phagocytosis enhancer TEP1. This protein is secreted into the hemocoel as a single chain molecule that is activated by proteolytic cleavage.[93] Cleaved TEP1 is then stabilized by forming a complex with the leucine rich repeat containing proteins LRIM1 and APL1C prior to binding bacteria in the hemocoel or ookinetes in the midgut, triggering their destruction.[36,38,93] The antiplasmodial activity of APL1 and TEP1 are further supported by studies showing that genetic variation at their respective loci has a profound effect on immune competence against *Plasmodium*.[4,95] A recent genome-wide mapping of reciprocal crosses of mosquito strains that are susceptible or resistant to *P. berghei* demonstrated that polymorphisms in Tep1 explain a portion of the variability in *P. berghei* killing efficiency observed among laboratory mosquito colonies.[95] While this study was published years after the discovery of Tep1, APL1 was initially characterized because it is coded within a *Plasmodium*-resistance island that explains naturally occurring resistance to *P. falciparum* in field-caught *An. gambiae*.[4] Dissection of the APL1 locus revealed that it is composed of three genes, all of which display major structural haplotypes.[96] Interestingly, APL1C is solely responsible for resistance to *P. berghei*, but distinct haplotypes in the neighboring APL1A gene are associated with various levels of resistance.[96]

C-type lectins are soluble or membrane bound proteins that bind carbohydrates in a calcium-dependent manner. In *An. gambiae*, C-type lectins are both positive and negative regulators of mosquito immune responses. In the midgut, CTL4 and CTLMA2 function as negative regulators of the melanization of *Plasmodium berghei* ookinetes.[37] However, in the hemocoel these same C-type lectins are present in the hemolymph as disulfide-linked heterodimers that function in the killing of *E. coli* in a melanization-independent manner.[97] Transcriptional knockdown of either of these lectins increases bacterial proliferation in the hemocoel and decreases mosquito survival, indicating that they are essential players in the antibacterial response.

Gram-negative binding proteins (GNBPs) were initially identified in *An. gambiae* because they share sequence similarities with GNBPs of other insects and because they are transcriptionally upregulated following infection with bacteria and *Plasmodium* parasites.[98] Six members of this gene family are expressed in *An. gambiae* and all presumably function as PRRs by binding ß-1,3-glucan and lipopolysaccharide on the surface of pathogens. GNBPs are transcribed in multiple tissues (hemocytes, midgut, salivary glands) and while they are all upregulated following an immune challenge, they vary in their antimicrobial specificities. GNBP4, for example, participates in the killing of *E. coli*, *Staphylococcus aureus* and *P. berghei*, but not *Plasmodium falciparum*. In contrast, GNBPA2 participates in the killing of *E. coli* and *P. falciparum*, exhibits mild activity against *P. berghei* and is ineffective against *S. aureus*.[99]

The immunoglobulin superfamily consists of 138 genes in *An. gambiae*, 85 of which are upregulated following an immune challenge. Six of these genes were recently extensively characterized: two are involved in the killing of *P. falciparum* ookinetes (IRID4 and 6), two control the growth of opportunistic bacteria (IRID3 and 4) and three are involved in the killing of exogenously introduced bacteria.[100] Another member of this gene family, AgDSCAM, has been shown to opsonize bacteria and to kill the midgut stages of *Plasmodium*.[101]

Fibrinogen-related proteins (FREPs) represent a PRR family that has experienced massive expansion in mosquitoes.[102] Fifty-nine and 37 FREPs have been identified in *An. gambiae* and *Ae. aegypti*, respectively, compared to 14 in *Drosophila melanogaster*.[91] Functional studies have shown that the majority of mosquito FREPs are upregulated following an immune challenge and that many are essential for the killing of bacteria and the maintenance of immune homeostasis.[11,103] Several also have antiplasmodial activity and one in particular (FBN9) binds ookinetes as they invade the midgut epithelium.[11]

Immune Signaling

Microorganism recognition by PRRs can lead to pathogen destruction through constitutive effector mechanisms and/or the activation of intracellular signaling pathways that activate the transcription of effector genes. The major immune signaling pathways in mosquitoes are Toll, Imd and JAK/STAT and the path that leads to activation of these pathways may be amplified or repressed by modulatory proteins such as serine proteases and serine protease inhibitors. Most of the components in these signaling and regulatory pathways are conserved among dipteran insects.[91]

In mosquitoes, genes regulated by the Toll pathway are controlled by the NF-κB transcription factor Rel1. This pathway is induced by fungi, Gram(+) bacteria, viruses and *Plasmodium*. Induction of the Toll pathway by silencing of the negative regulator of Rel1, Cactus, dramatically decreases *P. berghei* and *Plasmodium gallinaceum* infection intensity in the *Anopheles* and *Aedes* midgut, respectively.[104,105] Co-silencing Rel1 and Cactus renders mosquitoes susceptible to infection, indicating that Cactus-mediated susceptibility is due to repression of Rel1.[104] Co-silencing of Cactus and LRIM1 or Tep1 also renders mosquitoes susceptible to infection, suggesting that these two effector molecules are induced through the Toll pathway. In addition to the antiplasmodial activity of genes induced through Rel1, the Toll pathway is also involved in controlling infection against entomopathogenic fungi and dengue virus.[106,107]

The Imd pathway is controlled by the NF-κB transcription factor Rel2. Rel2 exists as short (Rel2S) and full-length (Rel2F) forms, both of which are involved in the immune response against bacteria and *Plasmodium*.[104,108] Interestingly, while both the Toll and Imd pathways are involved in immunity against *P. berghei*, immunity against *P. falciparum* is controlled primarily through the Imd pathway, as transcriptional knockdown of the Imd negative regulator Caspar and not the Toll negative regulator Cactus renders *An. gambiae* resistant to infection.[109]

The least studied immune pathway in mosquitoes is the JAK/STAT pathway. In *An. gambiae*, this pathway is controlled through two STAT transcription factors that are the result of a gene duplication event. In mosquitoes, activation of the STAT pathway requires STAT-B mediated activation of STAT-A, which is regulated by a negative feedback loop controlled by the signaling suppressor protein SOCS. Activation of this pathway leads to the induction of nitric oxide synthase transcription, which is a positive regulator of *Plasmodium* infection.[110] More recent studies have also implicated the JAK/STAT pathway in the immune response against dengue virus in *Ae. aegypti*. Here, inactivation of the JAK/STAT pathway by depletion of the receptor Domeless results in increased viral loads.[111] Conversely, hyperactivation of the pathway by depletion of the negative regulator PIAS results in a more resistant phenotype.

Antimicrobial Peptides

Antimicrobial peptides (AMPs) are secreted low molecular weight proteins that were initially identified for their antimicrobial activity in vitro.[74,75,112,113] Defensins, cecropins and gambicins comprise the three main AMP gene families in mosquitoes. In vitro analyses of their antimicrobial spectra showed that cecropins and gambicin are cytotoxic primarily against Gram(–) bacteria and defensins are cytotoxic primarily against Gram(+) bacteria.[74,75,112] Transcriptional regulation of AMPs occurs through the Toll and Imd pathways,[114,115] and all are transcriptionally upregulated in fat body following exposure to viruses, bacteria, *Plasmodium* and filarial nematodes.[16,107,116-118]

While defensins have antimicrobial activity in vitro, their function as essential components of the mosquito immune response continues to be debated. In *An. gambiae*, transcriptional knockdown of Defensin decreases mosquito survival following *S. aureus* infection,[119] but RNAi-based silencing of *Ae. aegypti* Defensin has no effect on mosquito survival following challenge with three bacterial species.[120,121] The role of cecropins in the antibacterial response in vivo is not known, but ectopic expression of a cecropin transgene results in increased killing of *P. berghei* ookinetes.[122] Lastly, Gambicin exhibits antiplasmodial activity in the midgut, as well as antibacterial activity in the hemocoel.[41]

Phenoloxidase-Based Melanization

Melanization in insects is essential for cuticle hardening, egg chorion tanning, wound healing and immunity. In mosquitoes, melanization (also known as melanotic encapsulation) is an immune effector mechanism involved in the killing of *Plasmodium*, filarial nematodes and bacteria and is visually manifested as a darkened proteinaceous capsule that surrounds invading pathogens (Fig. 3D).[5,8,35] Melanization involves a series of reactions that include the conversion of tyrosine to melanin precursors and the cross-linking of proteins to form a layer of melanin that surrounds and sequesters invading pathogens (Fig. 4).[10] Melanization often results in pathogen death but the killing mechanism remains unclear. It has been hypothesized that death may be caused by either oxidative damage brought on by unstable intermediates created during melanogenesis or by starvation, since the foreign agent becomes isolated from the nutrient-rich hemolymph.[123,124] Besides functioning as a killing mechanism, melanization is also involved in the clearing of already dead or dying pathogens.[37,40]

The process of melanization begins with the proteolytic cleavage of a pro phenoloxidase zymogen into its active form. The exact sequence of events that leads

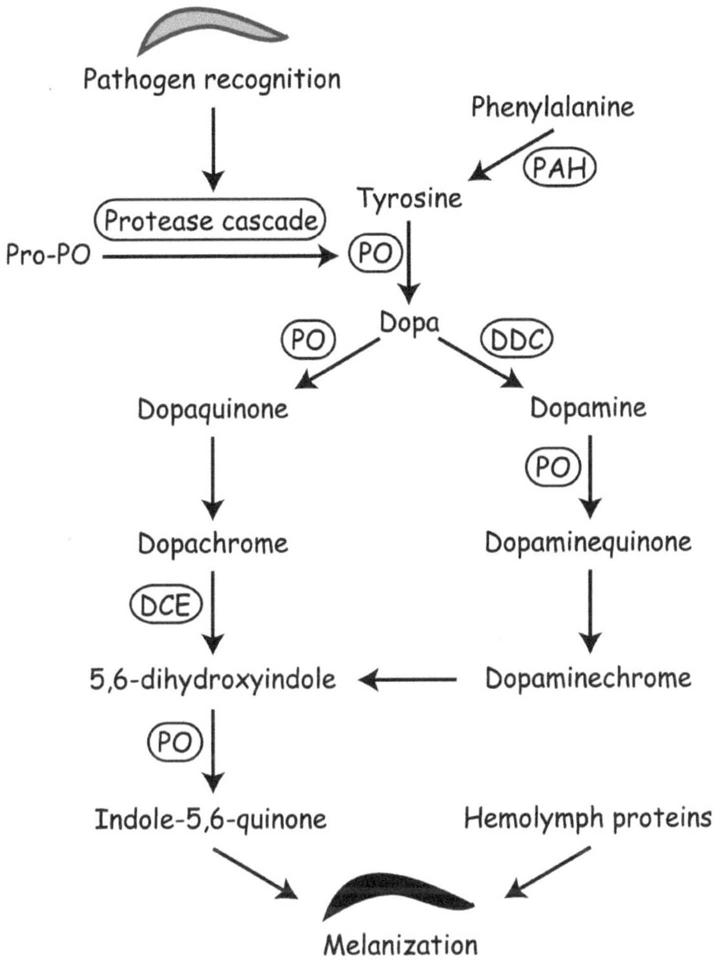

Figure 4. Proposed biochemical pathway leading to the melanization of pathogens. PAH, phenylalanine hydroxylase; PO, phenoloxidase; DDC, dopa decarboxylase; DCE, dopachrome conversion enzyme.

to this cleavage is not well understood, but involves the coordinated action of pattern recognition receptors, serine proteases and serine protease inhibitors. In the hemocoel, ß-1,3-glucan recognition protein is required for the melanization of filarial nematodes and bacteria by functioning as a pattern recognition receptor.[56,125] In the midgut, various proteins serve as promoters and inhibitors of melanization, with their exact role varying greatly between mosquito strains. For example, in *An. gambiae*, the *Plasmodium*-resistant L3-5 strain naturally melanizes *P. berghei* ookinetes but the G3 strain is susceptible to infection. Interestingly, mosquito C-type lectins function as repressors of melanization in the susceptible strain, as silencing CTL4 and CTLMA2 results in the LRIM1-dependent induction of ookinete melanization.[37] CLIP domain serine proteases also modulate the melanization response against *P. berghei* ookinetes and share structural similarities with pro-phenoloxidase activating enzymes from other insects.[126] However, once again, the genetic background of the mosquito has a profound effect on the function of CLIPs.[40,126] For example, CLIPA7 inhibits melanization in the susceptible strain but has no effect on the resistant strain. Conversely, CLIPB3 has no effect on the susceptible strain but promotes melanization in the resistant strain. Similar observations on the role of CLIPs as inhibitors and promoters of melanization have been made in the mosquito hemocoel after intrathoracic injection of sephadex beads.[127]

In addition to CLIPs, mosquito serine protease inhibitors (SRPNs) modulate melanization responses.[128] Both SRPN6 and SRPN2 inhibit the melanization of *P. berghei* ookinetes in the resistant strain and SRPN2 also inhibits the spontaneous formation of melanin-based pseudotumors in the hemocoel.[129,130] These pseudotumors drastically reduce mosquito fitness, underscoring the importance of tightly regulating this immune process. Moreover, the factors that trigger melanization in the midgut are at least partially dependent on the host-parasite combination: SRPN2, CTLs and LRIM1, molecules that regulate development of *P. berghei* in the midgut, have no detectable effect on the development of the human malaria parasite *P. falciparum*.[131,132] These findings are not surprising given that *P. berghei* infection triggers the transcriptional regulation of over twice as many genes as *P. falciparum*.[41]

Once phenoloxidase becomes activated, the formation of melanin is initiated by the phenoloxidase-mediated hydroxylation of tyrosine to form dopa (Fig. 4).[133] Dopa is then oxidized by phenoloxidase to form dopaquinone, which is then converted to dopachrome. Dopachrome conversion enzyme converts dopachrome to 5,6-dihydroxyindole, which is oxidized into indole-5,6-quinone by phenoloxidase and then cross-linked with hemolymph proteins to form melanotic capsules. In an alternative pathway, dopa formed by the hydroxylation of tyrosine is decarboxylated by dopa decarboxylase to form dopamine, which is then converted into melanin by phenoloxidase and other enzymes. Empirical testing has shown that pathogen-induced melanin formation is accomplished via both of these pathways, as transcriptional knockdown of phenoloxidase, dopachrome conversion enzyme and dopa decarboxylase all lead to impaired melanization responses.[62,134-136] Throughout these reactions, tyrosine remains the rate-limiting substrate and endogenous production of tyrosine is accomplished by the hydroxylation of phenylalanine by phenylalanine hydroxylase. This latter reaction is essential for the melanization of filarial nematodes but not sephadex beads,[61,136] with the observed difference possibly related to the larger surface area that needs to be melanized following filarial worm infection. Finally, while melanization events are extracellular, many melanin-producing enzymes (e.g., phenoloxidase, dopachrome conversion enzyme, dopa decarboxylase and phenylalanine hydroxylase) are produced by circulating hemocytes,[9,57,60,134,135] and melanized pathogens are often subsequently phagocytosed by granulocytes (Fig. 3G).[8,9,54]

Nitric Oxide and Other Reactive Species

Nitric oxide is a multifunctional free radical created during the oxidation of L-Arginine to L-Citrulline by the enzyme nitric oxide synthase.[137] In *Anopheles*, nitric oxide synthase is a single copy gene with 18-22 distinct transcripts. Three of these transcripts are induced by *Plasmodium* infection,[138,139] and at least one is induced by bacterial infection.[12] In the midgut of the mosquito, *Plasmodium* glycosylphosphatidylinositols and *Plasmodium*-derived hemozoin acquired with an infectious blood meal induce the transcription of nitric oxide synthase through the STAT pathway,[43,110,140] and the resultant nitric oxide kills *Plasmodium* ookinetes via lysis.[14,141,142] In the hemocoel, nitric oxide synthase is transcriptionally upregulated following bacterial infection and the production of nitric oxide is required for bacterial killing and mosquito survival during systemic infections with *E. coli*.[12]

Reactive oxygen species (ROS) kill *Plasmodium* ookinetes in the midgut and bacteria in the hemocoel.[15,143,144] While the exact mechanism of action against bacteria remains unknown, ROS kill *Plasmodium* through both lytic and melanization pathways.[15,143,144] The *P. berghei*-resistant L3-5 strain of *An. gambiae* lives in a constant state of oxidative stress that promotes melanization of ookinetes as they traverse the midgut epithelium.[143] Conversely, the susceptible G3 strain kills ookinetes via a lytic mechanism that is dependent on infection-induced oxidative stress that is maintained by the repression of catalase, an enzyme that breaks down hydrogen peroxide into oxygen and water.[15] Furthermore, reactive oxygen and reactive nitrogen species are intimately linked in mosquito immunity: peroxidases in the mosquito midgut use nitrite and hydrogen peroxide to synthesize highly reactive nitrogen dioxide and hydrogen peroxide triggers the transcriptional induction of nitric oxide synthase.[145,146]

Phagocytosis

Phagocytosis is an evolutionarily conserved immune process used for the killing and sequestration of small microorganisms. In this process, a particle is recognized, bound by proteins in the plasma membrane and internalized into a membrane-delimited phagosome. The phagosome then fuses with a lysosome and hydrolytic enzymes digest the particle. In mosquitoes, the granulocyte subpopulation of hemocytes uses this immune process to sequester and kill bacteria as early as 5 minutes after exposure.[8,9] Approximately 95% of circulating hemocytes are phagocytic and it has been estimated that individual hemocytes are capable of phagocytosing over 1,000 bacteria within 24 hours of infection.[58]

Several studies have investigated the molecular basis of phagocytosis by visualizing the uptake of fluorescently labeled dead bacteria by mosquito immortal cell lines or by low magnification fluorescence microscopy of whole mosquitoes.[93,94,147] Identified regulators of phagocytosis include pattern recognition receptors, transmembrane receptors and intracellular signaling proteins.

Thioester containing proteins TEP1, TEP3 and TEP4 are involved in the phagocytosis of Gram(+) and Gram(–) bacteria in the mosquito hemocoel.[93,94] The modes of action of TEP3 and TEP4 are not known, but proteolytically activated TEP1 opsonises bacteria by thioester-mediated binding, which in turn initiates phagocytosis. In addition to the above TEPs, the leucine rich repeat containing protein LRIM1 is also required for phagocytosis.[94] The exact mechanism by which LRIM1 functions has not been resolved, but TEP1-based antimicrobial activity in the midgut is dependent on the complexing of TEP1 with LRIM1 in the hemocoel.[36]

AgDSCAM is a hypervariable immunoglobulin that is encoded by 101 exons, which can be transpliced into over 31,000 variants.[101] AgDSCAM binds bacteria in vitro and triggers their phagocytosis by an immortal hemocyte-like cell line. In vivo transcriptional knockdown of AgDSCAM increases bacterial proliferation in the hemocoel and decreases mosquito survival, suggesting that AgDSCAM mediated phagocytosis or humoral killing is required for effective antibacterial responses in the hemocoel.

Several transmembrane receptors have also been implicated in phagocytosis and may function by either directly recognizing pathogens or by recognizing pathogens that have been opsonized by hemolymph proteins. Among these are a ß integrin (BINT2), a peptidoglycan recognition protein (PGRPLC) and a low-density lipoprotein receptor-related protein (LRP1).[94,147]

Lastly, several intracellular proteins trigger the internalization of bacteria. Transcriptional knockdown of *An. gambiae* CED2, CED5, or CED6 reduces phagocytosis efficacy by up to 80%.[94] Epistatic analyses then showed that TEP1, TEP3, LRIM1 and LRP1-mediated phagocytosis occurs through the CED6 pathway and TEP4 and BINT2-mediated phagocytosis occurs through the CED2/CED5 pathway. The involvement of mosquito CEDs in the phagocytosis of foreign bodies through two genetically independent pathways is significant but not unexpected. Cell death abnormal genes were initially discovered in *Caenorhabditis elegans* during screens aimed at identifying genetic factors required for the phagocytosis of apoptotic bodies in the developing worm and these factors function through two independent but partially redundant pathways: the CED-1/CED-6/CED-7 pathway and the CED-2/CED-5/CED-10/CED-12 pathway.[148-150]

CONCLUSION

The field of mosquito immunity has experienced unprecedented growth in the past decade. The publication of the *An. gambiae* and *Ae. aegypti* genomic sequences has led to the bioinformatic identification of numerous putative immunity genes.[88-91] Technical advances such as RNA interference, paratransgenesis and transgenesis have then allowed researchers to empirically test their function in effecting or regulating immune responses against diverse groups of pathogens.[119,151,152] As a result of these studies, it has become apparent that the mosquito immune system shares numerous similarities with vertebrate immune systems. Immune responses in both vertebrates and invertebrates are initiated by microbe recognition events that trigger signaling pathways and effector mechanisms. Moreover, similar to vertebrates, mosquitoes recognize pathogens using complement-like cascades, possess phagocytic cells that circulate with the blood and transcribe effector molecules through Toll and JAK/STAT pathways. Further studies on mosquito immunity will continue to shed light on the evolution of these complex and essential responses.

In addition to the evolutionary conservation of immune components, obtaining a better understanding of the mosquito immune system may translate into novel public health interventions. Mosquitoes are cosmopolitan pests and disease vectors.[153] Because of their global importance, these organisms are subjects of constant study in efforts to uncover mechanisms that reduce their population densities as well as their ability to transmit disease. One aspect of mosquito biology that continues to receive considerable attention is their ability to fight infectious agents, as it has been hypothesized that understanding how mosquitoes kill microbial pathogens may allow us to exploit weaknesses that increase the population-reduction effectiveness of biological control strategies, or to strengthen immune responses such that otherwise susceptible mosquitoes are rendered resistant to infection, halting disease transmission cycles.[19,20,154,155] Much must be done before the feasibility of these approaches can be further considered, including the continued expansion of work in immunity to wild mosquito populations and to a broader range of mosquito species in order to ensure that our current understanding of mosquito immunity is representative of natural host-pathogen interactions.

ACKNOWLEDGEMENTS

The author is funded by NSF grant IOS-0817644. Useful discussions with members of the Hillyer lab are greatly appreciated.

REFERENCES

1. Nasci RS, Miller BR. Culicine mosquitoes and the agents they transmit. In: Beaty BJ, Marquardt WC, eds. The Biology of Disease Vectors. Niwot: University Press of Colorado, 1996:85-97.
2. Gwadz R, Collins FH. Anopheline mosquitoes and the pathogens they transmit. In: Beaty BJ, Marquardt WC, eds. The Biology of Disease Vectors. Niwot: University Press of Colorado, 1996:73-84.
3. Niaré O, Markianos K, Volz J et al. Genetic loci affecting resistance to human malaria parasites in a West African mosquito vector population. Science 2002; 298(5591):213-216.
4. Riehle MM, Markianos K, Niaré O et al. Natural malaria infection in Anopheles gambiae is regulated by a single genomic control region. Science 2006; 312(5773):577-579.
5. Beerntsen BT, Luckhart S, Christensen BM. Brugia malayi and Brugia pahangi: inherent difference in immune activation in the mosquitoes Armigeres subalbatus and Aedes aegypti. J Parasitol 1989; 75(1):76-81.
6. Castillo JC, Robertson AE, Strand MR. Characterization of hemocytes from the mosquitoes Anopheles gambiae and Aedes aegypti. Insect Biochem Mol Biol 2006; 36(12):891-903.
7. Hernández-Martínez S, Lanz H, Rodríguez MH et al. Cellular-mediated reactions to foreign organisms inoculated into the hemocoel of Anopheles albimanus (Diptera: Culicidae). J Med Entomol 2002; 39(1):61-69.
8. Hillyer JF, Schmidt SL, Christensen BM. Rapid phagocytosis and melanization of bacteria and Plasmodium sporozoites by hemocytes of the mosquito Aedes aegypti. J Parasitol 2003; 89(1):62-69.
9. Hillyer JF, Schmidt SL, Christensen BM. Hemocyte-mediated phagocytosis and melanization in the mosquito Armigeres subalbatus following immune challenge by bacteria. Cell Tissue Res 2003; 313(1):117-127.
10. Christensen BM, Li J, Chen C-C et al. Melanization immune responses in mosquito vectors. Trends Parasitol 2005; 21(4):192-199.
11. Dong Y, Dimopoulos G. Anopheles fibrinogen-related proteins provide expanded pattern recognition capacity against bacteria and malaria parasites. J Biol Chem 2009; 284(15):9835-9844.
12. Hillyer JF, Estévez-Lao TY. Nitric oxide is an essential component of the hemocyte-mediated mosquito immune response against bacteria. Dev Comp Immunol 2010; 34(2):141-149.
13. Lowenberger C. Innate immune response of Aedes aegypti. Insect Biochem Mol Biol 2001; 31(3):219-229.
14. Luckhart S, Vodovotz Y, Cui L et al. The mosquito Anopheles stephensi limits malaria parasite development with inducible synthesis of nitric oxide. Proc Natl Acad Sci USA 1998; 95(10):5700-5705.
15. Molina-Cruz A, DeJong RJ, Charles B et al. Reactive oxygen species modulate Anopheles gambiae immunity against bacteria and Plasmodium. J Biol Chem 2008; 283(6):3217-3223.
16. Bartholomay LC, Cho WL, Rocheleau TA et al. Description of the transcriptomes of immune response-activated hemocytes from the mosquito vectors Aedes aegypti and Armigeres subalbatus. Infect Immun 2004; 72(7):4114-4126.
17. Baton L, Robertson A, Warr E et al. Genome-wide transcriptomic profiling of Anopheles gambiae hemocytes reveals pathogen-specific signatures upon bacterial challenge and Plasmodium berghei infection. BMC Genomics 2009; 10(1):257.
18. Walther CJ, Couche GA, Pfannenstiel MA et al. Analysis of mosquito larvicidal potential exhibited by vegetative cells of Bacillus thuringiensis subsp. israelensis. Appl Environ Microbiol 1986; 52(4):650-653.
19. Blanford S, Chan BHK, Jenkins N et al. Fungal pathogen reduces potential for malaria transmission. Science 2005; 308(5728):1638-1641.
20. Schnepf E, Crickmore N, Van Rie J et al. Bacillus thuringiensis and its pesticidal crystal proteins. Microbiol Mol Biol Rev 1998; 62(3):775-806.
21. Roberts LS, Janovy Jr. J. Foundations of parasitology. 8th ed. Boston: McGraw Hill Higher Education; 2009.
22. Ferdig MT, Beerntsen BT, Spray FJ et al. Reproductive costs associated with resistance in a mosquito-filarial worm system. Am J Trop Med Hyg 1993; 49(6):756-762.
23. Ferguson HM, Read AF. Genetic and environmental determinants of malaria parasite virulence in mosquitoes. Proc Biol Sci 2002; 269(1497):1217-1224.
24. Scott TW, Lorenz LH. Reduction of Culiseta melanura fitness by eastern equine encephalomyelitis virus. Am J Trop Med Hyg 1998; 59(2):341-346.

25. Styer LM, Meola MA, Kramer LD. West Nile virus infection decreases fecundity of Culex tarsalis females. J Med Entomol 2007; 44(6):1074-1085.
26. Schmidt O, Theopold U, Beckage NE. Insect and vertebrate immunity: key similarities versus differences. In: Beckage NE, ed. Insect Immunology. San Diego: Academic Press, 2008:1-23.
27. McGreevy PB, Bryan JH, Oothuman P et al. The lethal effects of the cibarial and pharyngeal armatures of mosquitoes on microfilariae. Trans R Soc Trop Med Hyg 1978; 72(4):361-368.
28. Abraham EG, Jacobs-Lorena M. Mosquito midgut barriers to malaria parasite development. Insect Biochem Mol Biol 2004; 34(7):667-671.
29. Gonzalez-Ceron L, Rodriguez MH, Chavez-Munguia B et al. Plasmodium vivax: impaired escape of Vk210 phenotype ookinetes from the midgut blood bolus of Anopheles pseudopunctipennis. Exp Parasitol 2007; 115(1):59-67.
30. Kato N, Mueller CR, Fuchs JF et al. Evaluation of the function of a type I peritrophic matrix as a physical barrier for midgut epithelium invasion by mosquito-borne pathogens in Aedes aegypti. Vector Borne Zoonotic Dis 2008; 8(5):701-712.
31. Ito J, Ghosh A, Moreira LA et al. Transgenic anopheline mosquitoes impaired in transmission of a malaria parasite. Nature 2002; 417(6887):452-455.
32. Santos JN, Lanfredi RM, Pimenta PFP. The invasion of the midgut of the mosquito Culex (Culex) quinquefasciatus Say, 1823 by the helminth Litomosoides chagasfilhoi Moraes Neto, Lanfredi and De Souza, 1997. J Invertebr Pathol 2006; 93(1):1-10.
33. Whitten MMA, Shiao S-H, Levashina EA. Mosquito midguts and malaria: cell biology, compartmentalization and immunology. Parasite Immunol 2006; 28(4):121-130.
34. Alavi Y, Arai M, Mendoza J et al. The dynamics of interactions between Plasmodium and the mosquito: a study of the infectivity of Plasmodium berghei and Plasmodium gallinaceum and their transmission by Anopheles stephensi, Anopheles gambiae and Aedes aegypti. Int J Parasitol 2003; 33(9):933-943.
35. Collins FH, Sakai RK, Vernick KD et al. Genetic selection of a Plasmodium-refractory strain of the malaria vector Anopheles gambiae. Science 1986; 234(4776):607-610.
36. Fraiture M, Baxter RHG, Steinert S et al. Two mosquito LRR proteins function as complement control factors in the TEP1-mediated killing of Plasmodium. Cell Host Microbe 2009; 5(3):273-284.
37. Osta MA, Christophides GK, Kafatos FC. Effects of mosquito genes on Plasmodium development. Science 2004; 303(5666):2030-2032.
38. Povelones M, Waterhouse RM, Kafatos FC et al. Leucine-rich repeat protein complex activates mosquito complement in defense against Plasmodium parasites. Science 2009; 324(5924):258-261.
39. Vernick KD, Fujioka H, Seeley DC et al. Plasmodium gallinaceum: a refractory mechanism of ookinete killing in the mosquito, Anopheles gambiae. Exp Parasitol 1995; 80(4):583-595.
40. Volz J, Müller HM, Zdanowicz A et al. A genetic module regulates the melanization response of Anopheles to Plasmodium. Cell Microbiol 2006; 8(9):1392-1405.
41. Dong Y, Aguilar R, Xi Z et al. Anopheles gambiae immune responses to human and rodent Plasmodium parasite species. PLoS Pathog 2006; 2(6):e52.
42. González-Lázaro M, Dinglasan RR, Hernández-Hernández FdlC et al. Anopheles gambiae Croquemort SCRBQ2, expression profile in the mosquito and its potential interaction with the malaria parasite Plasmodium berghei. Insect Biochem Mol Biol 2009; 39(5-6):395-402.
43. Lim J, Gowda DC, Krishnegowda G et al. Induction of nitric oxide synthase in Anopheles stephensi by Plasmodium falciparum: mechanism of signaling and the role of parasite glycosylphosphatidylinositols. Infect Immun 2005; 73(5):2778-2789.
44. Sanders HR, Foy BD, Evans AM et al. Sindbis virus induces transport processes and alters expression of innate immunity pathway genes in the midgut of the disease vector, Aedes aegypti. Insect Biochem Mol Biol 2005; 35(11):1293-1307.
45. Xu X, Dong Y, Abraham EG et al. Transcriptome analysis of Anopheles stephensi-Plasmodium berghei interactions. Mol Biochem Parasitol 2005; 142(1):76-87.
46. Blandin SA, Shiao S-H, Moita LF et al. Complement-like protein TEP1 is a determinant of vectorial capacity in the malaria vector Anopheles gambiae. Cell 2004; 116(5):661-670.
47. Pinto SB, Lombardo F, Koutsos AC et al. Discovery of Plasmodium modulators by genome-wide analysis of circulating hemocytes in Anopheles gambiae. Proc Natl Acad Sci USA 2009; 106(50):21270-21275.
48. Lai S-C, Chen C-C, Hou RF. Electron microscopic observations on wound-healing in larvae of the mosquito Armigeres subalbatus (Diptera: Culicidae). J Med Entomol 2001; 38(6):836-843.
49. Lai S-C, Chen C-C, Hou RF. Immunolocalization of prophenoloxidase in the process of wound healing in the mosquito Armigeres subalbatus (Diptera: Culicidae). J Med Entomol 2002; 39(2):266-274.
50. Bain O, Babayan S. Behaviour of filariae: morphological and anatomical signatures of their life style within the arthropod and vertebrate hosts. Filaria J 2003; 2(1):16.
51. Glenn JD, King JG, Hillyer JF. Structural mechanics of the mosquito heart and its function in bidirectional hemolymph transport. J Exp Biol 2010; 213(4):541-550.

52. Morahan BJ, Wang L, Coppel RL. No TRAP, no invasion. Trends Parasitol 2009; 25(2):77-84.
53. Vanderberg JP. Studies on the motility of Plasmodium sporozoites. J Protozool 1974; 21(4):527-537.
54. Hillyer JF, Schmidt SL, Christensen BM. The antibacterial innate immune response by the mosquito Aedes aegypti is mediated by hemocytes and independent of Gram type and pathogenicity. Microbes Infect 2004; 6(5):448-459.
55. Bartholomay LC, Mayhew GF, Fuchs JF et al. Profiling infection responses in the haemocytes of the mosquito, Aedes aegypti. Insect Mol Biol 2007; 16(6):761-776.
56. Wang X, Fuchs JF, Infanger L-C et al. Mosquito innate immunity: involvement of beta 1,3-glucan recognition protein in melanotic encapsulation immune responses in Armigeres subalbatus. Mol Biochem Parasitol 2005; 139(1):65-73.
57. Hillyer JF, Christensen BM. Characterization of hemocytes from the yellow fever mosquito, Aedes aegypti. Histochem Cell Biol 2002; 117(5):431-440.
58. Hillyer JF, Schmidt SL, Fuchs JF et al. Age-associated mortality in immune challenged mosquitoes (Aedes aegypti) correlates with a decrease in haemocyte numbers. Cell Microbiol 2005; 7(1):39-51.
59. Hillyer JF, Barreau C, Vernick KD. Efficiency of salivary gland invasion by malaria sporozoites is controlled by rapid sporozoite destruction in the mosquito haemocoel. Int J Parasitol 2007; 37(6):673-681.
60. Johnson JK, Rocheleau TA, Hillyer JF et al. A potential role for phenylalanine hydroxylase in mosquito immune responses. Insect Biochem Mol Biol 2003; 33(3):345-354.
61. Infanger L-C, Rocheleau TA, Bartholomay LC et al. The role of phenylalanine hydroxylase in melanotic encapsulation of filarial worms in two species of mosquitoes. Insect Biochem Mol Biol 2004; 34(12):1329-1338.
62. Shiao S-H, Higgs S, Adelman Z et al. Effect of prophenoloxidase expression knockout on the melanization of microfilariae in the mosquito Armigeres subalbatus. Insect Mol Biol 2001; 10(4):315-321.
63. Taft AS, Chen CC, Li J et al. Molecular cloning of two prophenoloxidase genes from the mosquito Aedes aegypti. Insect Mol Biol 2001; 10(1):97-103.
64. Bidla G, Dushay MS, Theopold U. Crystal cell rupture after injury in Drosophila requires the JNK pathway, small GTPases and the TNF homolog Eiger. J Cell Sci 2007; 120(Pt 7):1209-1215.
65. Shrestha S, Kim Y. Eicosanoids mediate prophenoloxidase release from oenocytoids in the beet armyworm Spodoptera exigua. Insect Biochem Mol Biol 2008; 38(1):99-112.
66. Charroux B, Royet J. Elimination of plasmatocytes by targeted apoptosis reveals their role in multiple aspects of the Drosophila immune response. Proc Natl Acad Sci USA 2009; 106(24):9797-9802.
67. Jones JC. The heart and associated tissues of Anopheles quadrimaculatus say (Diptera: Culicidae). J Morphol 1954; 94(1):71-124.
68. Pal R. Nephrocytes in some Culicidae-Diptera. Indian J Entomol 1944; 6:143-148.
69. Barillas-Mury C, Han YS, Seeley D et al. Anopheles gambiae Ag-STAT, a new insect member of the STAT family, is activated in response to bacterial infection. EMBO J 1999; 18(4):959-967.
70. Danielli A, Kafatos FC, Loukeris TG. Cloning and characterization of four Anopheles gambiae serpin isoforms, differentially induced in the midgut by Plasmodium berghei invasion. J Biol Chem 2003; 278(6):4184-4193.
71. Martins GF, Pimenta PFP. Structural changes in fat body of Aedes aegypti caused by aging and blood feeding. J Med Entomol 2008; 45(6):1102-1107.
72. Raikhel AS, Kokoza VA, Zhu J et al. Molecular biology of mosquito vitellogenesis: from basic studies to genetic engineering of antipathogen immunity. Insect Biochem Mol Biol 2002; 32(10):1275-1286.
73. Bartholomay LC, Farid HA, Ramzy RM et al. Culex pipiens pipiens: characterization of immune peptides and the influence of immune activation on development of Wuchereria bancrofti. Mol Biochem Parasitol 2003; 130(1):43-50.
74. Lowenberger C, Charlet M, Vizioli J et al. Antimicrobial activity spectrum, cDNA cloning and mRNA expression of a newly isolated member of the cecropin family from the mosquito vector Aedes aegypti. J Biol Chem 1999; 274(29):20092-20097.
75. Vizioli J, Bulet P, Hoffmann JA et al. Gambicin: a novel immune responsive antimicrobial peptide from the malaria vector Anopheles gambiae. Proc Natl Acad Sci USA 2001; 98(22):12630-12635.
76. Cao-Lormeau V-M. Dengue viruses binding proteins from Aedes aegypti and Aedes polynesiensis salivary glands. Virol J 2009; 6:35.
77. Ghosh AK, Devenport M, Jethwaney D et al. Malaria parasite invasion of the mosquito salivary gland requires interaction between the plasmodium TRAP and the anopheles saglin proteins. PLoS Pathog 2009; 5(1):e1000265.
78. Korochkina S, Barreau C, Pradel G et al. A mosquito-specific protein family includes candidate receptors for malaria sporozoite invasion of salivary glands. Cell Microbiol 2006; 8(1):163-175.
79. Arcà B, Lombardo F, Francischetti IMB et al. An insight into the sialome of the adult female mosquito Aedes albopictus. Insect Biochem Mol Biol 2007; 37(2):107-127.

80. Arcà B, Lombardo F, Valenzuela JG et al. An updated catalogue of salivary gland transcripts in the adult female mosquito, Anopheles gambiae. J Exp Biol 2005; 208(Pt 20):3971-3986.
81. Choumet V, Carmi-Leroy A, Laurent C et al. The salivary glands and saliva of Anopheles gambiae as an essential step in the Plasmodium life cycle: a global proteomic study. Proteomics 2007; 7(18):3384-3394.
82. Dixit R, Sharma A, Mourya DT et al. Salivary gland transcriptome analysis during Plasmodium infection in malaria vector Anopheles stephensi. Int J Infect Dis 2009; 13(5):636-646.
83. Ribeiro JMC, Charlab R, Pham VM et al. An insight into the salivary transcriptome and proteome of the adult female mosquito Culex pipiens quinquefasciatus. Insect Biochem Mol Biol 2004; 34(6):543-563.
84. Rosinski-Chupin I, Briolay J, Brouilly P et al. SAGE analysis of mosquito salivary gland transcriptomes during Plasmodium invasion. Cell Microbiol 2007; 9(3):708-724.
85. Serazin AC, Dana AN, Hillenmeyer ME et al. Comparative analysis of the global transcriptome of Anopheles funestus from Mali, West Africa. PLoS ONE 2009; 4(11):e7976.
86. Thangamani S, Wikel SK. Differential expression of Aedes aegypti salivary transcriptome upon blood feeding. Parasit Vectors 2009; 2(1):34.
87. Pinto SB, Kafatos FC, Michel K. The parasite invasion marker SRPN6 reduces sporozoite numbers in salivary glands of Anopheles gambiae. Cell Microbiol 2008; 10(4):891-898.
88. Christophides GK, Zdobnov E, Barillas-Mury C et al. Immunity-related genes and gene families in Anopheles gambiae. Science 2002; 298(5591):159-165.
89. Holt RA, Subramanian GM, Halpern A et al. The genome sequence of the malaria mosquito Anopheles gambiae. Science 2002; 298(5591):129-149.
90. Nene V, Wortman JR, Lawson D et al. Genome sequence of Aedes aegypti, a major arbovirus vector. Science 2007; 316(5832):1718-1723.
91. Waterhouse RM, Kriventseva EV, Meister S et al. Evolutionary dynamics of immune-related genes and pathways in disease-vector mosquitoes. Science 2007; 316(5832):1738-1743.
92. Das S, Dong Y, Garver L et al. Specificity of the innate immune system: a closer look at the mosquito pattern-recognition receptor repertoire. sws Rolff J, Reynolds SE, eds. Insect Infection and Immunity. Oxford, Oxford University Press 2009: 69-85.
93. Levashina EA, Moita LF, Blandin SA et al. Conserved role of a complement-like protein in phagocytosis revealed by dsRNA knockout in cultured cells of the mosquito, Anopheles gambiae. Cell 2001; 104(5):709-718.
94. Moita LF, Wang-Sattler R, Michel K et al. In vivo identification of novel regulators and conserved pathways of phagocytosis in A. gambiae. Immunity 2005; 23(1):65-73.
95. Blandin SA, Wang-Sattler R, Lamacchia M et al. Dissecting the genetic basis of resistance to malaria parasites in Anopheles gambiae. Science 2009; 326(5949):147-150.
96. Riehle MM, Xu J, Lazzaro BP et al. Anopheles gambiae APL1 is a family of variable LRR proteins required for Rel1-mediated protection from the malaria parasite, Plasmodium berghei. PLoS ONE 2008; 3(11):e3672.
97. Schnitger AKD, Yassine H, Kafatos FC et al. Two C-type lectins cooperate to defend Anopheles gambiae against Gram-negative bacteria. J Biol Chem 2009; 284(26):17616-17624.
98. Dimopoulos G, Richman A, Müller HM et al. Molecular immune responses of the mosquito Anopheles gambiae to bacteria and malaria parasites. Proc Natl Acad Sci USA 1997; 94(21):11508-11513.
99. Warr E, Das S, Dong Y et al. The Gram-negative bacteria-binding protein gene family: its role in the innate immune system of anopheles gambiae and in anti-Plasmodium defence. Insect Mol Biol 2008; 17(1):39-51.
100. Garver LS, Xi Z, Dimopoulos G. Immunoglobulin superfamily members play an important role in the mosquito immune system. Dev Comp Immunol 2008; 32(5):519-531.
101. Dong Y, Taylor HE, Dimopoulos G. AgDscam, a hypervariable immunoglobulin domain-containing receptor of the Anopheles gambiae innate immune system. PLoS Biol 2006; 4(7):e229.
102. Wang X, Zhao Q, Christensen BM. Identification and characterization of the fibrinogen-like domain of fibrinogen-related proteins in the mosquito, Anopheles gambiae and the fruitfly, Drosophila melanogaster, genomes. BMC Genomics 2005; 6:114.
103. Wang X, Rocheleau TA, Fuchs JF et al. A novel lectin with a fibrinogen-like domain and its potential involvement in the innate immune response of Armigeres subalbatus against bacteria. Insect Mol Biol 2004; 13(3):273-282.
104. Frolet C, Thoma M, Blandin SA et al. Boosting NF-kappaB-dependent basal immunity of Anopheles gambiae aborts development of Plasmodium berghei. Immunity 2006; 25(4):677-685.
105. Zou Z, Shin SW, Alvarez KS et al. Mosquito RUNX4 in the immune regulation of PPO gene expression and its effect on avian malaria parasite infection. Proc Natl Acad Sci USA 2008; 105(47):18454-18459.
106. Bian G, Shin SW, Cheon H-M et al. Transgenic alteration of Toll immune pathway in the female mosquito Aedes aegypti. Proc Natl Acad Sci USA 2005; 102(38):13568-13573.
107. Xi Z, Ramirez JL, Dimopoulos G. The Aedes aegypti toll pathway controls dengue virus infection. PLoS Pathog 2008; 4(7):e1000098.

108. Meister S, Kanzok SM, Zheng X-L et al. Immune signaling pathways regulating bacterial and malaria parasite infection of the mosquito Anopheles gambiae. Proc Natl Acad Sci USA 2005; 102(32):11420-11425.
109. Garver LS, Dong Y, Dimopoulos G. Caspar controls resistance to Plasmodium falciparum in diverse anopheline species. PLoS Pathog 2009; 5(3):e1000335.
110. Gupta L, Molina-Cruz A, Kumar S et al. The STAT pathway mediates late-phase immunity against Plasmodium in the mosquito Anopheles gambiae. Cell Host Microbe 2009; 5(5):498-507.
111. Souza-Neto JA, Sim S, Dimopoulos G. An evolutionary conserved function of the JAK-STAT pathway in anti-dengue defense. Proc Natl Acad Sci USA 2009; 106(42):17841-17846.
112. Lowenberger C, Bulet P, Charlet M et al. Insect immunity: isolation of three novel inducible antibacterial defensins from the vector mosquito, Aedes aegypti. Insect Biochem Mol Biol 1995; 25(7):867-873.
113. Steiner H, Hultmark D, Engström A et al. Sequence and specificity of two antibacterial proteins involved in insect immunity. Nature 1981; 292(5820):246-248.
114. Luna C, Hoa NT, Lin H et al. Expression of immune responsive genes in cell lines from two different Anopheline species. Insect Mol Biol 2006; 15(6):721-729.
115. Shin SW, Kokoza V, Lobkov I et al. Relish-mediated immune deficiency in the transgenic mosquito Aedes aegypti. Proc Natl Acad Sci USA 2003; 100(5):2616-2621.
116. Dimopoulos G, Seeley D, Wolf A et al. Malaria infection of the mosquito Anopheles gambiae activates immune-responsive genes during critical transition stages of the parasite life cycle. EMBO J 1998; 17(21):6115-6123.
117. Erickson SM, Xi Z, Mayhew GF et al. Mosquito infection responses to developing filarial worms. PLoS Negl Trop Dis 2009; 3(10):e529.
118. Kumar BA, Paily KP. Identification of immune-responsive genes in the mosquito Culex quinquefasciatus infected with the filarial parasite Wuchereria bancrofti. Med Vet Entomol 2008; 22(4):394-398.
119. Blandin SA, Moita LF, Köcher T et al. Reverse genetics in the mosquito Anopheles gambiae: targeted disruption of the Defensin gene. EMBO Rep 2002; 3(9):852-856.
120. Bartholomay LC, Fuchs JF, Cheng L-L et al. Reassessing the role of defensin in the innate immune response of the mosquito, Aedes aegypti. Insect Mol Biol 2004; 13(2):125-132.
121. Magalhaes T, Leandro DC, Ayres CFJ. Knock-down of REL2, but not defensin A, augments Aedes aegypti susceptibility to Bacillus subtilis and Escherichia coli. Acta Trop 2010; 113(2):167-173.
122. Kim W, Koo H, Richman AM et al. Ectopic expression of a cecropin transgene in the human malaria vector mosquito Anopheles gambiae (Diptera: Culicidae): effects on susceptibility to Plasmodium. J Med Entomol 2004; 41(3):447-455.
123. Chen CC, Chen CS. Brugia pahangi: effects of melanization on the uptake of nutrients by microfilariae in vitro. Exp Parasitol 1995; 81(1):72-78.
124. Nappi AJ, Christensen BM. Melanogenesis and associated cytotoxic reactions: applications to insect innate immunity. Insect Biochem Mol Biol 2005; 35(5):443-459.
125. Wang X, Rocheleau TA, Fuchs JF et al. Beta 1, 3-glucan recognition protein from the mosquito, Armigeres subalbatus, is involved in the recognition of distinct types of bacteria in innate immune responses. Cell Microbiol 2006; 8(10):1581-1590.
126. Volz J, Osta MA, Kafatos FC et al. The roles of two clip domain serine proteases in innate immune responses of the malaria vector Anopheles gambiae. J Biol Chem 2005; 280(48):40161-40168.
127. Paskewitz SM, Andreev O, Shi L. Gene silencing of serine proteases affects melanization of Sephadex beads in Anopheles gambiae. Insect Biochem Mol Biol 2006; 36(9):701-711.
128. Liu CT, Hou RF, Ashida M et al. Effects of inhibitors of serine protease, phenoloxidase and dopa decarboxylase on the melanization of Dirofilaria immitis microfilariae with Armigeres subalbatus haemolymph in vitro. Parasitology 1997; 115(Pt 1):57-68.
129. Abraham EG, Pinto SB, Ghosh A et al. An immune-responsive serpin, SRPN6, mediates mosquito defense against malaria parasites. Proc Natl Acad Sci USA 2005; 102(45):16327-16332.
130. Michel K, Budd A, Pinto S et al. Anopheles gambiae SRPN2 facilitates midgut invasion by the malaria parasite Plasmodium berghei. EMBO Rep 2005; 6(9):891-897.
131. Cohuet A, Osta MA, Morlais I et al. Anopheles and Plasmodium: from laboratory models to natural systems in the field. EMBO Rep 2006; 7(12):1285-1289.
132. Michel K, Suwanchaichinda C, Morlais I et al. Increased melanizing activity in Anopheles gambiae does not affect development of Plasmodium falciparum. Proc Natl Acad Sci USA 2006; 103(45):16858-16863.
133. Zhao X, Ferdig MT, Li J et al. Biochemical pathway of melanotic encapsulation of Brugia malayi in the mosquito, Armigeres subalbatus. Dev Comp Immunol 1995; 19(3):205-215.
134. Huang C-Y, Christensen BM, Chen C-C. Role of dopachrome conversion enzyme in the melanization of filarial worms in mosquitoes. Insect Mol Biol 2005; 14(6):675-682.
135. Huang C-Y, Chou S-Y, Bartholomay LC et al. The use of gene silencing to study the role of dopa decarboxylase in mosquito melanization reactions. Insect Mol Biol 2005; 14(3):237-244.

136. Paskewitz SM, Andreev O. Silencing the genes for dopa decarboxylase or dopachrome conversion enzyme reduces melanization of foreign targets in Anopheles gambiae. Comp Biochem Physiol B, Biochem Mol Biol 2008; 150(4):403-408.
137. Rivero A. Nitric oxide: an antiparasitic molecule of invertebrates. Trends Parasitol 2006; 22(5):219-225.
138. Luckhart S, Li K. Transcriptional complexity of the Anopheles stephensi nitric oxide synthase gene. Insect Biochem Mol Biol 2001; 31(3):249-256.
139. Luckhart S, Rosenberg R. Gene structure and polymorphism of an invertebrate nitric oxide synthase gene. Gene 1999; 232(1):25-34.
140. Akman-Anderson L, Olivier M, Luckhart S. Induction of nitric oxide synthase and activation of signaling proteins in Anopheles mosquitoes by the malaria pigment, hemozoin. Infect Immun 2007; 75(8):4012-4019.
141. Han YS, Thompson J, Kafatos FC et al. Molecular interactions between Anopheles stephensi midgut cells and Plasmodium berghei: the time bomb theory of ookinete invasion of mosquitoes. EMBO J 2000; 19(22):6030-6040.
142. Peterson TML, Gow AJ, Luckhart S. Nitric oxide metabolites induced in Anopheles stephensi control malaria parasite infection. Free Radic Biol Med 2007; 42(1):132-142.
143. Kumar S, Christophides GK, Cantera R et al. The role of reactive oxygen species on Plasmodium melanotic encapsulation in Anopheles gambiae. Proc Natl Acad Sci USA 2003; 100(24):14139-14144.
144. Lanz-Mendoza H, Hernández-Martínez S, Ku-López M et al. Superoxide anion in Anopheles albimanus hemolymph and midgut is toxic to Plasmodium berghei ookinetes. J Parasitol 2002; 88(4):702-706.
145. Herrera-Ortiz A, Lanz-Mendoza H, Martínez-Barnetche J et al. Plasmodium berghei ookinetes induce nitric oxide production in Anopheles pseudopunctipennis midguts cultured in vitro. Insect Biochem Mol Biol 2004; 34(9):893-901.
146. Kumar S, Gupta L, Han YS et al. Inducible peroxidases mediate nitration of anopheles midgut cells undergoing apoptosis in response to Plasmodium invasion. J Biol Chem 2004; 279(51):53475-53482.
147. Moita LF, Vriend G, Mahairaki V et al. Integrins of Anopheles gambiae and a putative role of a new beta integrin, BINT2, in phagocytosis of E. coli. Insect Biochem Mol Biol 2006; 36(4):282-290.
148. Ellis RE, Jacobson DM, Horvitz HR. Genes required for the engulfment of cell corpses during programmed cell death in Caenorhabditis elegans. Genetics 1991; 129(1):79-94.
149. Hedgecock EM, Sulston JE, Thomson JN. Mutations affecting programmed cell deaths in the nematode Caenorhabditis elegans. Science 1983; 220(4603):1277-1279.
150. Mangahas PM, Zhou Z. Clearance of apoptotic cells in Caenorhabditis elegans. Semin Cell Dev Biol 2005; 16(2):295-306.
151. Jasinskiene N, Coates CJ, Benedict MQ et al. Stable transformation of the yellow fever mosquito, Aedes aegypti, with the Hermes element from the housefly. Proc Natl Acad Sci USA 1998; 95(7):3743-3747.
152. Olson KE, Higgs S, Gaines PJ et al. Genetically engineered resistance to dengue-2 virus transmission in mosquitoes. Science 1996; 272(5263):884-886.
153. Roberts L. Mosquitoes and disease. Science 2002; 298(5591):82-83.
154. Curtis C. Possible use of translocations to fix desirable genes in insect pest populations. Nature 1968; 218(5139):368-369.
155. Terenius O, Marinotti O, Sieglaff D et al. Molecular genetic manipulation of vector mosquitoes. Cell Host Microbe 2008; 4(5):417-423.

CHAPTER 13

CRUSTACEAN IMMUNITY

Lage Cerenius,*[,1] Pikul Jiravanichpaisal,[2] Hai-peng Liu[3] and Irene Söderhäll[1]

[1]*Department of Comparative Physiology, Uppsala University, Uppsala, Sweden;* [2]*Molecular Aquatic Biology and Genetic Laboratory, National Center for Genetic Engineering and Biotechnology, National Science and Tecnology Development Agency, Rajdhevee, Bangkok, Thailand;* [3]*State Key Laboratory of Marine Environmental Science, College of Oceanography and Environmental Science, Xiamen University, Xiamen, Fujian, China*
**Corresponding Author: Lage Cerenius—Email: lage.cerenius@ebc.uu.se*

Abstract: This chapter provides a review of recent progress in the elucidation of innate immune mechanisms in crustaceans. Mainly due to the importance of crustacean aquaculture interest in this field is large and the subject for extensive research efforts. Here, we provide detailed data on the molecular characterisation of lectins, antiviral reactions, hemocyte formation and differentiation and on the regulation of innate immune pathways.

INTRODUCTION

Crustaceans are relatively well investigated with respect to immune reactions when compared to most other invertebrates except fruit flies. This is of course due to their large size and to the intensive fishing and aquaculture of shrimps and some other decapod crustaceans. Although they are (at least until now) less amenable for genetical experiments they are relatively easy to keep in aquaria and to bleed and, therefore, considerable amounts of plasma and hemocytes can be collected for work at molecular or cellular level. Most researchers in the field are using shrimps but many pioneering studies have been carried out on other crustaceans, in particular freshwater crayfish. In this chapter we will cover recent advances in crustacean immunity with emphasis on pattern recognition and lectins, hemocytes and hematopoiesis, prophenoloxidase activating system and on antiviral mechanisms. For earlier work in general we refer to references 1 and 2, and for reviews on crustacean antimicrobial peptides to references 3 and 4 and on antiviral immunity to reference 5.

Invertebrate Immunity, edited by Kenneth Söderhäll.

PATTERN RECOGNITION

Pathogens that manage to break through the outer protective parts of the animal such as the resistant chitin-containing cuticle (or through the midgut which lacks chitin) will encounter an array of hemocyte- and plasma-derived immune factors. The activities of these factors are triggered by molecular signatures typically present on or released by different microorganisms. Lipopolysaccharides, β-1,3-glucans and to some extent peptidoglycans, i.e., polysaccharides from the microbial cell-wall, are known to initiate immune reactions in crustaceans. Double-stranded RNA derived from some viruses is another inducer of such reactions. It is possible that enzymes (i.e., proteinases) produced by microorganisms, or the damage on host tissue caused by such enzymes are efficient triggers of the defence as has been shown in *Drosophila*,[6] but this is less known in crustaceans.

A number of pattern recognition proteins have been isolated from crustaceans and characterised in detail. Most known crustacean pattern recognition proteins were originally purified and cloned in freshwater crayfish but have subsequently been found in shrimps and other crustaceans. These include β-glucan-binding protein[7,8] (BGBP also abbreviated βGBP), lipopolysaccharide- and glucan-binding protein[9] (LGBP) some masquerade-like proteins/serine proteinase homologues[10,11] (SPHs) and a large number of lectins (see below). The LGBPs and the BGBPs will bind β-1,3-glucans and after this binding they will trigger immune reactions such as proPO-activation. LGBP is probably the main vehicle in crustaceans for the recognition of these glucans and thus for mediating defence reactions directed against fungi and oomycetes.[9] Also BGBP, which is not structurally related to LGBP, is capable of binding β-1,3-glucans and mediating immune reactions.[7,8] The affinity of binding to the glucans is lower for crayfish BGBP than for LGBP and the latter is as mentioned therefore likely to be more important for mediating glucan-triggered immune reactions. However, the plasma concentration of BGBP is high and it is therefore possible that this protein is important in removing excess glucans, if present. LGBP is also rendered active by gram-negative bacteria since it is capable of binding LPS and thereafter mediate the activation of the proPO-system. Insect homologues of LGBP (variously called β-1,3-glucan binding proteins or gram-negative bacteria binding proteins, i.e., GNBPs) have been shown to trigger the Toll pathway, melanisation and other immune reactions in the presence of bacteria or fungi.[12] This family of proteins appears to have gone through a large expansion in the crustacean *Daphnia pulex* since 11 genes coding for LGBP-like proteins were detected in its genome.[13]

No obvious peptidoglycan recognition protein (PGRP) candidates were found in the recently released *Daphnia* genome, nor has yet any PGRP gene been cloned from any crustacean. PGRPs are present in many vertebrate and invertebrate species so their apparent absence in crustaceans is a surprise. Still, peptidoglycans have been reported to stimulate immune responses from crustaceans[14] although the molecular mechanisms behind this need to be clarified.

SPHs are known as activators/regulators of proPO in some insects[15] but in crustaceans, they have mainly been implicated in pattern recognition and as opsonins so far. The first crustacean SPH characterised, the masquerade-like protein[10] is binding to gram-negative bacteria.[11] This binding is triggering a proteolytic processing of the protein that produces four different subunits.[11] Crayfish masquerade-like protein was demonstrated experimentally to be an efficient opsonin and important in the clearance of *E. coli*. A black tiger SPH that could bind lipopolysaccharides and intact *V. harveyi* and acting as opsonin has been described recently.[16]

LECTINS

Lectins are proteins or glycoproteins normally without catalytic activity that can recognise and noncovalently bind to specific sugar moieties and thereby agglutinate cells by binding to cell surface glycoproteins and glycoconjugates.[17] Lectins, therefore, are considered important pattern recognition proteins in innate immunity and play significant roles in nonself-recognition and clearance of invading microorganisms, either as cell surface receptors or as soluble proteins existing in circulating fluids.[18,19] C-type lectins are the most diverse and well studied among the lectin families. The term C-type lectin was originally used to distinguish a group of Ca^{2+}-dependent (C-type) carbohydrate-binding proteins from the other types of lectins.[20] The structures of C-type lectins were defined and found to contain a conserved single module of approximately 150 amino acid residues (carbohydrate recognition domain, CRD).[20-22] This domain contains a characteristic double-loop stabilised by two highly conserved disulphide bridges and four Ca^{2+}-binding sites where the Ca^{2+} binding site 2 is involved in carbohydrate binding.[23] The CRD´s usually have a key motif, either QPD (Gln-Pro-Asp) or EPN (Glu-Pro-Asn), which has been predicted to be ligand-binding specific for galactose or mannose, respectively.[23] Recently many C-type lectins containing nonstandard CRDs, which do not bind Ca^{2+} have been identified. These are considered to interact with noncarbohydrate ligands[24] and for these CRDs the term C-type lectin-like domain (CTLD) was introduced.[25]

Although C-type lectins have been well studied in vertebrates for many years, they have not been well characterised in invertebrates. Recently, genes containing CTLDs have been found to be abundant in the *Daphnia pulex* (6 genes) genome, *Drosophila melanogaster* genome (34 genes) and in *Caenorhabditis elegans* genome (278 genes), respectively.[13,26] This suggests that there is a high potential for generating many C-type lectins, perhaps with different ligand specificities. C-type lectins are the largest group of immune-function ESTs found in the hepatopancreas of the shrimps *Litopenaeus vannamei* and *L. setferus*.[27] It has become clear that vertebrate C-type lectins have a broad range of biological functions including cell adhesion, endocytosis, pathogen neutralisation, glycoprotein clearance, phagocytosis.[17,28,29] In invetebrates, lectins have been reported to contribute in innate immune responses, including prophenoloxidase activation,[30,31] enhancement of encapsulation,[19,32,33] nodule formation of hemocytes,[34] opsonin formation,[35] antibacterial activity,[36] antifungal activity[37] and maybe contribute to injury healing.[38]

A large number of natural lectins have been purified and characterised by biochemical methods from hemolymph of crustaceans (for reviews see refs. 39,40). Compared to vertebrate lectins, the molecular features and functions of lectins in crustaceans are just at the beginning of becoming understood. Here focus will be placed on those lectins which have been sequenced and whose functional properties have been determined using e.g., recombinant proteins (listed in Table 1).

Structure of the Shrimp C-Type Lectins

All lectins listed in the Table 1 contain a CTLD in the putative protein indicating that these lectins fall into the C-type lectin family. The four or six cysteine residues important in the formation of the CRD disulphide bonds are conserved.. PmAV, PmLec and Fc-Lec4 contain a single CRD with a QPD motif that has a predicted ligand-binding specificity for galactose, while Fc-hsL, LvLec and LvCTL1 contain a single CRD with an EPN motif with predicted ligand-binding affinity toward mannose. C-type lectin-1

Table 1. Shrimp lectins characterised by molecular methods

		Amino Acids	CRD	Sugar-Binding Motif	Biological Activity	GenBank Accession Number	Challenged	Tissue Specific	Reference
C-type lectin-1	*P. stylirostris*	168	1	EPK	-	-	WSSV	Hepatopancreas	141
PmAV	*P. monodon*	170	1	QPD	Antiviral activity	AY302750	WSSV	-	50
PmLec	*P. monodon*	182	1	QPD	Binding to LPS Agglutinating, opsonic effect	DQ078266	-	-	44
LvLT	*L. vannamei*	345	2	QPD and EPD		DQ871245	WSSV	Hepatopancreas	142
Fclectin	*F. chinensis*	287	2	2 QPD		AY871270	Bacteria/ WSSV	Hemocyte	43
PmLT	*P. monodon*	333	2	QPD and EPN	WSSV binding, Enhance encapsulation	DQ871244	Bacteria/ WSSV	Hepatopancreas	46
Fc-hsL	*F. chinensis*	159	1	EPN	Agglutinating, binding, antimicrobial activity	DQ167572	Bacteria/ WSSV	Hepatopancreas	47
Fc-Lec2	*F. chinensis*	333	2	QPD and EPN	Agglutinating	EU834289	Bacteria/ WSSV	Hepatopancreas	41
FcLec3	*F. chinensis*	158	1	EPS	Agglutinating		Bacteria/ WSSV	Hepatopancreas	49
FcLec4	*F. chinensis*	237	1	QPD	Agglutinating, binding activity to bacteria, clearance of bacteria	EU834293	Bacteria	Hepatopancreas, gills and stomach	42

continued on next page

Table 1. Continued

		Amino Acids	CRD	Sugar-Binding Motif	Biological Activity	GenBank Accession Number	Challenged	Tissue Specific	Reference
LvLec	*L. vannamei*	157	1	EPN	Agglutinating, Mannose binding	EF583939	-	Brain, hemocytes and hepatopancreas	45
LvCTL1	*L. vannamei*	156	1	EPN	Hemagglutinating, sugar binding, Binding to WSSV envelope proteins	DQ858900	WSSV	Hepatopancreas	48
PtLP	*P. trituberculatus*	164	1	-	-	ACC86854	-	-	143

and Fc-Lec3 also contain a single CRD, but with EPK and EPS, respectively instead of the usual EPN motif. PmLT, Fc-Lec2 and LvLT consist of two CRDs, the N-terminal CRD1 contains a QPD motif and the C-terminal contains an EPN motif, but LvLT has here an EPD instead. Those mutated motifs and their binding specificities need further studies. Fclectin contains two CRDs, each with a QPD motif. The number of CRDs identified in each species is variable and it seems likely that lectins with multiple CRDs have stronger affinity for binding to their ligands.[23]

According to the phylogenetic tree made by Zhang et al[41] lectins from shrimp, scallop and the arctic lamprey *Lethenteron japonicum* form one large cluster, those from mammals and fishes form the second cluster and those from insect belongs to the third cluster. Fc-Lec2, PmLT and LvLT are belonging to same subcluster, while Fc-hsL and LvLec are grouped in another subcluster. This could indicate that shrimp lectins may have a closer relationship with mammalian lectins than insect lectins. Thereafter was FcLec4, however, found to be closely related to insect lectins.[42]

Tissue-Specific Expression of Shrimp Lectins

Hepatopancreas and hemocytes of crustacean are regarded as the most important tissues involved in crustacean immunity.[1,27] Most shrimp lectins characterised to date have been isolated from hepatopancreas or hemocyte cDNA libraries. The Fclectin transcript was detected exclusively in hemocytes,[43] whereas the transcripts of LvLT, PmLT, Fc-hsL, Fc-lec2, Fclec3 and LvCTL1 were found specifically expressed in hepatopancreas. The detailed localisation of PmLT and Fc-Lec2 was demonstrated by immunohistochemistry to be in the F (fibrillar) cells of the hepatopancreas. In contrast, the FcLec4 transcripts were distributed in diverse tissues, mainly in the hepatopancreas, gill, stomach and a lower level could also be detected in intestine.[42] LvLec is unusual in having its highest expression in the brain; an interesting finding that needs further functional studies.

Functional Studies

By the use of recombinant proteins functional studies of several crustacean lectins have been carried out. The lipopolysaccharide-binding lectin PmLec can function as an opsonin that enhances hemocytic phagocytosis.[44] Recombinant protein of Fc-hsL has no hemagglutinating activity, but a Ca^{2+} dependent agglutinating activity against several Gram-positive and Gram-negative bacteria. Similarly, Fc-Lec2 and its two individual CRDs did not have hemagglutination activity, but had agglutinating activity and binding activity to some bacteria in a Ca^{2+}-dependent and Ca^{2+}-independent manner, respectively.[41] Their studies also suggest that two CRDs have synergistic effect. Recombinant LvLec has agglutinating activity to *E. coli* JM 109 depending on Ca^{2+} and the agglutination could be inhibited by mannose and EDTA.[45] Another role in immunity encapsulation, was demonstrated with PmLT by using agarose beads coated with the lectin.[46] Unlike other shrimp lectins, Fc-hsL has antimicrobial activity against several bacteria and fungi.[47]

A possible effect of LvCTL1 on virus defence is indicated by the binding of the lectin to WSSV virions and the interaction in a pull-down assay with several envelope proteins of WSSV including VP95, 28 26, 24, 19 and 14.[48] Also FcLec3 was shown to interact with a major envelope protein of WSSV, VP28.[49] PmAV, a C-type lectin identified from WSSV-resistant shrimp, has a strong antiviral activity in inhibiting virus-induced cytopathic effect in fish cell cultures. Surprisingly, neither recombinant nor native PmAV has agglutination activity[50] so the mechanism for its antiviral activity needs to be determined.

Expression Profile of Lectins after Challenge with Bacteria or WSSV

Almost all C-type lectins are synthesised in the hepatopancreas or hemocytes, tissues that are important tissues in immunity.[1,27] Thus, expression of mRNA and protein might be affected by bacterial and viral infection. *PmLec* was isolated from hepatopancreas libraries[44] and it was found to be highly expressed in the midgut of *P. monodon* challenged by an immersion with *V. harveyi*.[51] *Fc-hsL, Fc-Lec2* and *FcLec3* were constitutively expressed in the hepatopancreas of normal shrimp and were highly up-regulated following challenge with either bacteria or WSSV.[42,45,47] The expression pattern of *PmLT* mRNA, specifically expressed in hepatopancreas, was decreased initially and then gradually increased after treatment with WSSV extract either in vivo or in vitro using a hepatopancreas tissue fragment.[46] Zhao et al[48] also reported that *LvCTL1* specifically expressed in the hepatopancreas, was induced in the shrimp hemolymph after WSSV infection. Moreover, the binding of rLvCTL1 to WSSV could protect shrimp from viral infection and prolonged the survival of shrimp against WSSV infection.

FcLec4, which is distinct from other shrimp C-type lectins, is expressed in hepatopancreas, gills and stomach and intestine. A significant up-regulation of FcLec4 transcripts in gills and stomach and higher level of protein in gill stomach and hemolymph was observed after challenge with *V. anguillarum*.[49] Fclectin expression in hemocytes increased on exposure with inactive mixed bacteria of *V. anguillarum* and *S. aureus* as well as with WSSV. Similar results were observed with in vitro experiments, which showed that Fclectin expression was gradually increased in cultured hemocytes stimulated by LPS.[43]

Apart from shrimp, a C-type lectin (PtLP) was also isolated from the swimming crab *Portunus trituberculatus,* but with unknown function. PtLP is phylogenetically related to PmAV, but no perfect QPD motif was found and its mRNA levels were very high in hepatopancreas but lower in gills, hemocytes and ovary of unchallenged animals.

In addition to C-type lectins two isoforms of Tachylecin5-like genes (*PmTL5*) have been found in *P. monodon.* It is interesting that the first PmTL5 isoform was mainly expressed in the hindgut and was induced during immersion with *V. harveyi*, while the second was expressed at a very high level in all parts of shrimp intestine and hemocytes.[51]

There is a rising list of putative C-type lectin genes which have been successfully cloned and characterised in different shrimps. Many lectins are up-regulated during infection and since there are some data on lectins showing that they are promoting bacterial agglutination, phagocytosis, encapsulation and other immune reactions, a role for these proteins in defence seems likely. In future, more efforts need to be concentrated on biochemical characters, regulatory mechanisms, evolution and precise function of shrimp C-type lectins as well as searching for other types of lectins in addition to C-type lectins.

HEMOCYTES AND HEMATOPOIESIS

Phagocytosis is likely to be of great importance in crustacean immunity and there are numerous studies demonstrating efficient uptake of bacteria and other particles by the circulating hemocytes or fixed phagocytic cells in these animals. Compared to vertebrates little is known about which proteins are regulating and accomplishing this uptake. A few crustacean proteins with opsonic properties have been characterised, though. One example is the masquerade-like protein mentioned earlier. Another very important protein is peroxinectin originally purified[52] and cloned from freshwater crayfish[53] and since then found in shrimps, other crustaceans and many other organisms. Hemocytes, in the presence of e.g., β-1,3-glucans or other triggers of immune reactions, will release peroxinectin whereupon the protein, by limited proteolysis, will gain a strong cell adhesion activity. Peroxinectin acts as an opsonin during phagocytosis and in promoting cellular encapsulation of foreign objects. The *D. pulex* genome contains six putative scavenger receptors[13] hinting at the possible existence of a cellular uptake mechanism via such receptors in crustaceans.

In crustaceans, the circulating hemocytes are crucial in protecting the animal against invading microorganisms by participating in recognition, phagocytosis, melanisation and cytotoxicity.[1] In most crustacean's three morphologically different classes of hemocytes, hyaline cells (HC), semigranular cells (SGCs) and granular cells (GCs) are observed within the hemolymph and all of them are important in immobilising or destroying invasive pathogens.[54,55] The hemocyte separating technique developed by Söderhäll and Smith[56] made it possible to study the function of individual hemocyte types. In freshwater crayfish and shore crab HCs were then shown to be phagocytic, while SGCs act in early detection of pathogens.[55] The GCs contain within their granules several immune factors such as the proPO-activating system, the cell adhesion protein peroxinectin and crustin antimicrobial peptides.[57] Exocytosis is induced in both SGCs and GCs as a response to microbial polysaccharides, resulting in the release of these immune proteins. Similar hemocyte types are identified in other arthropods although with slight variations.[58]

The continuous formation of new hemocytes (hematopoiesis) is essential for survival of the animals and this process is tightly regulated by factors released from circulating hemocytes. Arthropod hemocyte development has mainly been studied in the fruit fly, *D. melanogaster*[59] and in the freshwater crayfish *Pacifastacus leniusculus*.[60-65] In *D. melanogaster* mature hemocytes are formed in the early embryonic head mesoderm and at larval stage in a specialised organ called the lymph gland, while no new hemocytes are produced in the adult flies. In contrast the crayfish hematopoietic development is an ongoing process throughout the animals whole life.

As early as 1891 the observation of a structure in the Atlantic ditch shrimp, *Palaemonetes varians,* named "the dorsal blood sinus" was reported by Weldon.[66] Later Allen[67] suggests in a more detailed study of this organ, "the dorsal sac", to be blood cell producing. Since then, several studies of crustacean hematopoietic tissues (HPTs) have been presented and are reviewed by Johnson.[68] In decapod crustaceans the HPT is a defined organ made up of lobules enveloped by a thin casing of connective tissue, that in noncaridean pleocyemata covers the dorsal and lateral walls of the foregut (Fig. 1A). The HPT consists of lobules of highly active proliferating cells (shown in Figure 1B-C by 5′-bromo-2′-deoxyuridine (BrdU) incorporation and mitotic figures in hematoxylin stained tissue). In dendrobranchiata, such as the penaeid shrimps, hemocytes are produced in paired nodules on the dorsolateral surface of the foregut and in some species supplementary HPTs are localised at the base of the maxillipeds, or adjacent to the antennal artery.[69-71] The lymphoid organ (LO) of penaeid shrimps consists of lobes of folded tubules located ventro-anterior to the hepatopancreas. This organ has been mistaken for a hematopoietic tissue, but according to detailed studies by Van de Braak et al[72] it is clear that the LO mainly has a role in bacterial clearance and is homologous in function to the phagocytic organ described by Cuenot[73] and fixed phagocytes associated with the hepatic artery in decapod crustaceans.[68]

The organisation of crustacean HPTs have been described in detail by electron microscopy studies in the crab *Carcinus maenas,*[74,75] the shrimp *Sicyonia ingentis,*[71,76] the lobster *Homarus americanus*[77] and in *P. leniusculus*.[60] These studies are in agreement with each other and we will in the further text adopt the nomenclature from Chaga et al[60] in classifying the HPT cells into five distinct categories based on morphological criteria. Type 1 cells have a large nuclei surrounded by small amount of cytoplasm which usually is characteristic of a stem cell while Type 2 also has large nuclei but larger cytoplasm containing cytoplasmic granules. Type 1 and 2 are the main proliferating cells in the HPT, whereas the other cell types in HPT can be categorised into precursors of granular hemocytes as Type 3 to 4, or as precursor of semigranular hemocytes as Type 5 (Fig. 1D).[64]

The formation and development of mature hemocytes involve proliferation, commitment and differentiation from undifferentiated HPT cells. Several transcription factors have been characterised as lineage specific markers in *Drosophila* and are conserved across taxonomic groups from flies to mammals.[78] Also in *P. leniusculus* the importance of a GATA transcription factor as well as a Runx protein homologue, during hematopoiesis has been revealed.[61] Apart from systematic detailed studies of hematopoietic transcription factors and signalling pathways associated with *Drosophila* hematopoiesis, little is known about the events regulating this process during development or an infection in insects.

In crustaceans, generally, hemocytes do not divide in the circulatory system and thus, new hemocytes need to be continuously and proportionally produced. Already experiments performed in the late 1800s revealed an increase in mitotic index in the HPT following experimental bleeding,[68] and since then several studies have confirmed that cell proliferation in the HPT can be influenced by moulting,[68] and different stress factors such as for example LPS injection,[72] and Mn-exposure.[79] In addition, the number of blood cells can be experimentally decreased by injection of microbial polysaccharides and then rapid recovery is stimulated, mainly due to production and release of new cells[61] form the HPT.

That new hemocytes are synthesised and partly differentiated in the HPT, but the final differentiation from stem cells not expressing proPO, into functional hemocytes expressing proPO is not completed until the hemocytes are released into the circulation are supported in several reports.[61,68] A method was developed to isolate the HPT cells from *P. leniusculus* in order to study their proliferation and differentiation in vitro.[62] The key to the successful stem cell culture was the isolation and characterisation of a new group of cytokines named astakines present in crayfish plasma. Two different astakines have been detected in crustaceans (Table 2) and astakine 1 is a small protein of 9 kDa containing a so-called prokineticin-domain present in several vertebrates. In *Penaeus monodon* astakine 2 contains a 13 amino acid insert as compared to astakine 1,[62,80] and this protein is also found in *P. leniusculus*. The importance of astakine 1 and astakine 2 in HPT cell proliferation, differentiation and release from the HPT has been shown by injection of recombinant protein as well as by in vivo and in vitro gene silencing by RNAi.[62] These experiments have revealed that astakine 1 supports differentiation of HPT cells into the SG cell lineage, since the addition of astakine 1 induces expression

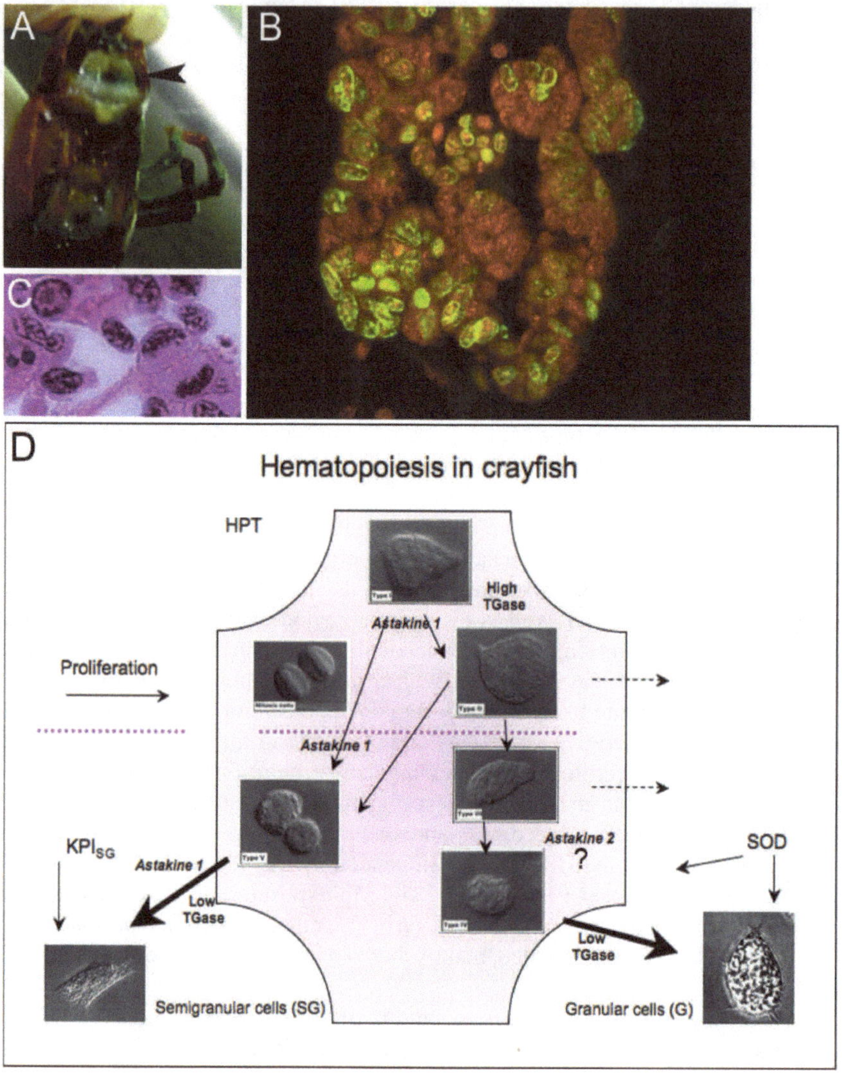

Figure 1. A) The hematopoietic tissue (HPT) of crayfish covers the dorsal and lateral walls of the foregut. B) Proliferating cells in HPT labelled by 5′-bromo-2′-deoxyuridine (BrdU) incorporation. C) Mitosis in HPT cells stained by hematoxylin. D) A hypothetical model for hemocyte development and release from the HPT in *P. leniusculus*. Based on data from references 9-12, 29. The cells in HPT follow two main cell lineages: one branch is from Type 1 via Type 5 cells to semigranular cells (SGC); the other is from Type 1 to Types 2, 3, 4 cells to granular cells (GC). Astakine 1 is involved in proliferation, differentiation through the SGC pathway and release of hemocytes into the circulation, while the definite role of astakine 2 still remains to be shown.

of the SGC specific Kazal type proteinase inhibitor.[62,80,81] Astakine 2 has been shown to increase proliferation in HPT cells in penaeid shrimp, whereas its role in crayfish is not fully understood although it has some stimulatory effect on GC differentiation.[81]

Table 2. Crustacean astakines

	Species	Accession Number
Astakine 1	*Pacifastacus leniusculus*	AY787656
Astakine 2	*Pacifastacus leniusculus*	EF568370-EF568371
	Penaeus monodon	AY787657
	Litopenaeus vannamei	FE148214
	Homarus americanus	FE535609
	Carcinus maenas	DW585080
	Daphnia pulex	FE329237

Transglutaminases (TGases) form a family of Ca^{2+}-dependent enzymes catalysing posttranslational remodelling of proteins by cross-linking and this enzyme acts as a clotting enzyme in crustacean hemolymph coagulation.[1] TGase is one of the most abundant proteins in crustacean HPTs[63,82] where its mRNA expression as well as enzyme activity is very high. TGase has been shown to play an important role in keeping the HPT cells in an undifferentiated stage inside the hematopoietic tissue and if expression of TGase mRNA is blocked, the cells start to differentiate and migrate out into the circulating hemolymph.[63] Interestingly astakine 1 seems to play a role in this process, since astakine by some unknown mechanism decrease extracellular TGase activity and induce cell migration.

ANTIVIRAL REACTIONS

Viruses remain a major obstacle to crustacean aquaculture. Among the viruses, the most intensively studied have been characterised from cultured penaeids such as the white spot syndrome virus (WSSV), yellow head virus (YHV) and Taura syndrome virus (TSV). We recently published[5] a review on antiviral reactions in crustaceans covering the literature up to 2008 so here mainly some recent developments are discussed. The molecular mechanisms that underlie the majority of crustacean antiviral immune responses are still unknown and are only starting to be addressed. Recently, high throughput identification of genes and proteins (e.g., EST, SSH, microarrays) have been taken into use in an attempt to solve this.[83]

Antiviral substances have been isolated from several crustaceans although the mechanism of this inhibitory activity remains unclear.[84] A well known cationic protein, antilipopolysaccharide factor (ALF) originally isolated from horseshoe crab[85] has been studied in crustaceans for its antibacterial activity.[86,87] Crayfish ALF was up-regulated by a WSSV challenge and was shown to be involved in antiviral response against WSSV. Silencing of ALF resulted in higher rates of WSSV propagation both in the animals and in an HPT cell culture.[88] In contrast, enhanced expression of ALF in the crayfish by the administration of UV-treated WSSV led to lower viral replication and a partial protection against a subsequent challenge with the active virus.[88] Silencing of *Lv*ALF1 resulted in a significant increase of mortality in *L. vannamei* challenged by *Vibrio penaeicida* and *Fusarium oxysporum* but no protection against WSSV.[86] However, a study on a *P. monodon* recombinant ALF*Pm*3, showed that this protein affected viral infection both in shrimp and in crayfish HPT cell cultures.[89] The mechanism for antiviral activity of crayfish ALF or shrimp ALF*Pm3* is still unknown.

Cytokine activation through JAK/STAT pathway of a number of genes has been suggested in counteracting viral infection in *Drosophila*.[12] Flies deficient in the JAK kinase Hopscotch show increased susceptibility to *Drosophila* C virus and contain a higher viral load. These data indicate that flies produce antiviral molecules in a JAK-STAT-dependent way.[90] However, the WSSV immediate early gene (*ie1*) was shown to employ a shrimp STAT as a transcription factor to enhance its expression.[91] Additional studies showed that shrimp STAT was activated in response to WSSV infection and the WSSV does not disrupt JAK-STAT pathway but benefits from STAT activation in the shrimp.[92,93] Also, some components of the Toll pathway (Toll and Dif) have been shown to be of importance for the resistance against *Drosophila* X virus.[94] However, in shrimp a recent study found that a Toll like receptor (lToll) was not involved in antiviral immunity.[95] Further work is needed to reveal if other Toll-like receptors are necessary for antiviral responses in crustaceans.

Apoptosis

Apoptosis is a critical cellular process for removing unnecessary or potentially harmful cells and possibly for limiting viral spread.[96] Caspases are central effectors in apoptosis and if the *M. japonicus Pj*caspase gene was silenced, the WSSV-induced apoptosis was significantly inhibited and the number of viral copies increased, indicating that apoptosis may play an antiviral role.[97] This proposal however needs to be ascertained by more experiments since knocking down caspase-3 reduces mortality in Pacific white shrimp challenged with a low dose of WSSV but not with a high-dose of WSSV. This suggests that apoptosis in some cases may increase rather than decrease mortality in WSSV-challenged shrimp.[98]

Similarly, the wide spread apoptosis in *P. monodon* infected with YHV is a major cause of dysfunction and death of the host. The expression of ribophorin I, a protein involved in apoptosis, was up-regulated and remained high until the moribund stage in YHV infected shrimp[99] whereas the defender against apoptotic death 1, a negative regulator of apoptosis, decreased dramatically after YHV challenge.[100] It ought to be stressed though that apoptosis can be triggered by a multitude of signals and much more need to be known about apoptosis in crustaceans before the mentioned findings can be properly evaluated.

Antiviral Activity Induced by RNA Interference or Injection of dsRNA

Injection of dsRNA/siRNA specific to viral genes can block viral disease progression. For instance, viral replication was efficiently suppressed with injection of WSSV-specific dsRNA/siRNA targeting VP19, VP28, VP281, or WSSV protein kinase in penaeid shrimp.[5,101]

Recent studies reveal the existence of both innate (nonsequence specific) and RNAi related (sequence specific) crustacean antiviral phenomena.[102] However, the protection induced by the innate pathway could be overwhelmed by a higher dose (8-fold) of infectious virus, suggesting it mediates a low degree of resistance. Two components of RNA silencing, Argonaute and Dicer, have been characterised from *P. monodon*[103,104] but the mechanism for this silencing is still not clear. Injection of

CpG oligodeoxynucleotides (CpG ODNs; these are actually typical for bacterial DNA) mediates a protection against WSSV propagation, possibly via Argonaute and Dicer.[105] Recently a protein homologous to HIV transactivating reponse RNA-binding protein was found to bind Dicer and to inhibit WSSV replication in *F. chinensis,*[106] These studies hint that the RNAi machinery may play an important role for antiviral activity in crustaceans, although much efforts remain to fully establish the mechanism(s) for this activity.

Studies on the mechanism of these antiviral responses have been hampered by absence of genome, tools for genetic manipulation and mutants and stable long-term cell lines for in vitro studies. We have succeeded in developing an HPT cell cultures from crayfish, which can be a useful tool for gene functional studies in crustaceans.[107] These HPT cultures can also be used to replicate WSSV and to study host-virus interactions.[108]

CLOTTING, SYNTHESIS OF ANTIMICROBIAL PROTEINS AND MELANISATION

Clotting

Clotting is an important reaction aimed at preventing hemolymph loss and microbial spread at sites of injury. The reaction has been extensively studied in crustaceans, in particular freshwater crayfish,[109] from which the first crustacean clotting protein was cloned.[110] For a review on the subject and a comparison with the corresponding reaction in other arthropods see ref 111. In the shrimp *Marsupenaus japonicus* RNAi silencing of clotting protein and a transglutaminase resulted in a defect clotting system.[112] Interestingly, challenging such animals with *V. penaceida* or WSSV resulted in higher mortalities. Although the reason for this effect is unknown it could mean that initiation of clotting also triggers the onset of other immune reactions and/or that the clotting reaction itself interferes with the propagation of these pathogens by e.g., entangling them.

Antimicrobial Proteins

Antimicrobial proteins are very important components of the immune system in many insects. They have received less attention in crustaceans, perhaps because their expression usually is not up-regulated as dramatically by the presence of microbial products as in holometabolous insects. In recent years a number of crustacean antimicrobial proteins (AMPs) have been purified and/or cloned. In some cases their effects on microbial growth in vitro have been investigated, but to what extent these activities reflect their importance in vivo is more difficult to assess. There are many kinds of crustacean AMPs that differ considerable in structure; two prominent groups with many members are the penaeidins and the crustins, thoroughly reviewed in Cuthbertson et al[3] and in Smith et al[4] respectively. Crustins have a wider occurrence among crustacean taxa than the penaedins. However, a recently characterised spider crab AMP, hyastatin may judged from its Cys bond pattern be related to the penaeidins.[113] The presence of a whey acidic protein (WAP) domain is a characteristic of crustins although WAP domains are present in many other types of proteins, e.g., proteinase inhibitors.[114] A third group of AMPs are the antilipolysaccharide factors that also have been implicated in antiviral reactions (see above). There are other potential sources of antimicrobial peptides and proteins. For example, some peptides released by limited proteolysis from hemocyanin exhibit antimicrobial properties.[115,116]

In many insects AMP expression is governed by either the Toll or the imd pathways. As mentioned earlier, crustacean AMP expression tends to be more or less constitutive although several cases of inducible AMPs have been recorded.[117] Some possible Toll pathway components were recently described in several shrimp species, e.g., Toll itself[95,117-120] spätzle,[121] relish,[121,123] dorsal.[124] *L. vannamei* Relish and dorsal were shown to regulate the expression of penaeidin-4 in transfected insect cells. To what extent this reflects in vivo AMP expression in the shrimp remains to establish. In one study, the recombinant protein from the spätzle-like gene in *F. chinensis* was injected into the crayfish *Procambarus clarkii* (shrimps died if injected with this product), which resulted in an increase of transcript levels for crustin-2 but not for the other tested AMP genes.[121] Also this finding needs confirmation by additional experiments. If transcription of one putative Toll-receptor was reduced by RNAi treatment in *L. vannamei* the animals became more susceptible to the bacterium *Vibrio harveyi* whereas the susceptibility to WSSV was unaffected.[117] Whether this is due to any possible Toll effects on AMP production or other Toll-mediated effects is unknown and, furthermore, the number of putative Toll receptors in shrimp is not known. However, it should be noticed that at least crustin expression was unaffected by the presence of LvToll dsRNA in this case. In another study specific RNA interference of *L. vannamei* crustin caused in increased susceptibility to a related bacterium, *V. penaeicida*.[125] Also a putative imd homologue has been reported from *L. vannamei*[126] that requires further functional studies. Silencing of a Relish homologue in the shrimp *F. chinensis* resulted in a lowered penaedin 5 transcription upon bacterial challenges, a result which could be interpreted as an imd pathway exists in shrimps. However, the extent (if any) to which AMP synthesis is regulated by Toll and imd homologues in crustaceans is still far from settled.

The Prophenoloxidase Activating System

The melanisation reaction is an important immune reaction and numerous studies in different types of animals have attested the crucial role of the phenoloxidase system in combating microbial infections. A large body of pioneering work (for reviews see refs. 127,128) on the proPO was carried out using freshwater crayfish as model. Recently several RNA interference studies aiming at the transcription of the proPO gene or genes whose products are involved in proPO activation have been carried out. In *P. leniusculus* reduced levels of proPO led to an increased susceptibility to the serious bacterial pathogen *Aeromonas hydrophila*.[129] Reducing the levels of pacifastin, a specific inhibitor of the proPO-activating proteinase, resulted into a higher melanisation capacity and increased survival to the pathogen. Also in vitro the products from an active crayfish PO are reducing the growth of several bacteria, both gram positive and gram negative species.[130] Two studies[131,132] using RNAi depletion of proPO transcripts in *Penaues monodon* have led to similar conclusions. Reduced transcription of the proPO genes or a gene for proPO-activating proteinase resulted into higher mortality upon challenge with *Vibrio harveyi*. An extensive study carried out with *M. japonicus* showed that after depletion of proPO, bacterial counts in hemolymph and other tissues increased.[133] Since these animals were not challenged either endogenous bacteria, or bacteria taken up from the rearing tank must be responsible for the increasing bacterial loads. In contrast to animals injected with control dsRNA the proPO-depleted animals exhibited increased mortality and reduced hemocyte numbers. The increased mortality could to some degree, although not completely, be counteracted by the administration of antibiotics. Interestingly, an array with more than 2000 shrimp genes showed that 28 genes were up-regulated and 49 down-regulated on the third day after starting the proPO gene interference. Thus, a consequence of the lowering of the PO levels could be that other immune factors are produced at a reduced rate, decreasing the immune capacity of the animal. Among the genes down-regulated after proPO depletion were some AMPs (penaedin and crustin) and two Kazal type inhibitors. Kazal type proteinase inhibitors have been shown to interfere with the growth of several bacteria. It is still not known to what extent the bacteriostatic activity is due to inhibition of microbial proteinases or to other effects (for a review see ref. 134). A single animal can produce a large number of different Kazal variants, a fact suggesting but not proving that these inhibitors is under selection pressure from various microbes.[135] Also other proteinase inhibitors, e.g., alpha-2-macroglobulin have been shown to be produced in a large number of sequence variants.[136] As is the case in the crustacean Kazal inhibitors the sequence variation is especially evident among those amino acid residues that takes part in the interaction with proteinases.

Once active PO has been produced, regulatory mechanisms to ensure that melanisation does not proceed uncontrolled for unlimited periods are likely to exist. Several such nonproteinaeous compounds and phenoloxidase inhibiting peptides are known from insects. One such control of excessive melanisation is likely to be carried out by the recently identified melanisation inhibition proteins (MIPs) from crayfish[137] and meal worms.[138] MIPs have been shown to prevent both proteolytic activation of proPO as well as to interfere with the melanin synthesis from quinones. The prevention of the melanin formation occurs at a late step(s) after the steps catalysed enzymatically by PO. Crayfish MIP is from an evolutionary viewpoint an interesting protein, since it is bearing significant sequence similarity with vertebrate ficolins, including a binding site for Ca^{2+}. In vitro mutagenesis of this site has demonstrated its importance for full inhibitory activity.[137]

CONCLUSION

Emerging evidence from other systems such as insects indicate that there is substantial cross-talk between different arms of the innate immune defence. For example, in several insect species some of the proteinases and regulatory serpins that are part of the Toll activation cascade and the proPO-activating system are shared.[139,140] This makes sense since it is likely that achieving an efficient response towards many pathogens will require that several parts of the immune system cooperate. It will be interesting to see whether examples of such cross-talks within crustacean innate immunity will be discovered. To conclude, large progress in elucidating innate immune pathways have been made in recent years. In the near future genomic information will be available for additional crustacean species and the speed with which progress in this area is made will increase further. However, we are still far from efficient therapies for the diseases that are plaguing crustacean aquaculture.

REFERENCES

1. Jiravanichpaisal P, Lee BL, Söderhäll K. Cell-mediated immunity in arthropods: Hematopoiesis, coagulation and opsonization. Immunobiol 2006; 211:213-236.
2. Lee SY, Söderhäll K. Early events in crustacean innate immunity. Fish Shellfish Immunol 2002; 12:421-437.
3. Smith VJ, Fernandes JMO, Kemp GD et al. Crustins: Enigmatic WAP domain-containing antibacterial proteins from crustaceans. Dev Comp Immunol 2008; 32:758-772.
4. Cuthbertson BJ, Deterding LJ, Williams JG et al. Diversity in penaeidin antimicrobial peptide form and function. Dev Comp Immunol 2008; 32:167-181.
5. Liu HP, Söderhäll K, Jiravanichpaisal P. Antiviral immunity in crustaceans. Fish Shellfish Immunol 2009; 27:79-88.
6. Chamy LE, Leclerc V, Caldelari I et al. Sensing of danger signals and pathogen-associated molecular patterns defines binary signaling pathways upstream of Toll. Nature Immunol 2008; 9:1165-1170.
7. Duvic B, Söderhäll K. Purification and characterization of a beta-1,3-Glucan binding-protein from plasma of the crayfish Pacifastacus leniusculus. J Biol Chem 1990; 265:9327-9332.
8. Cerenius L, Liang ZC, Duvic B et al. Structure and biological activity of a 1,3-beta-D-glucan-binding protein in crustacean blood. J Biol Chem 2004; 269:29462-29467.
9. Lee SY, Wang RG, Söderhäll K. A lipopolysaccharide- and beta-1,3-glucan-binding protein from hemocytes of the freshwater crayfish Pacifastacus leniusculus—Purification, characterization and cDNA cloning. J Biol Chem 2000; 275:1337-1343.
10. Huang TS, Wang HY, Lee SY et al. A cell adhesion protein from the crayfish Pacifastacus leniusculus, a serine proteinase homologue similar to Drosophila masquerade. J Biol Chem 2000; 275:9996-10001.
11. Lee SY, Söderhäll K. Characterization of a pattern recognition protein, a masquerade-like protein, in the freshwater crayfish Pacifastacus leniusculus. J Immunol 2001; 166:7319-7326.
12. Lemaitre B, Hoffmann J. The host defense of Drosophila melanogaster. Annu Rev Immunol 2007; 25:697-743.
13. McTaggart SJ, Conlon C, Colbourne JK et al. The components of the Daphnia pulex immune system as revealed by complete genome sequencing. BMC Genomics 2009; 10:175.
14. Fagutao FF, Yasuike M, Caipang CM et al. Gene expression profile of hemocytes of kuruma shrimp, Marsupenaeus japonicus following peptidoglycan stimulation Mar Biotech 2008; 10:731-740.
15. Cerenius L, Söderhäll K. The prophenoloxidase activating system in invertebrates. Immunol Rev 2004; 198:72-82.
16. Jitvaropas R, Amparyup P, Gross PS et al. Functional characterization of a masquerade-like serine proteinase homologue from the black tiger shrimp Penaeus monodon. Comp Biochem Physiol B 2009; 153:236-243.
17. Lis H, Sharon N. Lectins: Carbohydrate-specific proteins that mediate cellular recognition. Chem Rev 1998; 98:1283-1297.
18. Christophides GK, Zdobnov E, Barillas-Mury C et al. Immunity-related genes and genes families in Anopheles gambiae. Science 2002; 298:159-165.
19. Yu XQ, Kanost MR. Immulectin-2, a pattern recognition receptor that stimulates haemocyte encapsulation and melanization in the tobacco hornworm, Manduca sexta. Dev Comp Immunol 2004; 28:891-900.
20. Drickamer K. Two distinct classes of carbohydrate-recognition domains in animal lectins. J Biol Chem 1988; 263:9557-9560.
21. Weis WI, Kahn R, Fourme R et al. Structure of the calcium-binding protein determined by MAD phasing. Science 1991; 254:1608-1615.
22. Drickamer K, Taylor ME. Biology of animal lectins. Annu Rev Cell Biol 1993; 9:237-264.
23. Zelensky AN, Gready JE. The C-type lectin-like domain superfamily. FEBS J 2005; 272:6179-6217.
24. Vales-Gomez M, Reyburn H, Strominger J. Interaction between the human NK receptors and their ligands. Crit Rev Immunol 2000; 20:223-244.
25. Weis WI, Taylor ME, Drickamer K. The C-type lectin superfamily in the immune system. Immunol Rev 1998; 163:19-34.
26. Schulenburg H, Hoeppner MP, Weiner III J et al. Specificity of the innate immune system and diversity of C-type lectin domain (CTLD) proteins in the nematode Caenorhabditis elegans. Immunol 2008; 213:237-250.
27. Gross PS, Bartlett TC, Browdy L et al. Immune gene discovery by expressed sequence tag analysis of hemocytes and hepatopancreas in the Pacific White Shrimp, Litopenaeus vannamei and the Atlantic White shrimp, L. setiferus. Dev Comp Immunol 2001; 25:565-577.
28. Dodd RB, Drickamer K. Lectin-like proteins in model organisms: implications for evolution of carbohydrate-binding activity. Glycobiol 2001; 11:71R-79R.
29. Kerrigan AM, Brown GD. C type lectins and phagocytosis. Immunobiol 2009, 214:562-575.

30. Yu XQ, Kanost MR. Immulectin-2, a lipopolysaccharide-specific lectin from an insect, manduca sexta, is induced in response to gram-negative bacteria. J Biol Chem 2000; 275:37373-37381.
31. Yu XQ, Gan H, Kanost MR. Immulectin, an induction C-type lectin from an insect, Manduca sexta, stimulates activation of plasma prophenoloxidase. Insect Biochem Mol Biol 1999; 29:585-597.
32. Yu XQ, Tracy ME, Ling E et al. A novel C-type immulectin-3 from Manduca sexta is translocated from hemolymph into the cytoplasm of hemocytes. Insect Biochem Mol Biol 2005; 35:285-295.
33. Ling E, Yu XQ. Cellular encapsulation and melanization are enhanced by immulectins, pattern recognition receptors from the tobacco hornworm Manduca sexta. Dev Comp Immunol 2006; 30:289-299.
34. Koizumi N, Imamura M, Kadotani T et al. The lipopolysaccharide-binding protein participating in haemocyte nodule formation in the silkworm Bombyx mori is a novel member of the C-type lectin superfamily with two different tandem carbohydrate-recognition domains. FEBS Lett 1999; 443:139-143.
35. Jomori T, Natori S. Functional of the lipopolysaccharide-binding protein of Periplaneta americana as an opsonin. FEBS Lett 1992; 283-286.
36. Schroder HC, Ushijima H, Krasko A et al. Emergence and disappearance of an immune molecule, an antimicrobial lectin, in basal metazoan. A tachylectin-related protein in the sponge Suberites domuncula. J Biol Chem 2003; 278:32810-32817.
37. Willment JA, Brown GD. C-type lectin receptors in antifungal immunity. Trends Microbiol 2008; 16:27-32.
38. Zhu L, Song LS, Xu W et al. Molecular cloning and immune responsive expression of a novel C-type lectin gene from bay scallop Argopecten irradians. Fish Shellfish Immunol 2008; 25:231-238.
39. Marques MRF, Barracco MA. Lectins as nonself recognition factors, in crustaceans. Aquaculture 2000; 191:23-44.
40. Vazquez L, Alpuche J, Maldonado G et al. Immunity mechanisms in crustaceans. Innate Immunity 2009; 15:179-188.
41. Zhang X-W, Xu W-T, Wang X-W et al. A novel C-type lectin with two CRD domains from Chinese shrimp Fenneropenaeus chinensis functions as a pattern recognition protein. Mol Immunol 2009; 46:1626-1637.
42. Wang X-W, Zhang X-W, Xu W-T et al. A novel C-type lectin (FcLec4) facilitates the clearance of Vibrio anguillarum in vivo in Chinese white shrimp. Dev Comp Immunol 2009; 33:1039-1047.
43. Liu Y-H, Li F-H, Dong B et al. Molecular cloning, characterization and expression analysis of a putative C-type lectin (Fclectin) gene in Chinese shrimp Fenneropenaeus chinensis. Mol Immunol 2007; 44:598-607.
44. Luo T, Yang H, Li F et al. Purification, characterization and cDNA cloning of a novel lipopolysaccharide-binding lectin from the shrimp Penaeus monodon. Dev Comp Immunol 2006; 30:607-617.
45. Zhang Y, Qui L, Song L et al. Cloning and characterization of a novel C-type lectin gene from shrimp Litopenaeus vannamei. Fish Shellfish Immunol 2009; 26:183-192.
46. Ma TH-T, Benzie JAH, He JG et al. PmLT, a C-type lectin specific to hepatopancreas is involved in the innate defense of the shrimp Penaeus monodon. J Invertebr Pathol 2008; 99:332-341.
47. Sun Y-D, Fu L-D, Jia Y-P et al. A hepatopancreas-specific C-type lectin from the Chinese shrimp Fenneropenaeus chinensis exhibits antimicrobial activity. Mol Immunol 2008; 45:348-361.
48. Zhao Z-Y, Yin Z-X, Xu X-P et al. A novel C-type lectin from the shrimp Litopenaeus vannamei possesses anti-white spot syndrome virus activity. J Virol 2009; 83:347-356.
49. Wang X-W, Xu W-T, Zhang X-W et al. A C-type lectin is involved in the innate immune response of Chinese white shrimp. Fish Shellfish Immunol 2009; 27:556-562.
50. Luo T, Zhang X, Shao Z et al. PmAV, a novel gene involved in virus resistance of shrimp Penaeus monodon. FEBS Lett 2003; 551:53-57.
51. Soonthornchai W, Rungrassamee W, Karoonuthaisiri N et al. Expression of immune-related genes in the digestive organ of shrimp, Penaeus monodon, after an oral infection by Vibrio harveyi. Dev Comp Immunol 2010; 34:19-28.
52. Johansson MW, Söderhäll K. Isolation and purification of a cell adhesion factor from crayfish blood cells. J Cell Biol 1988; 106:1795-1803.
53. Johansson MW, Lind MI, Holmblad T et al. Peroxinectin, a novel cell-adhesion protein from crayfish blood. Biochem Biophys Res Com 1995; 216:1079-1097.
54. Bauchau AG. Crustaceans. s Ratcliffe NA, Rowley AF, eds. Invertebrate Blood Cells. London and New York: Academic Press 1981, 385-420.
55. Johansson MW, Keyser P, Sritunyalucksana K et al. Crustacean haemocytes and haematopoiesis. Aquaculture 2000; 191:45-52.
56. Söderhäll K, Smith VJ. Separation of hemocyte population of Carcinus maenas and other marine decapods and prophenoloxidase distribution. Dev Comp Immunol 1983; 7:229-239.

57. Sricharoen S, Kim JJ, Tunkijanukij S et al. Exocytosis and proteomic analysis of the vesicle content of granular hemocytes from a crayfish. Dev Comp Immunol 2005; 29:1017-1031.
58. Lavine MD, Strand MR. Insect hemocytes and their role in immunity. Insect Biochem Mol Biol 2002; 32:1295-1309.
59. Crozatier M, Meister M. Drosophila hematopoiesis. Cell Microbiol 2007; 5:1117-11126.
60. Chaga O, Lignell M, Söderhäll K. The haemopoietic cells of the freshwater crayfish, Pacifastacus leniusculus. Anim Biol 1995; 4:57-70.
61. Söderhäll I, Bangyeekhun E, Mayo S et al. Hemocyte production and maturation in an invertebrate animal: proliferation and gene expression in hematopoietic stem cells of Pacifastacus leniusculus. Dev Comp Immunol 2003; 27:661-672.
62. Söderhäll I, Kim YA, Jiravanichpaisal P et al. An ancient role for a prokineticin domain in invertebrate hematopoiesis. J Immunol 2005; 174:6153-6160.
63. Lin XH, Söderhäll K, Söderhäll I. Transglutaminase activity in the hematopoietic tissue of a crustacean, Pacifastacus leniusculus, importance in hemocyte homeostasis. BMC Immunol 2008; 9:58.
64. Wu CL, Söderhäll I, Kim YA et al. Hemocyte-lineage marker proteins in a crustacean, the freshwater crayfish, Pacifastacus leniusculus. Proteomics 2008; 8:4226-4235.
65. Lin XH, Kim YA, Lee BL et al. Identification and properties of a receptor for the invertebrate cytokine astakine, involved in hematopoiesis. Exp Cell Res 2009; 315:1171-1180.
66. Weldon WFR. The renal organs of certain decapod crustacea. Quart J Micr Sci 1891; 32:279-291.
67. Allen EJ. Nephridia and body-cavity of some decapod crustacea. Quart J Micr Sci 1893; 34:403-426.
68. Johnson, PT. Histology of the blue crab, Callinectes sapidus: A Model for the Decapoda. New York: Praeger, 1980: 440.
69. Bell T, Lightner DV. A handbook of normal penaeid shrimp histology. World Aquaculture Society. Kansas: Allen Press, 1988: 114.
70. Oka M. Studies on Penaeus orientalis Kishinouye. VIII. Structure of newly found lymphoid organ. Bull Jpn Soc Sei Fish 1969; 35:245-250.
71. Hose JE, Martin GG, Tiu S et al. Patterns of hemocyte production and release throughout the molt cycle in the penaeid shrimp Sicyonia ingentis. Biol Bull 1992; 183:185-199.
72. Van de Braak CB, Botterblom MH, Liu W et al. The role of the haematopoietic tissue in haemocyte production and maturation in the black tiger shrimp (Penaeus monodon). Fish Shellfish Immunol 2002; 12:253-272.
73. Cuénot L. L'organe phagocytaire des Crustacés Décapodes. Arch Zool Exp Gen 1905; 4:1-16.
74. Ghiretti-Magaldi A, Milanesi C, Tognon G. Hemopoiesis in crustacea decapoda: origin and evolution of hemocytes and cyanocytes of Carcinus maenas. Cell Differ 1977; 6:167-186.
75. Bazin F. Ultrastructure de l'organe hématopolétique chez le crabe Carcinus maenas (L.) Arch Anat Microsc Morphol Exp 1979; 68:142-157.
76. Martin GG, Hose JE, Kim JJ. Structure of hematopoietic nodules in the ridgeback prawn, Sicyonia ingentis: light and electron microscopic observations. J Morphol 1987; 192:193-204.
77. Martin GG, Hose JE, Chou M et al. Organization of hematopoietic tissue in the intermolt lobster, Homarus americanus. J Morphol 1993; 216:65-78.
78. Martinez-Agosto JA, Mikkola HKA, Hartenstein V et al. The hematopoietic stem cell and its niche: a comparative view. Genes Dev 2007; 21:3044-3060.
79. Hernroth B, Baden SP, Holm K et al. Manganese induced immune suppression of the lobster, Nephrops norvegicus. Aquat Toxicol 2004; 70:223-231.
80. Hsiao CY, Song YL. A long form of shrimp astakine transcript: Molecular cloning, characterization and functional elucidation in promoting hematopoiesis. Fish Shellfish Immunol 2010; 28:77-86.
81. Lin X, Novotny M, Söderhäll K et al. Function of crayfish astakines in hemocyte lineage determination. J Biol Chem 2010; Epub ahead of print, PMID 2059 2028.
82. Huang CC, Sritunyalucksana K, Söderhäll K et al. Molecular cloning and characterization of tiger shrimp (Penaeus monodon) transglutaminase. Dev Comp Immunol 2004; 28:279-294.
83. Robalino J, Carnegie RB, O'Leary N et al. Contributions of functional genomics and proteomics to the study of immune responses in the Pacific white leg shrimp Litopenaeus vannamei. Vet Immunol Immunopathol 2009; 128:110-118.
84. Pan J, Kurosky A, Xu B et al. Broad antiviral activity in tissues of crustaceans. Antiviral Res 2000; 48:39-47.
85. Aketagawa J, Miyata T, Ohtsubo S et al. Primary structure of limulus anticoagulant anti-lipopolysaccharide factor. J Biol Chem 1986; 261:7357-7365.
86. de la Vega E, O'Leary NA, Shockey JE et al. Anti-lipopolysaccharide factor in Litopenaeus vannamei (LvALF): A broad spectrum antimicrobial peptide essential for shrimp immunity against bacterial and fungal infection. Mol Immunol 2008; 45:1916-1925.

87. Robalino J, Almeida JS, McKillen D et al. Insights into the immune transcriptome of the shrimp Litopenaeus vannamei: tissue-specific expression profiles and transcriptomic responses to immune challenge. Physiol Genomics 2007; 29:44-56.
88. Liu H, Jiravanichpaisal P, Söderhäll I et al. Antilipopolysaccharide Factor interferes with white spot syndrome virus replication in vitro and in vivo in the crayfish Pacifastacus leniusculus. J Virol 2006; 80:10365-10371.
89. Tharntada S, Ponprateep S, Somboonwiwat K et al. Role of antilipopolysaccharide factor from the black tiger shrimp, Penaeus monodon, in protection from white spot syndrome virus infection. J Gen Virol 2009; 90:1491-1498.
90. Dostert C, Jouanguy E, Irving P et al. The Jak-STAT signaling pathway is required but not sufficient for the antiviral response of Drosophila. Nat Immunol 2005; 6:946-953.
91. Liu WJ, Chang YS, Wang AH et al. White spot syndrome virus annexes a shrimp STAT to enhance expression of the immediate-early gene ie1. J Virol 2007; 81:1461-1471.
92. Chen WY, Ho KC, Leu JH et al. WSSV infection activates STAT in shrimp. Dev Comp Immunol 2008; 32:1142-1150.
93. Liu WJ, Chang YS, Wang HC et al. Transactivation, dimerization and DNA-binding activity of White spot syndrome virus immediate early protein IE1. J Virol 2008; 82:11362-11373.
94. Zambon RA, Nandakumar M, Vakharia VN et al. The Toll pathway is important for an antiviral response in Drosophila. Proc Natl Acad Sci USA 2005; 102:7257-7262.
95. Labreuche Y, O'Leary NA, de la Vega E et al. Lack of evidence for Litopenaeus vannamei Toll receptor (lToll) involvement in activation of sequence-independent antiviral immunity in shrimp. Dev Comp Immunol 2009; 33:806-810.
96. Menze MA, Fortner G, Nag S et al. Mechanisms of apoptosis in Crustacea: what conditions induce versus suppress cell death? Apoptosis 2010; 15:293-312.
97. Wang L, Zhi B, Wu W et al. Requirement for shrimp caspase in apoptosis against virus infection. Dev Comp Immunol 2008; 32:706-715.
98. Rijiravanich A, Browdy CL, Withyachumnarnkul B. Knocking down caspase-3 by RNAi reduces mortality in Pacific white shrimp Penaeus (Litopenaeus) vannamei challenged with a low dose of white-spot syndrome virus. Fish Shellfish Immunol 2008; 24:308-313.
99. Molthathong S, Buaklin A, Senapin S et al. Up-regulation of ribophorin I after yellow head virus (YHV) challenge in black tiger shrimp Penaeus monodon. Fish Shellfish Immunol 2008; 25:40-46.
100. Molthathong S, Senapin S, Klinbunga S et al. Down-regulation of defender against apoptotic death (DAD1) after yellow head virus (YHV) challenge in black tiger shrimp Penaeus monodon. Fish Shellfish Immunol 2008; 24:173-179.
101. Krishnan P, Gireesh-Babu P, Rajendran KV et al. RNA interference-based therapeutics for shrimp viral diseases. Dis Aquat Org 2009; 86:263-272.
102. Robalino J, Bartlett TC, Chapman RW et al. Double-stranded RNA and antiviral immunity in marine shrimp: inducible host mechanisms and evidence for the evolution of viral counter-responses. Dev Comp Immunol 2007; 31:539-547.
103. Su J, Oanh DT, Lyons RE et al. A key gene of the RNA interference pathway in the black tiger shrimp, Penaeus monodon: identification and functional characterisation of Dicer-1. Fish Shellfish Immunol 2008; 24:223-233.
104. Unajak S, Boonsaeng V, Jitrapakdee S. Isolation and characterization of cDNA encoding Argonaute, a component of RNA silencing in shrimp (Penaeus monodon). Comp Biochem Physiol B Biochem Mol Biol 2006; 145:179-187.
105. Zhang Y, Song L, Zhao J et al. Protective immunity induced by CpG ODNs against white spot syndrome virus (WSSV) via intermediation of virus replication indirectly in Litopenaeus vannamei. Dev Comp Immunol 2010; 34:418-424.
106. Wang S, Liu N, Chen A-J et al. TRBP homolog interacts with eukaryotic initiation factor 6 (eIF6) in Fenneropenaeus chinensis. J Immunol 2009; 182:5250-5258.
107. Söderhäll I, Bangyeekhun E, Mayo S et al. Hemocyte production and maturation in an invertebrate animal; proliferation and gene expression in hematopoietic stem cells of Pacifastacus leniusculus. Dev Comp Immunol 2003; 27:661-672.
108. Jiravanichpaisal P, Söderhäll K, Söderhäll I. Characterization of white spot syndrome virus replication in vitro-cultured haematopoietic stem cells of freshwater crayfish, Pacifastacus leniusculus. J Gen Virol 2006; 87:847-854.
109. Kopácek P, Hall M, Söderhäll K. Characterization of a clotting protein, isolated from plasma of the crayfish Pacifastacus leniusculus. Eur J Biochem 1993; 213:591-597.
110. Hall M, Wang RG, van Antwerpen R et al. The crayfish plasma clotting protein: A vitellogenin-related protein responsible for clot formation in crustacean blood. Proc Natl Acad Sci USA 1999; 96:1965-1970.

111. Theopold U, Schmidt O, Söderhäll K et al. Coagulation in arthropods: defence, wound closure and healing. Trends Immunol 2004; 25:289-294.
112. Maningas MBB, Kondo H, Hirono I et al. Essential function of transglutaminase and clotting protein in shrimp immunity. Mol Immunol 2008; 45:1269-1275.
113. Spersted SV, Haug T, Vasskog T et al. Hyastatin, a glycine-rich muti-domain antimicrobial peptide isolated from the spider crab (Hyas araneus) hemocytes. Mol Immunol 2009; 46:2604-2612.
114. Du ZQ, Ren Q, Zhao XF et al. A double WAP domain (DWD)-containing protein with proteinase inhibitory activity in Chinese white shrimp, Fenneropenaeus chinensis. Comp Biochem Physiol B 2009; 154:203-210.
115. Destoumieux-Garzon D, Saulnier D, Garnier J et al. Crustacean immunity—Antifungal peptides are generated from the C terminus of shrimp hemocyanin in response to microbial challenge. J Biol Chem 2001; 276:47070-47077.
116. Lee SY, Lee BL, Söderhäll K. Processing of an antibacterial peptide from hemocyanin of the freshwater crayfish Pacifastacus leniusculus. J Biol Chem 2003; 278:7927-7933.
117. Wang KCHC, Tseng CW, Lin HY et al. RNAi knock-down of the Litopenaeus vannamei Toll gene (LvToll) significantly increases mortality and reduces bacterial clearance after challenge with Vibrio harveyi. Dev Comp Immunol 2010; 34:49-58.
118. Arts JAJ, Cornelissen FHJ, Cijsouw T et al. Molecular cloning and expression of a Toll receptor in the giant tiger shrimp, Penaeus monodon. Fish Shellfish Immunol 2007; 23:504-513.
119. Mekata T, Kono T, Yoshida T et al. Identification of cDNA encoding Toll receptor, MjToll gene from kuruma shrimp, Marsupenaeus japonicus. Fish Shellfish Immunol 2008; 24:122-133.
120. Yang, C, Zhang J, Li F et al. A Toll receptor from Chinese shrimp Fenneropenaeus chinensis is responsive to Vibrio anguillarum infection. Fish Shellfish Immunol 2008; 24:564-574.
121. Shi XZ, Zhang RR, Jia YP et al. Identification and molecular characterization of a Spätzle-like protein from Chinese shrimp (Fenneropenaeus chinensis). Fish Shellfish Immunol 2009; 27:610-617.
122. Huang XD, Yin ZX, Liao JX et al. Identification and functional study of a shrimp Relish homologue. Fish Shellfish Immunol 2009; 27:230-238.
123. Li FH, Yan H, Wang DD et al. Identification of a novel relish homolog in Chinese shrimp Fenneropenaeus chinensis and its function in regulating the transcription of antimicrobial peptides. Dev Comp Immunol 2009; 33:1093-1101.
124. Huang XD, Yin ZX, Jia XT et al. Identification and functional study of a shrimp Dorsal homologue. Dev Comp Immunol 2010; 34:107-113.
125. Shockey JE, O'Leary NA, de la Vega E et al. The role of crustins in Litopenaeus vannamei in response to infection with shrimp pathogens: An in vivo approach. Dev Comp Immunol 2009; 33:668-673.
126. Wang PH, Gu ZH, Huang XD et al. An immune deficiency homolog from the white shrimp, Litopenaeus vannamei, activates antimicrobial peptide genes. Mol Immunol 2009; 46:1897-1904.
127. Söderhäll K, Cerenius L. Role of the prophenoloxidase-activating system in invertebrate immunity. Cur Op Immunol 1998; 10:23-28.
128. Cerenius L, Lee BL, Söderhäll K. The proPO-system: pros and cons for its role in invertebrate immunity. Trends Immunol 2008; 29:263-271.
129. Liu HP, Jiravanichpaisal P, Cerenius L et al. Phenoloxidase is an important component of the defense against Aeromonas hydrophila infection in a crustacean, Pacifastacus leniusculus. J Biol Chem 2007; 282:33593-33598.
130. Cerenius L, Ramesh B, Söderhäll K et al. In vitro effects on bacterial growth of phenoloxidase reaction products. J Invertebr Pathol 2010; 103:21-23.
131. Charoensapsri W, Amparyup P, Hirono I et al. Gene silencing of a prophenoloxidase activating enzyme in the shrimp, Penaeus monodon, increases susceptibility to Vibrio harveyi infection. Dev Comp Immunol 2009; 33:811-820.
132. Amparyup P, Charoensapsri W, Tassanakajon A. Two prophenoloxidases are important for the survival of Vibrio harveyi challenged shrimp Penaeus monodon. Dev Comp Immunol 2009; 33:247-256.
133. Fagutao FF, Koyama T, Kaizu A et al. Increased bacterial load in shrimp hemolymph in the absence of prophenoloxidase. FEBS J 2009; 276:5298-5306.
134. Rimphanitchayakit V, Tassanakajon A. Structure and function of invertebrate Kazal-type serine proteinase inhibitors. Dev Comp Immunol 2010; 34:377-386.
135. Cerenius L, Liu HP, Zhang YJ et al. High sequence variability among hemocyte-specific Kazal-type proteinase inhibitors in decapod crustaceans. Dev Comp Immunol 2010; 34:69-75.
136. Ma H, Wang B, Zhan J et al. Multiple forms of alpha-2 macroglobulin in shrimp Fenneropenaeus chinensis and their transcriptional response to WSSV or Vibrio pathogen infection. Dev Comp Immunol 2010; 34:677-684.

137. Söderhäll I, Wu CL, Novotny M et al. A novel protein acts as a negative regulator of prophenoloxidase activation and melanization in the freshwater crayfish Pacifastacus leniusculus. J Biol Chem 2009; 284:6301-6310.
138. Zhao MY, Söderhäll I, Park, JW et al. A novel 43-kDa protein as a negative regulatory component of phenoloxidase-induced melanin synthesis. J Biol Chem 2005; 280:24744-24751.
139. Kan H, Kim CH, Kwon HM et al. Molecular control of phenoloxidase-induced melanin synthesis in an insect. J Biol Chem 2008; 283:25316-25323.
140. Zhou Z, Shin SW, Alvarez KS et al. Distinct melanization pathways in the mosquito Aedes aegyptii. Immunity 2010; 32:41-53.
141. Dhar AK, Dettori A, Roux MM et al. Identification of differentially expressed genes in shrimp (Penaeus stylirostris) infected with White spot syndrome virus by cDNA microarrays. Arch Virol 2003; 148:2381-2396.
142. Ma THT, Tiu SHK, He J-G, Chan S-M. Molecular cloning of a C-type lectin (LvLT) from the shrimp Litopenaeus vannamei: Early gene down-regulation after WSSV infection. Fish Shellfish Immunol 2007; 23:430-437.
143. Kong H-J, Park E-M, Nam B-H et al. A C-type like-domain (CTLD)-containing protein (PtLD) from the swimming crab Portunus trituberculatus. Fish Shellfish Immunol 2008; 25:311-314.

CHAPTER 14

ECHINODERM IMMUNITY

L. Courtney Smith,*[,1] Julie Ghosh,[1] Katherine M. Buckley,[2] Lori A. Clow,[3] Nolwenn M. Dheilly,[4] Tor Haug,[5] John H. Henson,[6] Chun Li,[5] Cheng Man Lun,[1] Audrey J. Majeske,[1] Valeria Matranga,[7] Sham V. Nair,[4] Jonathan P. Rast,[2] David A. Raftos,[4] Mattias Roth,[4] Sandro Sacchi,[1] Catherine S. Schrankel[1] and Klara Stensvåg[5]

[1]Department of Biological Sciences, George Washington University, Washington DC, USA; [2]Department of Medical Biophysics and Department of Immunology, Sunnybrook Health Sciences Centre, University of Toronto, Toronto, Ontario, Canada; [3]US Patent and Trademark Office, Alexandria, Virginia, USA; [4]Department of Biological Sciences, Macquarie University, Sydney, Australia; [5]Norwegian College of Fishery Science, University of Tromso, Tromso, Norway; [6]Department of Biology and the Biochemistry and Molecular Biology Program, Dickinson College, Carlisle, Pennsylvania, USA; [7]CNR, Institute of Biomedicine and Molecular Immunology "Albert Monroy", Palermo, Italy

**Corresponding Author: L. Courtney Smith—Email: csmith@gwu.edu*

Abstract: A survey for immune genes in the genome for the purple sea urchin has shown that the immune system is complex and sophisticated. By inference, immune responses of all echinoderms may be similar. The immune system is mediated by several types of coelomocytes that are also useful as sensors of environmental stresses. There are a number of large gene families in the purple sea urchin genome that function in immunity and of which at least one appears to employ novel approaches for sequence diversification. Echinoderms have a simpler complement system, a large set of lectin genes and a number of antimicrobial peptides. Profiling the immune genes expressed by coelomocytes and the proteins in the coelomic fluid provide detailed information about immune functions in the sea urchin. The importance of echinoderms in maintaining marine ecosystem stability and the disastrous effects of their removal due to disease will require future collaborations between ecologists and immunologists working towards understanding and preserving marine habitats.

Invertebrate Immunity, edited by Kenneth Söderhäll.

INTRODUCTION

Echinoderms are a phylum of ubiquitous benthic marine invertebrates, found in a continuous distribution from the intertidal zone to the deepest depths of the ocean explored to date. They are the sister group to the chordates and are the basal deuterostomes.[1] Approximately 7000 extant echinoderm species have been described, falling into five classes—echinoids (sea urchins and sand dollars), holothurians (sea cucumbers), asteroids (sea stars), crinoids (sea lilies and feather stars) and ophiuroids (brittle stars), as well as a number of extinct classes known from the fossil record. Many members of the phylum are large and long lived.[2]

Echinoderms are central players in many benthic ecosystems, variously as herbivores, carnivores and detritivores. They possess a simple gut which opens into an anus on the aboral surface, or, for crinoids, on the oral surface of the animal.[1] All echinoderms have basic pentameric radial symmetry in their adult forms and most metamorphose from bilateral larvae found in the plankton that function as the dispersal stage, feed and produce the adult rudiment. Adults lack obvious sensory organs or any kind of head and locomote largely by means of tube feet, which function by means of a water vascular system unique to the phylum. Activities are coordinated by a simple neural ring that communicates to five nerve bundles into the five ambulacral grooves that run longitudinally along each radial segment of the body. All echinoderms have some form of calcified skeleton composed of aragonite, which is more or less extensive in the various classes. At one extreme, echinoids are completely enclosed in a calcified test, with only a thin layer of muscular tissue lining the interior of the body cavity (or coelom) while holothurians possess a highly muscularized body wall in which the exoskeleton is reduced to microscopic ossicles dispersed throughout the dermis. The coelomic cavity contains the gonads and gut of the animal, bathed in coelomic fluid. This coelomic fluid is key to the immunological capabilities of echinoderms, being the medium in which the immunocytes or coelomocytes reside, and which also contains antimicrobial molecules.

A larval sea star was first used by Metchnikoff[3] to demonstrate that invertebrate immune cells could recognize the presence of and respond to (encapsulate) a rose prickle. Subsequent characterization of the echinoderm immune system in adults relied on allograft rejection assays in sea cucumbers, sea stars,[4,5] and sea urchins,[6-8] and demonstrated the ability of echinoderms to recognize self from nonself. Later evaluation of the sea urchin graft rejection kinetics illustrated the innate characteristics of echinoderm immunity.[9] Although the immune system is entirely innate, it has recently been found to be highly complex and sophisticated[10,11] and is clearly effective given that echinoderms have survived as a phylum for 450-500 million years and that individual echinoderms can survive the constant assault by pathogens in the marine environment (reviewed by ref. 12) for upwards of 100 years.[2]

COELOMOCYTES, THE IMMUNE MEDIATORS IN ECHINODERMS

The body cavity of echinoderms is filled with coelomic fluid,[13] which bathes the internal organs and forms the fluid medium in which the coelomocytes are suspended. The composition of coelomic fluid (CF) is similar to sea water in terms of minor dissolved salts and other minerals and contains proteins, the best-characterized of which are those involved in antipathogen responses. Indeed, whole CF (wCF) is a complex tissue that

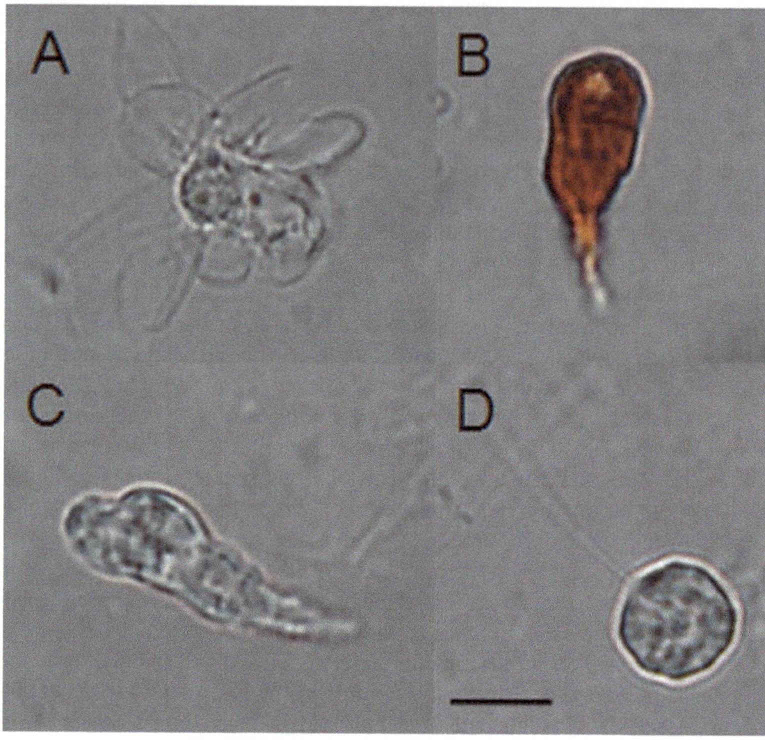

Figure 1. Live coelomocytes from the sea urchin, *Paracentrotus lividus*. A) petaloid phagocyte. B) red spherule cell. C) colorless spherule cell. D) vibratile cell. Scale bar = 5 microns. Images taken by R. Bonaventura.

mediates responses to wounding and microbial infections by undergoing reactions such as opsonization, coagulation, encapsulation and phagocytosis. Initial publications on echinoderm coelomocytes appeared in the late 19th century (e.g., ref. 14) and since then many reports have appeared describing the morphologies and functions of the various coelomocytes[9,15-17] (reviewed by refs. 18,19). Observations of live cells in CF from sea urchins suggest three basic categories of coelomocytes; phagocytes, spherule cells (also called amoebocytes,[20,21] or morula cells) and vibratile cells (Fig. 1); however further analyses suggest additional categories (Table 1) (reviewed in ref. 22). Within the spherule cell category, there are red and colorless spherules within the cells (Fig. 1B,C) and there are three different categories of phagocytes depending on cytoskeletal morphology and size (Fig. 2). Additional cell types have been noted in nonechinoid species including crystal cells, fusiform cells and progenitor cells, about which very little is known. The descriptions below refer to sea urchin coelomocytes, which are the best studied.

Phagocytes

Phagocytes have been variously referred to as leukocytes, or as bladder, petaloid or filoform phagocytes in the literature and constitute the most abundant type of coelomocyte (Table 1) in the coelomic fluid. In suspension the cells appear in a petaloid form (Fig. 1A;

Table 1. Coelomocytes in sea urchins

Cell Type	% in Coelomic Fluid	Function
Phagocyte Types Type 1—Discoidal cells Type 2—Polygonal cells Type 3—Small phagocytes	Total phagocytes *Sp** 40-80% *Sd* 67% *Pl* 80%	Encapsulation, Opsonization, Graft rejection, Chemotaxis, Phagocytosis, Antibacterial activity, Cellular clotting
Red spherule cells	*Sp* 7-40% *Sd* 8% *Pl* 4.7%	Oxygen transport. Antibacterial activity from echinochrome A.
Colorless spherule cells	*Sp* 3.7-25% *Sd* 6.5% *Pl* 7.8%	Cytotoxicity, Clotting?
Vibratile cells	*Sp* 11.9-20% *Sd* 18.5% *Pl* 7.5%	Movement or agitation of coelomic fluid? Associated with clotting.

**Sp, Strongylocentrotus purpuratus. Sd, Strongylocentrotus droebachiensis. Pl, Paracentrotus lividus*
Information collected from references 7, 16, 17, 19, 22, 24, 28, 33, 34, 38, 39, 50, 95, 222 and 223.

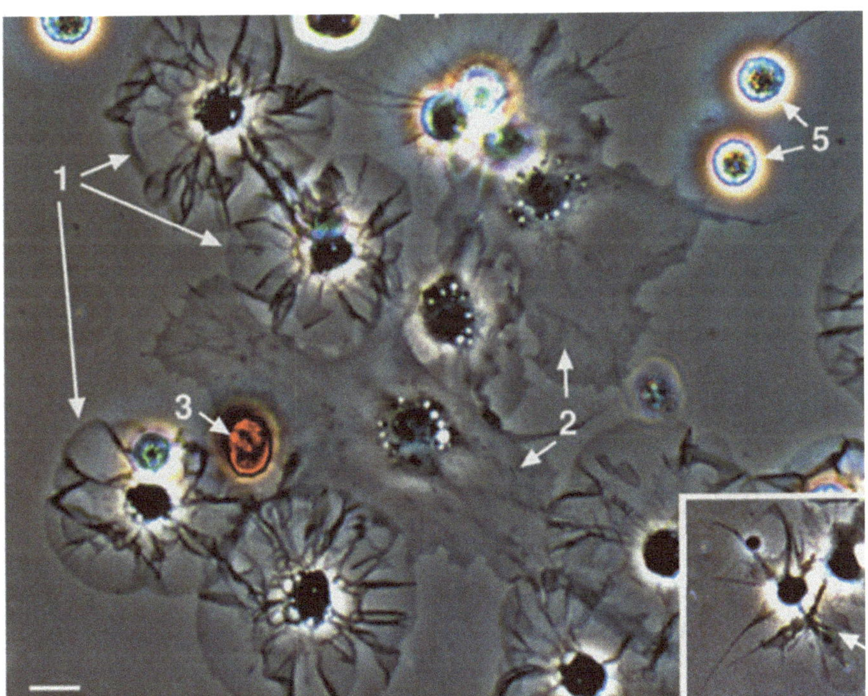

Figure 2. Coelomocytes from the sea urchin *S. droebachiensis*. Cells were withdrawn in anticoagulant and settled onto a glass coverslip. All cell types are shown and labeled with numbers. Large phagocytes; 1 = discoidal phagocyte; 2 = polygonal phagocyte; 3 = red spherule cell; 4 = colorless spherule cell; 5 = vibratile cell (the lower cell has lost the prominent flagellum seen in the upper cell). Inset; 6 = small phagocyte. Bar = 10 microns.

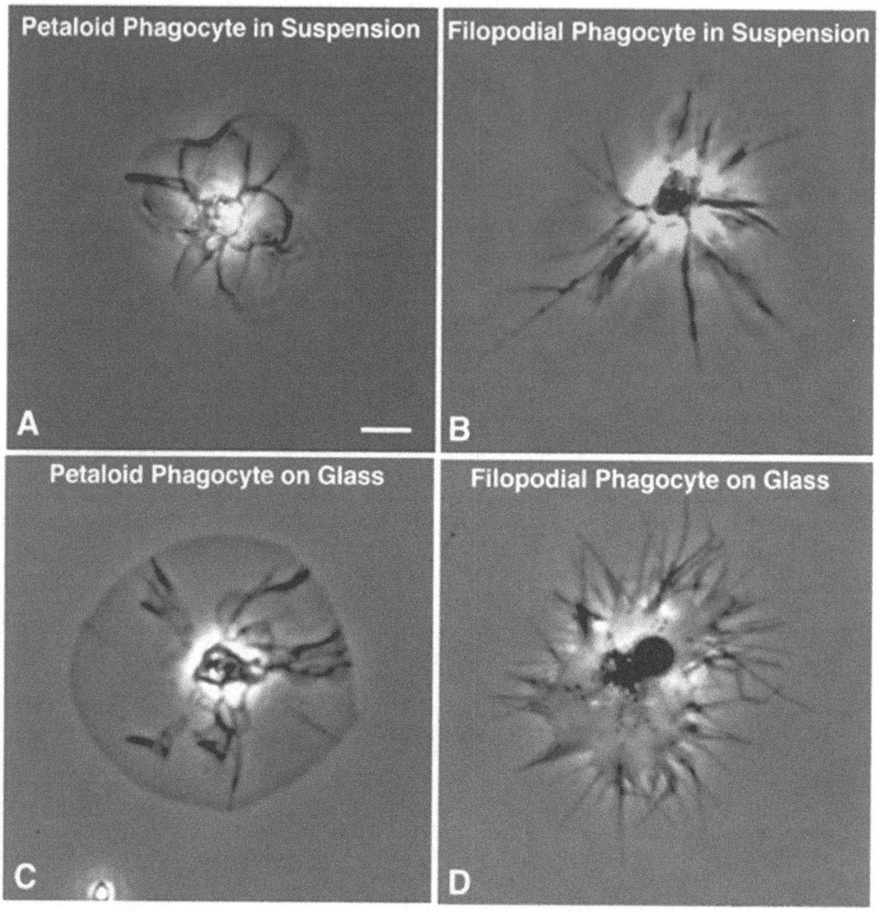

Figure 3. Phagocytes undergo an inducible petaloid/lamellipodial to filopodial/filiform shape change that is important in the coelomic fluid clotting process. This shape change occurs in cells in suspension; panel (A), petaloid cell; panel (B), filopodial cell. Shape change also occurs when cells are attached to glass substrate; panel (C), petaloid (lamellipodial) cell; panel (D), filopodial cell. Shape change was induced by hypotonic shock. Bar = 10 microns; magnifications of panels (A-D) are equivalent.

Fig. 3A) in which numerous cytoplasmic lamellipodia give the impression of the petals of a flower.[23] More recent work indicates that the phagocyte population actually consists of several different cell types that can be distinguished based on cytoskeletal morphology and organization, actin-based motility patterns,[24-27] and differential gene and protein expression.[19,28,29] Phagocytes are most easily distinguished as three morphologies in spread cells; two types of large phagocytes[24-27] and small phagocytes.[28,29] The discoidal-shaped phagocytes (Fig. 2) are large phagocytes that display rapid, actin-based retrograde/centripetal flow over their entire diameter, and contain a dense cortical actin meshwork rich in the Arp2/3 complex (Fig. 4C), radial actin bundles that connect the cortex to the perinuclear region, and a centralized distribution of myosin II (Fig. 4B).[25,27] The retrograde/centripetal flow activity in discoidal cells tends to restrict the distribution of microtubules (Fig. 4A), organelles (endoplasmic reticulum, mitochondria, lysosomes and granules) and

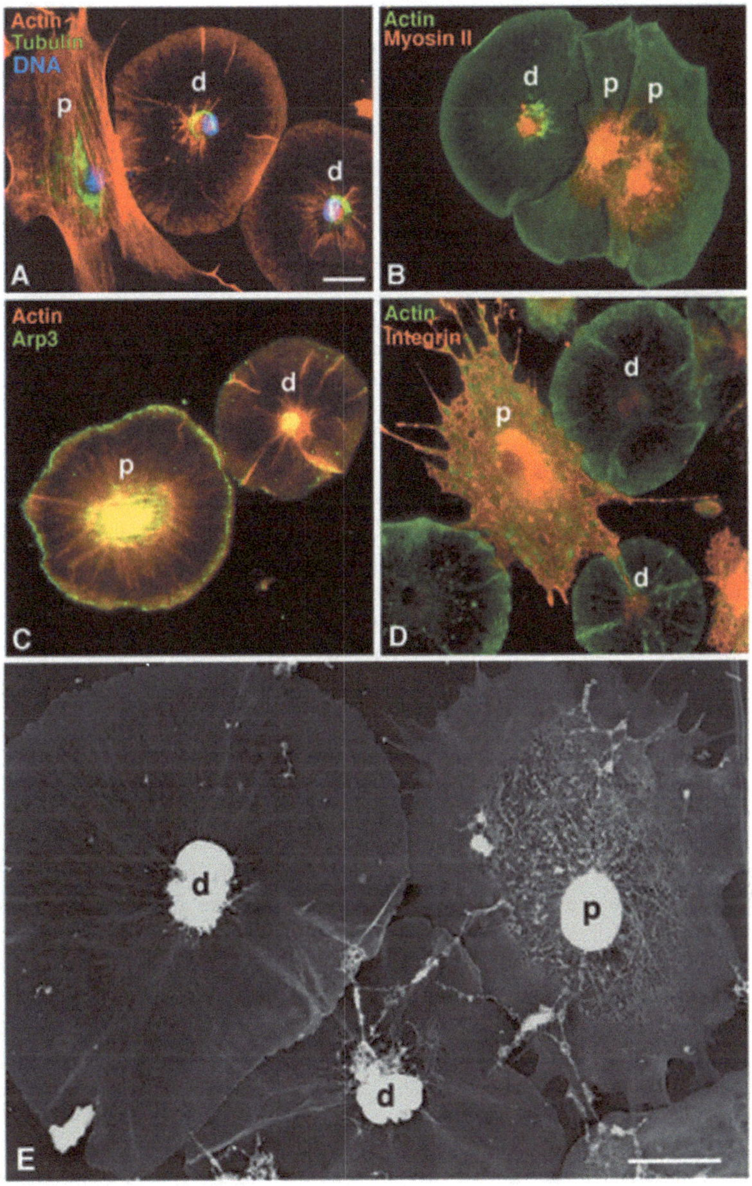

Figure 4. Large phagocytes of the discoidal (d) and polygonal (p) subpopulations can be differentiated based on cytoskeletal organization. Actin (panels A-D) in discoidal cells displays a broad cortical meshwork containing Arp3 (panel C) that feeds into a radial array of centralized bundles. The cortical actin and Arp3 meshwork is more limited in polygonal cells (panel C) and cells often contain prominent actin bundles oriented either radially in discoidal cells (panel C) or parallel to the long axis in polygonal cells (panels A,D). Both myosin II (panel B) and microtubules (panel A) are perinuclear in discoidal cells and more widespread in polygonal cells. β integrin expression is restricted to the polygonal cells (panel D). The significant difference between the structural organization of the cytoskeletons of the two large phagocytes is best appreciated in TEM images of critical point dried and rotary shadowed replicas of detergent extracted cells (panel E). Bar = 10 microns; magnifications of panels A-D are equivalent.

associated kinesin motor proteins to the perinuclear region.[26] The second type of large phagocyte is polygonal-shaped cells (Fig. 2, labeled 2), which display retrograde flow in a thin rim along the cell margin corresponding to more limited distribution of the Arp2/3 complex-containing actin meshwork (Fig. 4C). These large cells are integrin-positive (Fig. 4D) and the interior cytoplasm contains elongate actin bundles associated with myosin II (Fig. 4A-D),[25,27] as well as a broad array of microtubules (Fig. 4A) and associated organelles and kinesin (Fig. 4E).[26] Furthermore, both cell types undergo significant shape changes from a lamellipodial/petaloid to a filopodial form (Fig. 3B; see also ref. 20), inducible by the calcium-dependent clotting process, treatment with hypotonic shock[23] or calcium ionophore.[30]

Large phagocytes display phagocytic activity associated with clearance of bacteria, xenogeneic cells, foreign particles and proteins, both in vivo and in vitro (reviewed in ref. 18). Phagocytes have also been implicated in encapsulation and graft rejection responses, as well as cytolytic/cytotoxic reactions,[8,31] (reviewed in refs. 18,19). Phagocytes participate in the cellular clotting process, in which their filopodial form facilitates the cell-cell interactions leading to cellular clot formation and retraction.[16,23] Based on their size, preponderance in the CF and their phagocytic and encapsulation functions, the large phagocytes appear to be the major mediators of cellular immunity in echinoderms.

The third and least common form of phagocyte is the small phagocyte (Fig. 2, labeled 6)[28,29] of which some express the Sp185/333 family of highly diverse immune response-related proteins (see below). These cells are characterized by their small size relative to the large phagocytes, numerous small cytoplasmic granules and filopodial morphology (Fig. 5). The morphology of small phagocyte filopodia appears different from that of large phagocytes in that they are less numerous, thicker and often contain periodic knobs and microtubule bundles (Fig. 5D).[29]

Spherule Cells

Red spherule cells (Fig. 1B; Fig. 2, labeled 3; Table 1) have been called amoebocytes, spherulocytes, morula cells, pigment cells, granulocytes and eleocytes in the literature, and are small cells filled with spherical red granules containing echinochrome A, a naphthoquinone pigment with antibacterial properties. Red spherule cells have been reported to move towards bacteria to initiate an encapsulation response and to degranulate in the presence of bacteria.[32-35] They have been reported to migrate to and form a rim around the edge of wounds, infections and tissue grafts.[8,36,37] When settled onto a substrate, red spherule cells exhibit dynamic actin-based, amoeboid mobility, which may account for their ability to surround wounds and infections. Colorless spherule cells (Fig. 1C; Fig. 2, labeled 4; Table 1) contain granules and have therefore been referred to as morula cells (from Latin for 'mulberry'). Their properties and functions are not well known, however, a recent study suggests that these cells have potent cytolytic activity that is augmented by the presence of phagocytes.[38]

Vibratile Cells

Vibratile cells are round, highly motile, flagellated cells that contain large cytoplasmic granules (Fig. 1D; Fig. 2, labeled 5; Table 1). Exocytosis of these granules may trigger the initiation of the clotting reaction[32,39] (Sacchi and Smith, unpublished), however this hypothesis will require additional testing.

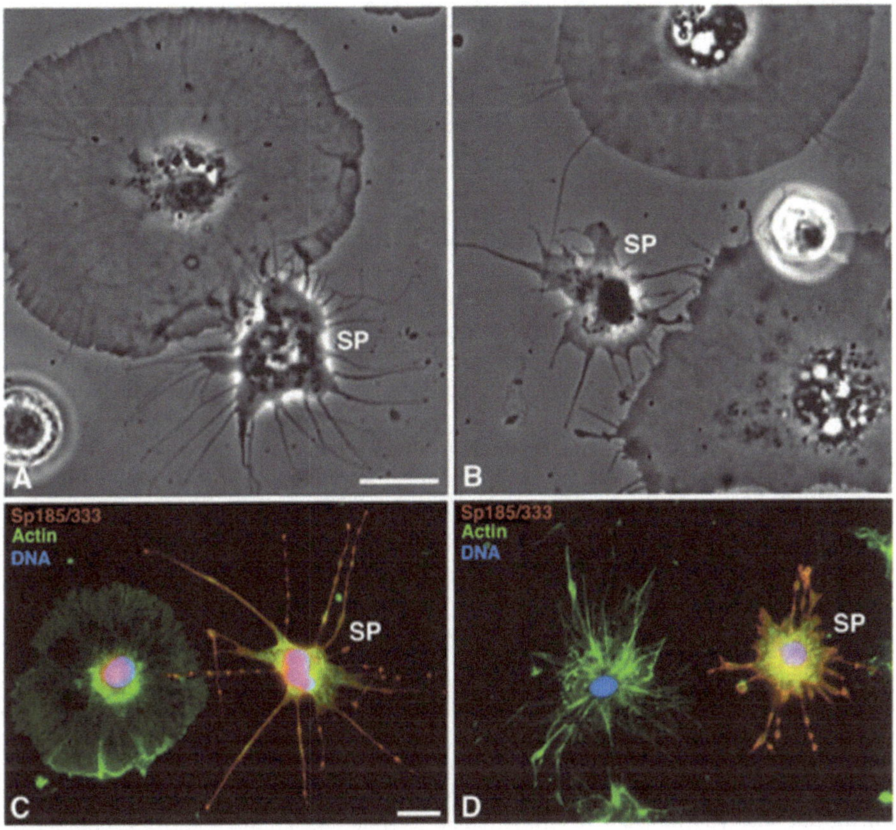

Figure 5. Small phagocytes in living preparations. Small phagocytes (SP) can be identified based on their relative small size, unusual filopodial morphology and granular cytoplasm (panels A,B). These cells are clearly labeled by antibodies against the Sp185/333 family of proteins (SP cell in panels C,D), whereas the petaloid and filopodial discoidal large phagocytes do not stain for Sp185/333. The Sp185/333 labeling of the small phagocytes highlights the atypical nature of their filopodia which are often thick and contain knobs, not common for filopodia of discoidal phagocytes (left-hand cell in panel D). Note that in some small phagocytes the filopodia become very elongate (panels A,C). Bar = 10 microns.

Coelomocytes in Other Echinoderms

Asteroids clear their coelomic cavities efficiently despite having only two major cell types, phagocytes and colorless spherule cells.[22,40] Rapid activation and differentiation of the phagocytes into multiple functional cell types upon bacterial infection is likely responsible for efficient microbial clearance.[41] The phagocytes in the sea star, *Asterias rubens*, are present in a continuum of small to large cells that change relative abundance upon immune stimulation. Larger cells become more numerous and exhibit strong phagocytic reactions to bacteria and the total number of phagocytes triple 24 hrs after challenge.[41] Increased numbers of total coelomocytes have also been noted in response to wounding and hypoxic conditions.[42] Swift responses, immune challenges, and the possible differentiation of phagocytes into cells with augmented phagocytic activity is an interesting approach to infection in sea stars that have fewer morphotypes of coelomocytes compared to echinoids.

Box 1. Larval immune cells.

While most investigations of echinoderm immunity have been carried out in the adult, the larvae of indirectly-developing species possess specialized mesodermal cells derived from embryonic blastocoelar cells that display a variety of immune behaviors. The immune activities of the larval immunocytes were recognized at the inception of cellular immunity by Metchnikoff and his infamous experiment of encapsulation of a rose prickle by blastodermal cells of a sea star larva.[3] This was the first demonstration of phagocytosis and encapsulation and lead to a nobel prize. More recently, recognition and phagocytic behavior of larval immune cells responding to bacteria or yeast injected into the blastocoel has been demonstrated.[218,219] The immune cells in the embryo and larvae are the blastocoelar cells located in the blastocoel[220] and the pigment cells in the ectoderm[19] that produce echinochrome A.[221] Embryonic and larval immune cells are also known to express immune factors, including the complement homologue SpC3[92] and the immune response genes *Sp185/333*.[10,70,92] Besides immune effector functions, these cells carry out a number of immune functions including surveillance-like behavior, wound healing and the expression of a complex suite of immune recognition, regulatory and effector genes[10] (Rast, unpublished). As more data are accumulated about this simple stage of the sea urchin life cycle, postgastrula embryos and larvae are becoming an increasingly valuable model system in which to investigate immunity.

Holothurians have lymphocytes, morula cells, amoebocytes, crystal cells, fusiform cells and vibratile cells.[22,40] In addition to phagocytosis, holothurian coelomocytes exhibit brown body formation in response to multicellular parasites. These are pigmented aggregates of phagocytes and spherule cells that encapsulate parasites that are too large to be phagocytized.[43] Brown bodies in the sea cucumbers, *Holothuria polii* and *H. tubulosa*, are tissue nodules containing entrapped parasites that are first surrounded by a fibrous, electron-dense, noncellular layer—probably melanin—and then covered by layers of specialized cells[44,45] including subsets of elongated phagocytes[44] and spherule cells.[46] Spherule cells likely degranulate to chemically kill and/or degrade the invader,[46] while other cells within the brown bodies possess active phenoloxidase resulting in melanization.[44] Brown bodies are progressively eliminated from the animal through coelo-rectal canaliculi.[44,45,47] Essentially, coelomocytes appear to have two means for clearing microbes and parasites; phagocytosis and encapsulation.

Crinoids have phagocytes, red spherule cells and colorless spherule cells (or morula cells), whereas ophiuroids have phagocytes, colorless spherule cells and possibly crystal and vibratile cells.[22,40] Because little work has been done on these classes of echinoderms, little is known about the functions of coelomocytes in these groups.

CLOT FORMATION

Clotting is mediated by a variety of agglutination factors. It is an important response to injury in echinoderms functioning to block loss of CF resulting from wounds, and to sequester pathogens and prevent their invasion throughout the body. A 220 kDa agglutinating factor thought to be involved with coagulation is secreted from coelomocytes of the sea cucumber *Holothuria polii*.[48] Coelomocytes from the purple sea urchin express amassin-1, a 75 kDa, multidomain protein with an olfactomedin domain that is stabilized by multiple disulfide

Table 2. ESTs from the purple sea urchin and a sea cucumber

Category	Examples
Defense	Sp185/333, DD104, DD186, Complement C3, Complement factor B, Galectin, SpEchinoidin and other C-type lectins, Thrombospondin, Amassin, Integrin-βC, Ferritin, Melanotransferrin, Annexin, Serum amyloid A, Echinonectin, Fibrinogen-like protein, Ficolin-like protein, Kazal-type serine proteinase inhibitors
Cytoskeleton	Actin, Gelsolin, Cofilin, Thymosin β, Protein tyrosine kinase-9, Fascin, Tubulin, Microtubule-associated protein
Signal Transduction	SpTie1/2, Steroid hormone receptor, LPS-induced TNFα, Protein tyrosine phosphatase receptor, Adenylyl cyclase, Receptor for activated protein kinase C, Guanine nucleotide binding protein β, Rho-GTPase, Rho-GDP dissociation inhibitor, GTPase-activating protein, FK Binding Protein-12, GF14/14-3-3
Nuclear Activities, Splicing	RNA/DNA binding protein, DNA methyl transferase-associated protein, Nonhistone nucleic acid-binding protein, Splicing factor 30, Paraspeckle protein, ET putative translation product, Heterogeneous nuclear ribonucleoprotein R, PolyA binding protein
Transcription factors	Steroid hormone receptor, LPS-induced tumor necrosis factor α, Immediate early-response protein, SpRunt, SpNFkB
Protein Metabolism, Synthesis, Degradation	Translation elongation factor 1α,1γ, 2, Translation initiation factor, Peptide chain release factor, Signal sequence receptor, Protein disulfide isomerase, Binding Protein, Heat shock protein, gp96, Presenilin, Proteosome subunit, Ubiquitin
Molecular Transport	Dynein, Kinesin, Vesicle trafficking protein, Coated vesicle membrane protein, ER transport protein, Vacuolar protein sorting protein, Rab7, Rab5-interacting protein, Sec22, vSNARE, Mannose-6 phosphate receptor
Endosomal System	Vacuolar H^+-ATPase, Cathepsins, Lysozyme, Arylsulfatase
Proliferation, Apoptosis	Polo-like kinase, Bax inhibitor-1, Allograft inflammatory factor-1
Metabolism	ATP synthase, Cytochrome C oxidase subunit 1, 4, Cytochrome b, NADH dehydrogenase subunit 2, 4, 5, Fatty acid desaturase, ATP/ADP translocase, Vacuolar ATP synthase subunit, Na^+/phosphate cotransporter, Na^+K^+-transporter, Citrate synthase, Malate dehydrogenase

From references 49, 52, 53, 88 and 165.

bonds (Table 2; Table 3;[49] Dheilly, Raftos and Nair, unpublished). Amasin-1 functions in cellular clot formation through homooligomerization and cross-linking coelomocytes.[50] A search of the sea urchin genome identified a putative receptor for amassin-1, colmedin, which also has an olfactomedin domain and is expressed in coelomocytes.[51]

Other molecules involved in clotting mechanisms, including transglutaminase homologues, were identified during an expressed sequence tag (EST) study from gut tissue of the sea cucumber *Holothuria glaberrima*.[52,53] Many of the genes referenced from

Table 3. Proteomic analysis of whole coelomic fluid from *Strongylocentrotus purpuratus*

Classification	Examples
Immune Response	Complement components (SpC3, SpBf), Scavenger receptors cysteine-rich, Sp185/333
Pathogen Destruction	Arylsulfatase and other lysosomal enzymes, α-2-macroglobulin
Clotting	Amassin, Annexin V, Von Willebrand factor
Metal binding	Major yolk protein, Transferrin, Ferritin, Ceruloplasmin
Cytoskeleton	Actin, Profilin, Fascin, Cofilin, Gelsolin, Myosin, Microtubule-associated protein, Arp 2/3 complex proteins, Coronin, Tubulin, α-actinin, Tetraspannin, Talin, Vinculin, Rab
Cell Adhesion	Integrin, NCAM, Selectin, Cadherin, Fibronectin
Signalling	Ras
Cytoplasmic Enzymes	Oxidative enzymes

From Dheilly, Raftos and Nair, unpublished.

these EST studies were originally studied in the context of the intestinal regeneration that occurs after sea cucumbers expel their internal organs as part of an antipredation response, and then regenerate the entire digestive tract in 30 days. As such, evisceration and regeneration exposes the coelomic cavity to microorganisms from the surrounding seawater and the animal's own enteric microflora. Thus, genes related to immune function and tissue regeneration are upregulated in intestinal tissues undergoing regeneration and are likely essential for the animal to survive the process.[52-55]

KEY MOLECULES IN ECHINODERM IMMUNITY

Proteins with Leucine Rich Repeats

Toll-Like Receptors

The Toll-like receptors (TLRs) are best characterized in insect and mammalian systems, in which they form small gene families of ~10 genes.[56,57] These receptors have an N-terminal solenoid-like leucine-rich repeat (LRR) ectodomain, a transmembrane region and a cytoplasmic Toll/interleukin 1 receptor (TIR) in the C-terminus.[58] The genes are often encoded in a single exon. There are more than 200 TLRs in the *Strongylocentrotus purpuratus* genome, far more than has been identified in any other species to date.[10] These genes fall into three categories: a large family of more than 200 genes with vertebrate-like structure; a set of three genes with structure similar to *Drosophila* Toll (which differs from that of the vertebrate TLRs and *Drosophila* Toll-9[59]) and a family of five genes with a short truncated ectodomain structure (Fig. 6). In vertebrates, the ectodomain consists of canonical leucine-rich repeats flanked by specialized, cysteine-rich domains (designated single cysteine cluster, sccTLR). In

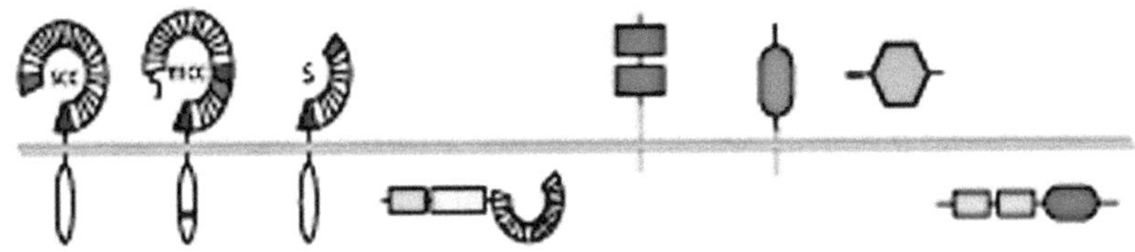

	TLR			NLR	SRCR	PGRP	GNBP	RLR
H.s.	10(+1Ψ)	0	0	22	81(16)	6	0	3
M.m.	13	0	0	33	118(27)	4	0	3
S.p.	**214**	**3**	**5**	**>200**	**1095(218)**	**5**	**3**	**12**
D.m.	1	8	0	0	14(7)	15	4	0
C.e.	0	1	0	0	3(1)	0	0	0

Figure 6. Pattern recognition receptor gene families are significantly expanded in the purple sea urchin with respect to other model organisms. The numbers of gene models encoding Toll-like receptors (TLR), Nod-like receptors (NLR), scavenger-receptor cysteine-rich (SRCR), peptidoglycan recognition proteins (PGRP), Gram negative binding proteins (GNBP) and RIG-I-like receptors (RLR) for human (*H.s.*), mouse (*M.m.*), purple sea urchin *S. purpuratus* (*S.p.*), fruit fly *Drosophila melanogaster* (*D.m.*) and nematode *Caenorhabditis elegans* (*C.e.*)[212] are shown. Given the complexity associated with identifying SRCR gene models accurately, the number of SRCR domains is shown, with the number of gene models that contain multiple SRCR domains indicated parenthetically. In the electronic version of this chapter, colors are used to illustrate protein domains and to emphasize the sea urchin gene model numbers. A color version of this image is available at www.landesbioscience.com/curie.

contrast, the *Drosophila* Toll has specialized domains that are located in the center of the ectodomain (multiple cysteine cluster; mccTLR).[58] The mccTLRs may be the ancestral form,[10] as mccTLR genes are present throughout eumetazoans, with the exception of the vertebrate lineage, in which it has apparently been lost. The sea urchin short TLRs have distant similarity within the TIR domain to the ancient Toll genes.[60]

The large family of sea urchin TLRs is unusual in both its multiplicity and the apparent rapid diversification of some of its subfamilies. The ~210 genes in this family can be divided into seven subfamilies based on phylogenetic analysis of the TIR domain. Some of these subfamilies are composed of many members that differ primarily within their ectodomains while the TIR domains show greater conservation. Divergence within the leucine-rich repeats takes a number of forms, including point mutations, insertion-deletions between LRRs and insertion-deletions of whole LRR units.[10] The extensive diversity of this class of sea urchin TLRs, along with a relatively large proportion of pseudogenes, differs from the more conserved picture seen in vertebrate TLR evolution[61] and suggests that this complex family of receptors function fundamentally differently.[60]

Many of the sea urchin TLR families are expressed most highly in coelomocytes in addition to gut tissue.[10] Expression of sea urchin TLRs is not detected in the embryo but many families are expressed in the feeding larvae. Thus, expression patterns and diversity of sea urchin TLRs are consistent with an immune rather than developmental function. Consistent with this is the fact that extensive investigations of vertebrate TLRs only show immune functions.

NOD-Like Receptors

The second family of expanded immune receptors in the sea urchin is the NOD-like receptors (NLRs).[10] NLRs are cytoplasmic pattern recognition receptors (PRRs) that recognize a variety of microbial signatures, including LPS, peptidoglycan, dsRNA and flagellin.[62] Structurally, NLRs are composed of C-terminal LRRs, a central NACHT domain and one of several N-terminal domains that function in protein-protein interactions. NLRs in mammals function in the immune response by serving as scaffolding proteins to assemble protein complexes that lead to the activation of the NFkB and MAPK signalling pathways. NLRs have also been shown to activate inflammatory caspases, including caspase-1, which is responsible for the processing of pro-interleukin-1β.[63] There is a significant expansion of the NLR gene family in the *S. purpuratus* genome, which has over 200 NLR gene models compared to vertebrates that typically have ~20 NLR genes (Fig. 6).[10] NLRs appear to be restricted to the deuterostome lineage and have not been identified in any of the sequenced protostome genomes, including *Drosophila* or *Caenorhabditis elegans*. As in vertebrates, the sea urchin NLR genes contain LRRs, a central NACHT domain and an N-terminal protein-protein interaction domain. However, unlike in the vertebrate system, this N-terminal domain is most commonly a DEATH domain, another member of the death-domain superfamily. A small number of sea urchin NLRs also contain CARD domains at the N-terminus, while the PYD domain is absent outside of the vertebrate lineage. Although many of the NLR gene models encode C-terminal LRR domains, these domains are lacking from some sea urchin gene models, which may be due to problems with accurate computational prediction of gene structure. The sea urchin NLR family appears to be the result of a sea urchin-specific gene expansion, which is seemingly more diverse than the similarly expanded sea urchin TLR family. The extent of this diversity is likely underrepresented given the incompleteness of the gene models, particularly within the LRR region. Although the function of the sea urchin NLRs is unknown, it is notable that they are most highly expressed in the gut,[10] and therefore may be involved in managing gut microflora. This mimics the role of the NLR NOD2 protein in mammalian systems, which when mutated, results in inappropriate inflammatory reactions in the gut tissue and leads to Crohn's disease.[64] The LRR gene families in the sea urchin are greatly expanded compared to families in vertebrates and insects and it is noteworthy that the LRR families in amphioxus are also expanded,[65] suggesting that these receptors play an important role in the innate immune functions of these deuterostome invertebrates.

Sp185/333—A DIVERSE FAMILY OF GENES AND PROTEINS EXPRESSED IN RESPONSE TO IMMUNE CHALLENGE

The *Sp185/333* Gene Family

The initial discovery of the *Sp185/333* family was the result of an EST analysis of transcripts that are upregulated in response to LPS challenge (see Box 2).[49] The diversity observed among the *Sp185/333* transcripts is intriguing for a putative immune response repertoire and resulted, in part, from an extraordinarily diverse gene family. The *Sp185/333* genes are small with two exons, of which the first is short and encodes a hydrophobic leader sequence, while the second encodes the remainder of the highly variable protein. The *Sp185/333* genes are atypical for *S. purpuratus* in three respects. (1) the second exon ranges in size from 771-1431 base pairs (bp), which is notably larger than the average exon length (100-115 bp) as characterized from all gene models in the genome; (2) the intron is smaller than average (~400 bp, compared to an average intron size of ~750 bp); and (3) the average *S. purpuratus* gene has 8.3 exons.[11,66] The most interesting aspect of the *Sp185/333* genes is the structure of the second exon. It is composed of contiguous blocks of sequence called *elements* that are defined from sequence alignments that require the insertions of large gaps (Fig. 7).[49,66,67] Elements are variably present or absent in different genes (and transcripts, see Box 2) in recognizable mosaic combinations that have been called *element patterns*. Each element is actually a set of sequences that are similar but not necessarily identical—elements in the genes and transcripts differ by single nucleotide polymorphisms (SNPs) and small insertions or deletions (indels). Much of the sequence diversity among the gene results from the element patterns in the second exon. Although there are only six *Sp185/333* genes assembled in the v. 2.1 build of the *S. purpuratus* genome, three independent lines of evidence suggest that the gene family is composed of 40 to 60 paralogous loci: quantitative PCR (qPCR) of genomic DNA,[68] statistical estimates based on the frequency with which unique genes were cloned,[69] and an estimate based on results from screening two BAC libraries for *Sp185/333* genes combined with assumptions about gene linkage.[70]

The *Sp185/333* genes have six types of repeats (Types 1-6)[69] in addition to elements (Fig. 7). Individual genes contain between two and four Type 1 repeats, which are located tandemly at the 5′ end of the second exon. In contrast, repeat types 2-6 are arranged as mixed, interspersed groups in the 3′ half of the second exon. The complexity of these repeats facilitates multiple sequence alignments, two of which have been analyzed in detail.[66] The "cDNA-based alignment" (Fig. 7A) was generated using the location of gaps in the *Sp185/333* transcript sequences as a guide.[49] Alternatively, because of the repeats within the sequences, genes (and transcripts) can also be aligned according to the boundaries of the repeats, which results in the "repeat-based alignment" (Fig. 7B). Regardless of the criteria used to align the gene sequences, they are characterized by a similar diversity in element pattern variation as is observed among the transcripts (see Box 2). To date, 171 genes have been isolated from four animals, of which 121 have unique coding sequences and have 33 different element patterns.[66] Furthermore, identical sequences have not been isolated from more than one individual, suggesting a level of diversity that exceeds the estimated 4-5% sequence divergence between *S. purpuratus* individuals.[71] Large, diverse gene families are a common theme in purple sea urchin immunity (see above) and understanding the mechanisms by which this diversity is generated and maintained is an important aspect of investigations of sea urchin immunity.

Gene-Level Diversification through Recombination

Despite the striking number of unique genes that have been isolated to date, the *Sp185/333* sequences are surprisingly similar. Overall, the genes share >88% pairwise identity. The number of different versions of each element is relatively low.[72] The complexity of the *Sp185/333* gene family, therefore, is the result of a mosaic pattern of these few element sequences distributed among the genes. Analysis of the evolutionary histories of five of the elements that were present in all of the genes revealed that each element evolved independently. This level of incongruence suggests that the extant *Sp185/333* gene family is the result of rapid and recent diversification events.[69] It is peculiar, given this high rate of diversification that pseudogenes have yet to be identified.

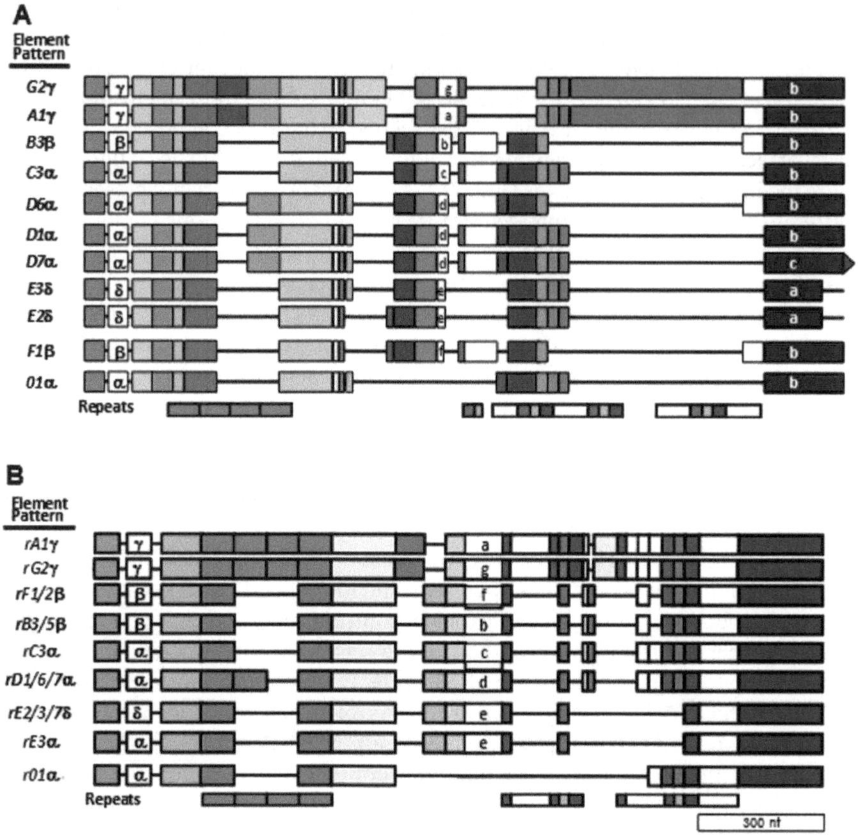

Figure 7. Two optimal alignments of the *Sp185/333* genes. The sequence complexity and repeats within the *Sp185/333* genes facilitate multiple alignments, two of which have been analyzed in detail.[66] The alignments differ in the pattern of elements in the second exon; the first exon and intron sequence do not have elements. Phylogenetic analysis of the intron sequences define five major types[66] (designated α-ε). The type of intron associated with each element pattern is indicated. A) The "cDNA-based alignment" results from the gaps used to optimize alignments of the *Sp185/333* ESTs and cDNAs.[49,68] This alignment divides the second exon into 25 elements. The terminal element is differentiated (designated a-c) based on the position of the first of three possible stop codons. The locations of each of the repeat types is indicated by the colored boxes under the alignments. B) An alternative alignment, the "repeat-based alignment", results from inserting gaps so that the repeats and elements correspond as much as possible.[66] This alignment divides the second exon into 27 elements based on the locations of gaps, as well as the locations of the repeats. The elements in this alignment that correspond to the repeats are color coded accordingly. The electronic version of this chapter shows the elements and repeats in color. A color version of this image is available at www.landesbioscience.com/curie.

With one exception, all but one of the cloned genes are predicted to encode intact open reading frames.[66] This may reflect a bias, however, in the primers that were designed in the untranslated regions of expressed genes,[49] so results do not confirm with certainty to a true lack of pseudogenes within the genome. It should be noted that, extensive computational analysis of elements and regions of the intron suggest that element boundaries do not necessarily serve as "recombination hotspots", but rather, that recombination appears to occur throughout the length of the genes.

The mechanisms that promote this frequent recombination are unknown. However, there are a few characteristics of the *Sp185/333* gene family that may contribute to its rapid diversification. Amplification of the regions between genes and preliminary analysis of a sequenced BAC insert with six *Sp185/333* genes show that many genes are closely linked (~3 kb). In addition to the six types of repeats found within the coding regions, the genes are flanked on either side by stretches of di- and trinucleotide microsatellites[66] (Miller, Buckley and Smith, unpublished). These repeats are closely associated with the boundaries of two types of large segmental duplications that include the *Sp185/333* genes (Miller, Buckley, Easley and Smith, unpublished). Microsatellites have been associated with genomic instability and increased recombination frequency[73,74] and have been implicated in mediating recombination of the variable surface glycoprotein (VSG) genes from *Trypanosma brucei*[75] and plant *R* genes.[76] The genomic organization, high sequence similarity and repeats within and surrounding the genes likely promote the diversification of the *Sp185/333* gene family through frequent recombination and thereby contributing to the complex protein repertoire.

Sp185/333 Transcript Editing

Given the diversity of the *Sp185/333* gene family, comparison of the gene and message sequences from individual animals yielded the surprising result that the two sets of sequences are very different.[72] Specifically, 148 messages and 53 genes were isolated from a single animal, of which only five of the sequences matched identically to another. Similarly, there was little or no overlap in gene and message sequence in two other animals. Furthermore, the gene and message repertoires were generally characterized by different element patterns, such that the predominantly expressed element pattern following immune challenge was *E2*, whereas the most common gene element pattern was *D1*. Although about half of the messages isolated from immunoquiescent animals had a truncated *E2* element pattern, called *E2.1*, the SNP that introduced an early stop codon was never found among the *Sp185/333* genes. In fact, no genes were identified with premature stop codons or indels resulting in frame shifts, which were both common features of *Sp185/333* transcripts.[67] When genes and message from individual animals were compared, the large majority of messages expressed both before and after immune challenge were the likely product of a few genes.[72] Conversely, most of the genes that comprise the large *Sp185/333* gene family were not transcribed. Notably, the pattern of nucleotide substitutions between the messages and the genes from which they were most likely transcribed indicated a bias towards transitions, specifically a uridine in the message at a position in which the gene contained a cytidine. This pattern of nucleotide substitution is consistent with cytidine deaminase activity. A number of cytidine deaminase-like molecules have been annotated within the sea urchin genome, but phylogenetic analysis of these sequences fails to identify homologues of activation-induced cytidine deaminase (AID).[10] AID, which has only been identified in vertebrates, is involved in class switch recombination and somatic hypermutation of immunoglobulins in B cells.[77] Alternatively, it is possible that the observed differences between *Sp185/333* gene and message sequences result from low-fidelity polymerase activity and the sea urchin genome does contain a homologue of terminal deoxytransferase and polymerase μ (Tdt/Polμ).[10,78] In higher vertebrates, this enzyme is also involved in immunoglobulin diversification, as well as low-fidelity DNA replication.[79,80]

Two Levels of Diversity

The *Sp185/333* gene family is an intriguingly diverse facet of the sea urchin immune response.[66] In response to immune challenge, this gene family is highly expressed and produces a diverse message repertoire.[49,67,68] The genes are believed to diversify through frequent recombination that does not appear to be limited to element boundaries and may be mediated by repeats within and flanking the coding sequences.[69] Given the diversity within the gene family, it is surprising that a second tier of diversification appears to affect the message sequences. That the majority of the messages appear to be derived from a few genes suggests that many of the *Sp185/333* genes may be nonexpressed pseudogenes and serve as a source of sequence diversification to the expressed genes. Alternatively, it may be that these genes are simply not expressed under the limited immunological challenges with which the animals have been presented.[67,68] Thus, the complex *Sp185/333* gene family represents a novel form of invertebrate immunological diversification both at the genomic and, also, possibly, at the posttranscriptional level.

In addition to diversity that appears to be generated by gene recombination, duplication, deletion, conversion,[69] and mRNA editing,[72] the array of Sp185/333 proteins show unexpectedly greater structural complexity than predicted from the genes and messages,[81] Not only are the arrays of Sp185/333 proteins different among different individual sea urchins, but the majority of sizes are at least twice as large as predicted and up to ≥ 200 kDa (Fig. 8).[29] The pI values range from 3-10 although the majority of isoforms have a pI more acidic than predicted. There are up to 260 discrete isoforms in

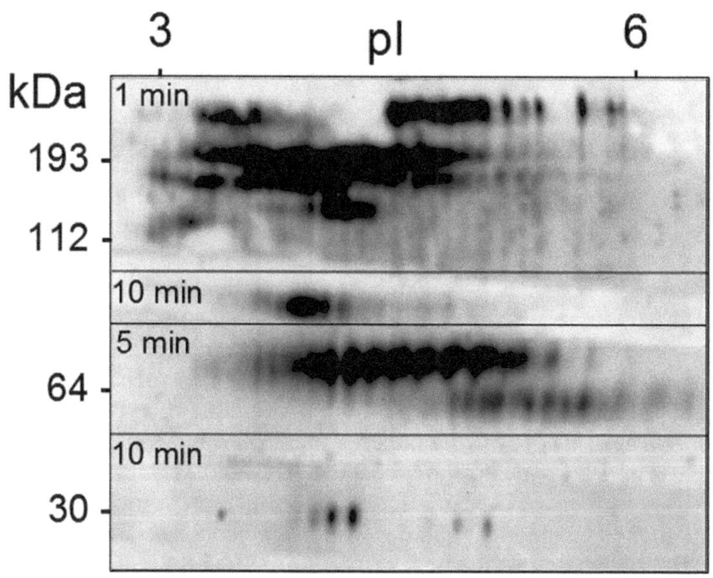

Figure 8. Over 260 spots for Sp185/333 proteins are present in CF from a single sea urchin. Total proteins from CF were separated by two dimensional gel electrophoresis and analyzed by Western blot with anti-185 antisera. The image is a composite of different regions of the blot that received different exposure times to optimize the spot intensities. pI units are shown at the top and molecular masses (kDa) are shown to the left. Reproduced from reference 81 with permission, ©2009, The American Association of Immunologists, Inc.

individual sea urchin coelomocytes and many of the expressed proteins are truncated, likely a result of mRNA editing.[67,81] Finally, the arrays of Sp185/333 proteins change in response to different pathogen-associated molecular patterns (PAMPs), which may result from a combination of variations in gene expression, mRNA editing and posttranslational processing of the proteins. How these several levels of diversification are coordinated and perhaps directed towards different types or species of pathogens will be the focus of future research.

COMPLEMENT IN ECHINODERMS

The complement system is a system of central importance in immunity for vertebrates and comprises over 30 known humoral and cell surface proteins.[82] Complement activation occurs by three major pathways: the classical pathway that is activated by antigen-antibody interaction, the lectin pathway that is activated by mannose binding lectin (MBL) or ficolins and the alternative pathway that is initiated through C3 autohydrolysis (reviewed in ref. 83). C3, a thioester containing protein, is the central component of the cascade and is activated by all three initiation pathways. It acts to coat the surfaces of pathogens, functioning as the initiator of the terminal pathway, an adjuvant for activating the adaptive immune response[84] and as an opsonin for direct pathogen recognition, phagocytosis and killing.

Investigations of opsonization and phagocytosis by coelomocytes from the green sea urchin *Strongylocentrotus droebachiensis* show that opsonizing target cells with mammalian C3 enhances the coelomocyte response,[39,85-87] suggesting that coelomocytes have receptors for mammalian C3 and consequently, sea urchins themselves may express C3-like proteins. The identification of two expressed sequence tags (ESTs) from the purple sea urchin, *S. purpuratus* with sequence homology to C3 and to factor B (Bf) called SpC3 and SpBf[88-90] was the first evidence that invertebrates have a complement system. Gene models annotated from the sea urchin genome reports several gene models encoding thioester proteins, including a second C3 homologue, Sp-C3-2, several factor B genes, MBL and ficolins (Table 4) (Rast, unpublished).[10]

SpC3

SpC3 has several conserved functional regions that are also present in other members of the thioester protein family, including a conserved thioester site, a histidine that regulates binding specificity, a putative C3-convertase site, a two-chain structure putatively disulfide bonded via cysteines in conserved positions, putative factor I cleavage sites and other conserved regions.[90] Immunoquiescent sea urchins responding to LPS generally show dramatic increases in the amounts of SpC3 in the CF[91] with localization to small transport vesicles in subpopulations of discoidal phagocytes and polygonal phagocytes.[28] Expression of *Sp064*, the gene which encodes SpC3, responds to immune challenge in adult animals and also in embryos cultured in the presence of heat-killed bacteria.[92] The function of SpC3 could be predicted from the deduced amino acid sequence and through comparison to mammalian C3 functions. The characteristics of the thioester, which forms covalent bonds with target molecules that lack protection against complement attack (i.e., microbial surfaces), suggests that SpC3 functions as an opsonin. Classic assays demonstrate that SpC3 binds methylamine, a small nucleophile that interacts covalently with thioester.[93] Furthermore, not only does SpC3 undergo autolysis, a reaction that, under appropriate

Table 4. Complement proteins in the sea urchin, *Strongylocentrotus purpuratus*

Gene Model or cDNA	Encoded Protein	Predicted Pathway
Sp-064	SpC3	Alternative
Sp-C3-2	SpC3-2	Alternative?
Sp-TCP1, Sp-TCP2	Thioester containing proteins	?
Sp-thioester containing protein-1, -2, -3/4	Thioester containing proteins	?
Sp-factor B	SpBf	Alternative
Sp-factor B-2, -3	SpBf-2, SpBf-3	Alternative?
SpSM30-F	Mannose-binding protein	Lectin
Sp-C1q-like (4*)	SpC1q	Lectin
Sp-MACPF (21*)	Perforin-like proteins	?
Sp-CD59, Sca2-like1, 2	CD59	Regulatory
Sp5	SpCRL	Regulatory?
Sp5013	SpCRS	Regulatory?

*Numbers of gene models.
From references 10 and 124.

conditions, results in the cleavage of the peptide bond between glutamic acid and glutamine within the thioester (reviewed in ref. 94), but methylamine binding blocks SpC3 autolysis.[93] When CF containing SpC3 is incubated with yeast, SpC3 can be detected on the yeast surface and augmented phagocytosis of the yeast by coelomocytes is inhibited by addition of anti-SpC3 antibody.[95]

SpBf

The deduced amino acid sequence and domain structure of SpBf show significant similarity to the vertebrate Bf/C2 family of proteins.[89] SpBf is a mosaic protein with short consensus repeat (SCR) domains, a Von Willebrand Factor (vWF) domain and a serine protease domain. SpBf has a conserved cleavage site for a putative factor D that is conserved compared to cleavage sites in other Bf/C2 proteins. Members of the Bf/C2 family are mosaic proteins and most have three SCRs, although some have more than three[96] including SpBf, which has five.[89] Sequence analysis of these small domains shows that the first two may be the result of a duplication event and that SCR4 may be the result of a recombination between SCR3 and SCR5.[89] Furthermore, alternative splicing produces some mRNAs with three or four SCRs.[97] The results suggest redundancies of SCRs in the SpBf protein and that the predicted function may be conserved in SpBf, even though additional SCRs are present, as the relative order of the SCRs in SpBf is maintained.[89] The gene encoding SpBf, *Sp152*, is expressed in the phagocyte fraction of coelomocytes with low levels detected in ovary, testes, gut and esophagus,[97] although it is not clear whether expression in tissues is actually due to coelomocytes present in the tissues rather than the tissue cells themselves. Unlike *Sp064* expression, *Sp152* expression is not induced by LPS and appears to be constitutive.

Complement Phylogeny and Evolution

The identification of a C3 homologue in the purple sea urchin[88] inspired searches for complement components in a wide range of invertebrates. Phylogenetic analysis of thioester proteins including homologues of C3 with α-2-macroglobulin homologues as the outgroup shows that vertebrate C3, C4 and C5 cluster into three well supported clades (Fig. 9). The invertebrate C3 homologues cluster in three paraphyletic clades at the base of the vertebrate complement clades. SpC3 clusters with other invertebrate C3 sequences, while the tunicate sequences form a chordate cluster with the vertebrate sequences (see also ref. 98). The structure of the tree suggests that the thioester complement proteins diverged after the separation of the major phyla. Phylogenetic analyses of members of the Bf/C2 family show that the sea urchin homologue, SpBf, is positioned near the base of the tree[89] with cnidarian Bf/C2 being more ancient.[98] A number of analyses of the complement family of proteins have suggested that they evolved from a restricted set of primordial genes[99,100] (reviewed in refs. 101, 102). Conserved sequence motifs, such as the thioester site (GCGEQ) and similar organization of domains suggest that the ancestral complement system may have included a thioester protein, a Bf/C2 protein and a mannose binding lectin. These are all present in a number of invertebrates including the purple sea urchin (Table 4).

LECTINS

Lectins are a large and heterogeneous group of proteins and glycoproteins present in plants, microorganisms and animals, that function to bind mono- and disaccharides.[103,104] Lectins can be soluble proteins or integral membrane proteins that often exist as oligomers, contain at least two carbohydrate recognition domains (CRDs) and are capable of agglutinating cells and/or precipitating glycoconjugates.[105] In immunity, lectins are key molecules that function in cell-cell interactions, self/nonself discrimination and interactions between cells and the extracellular matrix (ECM), among other functions.[106] Lectins in invertebrates fall into four major groups based on similarity of structure to vertebrate lectins. C-type lectins are the most common type in various invertebrates[107-110] and consist of both soluble and integral membrane proteins that require divalent cations (Ca^{2+}) to maintain the CRD structure to bind carbohydrates. A second major group of invertebrate lectins are S-type lectins that specifically bind β-galactosyl residues,[111,112] S-type lectins are predominantly intracellular and employ free thiols for binding carbohydrates. A third major group of invertebrate lectins resemble vertebrate pentraxins and share properties with C-reactive protein and serum amyloid protein.[113-115] Finally, a fourth group of invertebrate lectins includes all those that cannot be otherwise categorized based on lack of information on their primary structure.

Innate immune functions performed by lectins include recognition and specific binding of microbial surface carbohydrates through the CRDs,[116] which exhibit seven different structural folding patterns for binding to different carbohydrate motifs.[117] Both mannose-binding lectin (MBL) and ficolins have been identified in the sea urchin genome[10] and a homologue of MBL has been characterized in a sea cucumber, *Apostichopus japonicus*.[118] It is speculated that these homologues may initiate the lectin pathway of complement in echinoderms (Table 4). MBLs and ficolins selectively bind mannose,

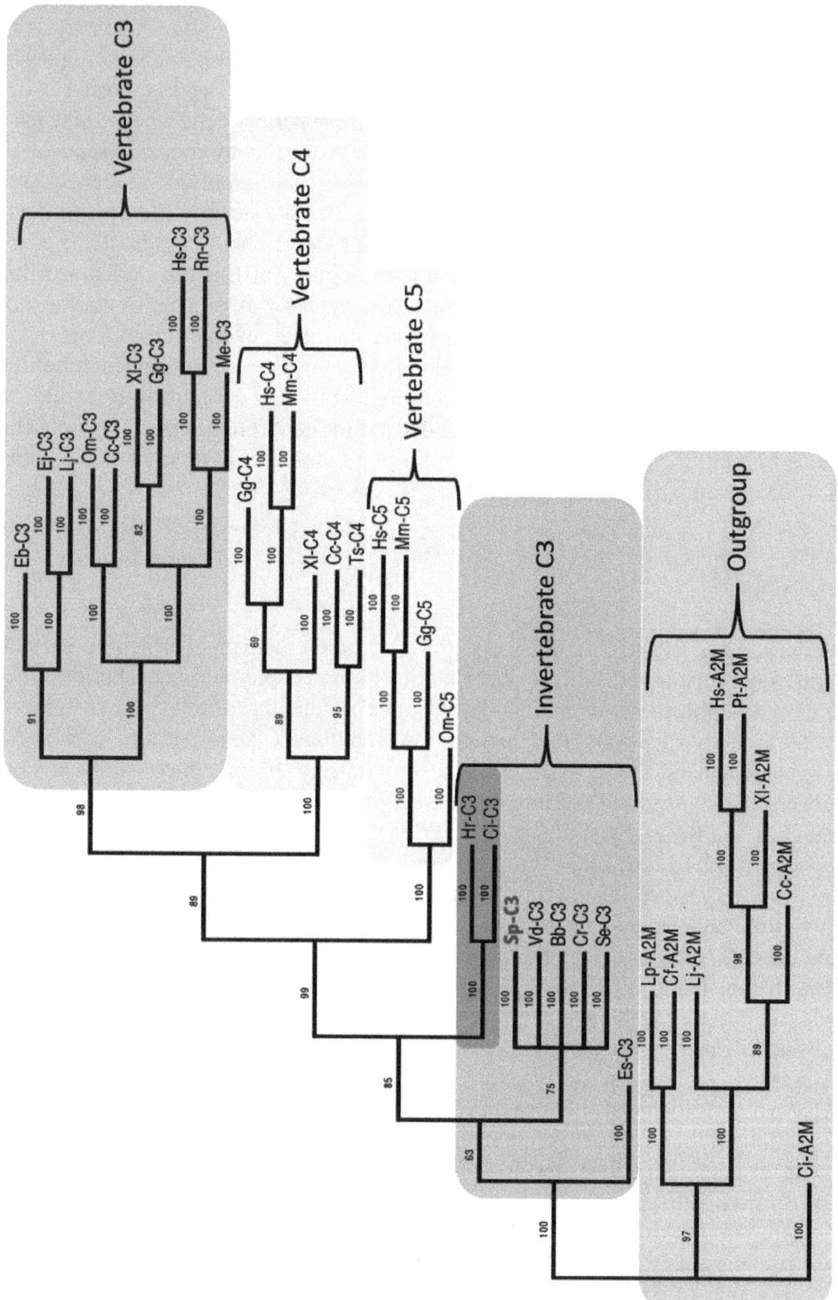

Figure 9. Please see the figure legend on the following page.

Figure 9, viewed on previous page. The phylogenetic relationships among members of the thioester protein family. Amino acid sequences for the thioester family were obtained from GenBank. The alignment was done in T-coffee[213] and manual editing was done with Mesquite.[214] A consensus tree was constructed in PAUP*[215] using maximum parsimony. Bootstrap support was generated with 10,000 iterations. Similar cladogram results were obtained by the maximum parsimony method in PAUP*, Neighbour-joining distance method in PAUP* and the Bayesian method in Mr. Bayes.[216] The Bayesian method was used with default priors and the GTR+G+I as the nucleotide substitution model suggested by jModelTest.[217] The α2macroglobulin sequence from horseshoe crab (*Limulus polyphemus*) was chosen as the outgroup. Sp-C3, *Strongylocentrotus purpuratus*, NP_999686; Hs-C3, *Homo sapiens*, AAR89906; Rn-C3, *Rattus norvegicus*, NP_058690.2; Me-C3, *Macropus eugenii*, AAW69835; Gg-C3, *Gallus gallus*, NP_990736; Xl-C3, *Xenopus laevis*, AAB60608; Lj-C3, *Lethenteron japonicum*, AAR13241; Ej-C3, *Entosphenus japonicus*, Q00685; Bb-C3, *Branchiostoma belcheri*, BAB47146; Cr-C3, *Carcinoscorpius rotundicauda*, AAQ08323; Se-C3, *Swiftia exserta*, AAN86548; Hr-C3, *Halocynthia roretzi*, BAA75069; Ci-C3, *Ciona intestinalis*, Q8WPD8; Eb-C3, *Eptatretus burgeri*, CAA77677; Om-C3, *Oncorhynchus mykiss*, I51339; Cc-C3, *Cyprinus carpio*, BAA36618; Es-C3, *Euprymna scolopes*, ACF04700; Vd-C3, *Venerupis decussates*, ACN37845; Gg-C4, *Gallus gallus*, T28153; Xl-C4, *Xenopus laevis*, BAA11188, Hs-C4, *Homo sapiens*, AAB67980; Cc-C4, *Cyprinus carpio*, BAB03284; Mm-C4, *Mus musculus*, CAA28936; Ts-C4, *Triakis scyllium*, BAC82347; Hs-C5, *Homo sapiens*, AAI13739; Mm-C5, *Mus musculus*, P06684 ; Gg-C5, *Gallus gallus*, XP_415405; Om-C5, *Oncorhynchus mykiss*, AAK82852 ; Lp-A2M, *Limulus polyphemus*, BAA19844; Lj-A2M, *Lethenteron japonicum*, BAA02762; Pt-A2M, *Pan troglodytes*, XP_001139559; Xl-A2M, *Xenopus laevis*, AAY98517; Cc-A2M, *Cyprinus carpio*, BAA85038; Cf-α2M, *Chlamys farreri*, AAR39412; Ci-A2M, *Ciona intestinalis*, NP_001027688; Hs-α2M, *Homo sapiens*, P01023.

fucose and other specific carbohydrates present exclusively on the surface of pathogens resulting in the direct activation of complement.[119-122]

Lectins were first demonstrated in echinoderms in the early 1980s.[123] Lectins have been identified in asteroids, echinoids and holothurians and characterized essentially from a functional and biochemical point of view. With improved molecular methods greater numbers of lectins and proteins with putative lectin function have been identified. Preliminary analysis of the sea urchin genome shows gene models encoding more than 100 small C-type lectins, over 400 mosaic proteins with lectin domains, 34 galectins, in addition to a few pentraxins and fucolectins[124] (Cohen and Smith, unpublished). This indicates that lectins in general are likely to have a variety of important functions in echinoderms, including recognition of foreign cells. Many, but not all of the lectins that have been identified in echinoderms have been C-type lectins (Table 5). Many show opsonin and agglutinin functions with the capability of binding carbohydrates on the surface of pathogens. It is generally accepted that lectins in echinoderms play an important role in the immune system, functioning as key molecules in immune responsiveness and to augment coelomocyte functions in host defense.

SCAVENGER RECEPTORS

Another expanded gene family in the sea urchin genome encodes a large repertoire of scavenger receptors containing multiple scavenger-receptor cysteine-rich (SRCR) domains that are both membrane-bound and secreted proteins.[125,126] Receptors of this structure are found throughout the animal kingdom but the sea urchin genome encodes more than 1000 SRCR domains in ~180 gene models, which greatly exceeds the multiplicity of these genes in other characterized species (Fig. 6).[10] These proteins are known to act as phagocytic receptors and some family members in mammals have been shown to bind bacteria.[127] These receptors are highly polymorphic in the population of purple sea urchins and show

Table 5. Echinoderm lectins

Class	Species	Name	Structure	Specificity	Agglutination	Reference
Asteroidea	*Asterina pectinifera*	(-)	C-type lectin	α-N-acetyl-galactosamine	yes	224
	Oreaster reticulatus	(-)	C-type lectin	galactosyl	(-)	225
Echinoidea	*Strongylocentrotus purpuratus*	SpEchinoidin	C-type lectin	galactose and derivates	(-)	88,124
	Anthocidaris crassispina	Echinoidin	C-type lectin	N-acetyl-galactosamine	yes	108
		SUEL	unique	D-galactoside	(-)	123
	Lytechinus variegatus	Echinonectin	(-)	galactoside	(-)	226
	Paracentrotus lividus	(-)	C-type lectin	(-)	yes	146
		*Pl*SL	(-)	D-glucose, L-rhamnose D-arabinose, L-fucose N-acetyl-D-glucosamine	yes	227
	Toxopneustes pileolus	SUL-I	(-)	D-galactose D-fucose	(-)	228
		SUL-II	(-)	D-galactose	(-)	228
	Tripneustes gratilla	TGL-I	C-type lectin	(-)	(-)	228

continued on next page

Table 5. Continued

Class	Species	Name	Structure	Specificity	Agglutination	Reference
Holothuroidea	*Apostichopus japonicus*	SJL-I	(-)	N-acetyl-D-galac-tosamine	yes	229
		SJL-II	(-)	simple carbohydrates	yes	229
		SPL-I	C-type lectin	D-glucuronic acid, D-galacturonic acid	(-)	230
		SPL-II	C-type lectin	D-galactosamine, D-galactose	(-)	230
		MLB-AJ	C-type lectin	α-D-mannans	yes	118
	Cucumaria echinata	CEL-I	C-type lectin	N-acetyl-galactosamine	yes	231
		CEL-II	C-type lectin	(-)	no	232
		CEL-III	C-type lectin	(-)	yes	231
		CEL-IV	C-type lectin	N-acetyl-galactosamine, α-galactose	yes	232
	Cucumaria japonica	(-)	(-)	branched α-D-mannans	yes	233
	Holoturia scabra	HSL	(-)	galctose derivates, T-antigen	yes	234

(-), not known, not done, not named.

pronounced variability in coelomocyte expression among individual animals suggesting a complex expression control system.[125] It is notable that representatives of all three of the most expanded families of sea urchin receptors (TLR, NLR and SRCR receptors) form a coregulated immune circuit that, in mammals, functions in gut immunity.[128]

ANTIMICROBIAL PEPTIDES

Antimicrobial peptides (AMPs) have been identified in a wide variety of species including bacteria, fungi, plants, insects, tunicates, amphibians, birds, fish and mammals.[129-132] Since the discovery of cecropins in insects[133] and defensins in mammals,[134] more than 1,200 different eukaryotic AMPs have been characterized.[135] In eukaryotes they form the first line of host defense against pathogenic infections and are a key component of the innate immune system. AMPs have an enormous variety of sequences and structures, but certain features are common. Most have a net positive charge and are 12-50 amino acids long, of which approximately half are hydrophobic.[136,137] However, a few peptides of up to 100 residues are also recognized as AMPs.[138]

Early work to document antimicrobial activities of crude extracts from echinoderms showed a wide range of activities against bacterial and fungal isolates[45,139] of which some functioned as antifoulants to deter the settlement of barnacle and bryozoan larvae in addition to bacterial colonization.[140] More recently, a variety of molecules with antimicrobial properties have been isolated from echinoderms, including steroidal glycosides,[141-143] polyhydroxylated sterols,[141] naphthoquinone pigments such as echinochrome A,[33,144] and complement homologues.[89,90] Lysozymes with antibacterial activity have also been detected,[145-147] and concentrations up to 15 μg/ml have been found in red spherule cells of the sea urchin *Paracentrotus lividus*.[34] Lysozyme and other antibacterial factors may act synergistically to provide effective defense against bacterial infections. In the Strongylocentrotids, the antimicrobial pigment echinochrome A is present in vesicles of red spherule cells and is bound to uncharacterized coelomocyte proteins.[148] A semi-purified coelomocyte fraction from *Paracentrotus lividus*, included fragments of beta-thymosin that were proposed to have antibacterial activity.[149] In extracts from coelomocytes of the sea star *Asterias rubens*, a number of partial peptide sequences were obtained and identified as fragments of actin, histone H2A and filamin A.[150,151] Antibacterial activity was detected in extracts of several tissues from the green sea urchin *Strongylocentrotus droebachiensis*, the common sea star *Asterias rubens*, and the sea cucumber *Cucumaria frondosa*,[152] with most activity in coelomocytes and body wall.

Strongylocins

Scans of the gene models in the purple sea urchin genome did not detect any sequences encoding recognizable AMPs.[10] This may be due to short exons that are difficult to recognize computationally, but may also be due to the divergent nature of these small proteins. However, two cysteine-rich AMPs, called strongylocins, have recently been isolated and characterized from the green sea urchin, *S. droebachiensis*.[153] Homologues are also present in the sister species *S. purpuratus*, called SpStrongylocins.[154] The strongylocin peptides are members of the cysteine-rich AMP family, which have six cysteines with three disulfide bonds involved in peptide conformation, stabilization and resistance to proteases,[155] and which are crucial for the antimicrobial activity.[156,157] The strongylocins have a novel cysteine pattern (Table 6) suggesting a different conformation than the other members of the group, perhaps to resist proteolysis within the coelomocytes and in the CF. Both the native and recombinant peptides show antibacterial activity against both Gram-positive and Gram-negative bacteria (Table 7). In addition, other peptides have been isolated and characterized from *S. droebachiensis* that appear as heterodimers and have strong activity against Gram-positive and Gram-negative bacteria (Stensvåg, unpublished). The heavy chain from one of the heterodimeric peptides also has strong activity against fungi and yeast.

The strongylocins are composed of three regions: a signal peptide, a prosequence and the mature peptide, and the strongylocin 1 peptides show high amino acid sequence similarity throughout (Fig. 10). However, SpStrongylocin 2 shares an identical signal peptide with strongylocin 1, instead of strongylocin 2. The prosequences are negatively charged, which may act to neutralize and stabilize the positive charge of the mature peptide[153,154] and presumably function as an intracellular steric chaperone during folding.[158-160] The peptides become active after the prosequences are cleaved during maturation. The first amino acid in the mature peptide of strongylocin 2 from *S. droebachiensis* is a tryptophan which is likely brominated. Although the recombinant SpStrongylocin 2 is not brominated, it shows equivalent antimicrobial activity to native peptides. Therefore, the posttranslational modification of tryptophan may affect the properties of the peptides by enhancing stability rather than mediating antimicrobial activity. The site of strongylocin activity is likely to be intracellular based on membrane integrity assays.[154]

Nonechinoids also synthesize a range of AMPs, of which many have been characterized. For example, the antibacterial activity in the CF of the orange-footed sea cucumber, *Cucumaria frondosa,* has been traced to small peptides ($\leq$6 kDa) that appear to be active at low pH (5.0-6.5) and which may be similar to the clavanins found in solitary tunicates.[161] Other immune-active chemical compounds with roles in maintaining antiseptic environments in nonechinoids include saponins and saponin-like compounds in sea stars and brittle stars, which are active against some Gram-positive bacteria.[141] Sea stars and brittle stars in particular express steroidal glycosides that exhibit antifungal activity and toxicity against brine shrimp.[162-164] It is likely that multitudes of molecules

Table 6. Cysteine patterns in AMPs containing six cysteines

Peptide Family	Cysteine Pattern[a]	Group of Organism
Strongylocins	C – C – C – **CC** – C	Echinoderms
Beta-defensins	C – C – C – C – **CC**	Mammals, birds
Alpha-defensins	C – C – C – C – **CC**	Mammals
Tachystatins	C – C – **CC** – C – C	Horseshoe crab
Knottin-type AMPs	C – C – **CC** – C – C	Plants
Thionins Type III and IV AMPs	**CC** – C – C – C – C	Plants
Insect defensins	C – C – C – C – C – C	Insects
Mytilus defensin	C – C – C – C – C – C	Molluscs

[a]Adjacent double cysteine residues are highlighted. Information regarding cysteine arrangements in the different peptides was obtained from the *Antimicrobial Peptide Database*.[135]

Table 7. Susceptibility of bacteria to strongylocins, recombinant SpStrongylocins and dimeric peptides from *Strongylocentrotus droebachiensis* and *S. purpuratus*

	Minimal Inhibitory Concentration (μM)					
Peptide	*Listonella anguillarum*	*Escherichia coli*	*Corynebacterium glutamicum*	*Staphylococcus aureus*	*Penicillium rogveforti*	*Candida albicans*
Strongylocin 1[a]	2.5	5.0	2.5	2.5	Nt[c]	Nt
Strongylocin 2[a]	1.3	5.0	2.5	2.5	Nt	Nt
Recombinant SpStrongylocin 1[b]	15.0	7.5	7.5	15.0	Nt	Nt
Recombinant SpStrongylocin 2[b]	15.0	7.5	3.8	15.0	Nt	Nt
Dimeric peptide 1[a]	2.5	1.3	1.3	2.5	6.25[d]	6.25[d]
Dimeric peptide 2[a]	2.5	2.5	1.3	5.0	Nt	Nt

[a]Minimal inhibitory concentration was determined as the lowest concentration of peptide causing an optical density less than 50% of the growth control without any peptide present.
[b]Minimal inhibitory concentration was determined as the lowest concentration of peptide causing 100% growth inhibition of the test organism compared to the growth control.
[c]Nt = Not tested.
[d]The peptide is the heavy chain of dimeric peptide 1. Growth inhibition was determined microscopically after 48 hrs of incubation.

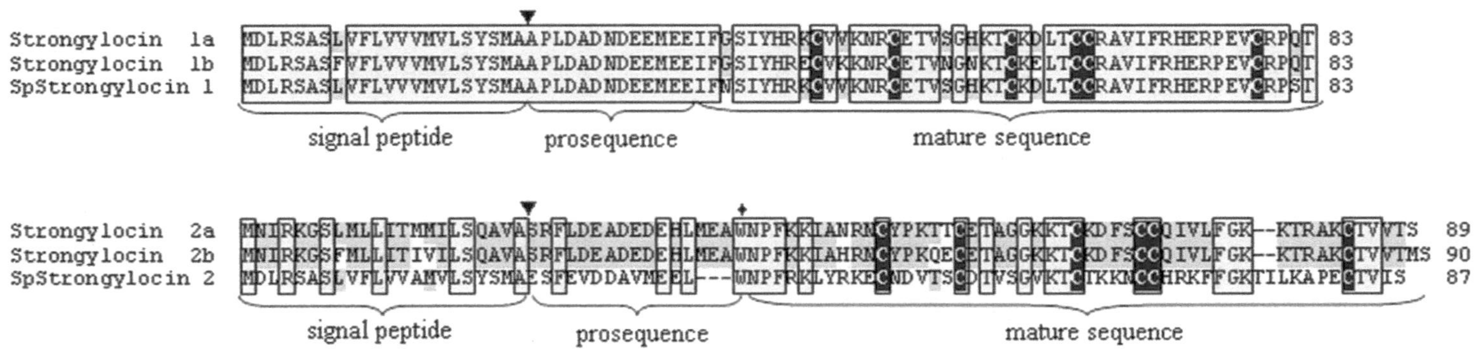

Figure 10. Alignment of strongylocins from *S. droebachiensis* and SpStrongylocins from *S. purpuratus*. The predicted cleavage site between the signal peptide and the proregion is indicated with a ▼. The first amino acid in active strongylocin 2 and SpStrongylocin 2 are likely a modified tryptophan, as indicated with a ♦. Identical amino acids are shown in boxes, similar amino acids are shown in gray and cysteines are highlighted in black. A color version of this image is available at www.landesbioscience.com/curie.

with activity against all types of pathogens function efficiently in echinoderms and have central importance for immune functions in this group.

GENE EXPRESSION IN SEA URCHIN COELOMOCYTES

Before it was possible to scan for gene models encoding proteins with putative immune function in the sea urchin genome, the only feasible approach for understanding the echinoderm molecular immunology was through EST profiling of messages in coelomocytes under challenged vs nonchallenged conditions.[49,88,165] ESTs encoding proteins with putative immune function have also been reported for the sea cucumber *Holothuria glaberrima*, undergoing gut regeneration after evisceration.[52,53] A variety of categories of proteins are expressed in coelomocytes that illustrate the activities of these cells in an assortment of functions (Table 2). The main focus of these studies was to identify immune response genes (see Box 2), which included protein matches suggesting several mechanisms for opsonization, clotting, iron sequestration plus the activity of a variety of signaling pathways. Matches were found to proteins with functions in the endosomal system (lysosomes) that may be involved in killing phagocytosed pathogens. Coelomocytes express a number of genes encoding proteins involved in splicing transcripts and translating proteins. They process, package, transport and secrete proteins as suggested from the transcripts encoding proteins involved in trafficking transport vesicles. Phagocytes have extensive cytoskeletons (as illustrated in Figs. 1-5), which is borne out by the number of ESTs matching proteins that function as cytoskeletal elements and others that act to modulate the cytoskeleton. One of the benefits of generating

Box 2. EST studies identify immune-related transcripts; discovery of the *Sp185/333* sequences.

Differential characterization of ESTs of coelomocytes from *Strongylocentrotus purpuratus* before and after immune challenge provide a picture of global changes in transcriptional activity. Genes encoding complement homologues,[88-90] transcription factors,[126] and a lectin,[67,88] among many others are induced by immune challenge (Table 6). Significantly upregulated transcripts include a large and diverse set of novel transcripts designated *Sp185/333* that were first discovered with the use of differential display[165] and subtracted probes followed by EST analysis.[49] Probes representing transcripts from LPS-activated coelomocytes were used to screen a high-density arrayed, conventional cDNA library made from bacterially-activated coelomocytes. About 4.5% of the clones in the library were positive and about 60% of clones selected for EST analysis showed significant sequence similarity to two previously uncharacterized cDNAs: *DD185* and *EST333*.[70,88,126] Given the considerable number of related *Sp185/333* sequences, the coding regions of the ESTs could be aligned. However, optimization of the alignment required the insertion of multiple, large artificial gaps[49] that defined 25 to 27 blocks of shared sequence called *elements*, which are variably present or absent in *element patterns* (Fig. 7).[66,67] The extraordinary diversity of the *Sp185/333* transcripts is based on the mosaic element patterns in addition to extensive single nucleotide polymorphisms (SNPs). The induction of *Sp185/333* gene expression in response to LPS and other pathogen-associated molecular patterns (PAMPs)[67,165] plus the diversity of the sequences suggest immune-related functions for the encoded proteins in *S.purpuratus*.

ESTs is that the level of expression rather than the level of match to known sequences can sometimes provide clues to putative immune function for unknown sequences. This has been the case for the *Sp185/333* sequences that are not known outside of the echinoids, but for which there is strong evidence of immune function.[70]

PROTEIN EXPRESSION PROFILE OF SEA URCHIN COELOMOCYTES

In addition to genome scans and EST analyses, high throughput methods in proteomics are being applied to analyzing proteins in echinoderms. In response to LPS challenge, 319 proteins were identified in the wCF of the purple sea urchin, *Strongylocentrotus purpuratus*, of which 284 were encoded by gene models in the genome and 48 were encoded by hypothetical open reading frames (Table 3) (Dheilly, Raftos and Nair, unpublished). The majority of the proteins were involved in modulating the cytoskeleton and linkage between the cytoskeleton and cell adhesion molecules, all of which are essential for intracellular transport and behaviors such as locomotion and phagocytosis. Cell adhesion molecules and intracellular signaling proteins were also identified, suggesting that coelomocytes respond to LPS with large-scale alterations to the cytoskeleton. Furthermore, the response to immune challenge implies the secretion of cellular products, as well as the endocytosis of extracellular fluids and phagocytosis of pathogens. Proteins involved in clotting and coagulation suggest another mechanism to sequester pathogens and to clear them from the CF, in addition to the prevention of CF loss after trauma. A number of proteins involved in opsonization are present in wCF including complement homologues. Metal-binding proteins are present, including ferritin, which has been identified in several echinoderm species under various conditions and using different analytical approaches.[49,52,88,166] Consequently, sequestration of iron must be an important mechanism for controlling the proliferation of invading microbes.

Proteomic analyses of coelomocytes from a second sea urchin species, *Heliocidaris erythrogramma*, demonstrated different expression profiles over time in response to sterile injections compared to injection of microbes (Dheilly, Raftos and Nair, unpublished). Results from shotgun proteomics showed two different profiles of proteomic changes. The proteins identified in the response to bacterial challenge were different from those identified in the controls. The proteomic profile changed in the coelomocytes from bacterially challenged animals within the first 6 hrs, largely due to the increased abundance of some proteins involved in cytoskeletal dynamics, while others decreased such as F-actin capping protein, advillin and α-actinin. Other cytoskeletal proteins that were absent in the controls, such as actins and annexin A7, were present in the challenged coelomocytes. Other proteins that appeared in the challenged coelomocytes included the complement homologue, C3 and SRCRs. The protein expression profile in sea urchin coelomocytes indicates that these cells show dynamic responses to wounding and immune challenge. Temporal analysis of proteomic changes in coelomocytes indicates that cellular responses to wounding and infection are biphasic. The initial phase, occurring within the first 24 hrs after treatment, appears to be a generalized response common to both types of insult. This phase involves reactions including CF coagulation and coelomocyte cytoskeletal remodelling, which resolves by 48 hrs to control levels. The second phase, which peaks at 48 hrs after injection, appears to be specific to microbial infections and the proteins expressed in this phase are involved in pathogen recognition and opsonization as well as the destruction of the invading microbes. A number of attendant sub-cellular pathways involved in signal transduction, endocytosis and exocytosis are also enhanced during this phase. These two phases thereby function in an integrated manner to repair wounds and to neutralize microbial infections.

REGENERATION

Many classes of echinoderms are remarkably plastic in their abilities for repair and regeneration resulting from both proliferation and transdifferentiation of circulating cells of mesenchymal origin (see Microscopy Research and Technique vol 55 no. 6, 2001). New evidence suggests that cells required for regeneration originate from coelomocytes.[167-169] Unfortunately, sea urchins have limited regenerative capabilities[170] compared to sea stars and crinoids, which have received more attention.[168] Sea stars are generally known for arm regeneration and levels of heat shock protein 70 (Hsp70) increase in the common sea star *Asterias rubens*, following arm tip amputation.[169] Furthermore, manganese treatment induces proliferation of coelomic epithelial cells that is coupled to traumatic stress responses including increased expression of Hsp70.[171] Red spherule cells (or amoebocytes) and polygonal phagocytes appear to be involved in regeneration.[169] These findings suggest a role for increased coelomocyte numbers and the expression of classic stress molecules in the early repair phase of tissue damage and regeneration.

Toposome, which is also referred to as the major yolk protein (MYP), is the most abundant protein in the CF (Table 3).[172] A monoclonal antibody to embryonic toposome also recognizes sea star coelomocytes and the coelomic epithelium.[169,173,174] Although previous efforts in cloning the MYP gene in *Strongylocentrotus purpuratus* suggested its relationship to vitellogenin,[175] protein fragments were too short for unambiguous identification. The cDNA sequences encoding MYP from the sea urchins, *Pseudocentrotus depressus*, *Paracentrotus lividus* and *Tripneustus gratilla*, show that they are members of the transferrin family, lack iron-binding sites, and are not homologous to vertebrate vitellogenins.[173,176] The toposome precursor, which has been postulated to serve multiple functions, is synthesized exclusively in the gut of the adult animal as a 180-190 kDa glycoprotein and the mature protein is found in the CF. The amount of toposome protein increases in response to traumatic stresses in agreement with increased gene expression as deduced from ESTs matching toposome in sea cucumbers regenerating gut tissue.[177] Toposome is expressed with other genes encoding proteins that function in wound healing, cell proliferation, differentiation, morphological plasticity, cell survival, stress response, immune challenge and neoplastic transformation.[55] It is emerging that circulating coelomocytes from some echinoderms originate from coelomic epithelia and are able to differentiate into a few tissue types, including nerve and muscle cells.[178] Accordingly, research on stem cells in marine organisms is becoming important for both comparative studies and for future applications.

ECOTOXICOLOGY AND THE ECHINODERM IMMUNE SYSTEM

Because coelomocytes are sensitive, stress-activated effectors of the echinoderm immune response,[19,179] they are good candidate biosensors for monitoring environmental stress in an environmental management context. Several examples of using coelomocytes as indicators employ both analyses of specific proteins and of cell function. Unfractionated coelomocytes from the sea urchin *Paracentrotus lividus* cultured at stress-inducing temperatures, (4°C or 35°C), or exposed to acidic pH, or high levels of cadmium, express high levels of Hsp70,[180,181] a well-recognized stress marker induced in response to a wide range of biological and physicochemical stresses.[106,182] In addition to serving as molecular chaperones, secreted and membrane-bound heat shock proteins (particularly Hsp60, Hsp70, Hsp90 and gp96) are potent activators of the innate immune system capable of inducing the production of proinflammatory cytokines by the monocyte-macrophage system.[183] Sp-gp96, which is expressed in sea urchin coelomocytes[88] is present on the surface of coelomocytes.[184] High levels of Hsp70 are also observed in coelomocytes obtained from sea urchins collected from waters heavily polluted with urban run-off and industrial waste[181] and, more recently, with the explosive 2,4,6-trinitrotoluene (TNT) from conventional weapons dumped at sea at the end of World War II.[185] Increased levels of Hsp70 levels in coelomocytes from specimens of the sea star *Asterias rubens*, collected along a transect from inland waters to the open sea along the Norwegian fiords correlate with a natural concentration gradient of heavy metals.[186] Increased production of reactive oxygen species (ROS) by coelomocytes from *A. rubens* is also observed after cadmium exposure in a dose-dependent manner.[187-189] Cadmium, on the other hand, leads to reduced phagocytic activity by coelomocytes.[190] Other heavy metals, including cadmium, as well as UV light can also result in single strand DNA breaks in coelomocytes.[20,191,192] Exposure to lead increases the phagocytic activity of sea star coelomocytes, though another pollutant, polychlorinated biphenyls (PCBs), do not have this effect.[190] Responses to temperature stress levels can also be detected in coelomocytes as increases in acetylcholine esterase (AChE) activity.[193] Overall, there are several approaches for monitoring protein levels and enzyme activities that can be employed as signs of environmental stress both in field and in laboratory studies.

Another approach for employing echinoderms as biosensors has been to evaluate the numbers of coelomocytes in the CF, which can be altered by environmental stressors. For example, the percentage of red spherule cells increase from 5% to 40% of total coelomocytes in animals collected from polluted seawaters or that are subjected to accidental injury.[181,185] Similarly, the numbers of total coelomocytes increase in immunoquiescent sea urchins after simple immune challenge in the lab.[29] Therefore, levels of red spherule cells and perhaps total coelomocytes may be used as a practical marker of environmental stress in animals collected in costal surveys for marine management.

CONCLUSION

Echinoderms are important members of marine ecosystems and are required for the stable maintenance of habitats. This has been illustrated by the aftermath of the disappearance of the long-spined black sea urchin, *Diadema antillarum*, the top herbivore on Caribbean coral reefs. The population crash in 1983-1984, with repeat crashes in 1985 and 1991-1992, reduced the population in the Caribbean and Western Pacific by 95-99%.[194,195,196] General destabilization of the coral reef ecology resulting from herbivore release that was exacerbated by hurricane damage, lead to a change in the trophic cascade resulting in a swift and sustained switch to an alternative ecological state; reef cover of mostly coral switched to mostly soft algae.[197-200] The effect of the disappearance of *D. antillarum* was predicted by Sammarco[201] and modeling these population changes show the same outcome.[202]

The population crash of *D. antillarum* progressed from west to east in the Caribbean and was speculated to have been due to a pathogen[198] that may also have been a commensal.[203] Disease outbreaks and mass mortalities have also been noted in the green sea urchin, *S. droebachiensis*, along the Atlantic coast of Nova Scotia.[204] Little is known of the bacterial pathogens of sea urchins[205-209] and much less is known of the fungal and viral pathogens of any class of echinoderm. This general lack of knowledge regarding pathogenesis in echinoderms is an important problem that will require coordinated efforts of both ecologists and immunologists. Understanding the population dynamics in a complex ecosystem such as coral reefs should include investigations of the host-pathogen interactions that are involved in the stability of the ecosystem or lack thereof.

Until recently, workers interested in the immune system of echinoderms were hobbled by identifying and analyzing one gene or protein at a time. Cross-phylum searches for genes encoding proteins involved in microbial recognition and immune effector functions were complicated by the rapid pace of immune gene evolution. However, the availability of the purple sea urchin genome[11] has greatly increased the sensitivity with which immune gene homologues can be identified, in addition to the efficient characterization of complex multigene families. Immune transcription regulators are generally very well conserved and BLAST type sequence identity searches are typically adequate to identify these factors. In the case of many immune receptors and effectors, primary sequence is poorly conserved but domain structure can be used as a unique identifier and combinatorial domain profile searches can be useful. A final class of immune mediators is encoded by genes that have novel structure relative to other phyla. These are typically identified in experimental surveys, but in some cases multiplicity of domains that are common to proteins with immune functions can be identified in purely bioinformatic surveys for candidates of novel immune mediators.

Analysis of the purple sea urchin genome has revealed a complex repertoire of immune receptors, regulators and effectors unlike those known in other phyla.[10,60,78,210] In addition to the LRR-containing proteins and the *Sp185/333* gene family described above, the genome encodes a virtually complete set of homologues of (i) vertebrate hematopoietic and immune transcription factors, (ii) candidate effector proteins and (iii) genes with distant homology to key adaptive immune mediators of the jawed vertebrates.[10] Transcription regulators of haematopoiesis include a nearly complete set with respect to the vertebrate homologues and include representatives of some subfamilies such as the PU.1/SpiB/SpiC Ets factors that are important regulators of vertebrate myeloid and lymphoid immunocyte development not found outside of deuterostomes.[211] Transcription factors involved in the regulation immune response genes include GATA1/2/3, SCL and NFkB have also been identified.

Future work on echinoderm immunity will focus on the purple sea urchin, but additional echinoderm genomes are needed. Preliminary 1X sequence coverage of the genomes for

S. franciscanus and *Allocentrotus fragilis*, sister species of *S. purpuratus* are available (http://www.hgsc.bcm.tmc.edu/project-species-o-Strongylocentrotus%20purpuratus. hgsc?pageLocation = Strongylocentrotus purpuratus) and preliminary analyses suggest a similar repertoire of immune genes. Additional genomes from species from other classes are needed and will not only promote phylogenetic and comparative evolutionary genomics, but will enable the characterization of the immune gene repertoire for a different echinoderm species. This will be of particular interest given that the immune genes and the mechanisms of immune gene diversification are dictated by the life history, pathogens and habitat of individual species.

ACKNOWLEDGEMENTS

The authors are grateful to Ms. Margaret Throckmorton for collecting and formatting the references. This work was supported in part by funding from the US National Science Foundation (MCB-0744999) to LCS, the US National Institutes of Health (GM60925) to JHH, the Australian Research Council (DP0556486) to SVN.

REFERENCES

1. Hyman LH. The Invertebrates: Echinodermata, The coelomate Bilateria. Vol IV. New Yok:McGraw-Hill, 1955.
2. Ebert TA. Negative Growth and longevity in the purple sea urchin Strongylocentrotus purpuratus (Stimpson). Science 1967; 157(3788):557-558.
3. Metchnikoff E. Lectures on the comparative pathology of inflammation: delivered at the Pasteur Institute in 1891: Kegan Paul, Trench, Trubner and Co. Ltd.; 1893.
4. Hildemann WH, Dix TG. Transplantation reactions of tropical Australian echinoderms. Transplantation 1972; 14(5):624-633.
5. Karp RD, Hildemann WH. Specific allograft reactivity in the sea star Dermasterias imbricata. Transplantation 1976; 22(5):434-439.
6. Coffaro KA. Memory and specificity in the sea urchin Lytechinus pictus [Doctoral Dissertation]. Santa Cruz, University of California; 1979.
7. Karp RD, Coffaro KA. Cellular defense systems of the Echinodermata. In: Manning ED, ed. Phylogeny of Immunological Memory. Amsterdam, The Netherlands: Elsevier/North Holland, 1980:257-282.
8. Coffaro KA, Hinegardner RT. Immune response in the sea urchin Lytechinus pictus. Science 1977; 197(4311):1389-1390.
9. Smith LC, Davidson EH. The echinoid immune system and the phylogenetic occurrence of immune mechanisms in deuterostomes. Immunol Today 1992; 13(9):356-362.
10. Hibino T, Loza-Coll M, Messier C et al. The immune gene repertoire encoded in the purple sea urchin genome. Dev Biol 2006; 300:349-365.
11. Sodergren E, Weinstock GM, Davidson EH et al. The genome of the sea urchin, Strongylocentrotus purpuratus. Science 2006; 314(5801):941-952.
12. Smith LC. Host responses to bacteria: innate immunity in invertebrates. In: McFall-Ngai M, Ruby N, Henderson B, eds. The Influence of Cooperative Bacteria on Animal Host Biology. Advances in Molecular and Cellular Microbiology, vol. 10. Cambridge: Cambridge University Press, 2005:293-320.
13. Castillo MG, Goodson MS, McFall-Ngai M. Identification and molecular characterization of a complement C3 molecule in a lophotrochozoan, the Hawaiian bobtail squid Euprymna scolopes. Dev Comp Immunol 2008; 33(1):69-76.
14. Geddes P. On the coalescence of amoeboid cells in plasmodia and on the so called coagulation of invertebrate fluids. Proc Royal Soc London 1880; (30):252-255.
15. Kindred JE. The cellular elements in the perivisceral fluid of echinoderms. Biol Bull 1924; (47):228-251.
16. Boolootian RA, Geise CA. Coelomic corpuscles of echinoderms. Biol Bull 1958; 15:53-56.

17. Johnson PT. The coelomic elements of sea urchins (Strongylocentrotus). I. The normal coelomocytes; their morphology and dynamics in hanging drops. J Invertebr Pathol 1969; 13:25-41.
18. Gross PS, Al-Sharif WZ, Clow LA et al. Echinoderm immunity and the evolution of the complement system. Dev Comp Immunol 1999; 23(4-5):429-442.
19. Smith LC, Rast JP, Brockton V et al. The sea urchin immune system. Invertebrate Survival Journal 2006; 3:25-39.
20. Matranga V, Pinsino A, Celi M et al. Impacts of UV-B radiation on short term cultures of sea urchin coelomocytes. Mar Biol 2006; 149(1):24-34.
21. Matranga V, Pinsino A, Celi M et al. Monitoring chemical and physical stress using sea urchin immune cells. Prog Mol Subcell Biol 2005; 39:85-110.
22. Smith VJ. The Echinoderms. In: Ratcliffe NA, Rowley AF, eds. Invertebrate Blood Cells. New York, NY: Academic Press; 1981:513-562.
23. Edds KT. Dynamic aspects of filopodial formation by reorganization of microfilaments. J Cell Biol 1977; 73(2):479-491.
24. Edds KT. Cell biology of echinoid coelomocytes. J Invertebr Pathol 1993; 61:173-178.
25. Henson JH, Kolnik SE, Fried CA et al. Actin-based centripetal flow: phosphatase inhibition by calyculin-A alters flow pattern, actin organization and actomyosin distribution. Cell Motil Cytoskeleton 2003; 56(4):252-266.
26. Henson JH, Nesbitt D, Wright BD et al. Immunolocalization of kinesin in sea urchin coelomocytes. Association of kinesin with intracellular organelles. J Cell Sci 1992; 103 (Pt 2):309-320.
27. Henson JH, Svitkina TM, Burns AR et al. Two components of actin-based retrograde flow in sea urchin coelomocytes. Mol Biol Cell 1999; 10(12):4075-4090.
28. Gross PS, Clow LA, Smith LC. SpC3, the complement homologue from the purple sea urchin, Strongylocentrotus purpuratus, is expressed in two subpopulations of the phagocytic coelomocytes. Immunogenetics 2000; 51(12):1034-1044.
29. Brockton V, Henson JH, Raftos DA et al. Localization and diversity of 185/333 proteins from the purple sea urchin—unexpected protein-size range and protein expression in a new coelomocyte type. J Cell Sci 2008; 121(3):339-348.
30. Henson JH, Schatten G. Calcium regulation of the actin-mediated cytoskeletal transformation of sea urchin coelomocytes. Cell Motil 1983; 3(5-6):525-534.
31. Bertheussen K. The cytotixic reaction in allogeneic mixtures of echinoid phagocytes. Exp Cell Res 1979; 120:373-381.
32. Johnson PT. The coelomic elements of the sea urchins (Strongylocentrotus) III. In vitro reaction to bacteria. J Invertebr Pathol 1969; 13:42-62.
33. Service M, Wardlaw AC. Echinochrome-A as a bactericidal substance in the coelomic fluid of Echinus esculentus (L). Comp Biochem Physiol Biochem Mol Biol 1984; 79(2):161-165.
34. Gerardi P, Lassegues M, Canicatti C. Cellular distribution of sea urchin antibacterial activity. Biol Cell 1990; 70(3):153-157.
35. Huang Y, Krein PM, Muruve DA et al. Complement factor B gene regulation: synergistic effects of TNF-alpha and IFN-gamma in macrophages. J Immunol 2002; 169(5):2627-2635.
36. Heatfield BM, Travis DF. Ultrastructural studies of regenerating spines of the sea urchin Strongylocentrotus purpuratus. II. Cells with spherules. J Morphol 1975; 145:51-72.
37. Hobaus E. Coelomocytes in normal and pathologically altered body walls of sea urchins. Paper presented at: Proceedings of the European Colloquium on Echinoderms, 1979:247-249.
38. Arizza V, Giaramita FT, Parrinello D et al. Cell cooperation in coelomocyte cytotoxic activity of Paracentrotus lividus coelomocytes. Comp Biochem Physiol A Comp Physiol 2007; 147(2):389-394.
39. Bertheussen K, Seljelid R. Echinoid phagocytes in vitro. Exp Cell Res 1978; 111(2):401-412.
40. Xing J, Yang HS, Chen MY. Morphological and ultrastructural characterization of the coelomocytes in Apostichopus japonicus. Aquat Biol 2008; 2:85-92.
41. Coteur G, DeBecker G, Warnau M et al. Differentiation of immune cells challenged by bacteria in the common European starfish, Asterias rubens (Echinodermata). Eur J Cell Biol 2002; 81(7):413-418.
42. Holm K, Hernroth B, Thorndyke M. Coelomocyte numbers and expression of HSP70 in wounded sea stars during hypoxia. Cell Tissue Res 2008; 334(2):319-325.
43. Hetzel HR. Studies on holothurian coelomocytes. II. The origin of coelomocytes and the formation of brown bodies. Biol Bull 1965; 128(1):102-111.
44. Canicatti C, Quaglia A. Ultrastructure of Holothuria polii encapsulating body. J Zool 1991; 224(3):419-429.
45. Dybas L, Fankboner PV. Holothurian survival strategies: mechanisms for the maintenance of a bacteriostatic environment in the coelomic cavity of the sea cucumber, Parastichopus californicus. Dev Comp Immunol 1986; 10(3):311-330.

46. Pagliara P, Carnevali C, Burighel P et al. The spherule cells of Holothuria polii Delle Chiaie, 1823 (Aspidochirota, Holothuroidea) during brown body formation: an ultrastructural study. J Submicrosc Cytol Pathol 2003; 35(3):295-301.
47. Jans D, Dubois P, Jangoux M. Defensive mechanisms of holothuroids (Echinodermata): Formation, role and fate of intracoelomic brown bodies in the sea cucumber Holothuria tubulosa. Cell Tissue Res 2005; (283):99-106.
48. Canicatti C, Rizzo A. A 220 kDa coelomocyte aggregating factor involved in Holothuria polii cellular clotting. Eur J Cell Biol 1991; 56:79-83.
49. Nair SV, Del Valle H, Gross PS et al. Macroarray analysis of coelomocyte gene expression in response to LPS in the sea urchin. Identification of unexpected immune diversity in an invertebrate. Physiol Genomics 2005; 22(1):33-47.
50. Hillier BJ, Vacquier VD. Amassin, an olfactomedin protein, mediates the massive intercellular adhesion of sea urchin coelomocytes. J Cell Biol 2003; 160(4):597-604.
51. Hillier BJ, Moy GW, Vacquier VD. Diversity of olfactomedin proteins in the sea urchin. Genomics 2007; 89(6):721-730.
52. Ramirez-Gomez F, Ortiz-Pineda PA, Rojas-Cartagena C et al. Immune-related genes associated with intestinal tissue in the sea cucumber Holothuria glaberrima. Immunogenetics 2008; 60(1):57-71.
53. Ramirez-Gomez F, Ortiz-Pineda PA, Rivera-Cardona G et al. LPS-induced genes in intestinal tissue of the sea cucumber Holothuria glaberrima. PLoS One 2009; 4(7):e6178.
54. Santiago-Cardona PG, Berrios CA, Ramirez F et al. Lipopolysaccharides induce intestinal serum amyloid A expression in the sea cucumber Holothuria glaberrima. Dev Comp Immunol 2003; 27(2):105-110.
55. Ortiz-Pineda PA, Ramirez-Gomez F, Perez-Ortiz J et al. Gene expression profiling of intestinal regeneration in the sea cucumber. BMC Genomics 2009; 10:262.
56. Takeda K, Kaisho T, Akira S. Toll-like receptors. Annual Review of Immunology 2003; 21:335-376.
57. Beutler B, Rehli M. Evolution of the TIR, Tolls and TLRs: functional inferences from computational biology. Curr Top Microbiol Immunol 2002; 270:1-21.
58. Leulier F, Lemaitre B. Toll-like receptors—taking an evolutionary approach. Nat Rev Genet 2008; 9:165-178.
59. Rock FL, Hardiman G, Timans JC et al. A family of human receptors structurally related to Drosophila Toll. Proc Natl Acad Sci U S A 1998; 95(2):588-583.
60. Messier-Solek C, Buckley KM, Rast JP. Highly diversificied innate receptor systems and new forms of animal immunity. Semin Immunol 2010; 22(1):39-47.
61. Roach JC, Glusman G, Rowen L et al. The evolution of vertebrate Toll-like receptors. Proc Natl Acad Sci USA 2005; 102(27):9577-9582.
62. Franchi L, Warner N, Viani K et al. Function of Nod-like receptors in microbial recognition and host defense. Immunol Rev 2009; 227(1):106-128.
63. Brodsky IE, Monack D. NLR-mediated control of inflammasome assembly in the host response against bacterial pathogens. Semin Immunol 2009; 21(4):199-207.
64. Schmucker D, Chen B. Dscam and DSCAM: complex genes in simple animals, complex animals yet simple genes. Genes Dev 2009; 23:147-156.
65. Huang S, Yuan S, Guo L et al. Genomic analysis of the immune gene repertoire of amphioxus reveals extraordinary innate complexity and diversity. Genome Res 2008; 18(7):1112-1126.
66. Buckley KM, Smith LC. Extraordinary diversity among members of the large gene family, 185/333, from the purple sea urchin, Strongylocentrotus purpuratus. BMC Mol Biol 2007; 8:68.
67. Terwilliger DP, Buckley KM, Brockton V et al. Distinctive expression patterns of 185/333 genes in the purple sea urchin, Strongylocentrotus purpuratus: an unexpectedly diverse family of transcripts in response to LPS, beta-1,3-glucan and dsRNA. BMC Mol Biol 2007; 8:16.
68. Terwilliger DP, Buckley KM, Mehta D et al. Unexpected diversity displayed in cDNAs expressed by the immune cells of the purple sea urchin, Strongylocentrotus purpuratus. Physiol Genomics 2006; 26(2):134-144.
69. Buckley KM, Munshaw S, Kepler TB et al. The 185/333 gene family is a rapidly diversifying host-defense gene cluster in the purple sea urchin, Strongylocentrotus purpuratus. J Mol Biol 2008; 379:912-928.
70. Ghosh JG, Buckley KM, Nair SV et al. Sp185/333: A novel family of genes and proteins involved in the purple sea urchin immune response. Dev Comp Immunol 2010; 34(3):235-245.
71. Britten RJ, Cetta A, Davidson EH. The single-copy DNA sequence polymorphism of the sea urchin Strongylocentrotus purpuratus. Cell 1978; 15(4):1175-1186.
72. Buckley KM, Terwilliger DP, Smith LC. Sequence variations between the 185/333 genes and messages from the purple sea urchin suggest posttranscriptional modifications. J Immunol 2008; 181(12):2203-2212.
73. Majewski J, Ott J. GT repeats are associated with recombination on human chromosome 22. Genome Res 2000; 10(8):1108-1114.

74. Rogaev EI. Simple human DNA-repeats associated with genomic hypervariability, flanking the genomic retroposons and similar to retroviral sites. Nucleic Acids Res 1990; 18(7):1879-1885.
75. Boothroyd CE, Dreesen O, Leonova T et al. A yeast-endonuclease-generated DNA break induces antigenic switching in Trypanosoma brucei. Nature 2009; 459(7244):278-281.
76. McDowell JM, Simon SA. Molecular diversity at the plant-pathogen interface. Dev Comp Immunol 2008; 32(7):736-744.
77. Hamilton CE, Papavasiliou FN, Rosenberg BR. Diverse functions for DNA and RNA editing in the immune system. RNA Biol 2010;7(2):in press.
78. Rast JP, Smith LC, Loza-Coll M et al. Genomic insights into the immune system of the sea urchin. Science 2006; 314(5801):952-956.
79. Kunkel TA, Bebenek K. DNA replication fidelity. Annu Rev Biochem 2000; 69:497-529.
80. Ruiz JF, Dominguez O, Lain de Lera T et al. DNA polymerase mu, a candidate hypermutase? Philos Trans R Soc Lond B Biol Sci 2001; 356(1405):99-109.
81. Dheilly NM, Nair SV, Smith LC et al. Highly variable immune response proteins from the sea urchin, Strongylocentrotus purpuratus: proteomic analysis of diversity within and between individuals. J Immunol 2009; 182:2203-2212.
82. Volanakis JE. Overview of the complement system. In: Volonakis JE, Frank MM, eds. The Human Complement System in Health and Disease. Vol 20. New York: Marcel Dekker; 1998:9-23.
83. Lambris JD, Ricklin D, Geisbrecht BV. Complement evasion by human pathogens. Nat Rev Microbiol 2008; 6(2):132-142.
84. Dempsey PW, Allison ME, Akkaraju S et al. C3d of complement as a molecular adjuvant: bridging innate and acquired immunity. Science 1996; 271(5247):348-350.
85. Kaplan G, Bertheussen K. The morphology of echinoid phagocytes and mouse peritoneal macrophages during phagocytosis in vitro. Scand J Immunol 1977; 6(12):1289-1296.
86. Bertheussen K. Endocytosis by echinoid phagocytes in vitro. II. Mechanisms of endocytosis. Dev Comp Immunol 1981; 5(4):557-564.
87. Bertheussen K. Receptors for complement on echinoid phagocytes. II. Purified human complement mediates echinoid phagocytosis. Dev Comp Immunol 1982; 6(4):635-642.
88. Smith LC, Chang L, Britten RJ et al. Sea urchin genes expressed in activated coelomocytes are identified by expressed sequence tags. Complement homologues and other putative immune response genes suggest immune system homology within the deuterostomes. J Immunol 1996; 156(2):593-602.
89. Smith LC, Shih CS, Dachenhausen SG. Coelomocytes express SpBf, a homologue of factor B, the second component in the sea urchin complement system. J Immunol. 1998; 161(12):6784-6793.
90. Al-Sharif WZ, Sunyer JO, Lambris JD et al. Sea urchin coelomocytes specifically express a homologue of the complement component C3. J Immunol 1998; 160(6):2983-2997.
91. Clow LA, Gross PS, Shih CS et al. Expression of SpC3, the sea urchin complement component, in response to lipopolysaccharide. Immunogenetics 2000; 51(12):1021-1033.
92. Shah M, Brown KM, Smith LC. The gene encoding the sea urchin complement protein, SpC3, is expressed in embryos and can be upregulated by bacteria. Dev Comp Immunol 2003; 27(6-7):529-538.
93. Smith LC. Thioester function is conserved in SpC3, the sea urchin homologue of the complement component C3. Dev Comp Immunol 2002; 26(7):603-614.
94. Sim RB, Sim E. Autolytic fragmentation of complement components C3 and C4 and its relationship to covalent binding activity. Ann N Y Acad Sci 1983; 421:259-276.
95. Clow LA, Raftos DA, Gross PS et al. The sea urchin complement homologue, SpC3, functions as an opsonin. J Exp Biol 2004; 207(Pt 12):2147-2155.
96. Nakao M, Fushitani Y, Fujiki K et al. Two diverged complement factor B/C2-like cDNA sequences from a teleost, the common carp (Cyprinus carpio). J Immunol 1998; 161(9):4811-4818.
97. Terwilliger DP, Clow LA, Gross PS et al. Constitutive expression and alternative splicing of the exons encoding SCRs in Sp152, the sea urchin homologue of complement factor B. Implications on the evolution of the Bf/C2 gene family. Immunogenetics 2004; 56(7):531-543.
98. Kimura A, Sakaguchi E, Nonaka M. Multi-component complement system of Cnidaria: C3, Bf and MASP genes expressed in the endodermal tissues of a sea anemone, Nematostella vectensis. Immunobiology 2009; 214(3):165-178.
99. Lachman PJ. An evolutionary view of the complement system. Behring Inst. Mitt 1979; 63:25-37.
100. Bentley DR. Structural superfamilies of the complement system. Exp Clin Immunogenet 1988; 5(2-3):69-80.
101. Smith LC, Azumi K, Nonaka M. Complement systems in invertebrates. The ancient alternative and lectin pathways. Immunopharmacology 1999; 42(1-3):107-120.
102. Smith LC, Clow LA, Terwilliger DP. The ancestral complement system in sea urchins. Immunol Rev 2001; 180:16-34.

103. Dodd RB, Drickamer K. Lectin-like proteins in model organisms: implications for evolution of carbohydrate-binding activity. Glycobiology 2001; 11(5):71R-79R.
104. Vasta GR, Ahmed H, Tasumi S et al. Biological Roles of lectins in innate immunity: molecular and structural basis for diversity and self/nonself recognition. Adv Exp Med Biol 2007; 598:389-406.
105. Goldstein IJ, Hughes RC, Monsigny M et al. What should be called a lectin? Nature 1980; 285(5760):66-66.
106. Srivastava P. Roles of heat-shock proteins in innate and adaptive immunity. Nat Rev Immunol 2002; 2:185-194.
107. Takahashi H, Komano H, Kawaguchi N et al. Cloning and sequencing of cDNA of Sarcophaga peregrina humoral lectin induced on injury of the body wall. J Biol Chem 1985; 260(22):12228-12233.
108. Giga Y, Ikai A, Takahashi K. The complete amino acid sequence of echinoidin, a lectin from the coelomic fluid of the sea urchin Anthocidaris crassispina. Homologies with mammalian and insect lectins. J Biol Chem 1987; 262(13):6197-6203.
109. Muramoto K, Kamiya H. The amino-acid sequence of multiple lectins of the acorn barnacle Megabalanus rosa and its homology with animal lectins. Biochim Biophys Acta 1990; 1039(1):42-51.
110. Suzuki T, Takagi T, Furukohri T et al. A calcium-dependent galactose-binding lectin from the tunicate Polyandrocarpa misakiensis. Isolation, characterization and amino acid sequence. J Biol Chem 1990; 265(3):1274-1281.
111. Hirabayashi J, Satoh M, Kasai K. Evidence that Caenorhabditis elegans 32-kDa beta-galactoside-binding protein is homologous to vertebrate beta-galactoside-binding lectins. cDNA cloning and deduced amino acid sequence. J Biol Chem 1992; 267(22):15485-15490.
112. Pfeifer K, Haasemann M, Gamulin V et al. S-type lectins occur also in invertebrates: high conservation of the carbohydrate recognition domain in the lectin genes from the marine sponge Geodia cydonium. Glycobiology 1993; 3(2):179-184.
113. Elola MT, Vasta GR. Lectins from the colonial tunicate Clavelina picta are structurally related to acute-phase reactants from vertebrates. Ann N Y Acad Sci 1994; 712:321-323.
114. Armstrong PB, Swarnakar S, Srimal S et al. A cytolytic function for a sialic acid-binding lectin that is a member of the pentraxin family of proteins. J Biol Chem 1996; 271(25):14717-14721.
115. Saito T, Hatada M, Iwanaga S et al. A newly identified horseshoe crab lectin with binding specificity to O-antigen of bacterial lipopolysaccharides. J Biol Chem 1997; 272(49):30703-30708.
116. Drickamer K, Fadden AJ. Genomic analysis of C-type lectins. Biochem Soc Symp 2002(69):59-72.
117. Vijayan M, Chandra N. Lectins. Curr Opin Struct Biol 1999; 9(6):707-714.
118. Bulgakov AA, Eliseikina MG, Petrova IY et al. Molecular and biological characterization of a mannan-binding lectin from the holothurian Apostichopus japonicus. Glycobiology 2007; 17(12):1284-1298.
119. Ikeda K, Sannoh T, Kawasaki N et al. Serum lectin with known structure activates complement through the classical pathway. J Biol Chem 1987; 262(16):7451-7454.
120. Endo Y, Takahashi M, Nakao M et al. Two lineages of mannose-binding lectin-associated serine protease (MASP) in vertebrates. J Immunol 1998; 161(9):4924-4930.
121. Stahl PD, Ezekowitz RA. The mannose receptor is a pattern recognition receptor involved in host defense. Curr Opin Immunol 1998; 10(1):50-55.
122. Zhang Y, Suankratay C, Zhang XH et al. Lysis via the lectin pathway of complement activation: minireview and lectin pathway enhancement of endotoxin-initiated hemolysis. Immunopharmacology 1999; 42(1-3):81-90.
123. Sasaki H, Aketa K. Purification and distribution of a lectin in sea urchin (Anthocidaris crassispina) egg before and after fertilization. Exp Cell Res 1981; 135(1):15-19.
124. Multerer KA, Smith LC. Two cDNAs from the purple sea urchin, Strongylocentrotus purpuratus, encoding mosaic proteins with domains found in factor H, factor I and complement components C6 and C7. Immunogenetics 2004; 56(2):89-106.
125. Pancer Z. Dynamic expression of multiple scavenger receptor cysteine-rich genes in coelomocytes of the purple sea urchin. Proc Natl Acad Sci USA 2000; 97(24):13156-13161.
126. Pancer Z, Rast JP, Davidson EH. Origins of immunity: transcription factors and homologues of effector genes of the vertebrate immune system expressed in sea urchin coelomocytes. Immunogenetics 1999; 49(9):773-786.
127. Bikker FJ, Ligtenberg AJ, End C et al. Bacteria binding by DMBT1/SAG/gp-340 is confined to the VEVLXXXXW motif in its scavenger receptor cysteine-rich domains. J Biol Chem 2004; 279(46):47699-47703.
128. Rosenstiel P, Sina C, End C et al. Regulation of DMBT1 via NOD2 and TLR4 in intestinal epithelial cells modulates bacterial recognition and invasion. J Immunol 2007; 178(12):8203-8211.
129. Boman HG. Peptide antibiotics and their role in innate immunity. Ann. Rev. Immunol 1995; 13:61-92.

130. Hancock JM, Simon M. Simple sequence repeats in proteins and their significance for network evolution. Gene 2005; 345(1):113-118.
131. Zasloff M. Antimicrobial peptides of multicellular organisms. Nature 2002; 415(6870):389-395.
132. Jenssen H, Hamill P, Hancock RE. Peptide antimicrobial agents. Clin Microbiol Rev 2006; 19(3):491-511.
133. Steiner H, Hultmark D, Engström Å et al. Sequence and specificity of two antibacterial proteins involved in insect immunity. Nature 1981; 292:246-248.
134. Selsted ME, Brown DM, DeLange RJ et al. Primary structures of MCP-1 and MCP-2, natural peptide antibiotics of rabbit lung macrophages. J Biol Chem 1983; 258(23):14485-14489.
135. Wang G, Li X, Wang Z. APD2: the updated antimicrobial peptide database and its application in peptide design. Nucleic Acids Res 2009; 37:D933-D937.
136. Scott MG, Hancock RE. Cationic antimicrobial peptides and their multifunctional role in the immune system. Crit Rev Immunol 2000; 20(5):407-431.
137. Bradshaw JP. Cationic antimicrobial peptides. Issues for potential clinical use. Biodrugs 2003; 17(4):233-240.
138. Lehrer RI, Ganz T. Endogenous vertebrate antibiotics. Defensins, protegrins and other cysteine-rich antimicrobial peptides. Ann N Y Acad Sci 1996; 797:228-239.
139. Rinehart J, K.L. Marine natural products as sources of antiviral, antimicrobial and dybasantineoplastic agents. Pure Appl Chem 1981; (53):795-817.
140. Bryan PJ, Rittschof D, McClintock JB. Bioactivity of echinoderm ethanolic body-wall extracts: an assessment of marine bacterial attachment and macroinvertebrate larval settlement. J Exp Mar Bio Ecol 1996; 196(1-2):79-96.
141. Andersson L, Bohlin L, Iorizzi M et al. Biological activity of saponins and saponin-like compounds from starfish and brittle-stars. Toxicon 1989; 27(2):179-188.
142. Chludil HD, Seldes AM, Maier MS. Antifungal steroidal glycosides from the Patagonian starfish Anasterias minuta: Structure-activity correlations. J Nat Prod 2002; 65(2):153-157.
143. Levina EV, Kalinovsky AI, Dmitrenok PV. Steroid compounds from two pacific starfish of the genus Evasterias. Russ J Bioorganic Chem 2009; 35(1):123-130.
144. Kuwahara R, Hatate H, Yuki T et al. Antioxidant property of polyhydroxylated naphthoquinone pigments from shells of purple sea urchin Anthocidaris crassispina. Lwt-Food Science and Technology 2009; 42(7):1296-1300.
145. Canicatti C, Roch P. Studies on Holothuria polii (Echinodermata) antibacterial proteins. I. Evidence for and activity of coelomocyte lysozyme. Experientia 1989; 45(8):756-759.
146. Canicatti C, Pagliara P, Stabili L. Sea urchin coelomic fluid agglutinin mediates coelomocyte adhesion. Eur J Cell Biol 1992; 58(2):291-295.
147. Shimizu M, Kohno S, Kagawa H et al. Lytic activity and biochemical properties of lysozyme in the coelomic fluid of the sea urchin Strongylocentrotus intermedius. J Invertebr Pathol 1999; 73(2):214-222.
148. Johnson PT. Infection with diatoms and other microorganisms in the sea urchin spines (Strongylocentrotus fanciscanus). J Invertebr Pathol 1970; 16:268-276.
149. Schillaci D, Arizza V, Parrinello N et al. Antimicrobial and antistaphylococcal biofilm activity from the sea urchin Paracentrotus lividus. J Appl Microbiol 2010, 108(1):17-24.
150. Maltseva AL, Aleshina GM, Kokryakov VN et al. Diversity of antimicrobial peptides in acidic extracts from coelomocytes of starfish Asterias rubens L. Izdatllstvo Sankt-Peterburgskogo Universiteta 2007; 1:85-94.
151. Maltseva AL, Aleshina GM, Kokryakov VN et al. New antimicrobial peptides from coelomocytes of sea star Asterias rubens L. Biologiya 2004; 4:101-108.
152. Haug T, Kjuul AK, Styrvold OB et al. Antibacterial activity in Strongylocentrotus droebachiensis (Echinoidea), Cucumaria frondosa (Holothuroidea) and Asterias rubens (Asteroidea). J Invertebr Pathol 2002; 81(2):94-102.
153. Li C, Haug T, Styrvold OB et al. Strongylocins, novel antimicrobial peptides from the green sea urchin, Strongylocentrotus droebachiensis. Dev Comp Immunol 2008; 32(12):1430-1440.
154. Li C, Blencke H-M, Smith LC et al. Two recombinant peptides, SpStrongylocins 1 and 2, from Strongylocentrotus purpuratus, show antimicrobial activity against Gram-positive and Gram-negative bacteria. Dev Comp Immunol 2010; 34:286-292.
155. Selsted ME, Ouellette AJ. Mammalian defensins in the antimicrobial immune response. Nat Immunol 2005; 6(6):551-557.
156. Daher KA, Selsted ME, Lehrer RI. Direct inactivation of viruses by human granulocyte defensins. J Virol 1986; 60(3):1068-1074.
157. Mandal M, Nagaraj R. Antibacterial activities and conformations of synthetic alpha-defensin HNP-1 and analogs with one, two and three disulfide bridges. J Pept Res 2002; 59(3):95-104.
158. Inouye M. Intramolecular chaperone: the role of the pro-peptide in protein folding. Enzyme 1991; 45(5-6):314-321.

159. Lemaitre B, Nicolas E, Michaut L et al. The dorsoventral regulatory gene cassette spatzle/Toll/cactus controls the potent antifungal response in Drosophila adults. Cell 1996; 86(6):973-983.
160. Reichhart JM, Achstetter T. Expression and secretion of insect immune peptides in yeast. Res Immunol 1990; 141(9):943-946.
161. Beauregard KA, Truong NT, Zhang H et al. The detection and isolation of a novel antimicrobial peptide from the echinoderm, Cucumaria frondosa. Adv Exp Med Biol 2001; 484:55-62.
162. Chludil HD, Muniain CC, Seldes AM et al. Cytotoxic and antifungal triterpene glycosides from the Patagonian sea cucumber Hemoiedema spectabilis. J Nat Prod 2002; 65(6):860-865.
163. Chludil HD, Seldes AM, Maier MS. Antifungal steroidal glycosides from the patagonian starfish anasteriasminuta: structure-activity correlations. J Nat Prod 2002; 65(2):153-157.
164. Lee J, Wang W, Hong J et al. A new 2.3-dimethyl butenolide from the brittle star Ophiomastix mixta. Chem Pharm Bull 2007; 55:459-461.
165. Rast JP, Pancer Z, Davidson EH. New approaches towards an understanding of deuterostome immunity. Curr Top Microbiol Immunol 2000; 248:3-16.
166. Beck G, Ellis TW, Habicht GS et al. Evolution of the acute phase response: iron release by echinoderm (Asterias forbesi) coelomocytes and cloning of an echinoderm ferritin molecule. Dev Comp Immunol 2002; 26(1):11-26.
167. García-Arrarás JE, Schenk C, Rodrígues-Ramírez R et al. Spherulocytes in the echinoderm Holothuria glaberrima and their involvement in intestinal regeneration. Dev Dyn 2006; 235(12):3259-3267.
168. Candia-Carnevali MD, Thorndyke MC, Matranga V. Regenerating echinoderms: a promise to understand stem cells potential. In: Rinkevich B, Matranga V, eds. Stem Cells in Marine Organisms: Springer; 2009:165-186.
169. Pinsino A, Thorndyke MC, Matranga V. Coelomocytes and posttraumatic response in the common sea star Asterias rubens. Cell Stress Chaperones 2007; 12(4):331-341.
170. Dubois P, Ameye L. Regeneration of spines and pedicellariae in echinoderms: a review. Microsc Res Tech 2001; 55:427-437.
171. Oweson C, Skold H, Pinsino A et al. Manganese effects on haematopoietic cells and circulating coelomocytes of Asterias rubens (Linnaeus). Aquat Toxicol 2008; 89(2):75-81.
172. Harrington FE, Easton DP. A putative precursor to the major yolk protein of the sea urchin. Dev Biol 1982; 94:505-508.
173. Noll H, Alcedo J, Daube M et al. The toposome, essential for sea urchin cell adhesion and development, is a modified iron-less caldium-binding transferrin. Dev Biol 2007; 310(1):54-70.
174. Cervello M, Arizza V, Lattuca G et al. Detection of vitellogenin in a subpopulation of sea urchin coelomocytes. Eur J Cell Biol 1994; 64:314-319.
175. Shyu AB, Raff RA, Bumenthal T. Expression of the vitellogenin gene in female and male sea urchin. Proc Natl Acad Sci U S A 1986; 83(1):3865-3869.
176. Unuma T, Konishi K, Kiyomoto M et al. The major yolk protein is synthesized in the digestive tract and secreted into the body cavities in sea urchin larvae. Mol Reprod Dev 2009; 76:142-150.
177. Rojas-Cartagena C, Ortiz-Pineda P, Ramirez-Gomez F et al. Distinct profiles of expressed sequence tags during intestinal regeneration in the sea cucumber Holothuria glaberrima. Physiol Genomics 2007; 31:203-215.
178. Rinkevich Y, Matranga V, Rinkevich B. Stem cells in aquatic invertebrates: Common premises and emerging unique themes. In: Rinkevich B, Matranga V, eds. Stem Cells in Marine Organisms: Springer Publishers; 2009:61-104.
179. Matranga V, Yokota Y. Responses of marine organisms to physical and chemical impacts. Cell Biol Toxicol 2008; 24(6):471-474.
180. Matranga V, Bonaventura R, Di Bella G. Hsp70 as a stress marker of sea urchin coelomocytes in short term cultures. Cell Mol Biol (Noisy-le-grand) 2002; 48(4):345-349.
181. Matranga V, Toia G, Bonaventura R et al. Cellular and biochemical responses to environmental and experimentally induced stress in sea urchin coelomocytes. Cell Stress Chaperones 2000; 5(2):113-120.
182. Becker J, Craig EA. Heat-shock proteins as molecular chaperones. Eur J Biochem 1994; 219:11-23.
183. Tsan MF, Gao B. Heat shock protein and innate immunity. Cell Mol Immunol 2004; 1(4):274-279.
184. Robert J. Evolution of heat shock protein and immunity. Dev Comp Immunol 2003; 27:449-464.
185. Pinsino A, Della Torre C, Sammarini V et al. Sea urchin coelomocytes as a novel cellular biosensor of environmental stress: a field study in the Tremiti Island Marine Protected Area, Southern Adriatic Sea, Italy. Cell Biol Toxicol 2008; 24(6):541-552.
186. Coteur G, Gillan D, Joly G et al. Field contamination of the starfish Asterias rubens by metals. Part 2: Effects on cellular immunity. Environ Toxicol Chem 2003; 22(9):2145-2151.
187. Coteur G, Danis B, Dubois P. Echinoderm reactive oxygen species (ROS) production measured by peroxidase, luminol-enhanced chemiluminescence (PLCL) as an immunotoxicological tool. Prog Mol Subcell Biol 2005; 39:71-83.

188. Danis B, Goriely S, Dubois P et al. Contrasting effects of coplanar versus noncoplanar PCB congeners on immunomodulation and CYP1A levels (determined using an adapted ELISA method) in the common sea star Asterias rubens L. Aquat Toxicol 2004; 69(4):371-383.
189. Danis B, Wantier P, Flammang R et al. Bioaccumulation and effects of PCBs and heavy metals in sea stars (Asterias rubens, L.) from the North Sea: a small scale perspective. Sci Total Environ 2006; 356(1-3):275-289.
190. Coteur G, Danis B, Wantier P et al. Increased phagocytic activity in contaminated seastars (Asterias rubens) collected in the Southern Bight of the North Sea. Mar Pollut Bull 2005; 50(11):1295-1302.
191. Schroder HC, Di Bella G, Janipour N et al. DNA damage and developmental defects after exposure to UV and heavy metals in sea urchin cells and embryons compared to other invertebrates. In: Matranga V, ed. Echinodermata. Heidelberg, Germany: Springer Publishers; 2005:111-137.
192. Hartwig A. Role of DNA repair inhibition in lead- and cadmium-induced genotoxicity: a review. Environ Health Perspect 1994; 102(Suppl 3):45-50.
193. Angelini C, Amaroli A, Falugi C et al. Acetylcholinesterase activity is affected by stress conditions in Paracentrotus lividus coelomocytes. Mar Biol 2003; 143:623-628.
194. Lessios HA. Population dynamics of Diadema antillarum (Echinodermata: Echinoidea) following mass mortality in Panama. Mar Biol 1988; 99:515-526.
195. Hughes TP, Keller BD, Backson JBC et al. Mass mortality of the echinoid Diadema antillarum Philippi in Jamaica. Bull Mar Sci 1985; 36:377-384.
196. Forcucci D. Population density, recruitment and 1991 mortality event of Diadema antillarum in the Florida Keys. Bull Mar Sci 1994; 53:917-928.
197. Hughes TP, Reed DC, Boyle MJ. Herbivory on coral reefs: Community structure following mass mortalities of sea urchins. J Exp Mar Bio Ecol 1987; 113:39-59.
198. Liddell WD, Ohlhorst SL. Changes in benthic community composition following the mass mortality of Diadema at Jamaica. J Exp Mar Bio Ecol 1986; 95:271-278.
199. Williams SL, Carpenter RC. Nitrogen-limited primary productivity of coral reef algal turfs: potential contribution of ammonium excreted by Diadema antillarum. Mar Ecol Prog Ser 1988; 47(2):145-152.
200. Carpenter RC. Mass mortality of Diadema antillarum. 1. Long-term effects on sea urchin population-dynamics and coral reef algal communities. Mar Biol 1990; 104:67-77.
201. Sammarco PW. Diadema and its relationship to coral spat mortality: grazing, competition and biological disturbance. J Exp Mar Bio Ecol 1980; 45:245-272.
202. Mumby PJ, Hastings A, Edwards HJ. Thresholds and the resilience of Caribbean coral reefs. Nature 2007; 450(7166):98-101.
203. Bauer JC, Agerter CJ. Isolation of potentially pathogenic bacterial flora from tropical sea urchins in selected West Atlantic and East Pacific sites. Bull Mar Sci. 1994; 55:142-150.
204. Scheibling RE, Hennigar AW. Recurrent outbreaks of disease in sea urchins Strongylocentrotus droebachiensis in Nova Scotia: evidence for a link with large-scale meteorologic and oceanographic events. Mar Ecol Prog Ser 1997; 152:155-165.
205. Shimizu M. Histopathological investigation of the spotted gonad disease in the sea urchin, Strongylocentrotus intermedius. J Invertebr Pathol 1994; 63:182-187.
206. Shimizu M, Takaya Y, Ohsake S et al. Gross and histopathological signs of the spotting disease in the sea urchin Strongylocentrotus intermedius. Fisheries Sci 1995; 61:608-661.
207. Tajima K, Hiranno R, Shimizu M et al. Isolation and pathogenicity of the causative bacterium of spotting disease of sea urchin Strongylocentrotus intermedius. Fisheries Sci 1997; 63:249-252.
208. Tajima K, Takeuchi K, Iqbal MM et al. Studies on a bacterial disease of sea urchin Strongylocentrotus intermedius occurring at low water temperatures. Fisheries Sci (Japan) 1998; 64(6):918-920.
209. Tajima K, Shimizu M, Miura M et al. Seasonal fluctuations of Flexibacter sp. the causative bacterium of spotting disease of sea urchin Strongylocentrotus intermedius in the culturing facilities and coastal area. Fisheries Sci (Japan) 1998; 64(1):6-9.
210. Rast JP, Messier-Solek C. Marine invertebrate genome sequences and our evolving understanding of animal immunity. Biol Bull 2008; 214:274-283.
211. Rizzo R, Fernandez-Serra M, Squarzoni P et al. Identification and developmental expression of the ets gene family in the sea urchin (Strongylocentrotus purpuratus). Dev Biol 2006; 300(1):35-48.
212. Livingston BT, Killian CE, Wilt F et al. A genome-side analysis of biomineralization-related proteins in the sea urchin Strongylocentrotus purpuratus. Dev Biol 2006; 300:335-348.
213. Notredame C, Higgins DG, Heringa J. T-Coffee: A novel method for multiple sequence alignments. JMB 2000; (302):205-217.
214. Mesquite: a modular system for evolutionary analysis. [computer program]. Version 2.72; 2009.
215. PAUP*. Phylogenetic Analysis Using Parsimony (*and Other Methods). [computer program]. Version 4. Sunderland, Massachusetts: Sinauer Associates; 2003.

216. Ronquist F, Huelsenbeck JP. MrBayes 3: Bayesian phylogenetic inference under mixed models. Bioinformatics 2003; 19(12):1572-1574.
217. Posada D, Crandall KA. MODELTEST: testing the model of DNA substitution. Bioinformatics 1998; 14(9):817-818.
218. Silva JR. The onset of phagocytosis and identity in the embryo of Lytechinus variegatus. Dev Comp Immunol 2000; 24(8):733-739.
219. Furukawa R, Takahashi Y, Kanajima Y et al. Defense system by mesenchyme cells in bipinnaria larvae of the starfish, Asterina pectinifera. Dev Comp Immunol 2009; 33(2):205-215.
220. Tamboline CR, Burke RD. Secondary mesenchyme of the sea urchin embryo: ontogeny of blastocoelar cells. J Exp Zool 1992; 262(1):51-60.
221. Castellani C, Rast JP, Davidson EH. Isolation of pigment cell specific genes in the sea urchin embryo by differential macroarray screening. Development 2003; 130(19):4587-4596.
222. Matranga V, Bonaventura R. Sea urchin coelomocytes, the progenitors of vertebrate immune effectors, as bio-indicators of stress and pollution. In: Yokota Y, Matranga V, Smolenicka Z, eds. The Sea Urchin: from Basic Biology to Aquaculture. Lisse, the Netherlands: Balkema; 2002:161-176.
223. Smith LC, Britten RJ, Davidson EH. SpCoel1: a sea urchin profilin gene expressed specifically in coelomocytes in response to injury. Mol Biol Cell 1992; 3(4):403-414.
224. Kakiuchi M, Okino N, Sueyoshi N et al. Purification, characterization and cDNA cloning of alpha-N-acetylgalactosamine-specific lectin from starfish, Asterina pectinifera. Glycobiology 2002; 12(2):85-94.
225. Snowden AM, Vasta GR. A dimeric lectin from coelomic fluid of the starfish Oreaster reticulatus cross-reacts with the sea urchin embryonic substrate adhesion protein, echinonectin. Ann N Y Acad Sci 1994; 712:327-329.
226. Alliegro MC, Ettensohn CA, Burdsal CA et al. Echinonectin: a new embryonic substrate adhesion protein. J Cell Biol 1988; 107(6 Pt 1):2319-2327.
227. Drago F, Malagoli D, Pezzino FM et al. Presence of a low molecular weight lectin in the coelomic fluid of the sea urchin Paracentrotus lividus. Inv Surv Journal 2009; 6(1):15-20.
228. Nakagawa H, Hashimoto T, Hayashi H et al. Isolation of a novel lectin from the globiferous pedicellariae of the sea urchin Toxopneustes pileolus. Adv Exp Med Biol 1996; 391:213-223.
229. Hatakeyama T, Himeshima T, Komatsu A et al. Purification and characterization of two lectins from the sea cucumber Stichopus japonicus. Biosci Biotechnol Biochem 1993; 57(10):1736-1739.
230. Matsui T, Ozeki Y, Suzuki M et al. Purification and characterization of two $Ca^{(2+)}$-dependent lectins from coelomic plasma of sea cucumber, Stichopus japonicus. J Biochem 1994; 116(5):1127-1133.
231. Hatakeyama T, Kohzaki H, Nagatomo H et al. Purification and characterization of four Ca^{2+}-dDependent lectins from the marine invertebrate, Cucumaria echinata. J Biochem 1994; 116:209-214.
232. Hatakeyama T, Ohuchi K, Kuroki M et al. Amino acid sequence of a C-type lectin CEL-IV from the marine invertebrate Cucumaria echinata. Biosci Biotechnol Biochem 1995; 59:1314-1317.
233. Bulgakov AA, Nazarenko EL, Petrova IY et al. Isolation and properties of a mannan-binding lectin from the coelomic fluid of the holothurian Cucumaria japonica. Biochemistry (Mosc) 2000; 65(8):933-939.
234. Gowda NM, Goswami U, Khan MI. Purification and characterization of a T-antigen specific lectin from the coelomic fluid of a marine invertebrate, sea cucumber (Holothuria scabra). Fish Shellfish Immunol 2008; 24(4):450-458.

CHAPTER 15

UROCHORDATE IMMUNITY

Masaru Nonaka*,[1] and Honoo Satake[2]

[1]*Department of Biological Sciences, Graduate School of Science, The University of Tokyo, Hongo, Tokyo, Japan*
[2]*Suntory Institute for Bioorganic Research, Mishima, Osaka, Japan*
**Corresponding Author: Masaru Nonaka—Email: mnonaka@biol.s.u-tokyo.ac.jp*

Abstract: This chapter provides a short review of the immune system of urochordates, the closest living relative of vertebrates. Since adaptive immunity is a unique property of vertebrates, urochordates rely exclusively on innate immunity to recognize and eliminate pathogens. Here we discuss three immune systems of urochordates which show different evolutionary relationship with the vertebrate immune system. Urochordate Toll-like receptors (TLR) show a clear orthologous relationship with vertebrate counterparts, although they show unique characteristics most likely gained in the urochordate lineage. The urochordate complement system also shows orthologous relationship with the vertebrate complement system. From the structural and functional viewpoints, it seems to represent a more primitive state of the vertebrate complement system without any major deviation. In contrast, the allorecognition systems of urochordates show no evolutionary relationship with any invertebrate or vertebrate systems, suggesting that they were invented in the urochordate lineage.

INTRODUCTION

Vertebrates developed two types of the adaptive immune system, a well-defined one based on conventional lymphocytes and the major histocompatibility complex (MHC) of jawed vertebrates and an emerging one based on novel lymphocytes of jawless fish.[1] No evidence for the presence of adaptive immunity has been reported from invertebrates thus far, whereas various types of innate immunity, some common with vertebrates and others specific to certain phylogenetic groups, have been reported. To understand the origin and evolution of vertebrate immune system, it is essential to analyze the immune system of the closest relatives of vertebrates. Vertebrates, urochordates and cephalochordates constitute the phylum Chordata. Traditionally, cephalochordates are considered as the closest living

Invertebrate Immunity, edited by Kenneth Söderhäll.

relatives of vertebrates, with urochordates representing the earliest chordate lineage. This view is mainly justified by overall morphological similarities and an apparently increased complexity in cephalochordates and vertebrates relative to urochordates. However, recent molecular phylogenetic analyses provided compelling evidence that urochordates and not cephalochordates, represent the closest living relatives of vertebrates.[2] Comprehensive search for immune-related genes from draft genome information of one urochordtae species, *Ciona intestinalis*, indicated that the pivotal genes for adaptive immunity, such as the MHC class I and II genes, T-cell receptors, or dimeric immunoglobulin molecules, are missing from the *Ciona* genome.[3] In contrast many genes possibly involved in innate immunity are identified. Among them, here we discuss three well-characterized systems, TLR (Toll-like receptor), complement system and allorecognition. The former two are evolutionary related to their vertebrate counterparts, whereas the last one is unique to urochordate lacking the vertebrate counterpart.

TOLL-LIKE RECEPTORS (TLR)

What Are Toll-Like Receptors?

Toll-like receptors (TLRs) play pivotal roles in host defenses via the innate immune system. All TLRs are Type I transmembrane proteins which harbor an intracellular Toll/Interleukin-1 receptor (TIR) domain and extracellular leucine rich repeat (LRR) motifs.[4,5] LRRs exhibit specific pathogenic ligand recognition and TIR participates in the activation of downstream signaling pathways. Nine functional human TLRs (hTLRs) have been identified. As summarized in Table 1, each hTLR directly recognizes their specific ligands (or pathogen-associated molecular patterns, PAMPs). Molecular diversity in the number and organization of LRRs enables the specific and sensitive recognition of PAMPs by respective TLRs. TLRs are expressed not only in immune cells such as lymphocytes, macrophages and dendritic cells but also nonimmune tissues including lung, small intestine, stomach and testis. TLR1, TLR2, TLR4, TLR5 and TLR6 recognize extracellular microbial pathogenic components on plasma membranes, whereas TLR3, TLR7, TLR8 and TLR9 respond to viral DNA or RNA on endosomes. In addition, TLR4 requires an extracellular associated protein, MD2, to recognize lipopolysaccharide (LPS).[4-6] Interaction of TLRs with specific PAMPs triggers signal transduction pathways via adaptor proteins (MyD88, TIRAP, TRIF and TRAM) followed by activation of a wide range of inducible transcriptional factors such as NF-κB, AP-1 and IRF, leading to production of a inflammatory cytokine TNFα, chemokines and/or Type I interferon.[4,5] TLRs or their related genes have also been detected in fish,[7-9] cyclostomes,[10] amphioxus,[11] sea urchin,[12] annelid[13] and cnidarian,[14] although their functions, except for several fish TLRs, have yet to be elucidated.

TLRs of *C. intestinalis*

A *Ciona* genome survey and molecular cloning revealed the presence of two TLRs in *C. intestinalis*, namely, Ci-TLR1 and -2.[3,15] Ci-TLR1 and Ci-TLR2 are composed of a TIR, transmembrane and LRR domain, which is typical of TLRs. Moreover, 7 and 13 LRRs are found in Ci-TLR1 and Ci-TLR2, respectively. Ci-TLR1 and -2 were most homologous to hTLR7 (26%) and hTLR8 (26%), respectively.[15] However, the sequence homology is inconsistent with PAMP recognition and intracellular localization of Ci-TLRs (Table 1).

Table 1. Ligands of human and *Ciona* TLRs

TLR	Ligands	Intracellular Localization
TLR1/2	triacylated lipoprotein	PM
TLR2	Zymosan (yeast cell wall 1, 3-β-glucan Lipoarabinomannan Heat-killed *Legionella pneumophila* (HKLP, Gram-negative) Heat-killed *Staphylococcu aereus* (HKSA, Gram-positive) Glycosylphosphatidylinositol (GPI)-anchored glycoprotein	PM
TLR3	poly(I:C) (double-stranded RNA)	ES
TLR4 (with MD2)	LPS (lipopolysaccharide from Gram-negative bacteria) Lipid A (lipid component of LPS)	PM
TLR5	Flagellin (bacterial flagellar filament)	PM
TLR6	MALP-2 (mycoplasma-derived macrophage-activating lipopeptide) FSL1 (micoplasma-dereived lipoprotein)	PM
TLR7	Imidazoquimod (imidazoquinolone amino acid analog), single-stranded RNA	ES
TLR8	Single-stranded RNA	ES
TLR9	Unmethylated CpG DNA	ES
Ci-TLR1	Zymosan (yeast cell wall)	PM and ES
Ci-TLR2	Heat-killed *Legionella pneumophila* (HKLP, Gram-negative) poly(I:C) (double-stranded-RNA) Flagellin (bacterial flagellar filament)	

PM, plasma membrane; ES, endosome.

The Ci-TLR1 and Ci-TLR2 genes were expressed intensively in the stomach, intestine and numerous hemocytes and, to a lesser degree, the central nervous system.[15] These findings indicate that Ci-TLRs function mainly in the alimentary tracts and hemocytes. Intriguingly, both of Ci-TLRs, unlike any vertebrate TLRs, were present on both the plasma membrane and a number of late endosomes.[15] Moreover, Ci-TLR1 and Ci-TLR2 activated NF-κB in response to multiple TLR ligands (Table 1), which are recognized by different mammalian TLRs. Zymosan (*Saccharomyces cerevisiae* cell wall) for hTLR2, heat-killed *Legionella pneumophila* (HKLP, a Gram-negative bacterium) for hTLR2, double-stranded RNA, poly(I:C) for hTLR3, *Salmonella typhimurium* Flagellin (the major component of the bacterial flagellar filament) for hTLR5 elicited a dose-dependent transactivation of NF-κB in the *ci-tlr1-* or *ci-tlr2*-expressing cells.[15] Poly(I:C) also elicited approximately 4-fold and 10-fold Ci-TNFα expression in the anterior and middle intestine, respectively.[15] Likewise, induction of 4-fold and 10-fold Ci-TNFα expression by Flagellin was observed in the stomach and middle intestine, respectively.[15] In contrast, no Ci-TNFα induction was detected in the posterior intestine.[15] These profiles of the Ci-TNFα induction are compatible with the tissue-distribution of *ci-tlr* expression; *ci-tlr1* and *ci-tlr2* are abundantly expressed in the stomach, anterior and middle intestine, but not in the posterior intestine.[15] These data lead to two important conclusions. Firstly, Ci-TLRs, like vertebrate TLRs, directly recognize their PAMPs and trigger the transactivation of NF-κB. Secondly, Ci-TLRs are 'functionally hybrid TLRs' of vertebrate cell-surface TLRs and endosome TLRs;

poly(I:C) is recognized by hTLR3 on endosomes TLR, whereas hTLR2 and hTLR5 respond to Zymosan, HKLP and Flagellin on the cell surface, respectively (Table 1). In addition, the PAMPs of Ci-TLRs are in good agreement with their cellular localization to both the plasma membrane and endosomes.

C. intestinalis possesses only two TLRs,[15] whereas other deuterostome invertebrates, amphioxus and sea urchin, were found to possess a great number of TLRs or their related genes: 72 genes in amphioxus[11] and 222 genes in sea urchin.[12] Furthermore, the molecular phylogenetic analyses demonstrated that most of these genes were generated via species-specific gene duplication, suggesting that these deuterostome invertebrates expand TLRs or their related genes in unique lineages of innate immunity, if most of the genes are functional. These findings lead to two scenarios of evolution of TLR or their related genes. First, only a few TLR or their related genes might have existed in a common deuterostome antecedent and *C. intestinalis* conserves the ancestral characteristics. Alternatively, a common deuterostome antecedent might have numerous TLR family genes. If this is true, *C. intestinalis* should have lost a large part of ancestral TLR family genes. Instead of such a gene deletion, Ci-TLRs are highly likely to have acquired multiple PAMP recognition and intracellular localization as mentioned above. Unfortunately, Ci-TLRs are at present the only invertebrate TLRs of which intracellular localization, PAMP recognition and signaling have been investigated. Elucidation of PAMPs and intracellular localization of sea urchin, amphioxus and cyclostome TLRs is expected to contribute not only to understanding of their biological roles but also to the investigation of molecular and functional divergence of the invertebrate TLR family.

COMPLEMENT SYSTEM

The mammalian complement system is a powerful defense mechanism consisting of more than 30 plasma and cell-surface proteins interacting in the recognition and elimination of pathogens.[16] Three major physiological functions of the mammalian complement system are; opsonization of the foreign particles, induction of the inflammatory reactions and cytolysis. Evolutionary studies revealed that the origin of the multi-component complement system consisting of C3, Bf (factor B) and MASP (mannan-binding lectin associated serine protease) is traced back to the common ancestor of Eumetazoa.[17,18] In addition, marked development of the complement system by gene duplication of the key components and subsequent functional differentiation likely occurred at the early stage of vertebrate evolution.[18] Thus, the urochordate complement system represents the evolutionary stage just before this development and accumulating analyses made it the best-analyzed invertebrate complement system. Several complement genes have been identified mainly from two species, *Halocynthia roretzi* and *Ciona intestinalis*. Those genes are; C3,[19,20] Bf,[3,21] MASPs,[3,22] mannan-binding lectin (MBL),[23] ficolin[24] and CR3 alpha[25] and beta.[26] In addition, a glucose binding lectin (GBL) lacking the collagen domain was reported from *H. roretzi* as a possible functional substitute for MBL.[27] For the functional aspect, *H. roretzi* C3, ficolin and GBL proteins were isolated from the body fluid and were shown to act as a component of the opsonic complement system. Moreover, the C3a fragment of *C. intestinalis* C3 was shown to have a chemotactic activity,[28] indicating that the role of the complement system in inflammation is also conserved between mammals and urochordates. In contrast, the third activity of the mammalian complement system, cytolytic activity, has not been recognized in the urochordate complement system.

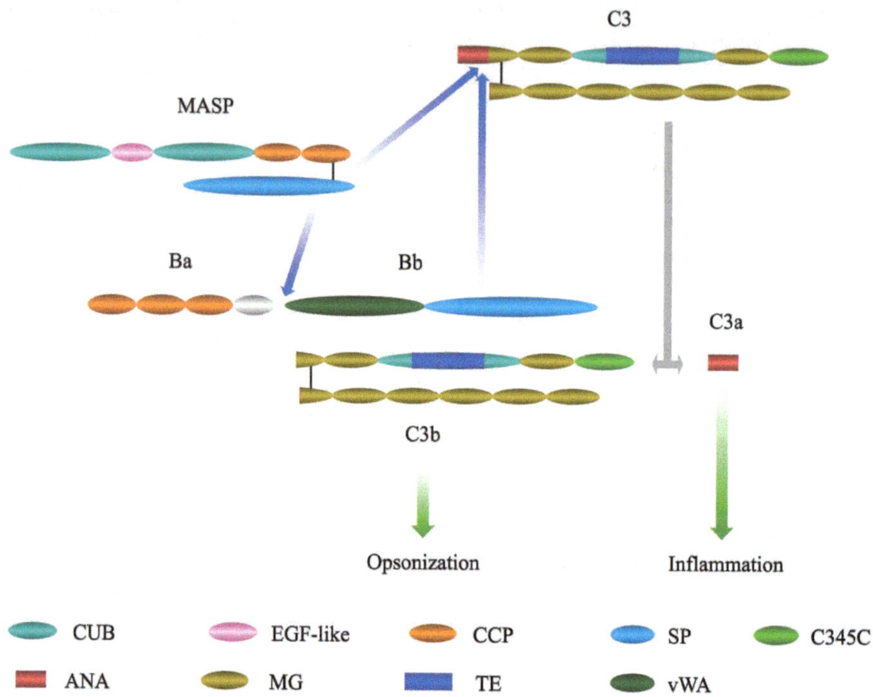

Figure 1. Schematic view of the core part of the urochordate complement system comprising C3, Bf and MASP. Conservation of the domain structure and functionally important residues of these components between mammals and urochordates suggests that the basic activation mechanism is also conserved. However, there is still no direct experimental evidence for proteolytic activation processes shown in blue arrows. The grey arrow indicates that C3 is cleaved by the C3 convertase (C3bBb) into two fragments, C3a and C3b and green arrows show biological functions of C3a and C3b. Abbreviations of domain names are: CUB, C1r, C1s, uEGF and bone morphogenetic protein; EGF-like, epidermal growth factor-like; CCP, complement control protein; SP, serine protease; C345C, C-terminal of C3, C4 and C5; ANA, anaphylatoxin; MG, macroglobulin; TE, thioester; vWA, von Willebrand factor Type A.

Although there are several C6-like genes with the membrane attack complex/perforin (MACP) domain in the *C. intestinalis* genome,[3] all of them lack the C-terminal short consensus repeat (SCR) and factor I/membrane attack complex (FIM) domains reported to be essential for interaction with other complement components. Thus, it is unlikely that these C6-like molecules are integrated in the urochordate complement system. All these results indicate that the urochordate complement system represents the primitive evolutionary stage just before the development occurred in the common ancestor of vertebrates. It lacks some components and functions of the mammalian complement system and shows no sign of acquisition of unique function. Figure 1 shows a schematic representation of the putative activation mechanism of the core part of the urochordate complement system comprising C3, Bf and MASP. Conservation of most structural motifs involved in proteolytic activation and C3 convertase formation of these complement components between urochordates and mammals strongly suggests that the activation mechanism of the urochordate complement system is the same as that of the mammalian complement system, although direct experimental evidence is still missing.

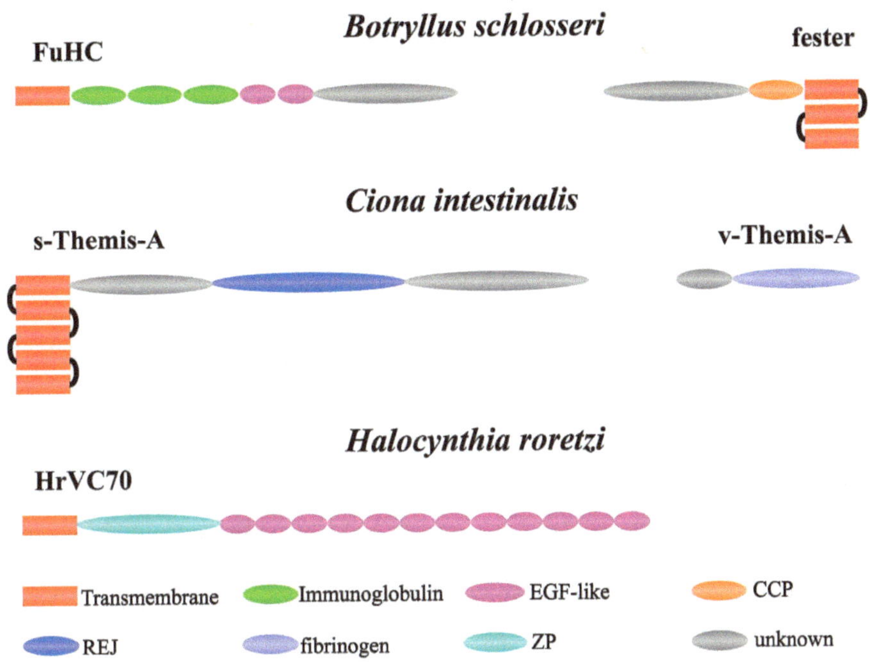

Figure 2. Domain structure of putative allorecognition molecules of three ascidian species. As shown here there is no orthologous relationship among these putative allorecognition molecules. Mutual interaction between FuHC and fester or between S-Themis and v-Themis is postulated, although it is still to be demonstrated directly. The *C. intestinalis* genome contains another set of the s-Themis and v-Themis genes containing similar domain structure as shown in this figure. Abbreviations of domain names not described in the legend to Figure 1 are: REJ, receptor for egg jelly; ZP, zona pellucida.

ALLORECOGNITION

Allorecognition is well known in vertebrates in the context of tissue transplantation where self or isogenic grafts are accepted whereas allogenic grafts are rejected. Although multiple genetic loci are involved in vertebrate allorecognition, by far the most important locus is the MHC. The evolutionary analyses indicated that the MHC was established in the common ancestor of jawed vertebrates[1] and urochordates completely lack the MHC. However, urochordates have two famous allorecognition systems working at colony fusion of the colonial ascidian[29] and fertilization of the solitary ascidian.[30] In both systems, the candidate genes for the key recognition molecules have been identified recently.

It has been known for 50 years that when two individuals of *Botryllus schlosseri*, a colonial ascidian, come into contact, they show histocompatibility reaction based on their genetic background.[29] If they share one or both alleles at a single histocompatibility locus, they will fuse. If they share no alleles, the colonies will reject each other. The candidate histocompatibility gene was isolated recently by positional cloning as described below and was termed *FuHC*.[31] About 1 Mb region identified by segregation analysis was sequenced to identify a candidate gene which showed polymorphism correlating with defined histocompatibility alleles in a fusion assay. The predicted open reading frame of this gene encoded a Type I transmembrane protein of 1007 amino acids in length. The amino terminus begins with a signal sequence, followed by an extracellular epidermal growth factor (EGF) repeats, two tandem immunoglobulin domains and the transmembrane domain and intracellular tail (Fig. 2). The high degree of polymorphism was demonstrated by identifying 18 alleles from 10 wild individuals. Most of allelic differences are single amino acid substitutions spread throughout the extracellular domains, with no obvious highly variable regions. Expression pattern analyzed by RT-PCR and in situ hybridization indicated that strong expression is observed in epithelia of ampullae and in a subset of blood cells, intimately associated with histocompatibility.

Another candidate for possible histocompatibility components is *fester* encoded near the *FuHC* locus.[32] The *fester* locus is highly polymorphic although this polymorphism does not contribute to histocompatibility, since it is not correlated with defined histocompatibility alleles. The *fester* is a Type I membrane protein having a signal peptide and several extracellular domains containing a single SCR domain. Alternative splicing generates several forms both membrane bound and secreted, all expressed in tissues intimately associated with histocompatibility. SiRNA-mediated knockdown of fester resulted in no histocompatibility reaction in both compatible and incompatible pairs, suggesting that *fester* is a receptor involved in histocompatibility. These data suggest that *FuHC* and *fester* are involved in allorecognition of *B. schlosseri*, although the underlying molecular mechanism including the possibility that *FuHC* and *fester* bind to each other is still to be clarified.

Ascidians are hermaphroditic and exhibit self-incompatibility (SI) at fertilization, self-sterility. Two species, *Ciona intestinalis* and *Halocynthia roretzi*, have been studied in detail for their SI system.[30] The SI system of *C. intestinalis* is genetically determined by multiple loci and takes place in the interaction between sperm and vitelline coat (VC), since removal of VC by acid treatment results in the loss of self-sterility. Recently, positional cloning of the SI loci was carried out using acid-induced self-fertilized siblings, the draft genome sequences and the detailed physical map.[33] Two loci A and B were identified in chromosome 2q and 7q, respectively. At both loci, a pair of genes termed s-Themis and v-Themis are present with a curious configuration that the v-Themis gene is located in the first intron of the s-Themis gene in opposite transcriptional direction. s-Themis is a polycystin-1 receptors and is expressed in testis. On the other hand, v-Themis is fibrinogen-like molecule and is a component of VC. Both s-Themis and V-Themis are highly polymorphic and autologous interaction between them is believed to reduce the binding ability of sperms. *H. roretzi* has much more strict SI system than *C. intestinalis*. In addition, *H. roretzi* has another allorecognition system termed "contact reaction", in which allogenic hemocytes show cytotoxic reactions. Although the common underlying mechanism is suggested for the SI system and contact reaction, their molecular basis is still to be clarified. Recently, a candidate for the VC molecule responsible for the SI system was reported.[34] A 70 kDa VC protein termed HrVC70 consists of 12 EGF-like repeats and show a high degree of polymorphism. HrVC70-agarose beads binds more nonself-sperms than self-sperms and pretreatment of sperm with nonself-HrVC70 more strongly inhibited fertilization than the pretreatment of sperm with self-HrVC70. These results suggest that HrVC70 is involved in the SI system of *H. roretzi*. Therefore, at least two different SI systems seem to be present in ascidians, although detailed molecular mechanism and evolution are still to be clarified.

CONCLUSION

Two major arms of innate immunity, TLRs and complement, show clear orthologous relationship between urochordates and vertebrates. Although functional information is still missing, Cnidaria also has the orthologous genes of vertebrate TLRs and complement, suggesting that their evolutionary origin can be traced back to the common ancestor of eumetazoa. In contrast, allorecognition systems of ascidians seem to be innovated in the urochordate lineage and use totally different genes from vertebrate MHC. Although vertebrate MHC has evolved as the antigen presentation system of adaptive immunity and its involvement in allorecognition is an accidental side effect, allorecognition seems to be the original purpose of the urochordate allorecognition systems.

REFERENCES

1. Flajnik MF, Kasahara M. Origin and evolution of the adaptive immune system: genetic events and selective pressures. Nat Rev Genet 2010; 11:47-59.
2. Delsuc F, Brinkmann H, Chourrout D et al. Tunicates and not cephalochordates are the closest living relatives of vertebrates. Nature 2006; 439:965-968.
3. Azumi K, De Santis R, De Tomaso A et al. Genomic analysis of immunity in a Urochordate and the emergence of the vertebrate immune system: "waiting for Godot". Immunogenetics 2003; 55:570-581.
4. Akira S, Takeda K. Toll-like receptor signalling. Nat Rev Immunol 2004; 4:4499-4511.
5. Dunne A, O'Neill LA. Adaptor usage and Toll-like receptor signaling specificity. FEBS Lett 2005; 579:3330-3335.
6. Shimazu R, Akashi S, Ogata H et al. MD-2, a molecule that confers lipopolysaccharide responsiveness on Toll-like receptor 4. J Exp Med 1999; 189:1777-1782.
7. Matsuo A, Oshiumi H, Tsujita T et al. Teleost TLR22 recognizes RNA duplex to induce IFN and protect cells from birnaviruses. J Immunol 2008; 181:3474-3485.
8. Takano T, Kondo H, Hirono I et al. Molecular cloning and characterization of Toll-like receptor 9 in Japanese flounder, Paralichthys olivaceus. Mol Immunol 2007; 44:1845-1853.
9. Yu Y, Zhong Q, Li C et al. Isolation and characterization of Toll-like receptor 9 in half-smooth tongue sole Cynoglossus semilaevis. Fish Shellfish Immunol 2009; 26:492-499.
10. Ishii A, Matsuo A, Sawa H et al. Lamprey TLRs with properties distinct from those of the variable lymphocyte receptors. J Immunol 2007; 178:397-406.
11. Huang S, Yuan S, Guo L et al. Genomic analysis of the immune gene repertoire of amphioxus reveals extraordinary innate complexity and diversity. Genome Res 2008; 18:1112-1126.
12. Hibino T, Loza-Coll M, Messier C et al. The immune gene repertoire encoded in the purple sea urchin genome. Dev Biol 2006; 300:349-365.
13. Davidson CR, Best NM, Francis JW et al. Toll-like receptor genes (TLRs) from Capitella capitata and Helobdella robusta (Annelida). Dev Comp Immunol 2008; 32:608-612.
14. Bosch TC, Augustin R, Anton-Erxleben F et al. Uncovering the evolutionary history of innate immunity: the simple metazoan Hydra uses epithelial cells for host defence. Dev Comp Immunol 2009; 33:559-569.
15. Sasaki N, Ogasawara M, Sekiguchi T et al. Toll-like receptors of the ascidian Ciona intestinalis: prototypes with hybrid functionalities of vertebrate Toll-like receptors. J Biol Chem 2009; 284:27336-27343.
16. Volanakis JE. Overview of the complement system. In: Volanakis JE, Frank MM, eds. The Human Complement System in Health and Disease. New York: Marcel Dekker, Inc., 1998:9-32.
17. Kimura A, Sakaguchi E, Nonaka M. Multi-component complement system of Cnidaria: C3, Bf and MASP genes expressed in the endodermal tissues of a sea anemone, Nematostella vectensis. Immunobiology 2009; 214:165-178.
18. Nonaka M, Kimura A. Genomic view of the evolution of the complement system. Immunogenetics 2006; 58:701-713.
19. Nonaka M, Azumi K, Ji X et al. Opsonic complement component C3 in the solitary ascidian, Halocynthia roretzi. J Immunol 1999; 162:387-391.
20. Marino R, Kimura Y, De Santis R et al. Complement in urochordates: cloning and characterization of two C3-like genes in the ascidian Ciona intestinalis. Immunogenetics 2002; 53:1055-1064.

21. Yoshizaki FY, Ikawa S, Satake M et al. Structure and the evolutionary implication of the triplicated complement factor B genes of a urochordate ascidian, Ciona intestinalis. Immunogenetics 2005; 56:930-942.
22. Ji X, Azumi K, Sasaki M et al. Ancient origin of the complement lectin pathway revealed by molecular cloning of mannan binding protein-associated serine protease from a urochordate, the Japanese ascidian, Halocynthia roretzi. Proc Natl Acad Sci USA 1997; 94:6340-6345.
23. Bonura A, Vizzini A, Salerno G et al. Isolation and expression of a novel MBL-like collectin cDNA enhanced by LPS injection in the body wall of the ascidian Ciona intestinalis. Mol Immunol 2009; 46:2389-2394.
24. Kenjo A, Takahashi M, Matsushita M et al. Cloning and characterization of novel ficolins from the solitary ascidian, Halocynthia roretzi. J Biol Chem 2001; 276:19959-19965.
25. Miyazawa S, Azumi K, Nonaka M. Cloning and characterization of integrin alpha subunits from the solitary ascidian, Halocynthia roretzi. J Immunol 2001; 166:1710-1715.
26. Miyazawa S, Nonaka M. Characterization of novel ascidian beta integrins as primitive complement receptor subunits. Immunogenetics 2004; 55:836-844.
27. Sekine H, Kenjo A, Azumi K et al. An ancient lectin-dependent complement system in an ascidian: novel lectin isolated from the plasma of the solitary ascidian, Halocynthia roretzi. J Immunol 2001; 167:4504-4510.
28. Pinto MR, Chinnici CM, Kimura Y et al. CiC3-1a-mediated chemotaxis in the deuterostome invertebrate Ciona intestinalis (Urochordata). J Immunol 2003; 171:5521-5528.
29. McKitrick TR, De Tomaso AW. Molecular mechanisms of allorecognition in a basal chordate. Semin Immunol 2009; 22:34-38.
30. Harada Y, Sawada H. Allorecognition mechanisms during ascidian fertilization. Int J Dev Biol 2008; 52:637-645.
31. De Tomaso AW, Nyholm SV, Palmeri KJ et al. Isolation and characterization of a protochordate histocompatibility locus. Nature 2005; 438:454-459.
32. Nyholm SV, Passegue E, Ludington WB et al fester, A candidate allorecognition receptor from a primitive chordate. Immunity 2006; 25:163-173.
33. Harada Y, Takagaki Y, Sunagawa M et al. Mechanism of self-sterility in a hermaphroditic chordate. Science 2008; 320:548-550.
34. Sawada H, Tanaka E, Ban S et al. Self/nonself recognition in ascidian fertilization: vitelline coat protein HrVC70 is a candidate allorecognition molecule. Proc Natl Acad Sci USA 2004; 101:15615-15620.

INDEX

A

B

C

D

E

I

K

L

M

SPRINGER NATURE

GPSR Compliance

The European Union's (EU) General Product Safety Regulation (GPSR) is a set of rules that requires consumer products to be safe and our obligations to ensure this.

If you have any concerns about our products, you can contact us on ProductSafety@springernature.com

In case Publisher is established outside the EU, the EU authorized representative is:

Springer Nature Customer Service Center GmbH
Europaplatz 3
69115 Heidelberg, Germany

FSC
www.fsc.org
MIX
Papier | Fördert gute Waldnutzung
FSC® C083411

Kolibri 360 GmbH
Ferdinand-Jühlke-Straße 7
99095 Erfurt, Deutschland
produktsicherheit@kolibri360.de